Epigenetics and Proteomics of Leukemia

Rūta Navakauskienė • Dalius Navakauskas
Veronika Borutinskaitė • Dalius Matuzevičius

Epigenetics and Proteomics of Leukemia

A Synergy of Experimental Biology and Computational Informatics

Springer

Rūta Navakauskienė
Department of Molecular Cell Biology
Institute of Biochemistry
Life Sciences Center
Vilnius University
Vilnius, Lithuania

Veronika Borutinskaitė
Department of Molecular Cell Biology
Institute of Biochemistry
Life Sciences Center
Vilnius University
Vilnius, Lithuania

Dalius Navakauskas
Department of Electronic Systems
Faculty of Electronics
Vilnius Gediminas Technical University
Vilnius, Lithuania

Dalius Matuzevičius
Department of Electronic Systems
Faculty of Electronics
Vilnius Gediminas Technical University
Vilnius, Lithuania

ISBN 978-3-030-68707-6 ISBN 978-3-030-68708-3 (eBook)
https://doi.org/10.1007/978-3-030-68708-3

This Springer imprint is published by the registered company Springer Nature Switzerland AG
The registered company address is: Gewerbestrasse 11, 6330 Cham, Switzerland

Dedication of Rūta and Dalius Navakauskai—
to our daughter, Aistė, Rūta's parents, Genovaitė and Povilas Petniūnai.

Dedication of Veronika Borutinskaitė—
to beloved family and dear friends.

Dedication of Dalius Matuzevičius—
to my wife, Greta, and my parents.

Preface

The decision to prepare the monograph “Epigenetics and Proteomics in Leukemia” was influenced by several factors. First of all, nowadays high technologies develop very fast and are used on a massive scale, but one of the world’s largest health problems—cancer—still exists. Second, we set ourselves the goal of describing our achievements after more than 20 years of active and productive research in the field of cancer treatment. Our innovative research in the field of cancer focuses on leukemia, a blood cancer.

Aging of the hematopoietic system is associated with various changes at the molecular, cellular, and physiological levels. Changes in the function and expression of molecular factors and age are the main prognostic factors for the development of leukemia. Genetic and/or epigenetic changes may be very different in patients with the same diagnosis of leukemia. Therefore, treatment of leukemia should be personalized based on the patient’s cytogenetic and epigenetic changes, biological age, drug tolerance, disease stage, etc. Thus, it is necessary to select a treatment strategy with maximum efficacy and minimal toxicity to the patient.

Epigenetic regulation plays a special role in the normal development of cells. As a result, the number of clinical trials using epigenetic regulators as drugs in the treatment of leukemia has increased significantly in recent years. Epigenetic-based therapies open up new possibilities for the successful treatment of leukemia. Nevertheless, it is necessary to select a treatment strategy with maximum efficacy and minimal toxicity to the patient. In our study, special attention has been paid to epigenetic differentiation treatment for leukemia. New drugs and their combinations are being examined and proposed for optimal treatment of leukemia with minimal side effects. In addition, epigenetic therapy also contributes to the prevention of leukemia recurrence. Therefore, epigenetic reprogramming of leukemic and cancer cells, in general, is a new strategy for cancer therapy.

The main motive of targeted epigenetic therapy is to detect epigenetically silenced genes and epigenetic effector proteins, activate them, and induce therapeutic effects such as differentiation, growth inhibition, and apoptosis. In this monograph, we present our achievements that contribute to the elucidation of molecular factors of myeloid leukemia involved in cell regeneration, differentiation, and aging processes, which would make a significant input to the effective treatment of leukemia. For visualization, characterization, and evaluation of protein-marker changes, we generate and present the computational methods necessary for proteome analysis and estimation of protein localization in the cell.

A number of our colleagues participated in the research that was presented and summarized in this monograph. We are especially grateful to current and former colleagues at the Department of Molecular Cell Biology at the Institute of Biochemistry, Life Sciences Center, Vilnius University. We want to give our special thanks to colleagues at the Electronic Systems Department, Faculty of Electronics, Vilnius Gediminas Technical University.

We are grateful for the fruitful collaboration within Visby Program project with Division of Medical Microbiology and Division of Automatic Control at the Linköping University. We are particularly grateful for the collaboration with hematologists of Vilnius University Hospital Santaros Klinikos. Our former student Monika Jasnauskaitė who studied epigenetics also kindly contributed to the excellent painted illustrations of the monograph; for this, we are very appreciative.

Also, our studies cannot be done without financial support from the international Visby Program and Lithuanian Research Council; for financial support through personal scholarships and funding of our research and experimental development projects, we are thankful to the Lithuanian Academy of Sciences, Agency for Scientific Innovation and Technology, Lithuanian State Science and Studies Foundation, and Forum Scientium program.

Our extensive research includes not only elucidating the contribution of epigenetic reprogramming in the treatment of leukemia but also the influence of various molecular factors on the development of leukemia and the management of this form of cancer. We made an effort to explain the elucidation problem in each chapter of the monograph and to present smoothly the obtained results that might have wide applicability in clinical practice.

Preparing the monograph "Epigenetics and Proteomics in Leukemia," we aim to make it useful for advanced biomedical scientists and students, medical doctors and students, bioinformatics specialists, health informatics, computational biologists, structural biologists, systems biologists, bioengineers, various societies (biochemistry, cell biology, hematology, bioengineers, etc.), and personnel in industry. This monograph would deepen the knowledge and understanding and expand the possibilities of using the results of these studies in clinics and pharmacy.

Vilnius, Lithuania
February 27, 2021

Rūta Navakauskienė

Introduction

Hematopoiesis is a tightly regulated system and is closely related to human immunity. Blood is one of the best recovering and most plastic tissues in the body. Mature blood cells are formed in bone marrow from hematopoietic stem cells during hematopoiesis. The precursors of this system are hematopoietic stem cells, which are characterized by regeneration and the possibility of differentiation into different blood cell lines. There are more than ten different types of cells in the blood that perform different functions in the body. There is a constant regeneration of the hematopoietic system because the differentiated blood cells function only for a defined period of time. However, hematopoietic stem cells are not bypassed by aging processes, their functions disappear with age, and the volume of bone marrow decreases. The aging hematological system causes anemia, decreased immunity, and increased incidence of blood malignancies. Aging of the hematopoietic system is associated with a variety of changes at the molecular, cellular, and physiological levels that lead to carcinogenesis and cancer progression. Changes in the function and expression of molecular factors and age are the main prognostic factors for the development of myeloid leukemia. The treatment of myeloid leukemia would be more effective if the molecular factors (epigenetic markers, transcription factors, etc.) involved in aging and the onset and progression of blood cancer were known. Genetic and epigenetic changes in cells involved in hematopoiesis can lead to leukemia, a blood cancer. Mutations can occur in both the early and late stages of hematopoiesis. As a result of the changes, the cells start to multiply uncontrollably, so leukemia is characterized by an increased number of malignant blood cells in the blood and/or bone marrow. Not only malignant blood cells do not perform their function, but their large numbers begin to interfere with the functioning of healthy cells. Genetic and/or epigenetic changes may be very different in patients with the same diagnosis of leukemia. Therefore, treatment of leukemia should be personalized based on the patient's cytogenetic and epigenetic changes, biological age, drug tolerance, disease stage, and so on. Thus, it is necessary to select a treatment strategy with maximum efficacy and minimal toxicity to the patient. In recent years, special attention has been paid to targeted epigenetic differentiation therapy for leukemia. Cancer cells are characterized by different genetic alterations (mutations, deletions, and translocations of genes required for normal development) that do not substantially disappear during treatment. However, epigenetic changes that occur during leukemic cell-induced differentiation prevent the onset of genetic disorders and arrest the development of cancer. Therefore, epigenetic reprogramming of leukemic and cancer cells, in general, is a new cancer therapy strategy for the targeted treatment of leukemias.

We are probably not mistaken in claiming that epigenetic cancer therapy is receiving increasing interest and is slowly gaining ground in conventional cancer treatment practices. Several epigenetic inhibitors are already officially approved and routinely used in the treatment of cancer. It should be mentioned that epigenetic cancer therapy serves several levels. By using epigenetic modifiers' inhibitors and their combinations with other drugs, the epigenetic state of cancer cells alters. The relationship between epigenetics and the immune system is also shown. A promising way to treat cancer is to stimulate the body's own immune system to fight cancer. Epigenetic mechanisms are known to be critical for the development and control of the immune system. For example, EZH2 (a PRC2 subunit of a polycomb repressive protein complex) has been shown to regulate T cell differentiation and activity. In addition, inhibition of EZH2 and DNA methyltransferases (DNMT) is known to be able to attract T cells to tumors and promote T cell infiltration. It should also be noted that HDAC (histone deacetylase) inhibitors may also serve to enhance the efficacy of chemotherapy and ionization therapy. The epigenetic therapy is more widely used in the treatment of hematologic diseases. For some reason, solid tumors are more resistant to this therapy.

This monograph presents research on the elucidation of molecular targets of leukemia and their applicability to targeted therapy. In Chap. 1, we describe the normal process of human hematopoiesis, genetic and epigenetic (histone and DNA modifications) alterations characteristic for human leukemia cells, types and pathogenesis of acute myeloid leukemia (AML). Also, main strategies of AML treatment are presented. The main motive of targeted epigenetic therapy is to detect epigenetically silenced genes and epigenetic effector proteins, activate them, and induce therapeutic effects such as differentiation, growth inhibition, and apoptosis.

Chapter 2 introduces main epigenetic modifiers and their inhibitors used in leukemia treatment and leukemia cell viability differentiation and apoptosis evaluation after the treatments. Epigenetic studies of leukemic cell mechanisms were performed in a model *in vitro* system (different cell lines of human myeloid leukemia) using the differentiation inducer ATRA and its combinations with epigenetic modifiers—HDAC, DNMT, and HMT inhibitors and other agents such as cell signaling molecules—protein kinases and their inhibitors, nucleoside analogues, and low-dose ionizing radiation. Molecular and epigenetic mechanisms that regulate the development of cancer cells have been purposefully investigated.

Chapter 3 delivers evaluation of genes expression and methylation in response to chromatin remodeling during leukemia cell proliferation and differentiation. Conditions and molecular mechanisms and therapeutic options have been studied in various leukemic cell lines exposed to different structures of HDAC (BML-210, FK228, sodium phenylbutyrate, vitamin B3, belinostat) and HMT inhibitors (BIX-

01294), DNMT inhibitors with different effects (zebularine, RG108) and their effective combinations with ATRA. The methylation status of cell cycle regulators and genes involved in cell differentiation processes can serve as a diagnostic and prognostic marker of cancer. The effect of DNMT, HDAC, and HMT inhibitors on the activation of epigenetically silenced genes (p15, p16, p21, E-cadherin) and the expression of epigenetic protein markers in proliferating leukemic and induced differentiation into granulocytes were evaluated. The role of many transcription factors (Sp1, p53, NFκB, PU.1s, C/EBPβ) in the genes responsible for cancer cell proliferation (p15, p21, p53) and apoptosis (FasL, NFκB) in cellular regulation and functioning is described.

Chapter 4 represents our research that focuses on the analysis of changes in protein expression, and in particular protein expression, in relation to disease status, drug action, or changes in extracellular conditions. Often, the level of gene expression does not match the level of expression of their products, proteins, in a cell, tissue, or organism, and the level of protein expression does not necessarily reflect protein activity. Using high-throughput proteomic methods (two-dimensional electrophoresis and mass spectrometry), proteomic changes in leukemic and induced differentiation cells were investigated, as well as analysis, characterization, and identification of regulatory and important proteins.

Chapter 5 describes newly tyrosine phosphorylated proteins that are modified after induction of differentiation of leukemia cell and followed apoptosis. This includes transcription factors, structural and signaling proteins, as well as the dystrobrevin as newly tyrosine phosphorylated protein immediately after induction of granulocytic differentiation with ATRA was identified and described.

Chapter 6 presents the assessments of the proteome in leukemia cells, primary and mature blood cells, and the leukemia cells induced by epigenetically acting agents. The proteome changes are characterized in different cell subfractions using bioinformatics methods developed by the authors. Bioinformatics and computational methods are essential for the analysis and comparative analysis of large-scale data obtained by protein fractionation in two-dimensional electrophoresis, visualization, and analysis. Depending on the specifics of the research object and the large scale of the proteomic research of leukemia, the authors have developed the tools for the proper analysis of these samples. Here, we ground the efficiency and reliability of evaluation criteria and present automatic 2DEG image analysis strategy, which consists of two essential steps: image matching and protein expression analysis. Study of several new 2DEG image matching algorithms results in original technique that works with natural and semi-artificial gel images more efficiently than the analogous MIR and RAIN methods. Three new 2DEG image segmentation into meaningful areas algorithms and their research results are presented. We also present protein spot modeling studies involving protein spot reconstruction and parameterization. Eleven protein spot models are formulated, implemented, and experimentally compared. Their use for protein spot parameterization is also investigated.

Chapter 7 represents the method for establishment of protein translocation from the cytoplasm into the nucleus during granulocytic differentiation of leukemic cells.

Also, here we describe the identified proteins that translocate into the nucleus after induction of leukemic cell differentiation with ATRA.

Chapter 8 presents the protein visualization in human leukemic cells, in cells during induced granulocytic differentiation, and apoptotic cells. The comparative analysis of histones and their variants and modifications distribution during hematopoietic cell granulocytic differentiation is presented. Also, the visualization of dystrobrevin-α and co-localization with other proteins in leukemic cells induced to granulocytic differentiation or apoptosis is defined.

Chapter 9 describes the computational methods used in our research to evaluate fluorescent images of leukemic cells and to define the quantified localization and of specific proteins in different cellular organs during leukemia cell differentiation, apoptosis, or in normal blood cell development (primary and mature blood cells). We describe development of the system for automated fluorescence image analysis, which would require minimal amount of manual operations at stages of initial parameter tuning and inspection of segmentation results. Before presenting the algorithms that make up the system, the chapter gives an overview of typical workflow of automated fluorescence image analysis and highlights its challenges.

The obtained data show that the differentiation efficiency due to chromatin remodeling modifiers depended on the sequence of their use and on the duration of exposure. The sequence of use of DNMT/HDAC/HMT inhibitors and ATRA significantly accelerated and improved the efficiency of leukemic cell differentiation into normal granulocytes, resulting in their death. The principle of action of other inhibitors is used not only in laboratory but also in clinical practice. However, the drugs currently used in the clinic do not yet fully meet the expectations of both patients and physicians regarding cytotoxicity, side effects, or acquired resistance. The studies identified new molecular targets important for the management of molecular mechanisms in leukemic cells, and selected schemes for the most effective combinations of new therapeutic agents are applicable for targeted leukemia therapy.

The Authors' Publications Are Presented in the Monograph

Borutinskaitė V, Virkšaitė A, Gudelytė G, Navakauskienė R (2018) Green tea polyphenol EGCG causes anti-cancerous epigenetic modulations in acute promyelocytic leukemia cells. Leuk Lymphoma 59(2):469–478. https://doi.org/10.1080/10428194.2017.1339881

Borutinskaite V, Navakauskiene R (2015) The histone deacetylase inhibitor BML-210 influences gene and protein expression in human promyelocytic leukemia NB4 cells via epigenetic reprogramming. Int J Mol Sci 16(8):18252–18269. https://doi.org/10.3390/ijms160818252

Borutinskaite V, Magnusson KE, Navakauskiene R (2005) Effects of retinoic acid and histone deacetilase inhibitor Bml-210 on protein expression in NB4 cells. Biologija 4:88–93

Borutinskaite V, Bauraite-Akatova J, Navakauskiene R (2016) Anti-leukemic activity of DNA methyltransferase inhibitor procaine targeted on human leukaemia cells. Open Life Sci 11(1):322–330. https://doi.org/10.1515/biol-2016-0044

Borutinskaite VV, Savickiene J, Navakauskiene R (2008) Apoptotic effects of the novel histone deacetylase inhibitor BML-210 on HeLa cells. Biologija (3):217–220

Borutinskaite VV, Magnusson KE, Navakauskiene R (2011) α-Dystrobrevin distribution and association with other proteins in human promyelocytic NB4 cells treated for granulocytic differentiation. Mol Biol Rep 38(5):3001–3011. https://doi.org/10.1007/s11033-010-9965-9

Kulyte A, Navakauskiene R, Treigyte G, Gineitis A, Magnusson KE (2001) Parallel assessment of tyrosine phosphorylation and nuclear targeting of proteins. Biotechniques 31(3):510+

Kulyte A, Navakauskiene R, Treigyte G, Gineitis A, Bergman T, Magnusson KE (2002) Characterization of human alpha-dystrobrevin isoforms in HL-60 human promyelocytic leukemia cells undergoing granulocytic differentiation. Mol Biol Cell 13(12):4195–4205. https://doi.org/10.1091/mbc.E02-03-0128

Laptik R, Navakauskas D (2005) Application of artificial neural networks for the recognition of car number plates. Elektronika ir elektrotechnika 8(64):27–31. In Lithuanian

Laptik R, Navakauskas D (2007) Application of ant colony optimization for image segmentation. Elektronika ir elektrotechnika 8(80):13–18

Laptik R, Navakauskas D (2009) MAX-MIN Ant System in image processing. Elektronika ir elektrotechnika 1(89):21–24

Matuzevičius D, Zurauskas E, Navakauskiene R, Navakauskas D (2008) Improved proteomic characterization of human myocardium and heart conduction system by using computational methods. Biologija 4:283–289

Matuzevičius D (2010a) Automatinė dvimatės elektroforezės gelių su mažiausiais geometriniais iškraipymais atranka. Mokslas – Lietuvos ateitis: Elektronika ir elektrotechnika 2(1):9–13

Matuzevičius D (2010b) Dvimatės elektroforezės gelių vaizdų analizė taikant intelektualiuosius metodus. daktaro disertacija, Vilniaus Gedimino technikos universitetas

Matuzevičius D, Navakauskas D (2005) Investigation of segmentation methods for proteomics. Elektronika ir elektrotechnika 7(63):66–70. In Lithuanian

Matuzevičius D, Navakauskas D (2008) Feature selection for segmentation of 2-D electrophoresis gel images. In: Proceedings of the 11th international biennial baltic electronics conference, BEC 2008, Tallinn, pp 341–344

Matuzevičius D, Navakauskas D (2010) Comparison of distance measures according to suitability for 2D electrophoresis image registration using synthetic image data and SOFN. In: Romaniuk RS, Kulpa KS (eds) Photonics applications in astronomy, communications, industry, and high-energy physics experiments, Bellingham, WA: SPIE, proceedings of SPIE, vol 7745, p CID 7745 16

Matuzevičius D, Serackis A, Navakauskas D (2007) Mathematical models of oversaturated protein spots. Elektronika ir elektrotechnika 1(73):63–68

Matuzevičius D, Serackis A, Navakauskas D (2010) Application of K-Means and MLP in the automation of matching of 2DE gel images. In: Lecture notes in computer science: proceedings of the 20th international conference on artificial neural networks, ICANN 2010. Springer, Thessaloniki, vol 1, pp 541–550

Merzvinskyte R, Treigyte G, Savickiene J, Magnusson KE, Navakauskiene R (2006) Effects of histone deacetylase inhibitors, sodium phenyl butyrate and vitamin B3, in combination with retinoic acid on granulocytic differentiation of human promyelocytic leukemia HL-60 cells. In: Diederich M (ed) Signal transduction pathways, PT B: stress signaling and transcriptional control, Annals of the New York Academy of Sciences, vol 1091, pp 356–367. https://doi.org/10.1196/annals.1378.080. Cell Signaling World 2006 Conference, Luxembourg, Jan 25–28, 2006

Navakauskas D (2005) Grid computing for proteomics. Notes of Seminar at The Lithuanian Academy of Sciences

Navakauskiene R (2017) Combination epigenetic therapy. In: Tollefsbol T (ed) Handbook of Epigenetics: the new molecular and medical genetics, 2nd edn. Academic Press, Cambridge; Elsevier, Amsterdam, chap 41, pp 623–632

Navakauskiene R, Savickiene J, Treigyte G (2000) Protein tyrosine phosphorylation modulates apoptosis and differentiation of human promyelocytic leukemia HL-60 cells. Biomedicine 2(1):23–29

Navakauskiene R, Treigyte G, Pivoriunas A, Savickiene J (2002) Cell cycle inhibitors in retinoic acid- and etoposide-mediated biological responses. Biologija 2:64–67

Navakauskiene R, Kulyte A, Treigyte G, Gineitis A, Magnusson KE (2003a) Translocation of transcription regulators into the nucleus during granulocyte commitment of HL-60 cells. Biochem Cell Biol 81(4):285–295. https://doi.org/10.1139/O03-055

Navakauskiene R, Treigyte G, Kulyte A, Magnusson KE (2003b) Proteomic analysis by MALDI-TOf mass spectrometry and its application to HL-60 cells. Biologija 3:63–65

Navakauskiene R, Treigyte G, Gineitis A, Magnusson KE (2004a) Identification of apoptotic tyrosine-phosphorylated proteins after etoposide or retinoic acid treatment of HL-60 cells. Proteomics 4(4):1029–1041. https://doi.org/10.1002/pmic.200300671

Navakauskiene R, Treigyte G, Savickiene J, Gineitis A, Magnusson KE (2004b) Alterations in protein expression in HL-60 cells during etoposide-induced apoptosis modulated by the caspase inhibitor ZVAD.fmk. In: Diederich M (ed) Signal transduction pathways, chromatin structure, and gene expression mechanisms as therapeutic targets, Fdn Rech Canc & Sang; Novartis Luxembourg; Q8 Petr, Annals of the New York Academy of Sciences, vol 1030, pp 393–402, https://doi.org/10.1196/annals.1329.0049

Navakauskiene R, Treigyte G, Borutinskaite VV, Matuzevicius D, Navakauskas D, Magnusson KE (2012) Alpha-dystrobrevin and its associated proteins in human promyelocytic leukemia cells induced to apoptosis. J Proteomics 75(11):3291–3303. https://doi.org/10.1016/j.jprot.2012.03.041

Navakauskiene R, Borutinskaite VV, Treigyte G, Savickiene J, Matuzevicius D, Navakauskas D, Magnusson KE (2014) Epigenetic changes during hematopoietic cell granulocytic differentiation – comparative analysis of primary CD34+cells, KG1 myeloid cells and mature neutrophils. BMC Cell Biol 15:4. https://doi.org/10.1186/1471-2121-15-4

Pivoriūnas A, Surovas A, Borutinskaitė V, Matuzevičius D, Treigytė G, Savickienė J, Tunaitis V, Aldonytė R, Jarmalavičiūtė A, Suriakaitė K, Liutkevičius E, Venalis A, Navakauskas D, Navakauskienė R, Magnusson KE (2010) Proteomic analysis of stromal cells derived from the dental pulp of human exfoliated deciduous teeth. Stem Cells Dev 19(7):1081–1093

Pivoriunas A, Navakauskiene R, Gineitis A (2004) Inhibition of phosphatidylinositol 3-kinase activity blocks nuclear accumulation of protein kinase zeta during granulocytic differentiation of HL-60 cells. Biologija 2:46–48

Pivoriunas A, Savickiene J, Treigyte G, Tunaitis V, Navakauskiene R, Magnusson KE (2007) PI3-K signaling pathway suppresses PMA-induced expression of p21WAF1/Cip1 in human leukemia cells. Mol Cell Biochem 302(1–2):9–18. https://doi.org/10.1007/s11010-007-9419-4

Savickiene J, Kazlauskaite N, Characiejus D, Treigyte G, Navakauskiene R (2002) Combined differentiating effects of retinoic acid and nucleoside analogues on acute promyelocytic leukemia. Acta Medica Lituanica 9:22–27

Savickiene J, Treigyte G, Magnusson KE, Navakauskiene R (2005) p21 (Waf1/Cip1) and FasL gene activation via Sp1 and NFκb is required for leukemia cell survival but not for cell death induced by diverse stimuli. Int J Biochem Cell Biol 37(4):784–796

Savickiene J, Borutinskaite VV, Treigyte G, Magnusson KE, Navakauskiene R (2006a) The novel histone deacetylase inhibitor BML-210 exerts growth inhibitory, proapoptotic and differentiation stimulating effects on the human leukemia cell lines. Eur J Pharmacol 549(1–3):9–18. https://doi.org/10.1016/j.ejphar.2006.08.010

Savickiene J, Treigyte G, Borutinskaite V, Navakauskiene R, Magnusson KE (2006b) The histone deacetylase inhibitor FK228 distinctly sensitizes the human leukemia cells to retinoic acid-induced differentiation. In: Diederich M (ed) Signal transduction pathways, PT B: stress signaling and transcriptional control, Annals of the New York Academy of Sciences, vol 1091, pp 368–384. https://doi.org/10.1196/annals.1378.081. Cell Signaling World 2006 Conference, Luxembourg, JAN 25–28, 2006

Savickiene J, Treigyte G, Magnusson KE, Navakauskiene R (2009) Response of retinoic acid-resistant KG1 cells to combination of retinoic acid with diverse histone deacetylase inhibitors. In: Diederich M (ed) Natural compounds and their role in apoptotic cell signaling pathways, Annals of the New York Academy of Sciences, vol 1171, pp 321–333. https://doi.org/10.1111/j.1749-6632.2009.04718.x

Savickiene J, Treigyte G, Aleksandraviciene C, Navakauskiene R (2010a) Low-dose ionizing radiation effects on differentiation of HL-60 cells. Cent Eur J Biol 5(5):600–612. https://doi.org/10.2478/s11535-010-0085-2

Savickiene J, Treigyte G, Gineitis A, Navakauskiene R (2010b) A critical role of redox state in determining HL-60 cell granulocytic differentiation and apoptosis via involvement of PKC and NF-kappa B. In Vitro Cell Dev Biol Anim 46(6):547–559. https://doi.org/10.1007/s11626-010-9296-0

Savickiene J, Treigyte G, Vistartaite G, Tunaitis V, Magnusson KE, Navakauskiene R (2011) C/EBPalpha and PU.1 are involved in distinct differentiation responses of acute promyelocytic leukemia HL-60 and NB4 cells via chromatin remodeling. Differentiation 81(1):57–67. https://doi.org/10.1016/j.diff.2010.08.003

Savickiene J, Treigyte G, Borutinskaite VV, Navakauskiene R (2012a) Antileukemic activity of combined epigenetic agents, DNMT inhibitors zebularine and RG108 with HDAC inhibitors, against promyelocytic leukemia HL-60 cells. Cell Mol Biol Lett 17(4):501–525. https://doi.org/10.2478/s11658-012-0024-5

Savickiene J, Treigyte G, Jazdauskaite A, Borutinskaite VV, Navakauskiene R (2012b) DNA methyltransferase inhibitor RG108 and histone deacetylase inhibitors cooperate to enhance NB4 cell differentiation and E-cadherin re-expression by chromatin remodeling. Cell Biol Int 36(11):1067–1078

Savickiene J, Treigyte G, Jonusiene V, Bruzaite R, Borutinskaite VV, Navakauskiene R (2012c) Epigenetic changes by zebularine leading to enhanced differentiation of human promyelocytic leukemia NB4 and KG1 cells. Mol Cell Biochem 359(1–2):245–261. https://doi.org/10.1007/s11010-011-1019-7

Savickiene J, Treigyte G, Stirblyte I, Valiuliene G, Navakauskiene R (2014a) Euchromatic histone methyltransferase 2 inhibitor, BIX-01294, sensitizes human promyelocytic leukemia HL-60 and NB4 cells to growth inhibition and differentiation. Leuk. Res. 38(7):822–829

Savickiene J, Treigyte G, Valiuliene G, Stirblyte I, Navakauskiene R (2014b) Epigenetic and molecular mechanisms underlying the antileukemic activity of the histone deacetylase inhibitor belinostat in human acute promyelocytic leukemia cells. Anti-Cancer Drugs 25(8):938–949. https://doi.org/10.1097/CAD.0000000000000122

Serackis A, Navakauskas D (2008) Reconstruction of overlapped protein spots using RBF networks. Elektronika ir elektrotechnika 1(81):83–88

Serackis A, Matuzevičius D, Navakauskas D (2006) Reconstruction of protein spots using DSP modules. In: Proceedings of 29th international conference on fundamentals of electrotechnics and circuit theory, IC-SPETO 2006, Gliwice-Ustron, vol 2, pp 573–576

Serackis A, Matuzevičius D, Navakauskas D (2010) 2DE gel image preprocessing using self-organizing maps. In: Romaniuk RS, Kulpa KS (eds) Photonics applications in astronomy, communications, industry, and high-energy physics experiments, Washington: SPIE, proceedings of SPIE, vol 7745, p CID 7745 1N

Treigytė G, Zaikova I, Matuzevičius D, Čeksterytė V, Dabkevičienė G, Kurtinaitienė B, Navakauskienė R (2014) Comparative proteomic analysis of pollen of trifolium pratense, t. alexandrinum and t. repens. Zemdirbyste-Agriculture 101(4):453–460

Treigyte G, Navakauskiene R, Kulyte A, Gineitis A, Magnusson KE (2000a) Characteristics of cytosolic proteins and changes in their tyrosine phosphorylation during HL-60 cell differentiation. Biologija 2:32–35

Treigyte G, Navakauskiene R, Kulyte A, Gineitis A, Magnusson KE (2000b) Tyrosine phosphorylation of cytoplasmic proteins in proliferating, differentiating, apoptotic HL-60 cells and blood neutrophils. Cell Mol Life Sci 57(13–14):1997–2008. https://doi.org/10.1007/PL00000681

Treigyte G, Navakauskiene R, Kulyte A, Gineitis A, Magnusson KE (2000c) Tyrosine phosphorylation of cytoplasmic proteins in proliferating, differentiating, apoptotic HL-60 cells and blood neutrophils. Cell Mol Life Sci 57(13–14):1997–2008. https://doi.org/10.1007/PL00000681

Treigyte G, Savickiene J, Navakauskiene R (2003) Changes in O- and N- glycosylation of cytoplasmic proteins in proliferating HL-60 cells and in those induced to granulocytic differentiation. Biologija 3:36–38

Treigyte G, Savickiene J, Navakauskiene R (2004) Identification of O- and N-glycosylated nuclear proteins of HL-60 cells induced to granulocytic differentiation. Biologija 2:49–51

Treigyte G, Savickiene J, Pivoriunas A, Navakauskiene R (2006) Effects of inhibition of phosphoinositide 3-kinase and p53 on monocytic differentiation driven leukaemia cells with different p53 status. Biologija 2:93–98

Valiulienė G, Treigytė G, Savickienė J, Matuzevičius D, Alksnė M, Jarašienė-Burinskaja R, Bukelskienė V, Navakauskas D, Navakauskienė R (2016) Histone modifications patterns in tissues and tumours from acute promyelocytic leukemia xenograft model in response to combined epigenetic therapy. Biomed Pharmacother 79:62–70. https://doi.org/10.1016/j.biopha.2016.01.044

Valiuliene G, Stirblyte I, Cicenaite D, Kaupinis A, Valius M, Navakauskiene R (2015) Belinostat, a potent HDACi, exerts antileukaemic effect in human acute promyelocytic leukaemia cells via chromatin remodelling. J Cell Mol Med 19(7):1742–1755. https://doi.org/10.1111/jcmm.12550

Valiuliene G, Stirblyte I, Jasnauskaite M, Borutinskaite V, Navakauskiene R (2017) Anti-leukemic effects of HDACi belinostat and HMTi 3-Deazaneplanocin A on human acute promyelocytic leukemia cells. Eur J Pharmacol 799:143–153. https://doi.org/10.1016/j.ejphar.2017.02.014

Vitkeviciene A, Baksiene S, Borutinskaite V, Navakauskiene R (2018) Epigallocatechin-3-gallate and BIX-01294 have different impact on epigenetics and senescence modulation in acute and chronic myeloid leukemia cells. Eur J Pharmacol 838:32–40. https://doi.org/10.1016/j.ejphar.2018.09.005

Vitkeviciene A, Skiauteryte G, Zucenka A, Stoskus M, Gineikiene E, Borutinskaite V, Griskevicius L, Navakauskiene R (2019) HDAC and HMT inhibitors in combination with conventional therapy: a novel treatment option for acute promyelocytic leukemia. J Oncol 2019:11. https://doi.org/10.1155/2019/6179573

Vitkevičienė A, Janulis V, Žučenka A, Borutinskaitė V, Kaupinis A, Valius M, Griškevičius L, Navakauskienė R (2019) Oxidative phosphorylation inhibition induces anticancerous changes in therapy-resistant–acute myeloid leukemia patient cells. Mol Carcinog 58(11):2008–2016

Contents

Notations

Text Parts

The following markup is used in order to outline the context of word or their groups:

- *important text*;
- cell (process);
- proteins and protein Uniprot names;
- gene;
- chromosome alterations;
- histone;
- drug or chemical;
- `software or program code`.

Models are numbered chapterwise and have their titles (cf. Model 0.0).

Model 0.0 (Title)

Model description.

Algorithms are numbered chapterwise, have their titles, and are outlined by horizontal rules (cf. Algorithm 0.0).

Algorithm 0.0

Title

Algorithm description.

General Setup

a —Scalar (number)
$\boldsymbol{a}$ —Vector (column of numbers)
$\boldsymbol{A}$ —Matrix (multidimensional number array)
$\mathcal{A}$ —Dataset (set of numbers)
$\mathfrak{a}$ —Class (criteria)
$\cdot^{\text{feat.}}$ —Additional feature indicator
$\cdot^{(i)}$ —Main index
$\cdot_{\text{feat.}}$ —Principal feature indicator
$\cdot_{i}$ —Additional index

Generalized Notations

$\cdot^{\|}, \cdot^{\angle}$ —(Vector) length and direction
$\cdot^{\rightarrow]}, \cdot^{\mapsto}$ —Input and output
$\cdot^{\bullet(r)}, \cdot^{\circ(r)}$ —rth Region and contour
$\cdot^{\text{feat.}+}, \cdot^{\text{feat.}-}$ —With and without the feature
$\cdot^{\diamond}$ —Aimed (true) value
$\cdot^{\star}$ —The best (calculated) value
$\cdot^{\star}$ —Important image region and point
$\cdot^{\top}, \cdot^{\perp}$ —Maximum and minimum value

Symbols

$\boldsymbol{C}^{\mathcal{W}}_{\mathrm{G}r}, \mathcal{C}^{\mathcal{W}}_{\mathrm{G}}$ —2DEG image watershed transformation rth contour and contour set
d_{tr} —Image warp diameter
d_{R} —Correlation matrix threshold
$\boldsymbol{D}$ —Image structural element
$e_h(n)$ —Instantaneous hth output error or uncertainty
E —Mean square error
$E(n)$ —Instantaneous mean square error
$\mathscr{E}_t$ —Objective function (validity index), here, t: S—Silhouette, CH—Calinski and Harabasz, D—Dunn, A—separation, C—Hubert and Levin (C index), STV—weighted inter–intra
$\mathcal{E}$ —Cumulative mean square error
$\Phi^{(l)}_t(\cdot)$ —Artificial neuron in lth layer activation function, here t: TS—hyperbolic tangent, LS—logistic sigmoid, T—linear, S—binary step, G—Gaussian function, 2G—2-D Gaussian function, L—logarithmic distribution, K—quadratic, IK—inverse quadratic
$\boldsymbol{\phi}(\cdot)$ —Artificial neuron neighborhood function
$H(\cdot), H(\cdot,\cdot)$ —Marginal and joint entropy
$\boldsymbol{H}$ —Hessian matrix

i, j, h	—Indexes of inputs, weights (coefficients) and outputs
$I(x, y)$	—Image pixel intensity
I_{R}	—Intensity value in specified image area R
$I_{\mathrm{p}}(x, y)$	—Protein spot intensity
$I^{\perp}$	—Image minimal intensity value
$I_{\mathrm{OT}}^{(r)}$	—Protein spot optical density
$\boldsymbol{I}, \mathcal{I}$	—Natural (digital) image and image set
$\boldsymbol{I}_{\mathrm{B}}, \boldsymbol{I}_{\mathrm{G}}$	—Binary and 2DEG image
$\boldsymbol{I}^{\mathrm{F+}}, \boldsymbol{I}^{\mathrm{F-}}$	—Only image background and image without background intensity
$\boldsymbol{I}^{\mathrm{rot}}$	—Image rotational symmetry intensity map
$\boldsymbol{I}^{\mathcal{F}}$	—Image complex spectrum
$\boldsymbol{I}$	—Matrix of ones
$\boldsymbol{J}$	—Jacobian matrix
$\boldsymbol{K}$	—Image filter mask
L	—Number of ANN layers
$L(\theta \| \boldsymbol{x})$	—Probability function
λ	—Eigenvalue
M	—Order (model, filter, etc.)
M_{M}	—Polypeptide (protein) molecular mass
μ	—ANN training step
n	—Time (sample) index
N	—Total number (samples, neurons, etc.)
$N^{\star}$	—Successful experiment's number
$N_{\mathrm{y}}^{\mathrm{A}}, N_{\mathrm{x}}^{\mathrm{A}}$	—Gel active part height and width
$\nabla^{\parallel}\boldsymbol{I}, \nabla^{\angle}\boldsymbol{I}$	—Image intensity gradient and direction
ν	—ANN training inertion constant
$\nu^{(r)}$	—Protein spot relative migration distance
$p(i), p(i, j)$	—Probability and mutual probability
$p^{\top}, p^{\perp}$	—Probability upper and lower limits
$\Psi_t(x, y; \boldsymbol{\theta}_t)$	—Protein spot models, here t: M1G—circularly symmetric 2-D Gaussian, M2G—two-way adapting 2-D Gaussian, M3G—three Gaussian function, M4G—four Gaussian functions, M4GFT—four Gaussian functions with flat top, MD—diffusion, MP—simple Π-shaped, MPFT—Π-shaped with flat top, MBS—symmetric bell shape, MBAS—asymmetric bell shape, MSS—two-way symmetric sigmoid-based, MSAS—asymmetric sigmoid-based, MD—diffusion
r	—Area or iteration index
$\boldsymbol{R}$	—Correlation matrix
$S_{\mathrm{tr}}^{\top}$	—Maximal image warping area
$\boldsymbol{S}_{xy}$	—(Rectangular) area with center coordinates x and y
$\boldsymbol{S}_{\mathrm{G}ij}^{\star}, \mathcal{S}_{\mathrm{G}i}^{\star}$	—Important in 2DEG ith image jth area and area set
$\boldsymbol{S}_{\mathrm{G}r}^{\mathcal{W}}, \mathcal{S}_{\mathrm{G}}^{\mathcal{W}}$	—2DEG image watershed transformation rth area and area set
σ	—Standard deviation

$\overline{\sigma}$ —Mean absolute deviation
θ_{r} —Protein radial diffusion area radius
θ_{R} —Diffusion coefficient
$\theta_{\mathrm{C_0}}$ —Initial protein concentration
$\theta_{\mathrm{Dx}}, \theta_{\mathrm{Dy}}$ —Diffusion constants in specified directions
$\theta_{\mathrm{x}}^{\mathrm{a}}, \theta_{\mathrm{x}}^{\mathrm{b}}, \theta_{\mathrm{x}}^{\mathrm{c}}, \theta_{\mathrm{x}}^{\mathrm{d}}$ —Protein spot model shape parameters
$V^{(r)}$ —Protein spot total optical density (volume)
$\boldsymbol{w}$ —System parameters
(x, y) —Image pixel coordinates
$(x_{\mathrm{m}}, y_{\mathrm{m}})$ —Binary image area mass center coordinates
$(x_{\mathrm{c}}, y_{\mathrm{c}})$ —Protein spot center coordinates
$\left(x_{\mathrm{c}}^{(r)}, y_{\mathrm{c}}^{(r)}\right)$ —(rth) Protein spot center coordinates
$\boldsymbol{x}, \boldsymbol{z}, \boldsymbol{y}$ —Multidimensional system input, state, and output
$\boldsymbol{x}_{\mathrm{m}}$ —Areas mass center vector calculated from a set of binary images
ξ_r —Random numbers in interval [0, 1]
$\mathbf{Z}, \mathcal{Z}$ —Simulated (synthetic) image and image set

Operators

$*$ —Convolution
$\cdot$ —Scalar multiplication
$\boxdot$ —Elementwise multiplication (array multiplication)
$\oplus$ —Morphological dilation
$\ominus$ —Morphological erosion
$\circ$ —Morphological opening
$\bullet$ —Morphological closing
$\equiv$ —Equivalence
$\Rightarrow$ —Implication
$\rightarrow$ —Substitution
$\longrightarrow$ —Mapping
$\triangleq$ —Definition
$\langle \cdot, \cdot \rangle$ —Correspondence
$|\cdot|$ —Absolute value (modulus)
$\|\cdot\|$ —Norm (length)
$\widehat{\cdot}$ —Estimate
$\overline{\cdot}$ —Mean value (constant constituent)
$\cdot^{\mathrm{T}}$ —Transposition
$\lfloor \cdot \rfloor, \lceil \cdot \rceil$ —Rounding toward $-\infty$ and $+\infty$
$\cdot', \cdot''$ —Derivative: first and second order
$\partial \cdot$ —Partial derivative
$D_r \cdot = \partial \cdot / \partial \Theta_r$ —Partial derivative with respect to rotation angle Θ_r
$\mathbb{E}(\cdot)$ —Mathematical expectation
$\nabla_x(n)$ —Instantaneous gradient with respect to x
z, z^{-1} —Unit delay and advancement

dist(.)	—Distance function
$(x \text{ div } y)$	—Remainder of x division by y
$\varepsilon(.)$	—Step function
entropy(.)	—Entropy function
$\max\limits_{T} f(n)$	—Maximum of function $f(n)$ in period T
median(.)	—Median function
$(x \bmod y)$	—Modulus of x division by y
mode(.)	—Mode (most frequent value) function
std(.)	—Standard deviation function
rank(.)	—Ranking function
$\mathcal{F}(.)$, $\mathcal{F}^{1}(.)$	—Direct and inverse Fourier transform
Re(.), Im(.)	—Real and imaginary parts of complex number

Abbreviations

2DE	– two-dimensional electrophoresis;
2DEG	– two-dimensional electrophoresis gel;
ALL	– acute lymphoid leukemia;
AML	– acute myeloid leukemia;
AMP	– adenosine monophosphate;
ANN	– artificial neural network;
APL	– acute promyelocytic leukemia;
AraC	– cytarabine;
ATO	– arsenic trioxide;
AU	– acid-urea;
BET	– bromodomain and extra-terminal motif (proteins);
CD	– cluster of differentiation (membrane antigens);
ChIP	– chromatin immunoprecipitation;
CLL	– chronic lymphoid leukemia;
CML	– chronic myeloid leukemia;
CMML	– chronic myelomonocytic leukemia;
CpG	– regions of DNA where a cytosine nucleotide is followed by a guanine nucleotide in the linear sequence of bases along its 5' → 3' direction;
CS	– central symetry;
DAG	– diacylglycerol;
DNA	– deoxyribonucleic acid;
DNMT	– DNA methyltransferase;
DNMTi	– DNA methyltransferase inhibitor;
EMSA	– electrophoretic mobility shift assay (method);
ERK	– extracellular signal-regulated kinases;
FAB	– French-American-British (organization);
FDA	– (United States) Food and Drug Administration;
FITC	– fluorescein isothiocyanate;
HDAC	– histone deacetylase;

HDACi	– histone deacetylase inhibitor;
HMT	– histone methyltransferase;
HMTi	– histone methyltransferase inhibitor;
HNE	– human neutrophil elastase;
HSC	– hematopoietic pluripotent stem cell;
IPG	– imobilized pH gradient (bar);
LoG	– Laplassian and Gaussian (filter or kernel);
LS	– logistic sigmoid (neuron activation function);
LSC	– leukemic stem cell;
M1G	– circularly-symmetric 2-D Gaussian model;
M2G	– 2-way adapting 2-D Gaussian model;
M3G	– three Gaussian function model;
M4G	– four Gaussian functions model;
M4GFT	– definition;
MALDI-MS	– matrix-assisted laser desorption/ionization mass spectrometry;
MBAS	– asymmetric bell shape model;
MBS	– two-way symmetric bell shape model;
MD	– diffusion model;
MIR	– multiresolution image registration;
MLP	– multilayer perceptron;
MM	– molecular mass;
MP	– simple Π-shaped model;
MPft	– Π-shape model with flat top;
MRP	– multidrug resistance proteins;
MS	– mass spectrometry;
MSAS	– asymmetric sigmoid-based model;
MSS	– symmetric sigmoid-based model;
MW	– molecular weight;
PB	– phenylbutyrate;
PDGFR	– platelet derived growth factor receptor;
PFT	– pifithrin;
pI	– isoelectric point;
PI2	– propidium iodide;
PKA	– protein kinase A;
PKC	– protein kinase C;
PLD	– phospholipase D;
PMA	– phorbol myristate acetate;
PML	– promyelocytic leukemia;
PTK	– protein tyrosine kinase;
RAIN	– robust automated image normalization;
RANSAC	– random sample consensus (algorithm);
RAR	– retinoic acid receptor;
RARE	– all-trans retinoic response elements;
RNA	– ribonucleic acid;
SAHA	– suberoylanilide hydroxamic acid;

SEM	– standard error of the mean;
SOM	– self-organizing map;
TS	– hyperbolic tangent (neuron activation function);
WHO	– World Health Organization;
ZEB	– zebularine.

List of Figures

List of Tables

Part I
Epigenetics in Leukemia

Chapter 1
Epigenetic and Genetic Alterations in Leukemia

Cancer develops due to genetic and epigenetic changes. Genetic alterations are caused by aging, ultraviolet light, radiation, chemicals, and other factors. Meanwhile, epigenetic changes are mainly caused by aging and chronic inflammation. The accumulation of cancer cells and the emergence of genetic and epigenetic changes reflect our previous level of exposure and life history. Most of the accumulated lesions are considered passengers, but their accumulation is associated with the development of cancer. Genetic alterations (mutations, deletions, and translocations of genes required for normal development) in cancer cells do not substantially disappear during cancer therapies. In cancers, including leukemia, genetic changes are very often accompanied by epigenetic changes. However, unlike genetic mutations, epigenetic changes can be regulated by the use of inhibitors of epigenetic modifiers. Managing epigenetic changes that occur during cancer "bypasses" genetic disorders and is a promising way to control cancer. In addition, epigenetic therapy may also help prevent the recurrence of leukemia. Therefore, reprogramming leukemia and cancer cells in general using inhibitors of epigenetic modifiers is a new strategy for cancer therapy.

Leukemia is a disease of hematopoietic cells in which the balance between cell proliferation, differentiation, and apoptosis is disturbed. Usually, the term leukemia (formed from the two Greek words leukos (λευκός), meaning—white, and haima (αἷμα), meaning—blood) refers to cases of cancer in which the production and/or function of white blood cells (leukocytes) is impaired. Depending on whether the malignancies occurred in the myeloid or lymphoid lineage, leukemia is distinguished into myeloid or lymphoid, respectively, depending on how rapidly it develops—into chronic or acute. The latter is more dangerous due to the faster proliferation of malignant cells. In acute leukemia, the altered blood cells are immature and unable to perform the functions typical of healthy blood cells. Meanwhile, in chronic leukemia, most cells are partially functional.

R. Navakauskienė et al., *Epigenetics and Proteomics of Leukemia*,
https://doi.org/10.1007/978-3-030-68708-3_1

1.1 Hematopoiesis

The process of formation of blood cellular components is called hematopoiesis. During hematopoiesis mature blood cells develop from immature precursor cells. Blood cell development includes erythropoiesis, granulopoiesis, lymphopoiesis, and monocytopoiesis. Each day in adult individuals, 2.5 billion erythrocytes, 2.5 billion platelets, and 50–100 billion granulocytes develop. The hematopoiesis process in adult individuals occurs predominantly in the bone marrow (the extravascular compartment), which consists of parenchyma of developing plasma cell, blood cells, fat cells, and macrophages, and stroma of reticular connective tissue. In bodies of adults, it occurs mainly in the pelvis, vertebrae, sternum, and cranium (Fig. 1.1a). Hematopoiesis in children occurs in the marrow of the long bones (e.g., femur and tibia). In the bone marrow, a tightly controlled local microenvironment/niche exists, which emanates regulatory signals (bound or secreted molecules). These signals regulate the quiescence, differentiation, and proliferation of hematopoietic pluripotent stem cells (HSCs) (Morrison and Scadden 2014).

HSCs can differentiate into cells of lower potency and, in turn, into blood components (Fig. 1.1b). All lymphoid, myeloid, megakaryocytic, and erythrocyte blood cells originate from the HSC. The remainder of the HSC progeny becomes myeloid and lymphoid progenitor cells; these cells begin their alternative pathways of differentiation into certain blood cells. Figure 1.1b shows the formation of the major hematopoietic elements. As mentioned earlier, the division of HSC cells results in the formation of so-called lymphoid and myeloid primary cells. In particular, HSC cells produce multipotent precursors, which further differentiate into primary lymphoid multipotent precursors and total myeloid precursors (some sources also indicate that multipotent precursors develop into lymphoid-primed multipotent progenitor) and only then into primary lymphoid multipotent precursors and granulocytic-monocytes precursors. Granulocytic-monocytes precursors are formed from primary lymphoid multipotent precursors or total myeloid precursors, but only total myeloid precursors can give rise to megakaryocytic-erythrocyte precursors. These stem cells can further divide and form differentiated cells. Primary lymphoid multipotent precursor cells through few intermediate stages differentiate into natural killer cells, T and B lymphocytes, and dendritic cells. Cells in the myeloid lineage include erythrocytes, platelets, monocytes/macrophages, and granulocytes (basophils, neutrophils, and eosinophils). Megakaryocytic-erythrocyte precursors further develop into megakaryocytes (further into platelets (Plt) and erythrocytes (E)); myeloid precursors develop into dendritic cells (DC), eosinophils, basophils, and granulocytic-monocytes precursors (further into neutrophils and monocytes). So far, there is no detailed evidence, but two lines of dendritic cell development are distinguished: myeloid lines are shared with phagocyte and lymphoid lines shared with T cells. The life span of already differentiated cells is very variable, ranging from several years for T and B cells and 3 months for erythrocytes to several days for granulocytes (Fiedler and Brunner 2012; Borutinskaite et al. 2008; Wilson et al. 2007).

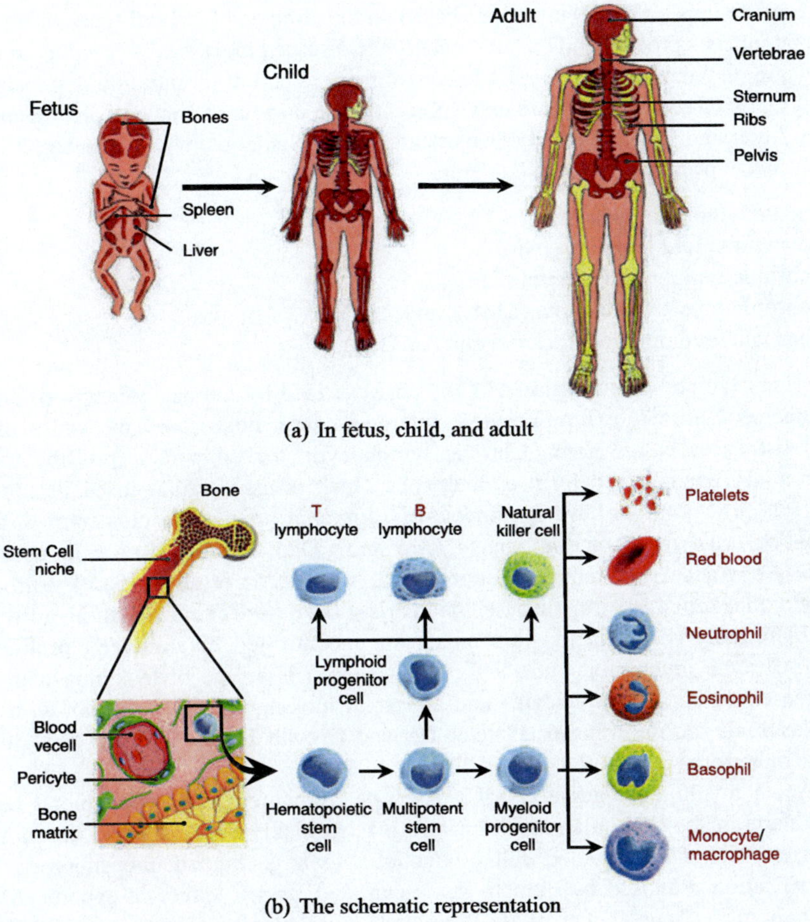

Fig. 1.1 The hematopoiesis process

1.2 Types of Leukemia

Leukemia is defined as the non-controllable proliferation of hematopoietic cells that are unable to develop normally into mature blood cells (Sawyers et al. 1991). There were 18 million cancer cases estimated around the world in 2018 according to World Cancer Research Fund information. And leukemia cases account for 2.6% of all types of cancers (except non-melanoma skin cancer). The disease can be acute when it progresses rapidly and, if untreated, is fatal within months or chronic, when the number of malignant cells increases slowly and the patient usually does not experience any symptoms for several years.

Leukemia is classified into types based on the affected blood cell type during the hematopoiesis process and on how quickly the disease progresses. According to this division, leukemia can be myeloid—when myeloid cell differentiation is impaired, or lymphoid—when lymphoid cell differentiation into functional cells is impaired. The American Cancer Society (American Cancer Society, www.cancer.org, 2019) identifies 5 principal types of leukemia:

- acute lymphoid leukemia (ALL),
- acute myeloid leukemia (AML),
- chronic lymphoid leukemia (CLL),
- chronic myeloid leukemia (CML), and
- chronic myelomonocytic leukemia (CMML).

Acute lymphoid leukemia (ALL) is characterized by the development of large numbers of immature lymphocytes, enlarged lymph nodes, and the cause of it in most cases is unknown. Chronic lymphocytic leukemia (CLL) is the most commonly encountered form of leukemia which is usually diagnosed in adults. Patients are found to have lymphocytosis, lymphocyte-specific cluster of differentiation (CD) membrane antigens—CD5 and CD23, and therefore cells usually have a characteristic immunophenotype (Grigoropoulos et al. 2013a,b). Chronic myeloid leukemia is a rare disease which arises from t(9;22) translocation, a fusion of BCR and ABL proteins that create an uncontrolled myeloid cell proliferation. Chronic myelomonocytic leukemia (CMML) is a type of leukemia with an increased number of monocytes and abnormal looking cells (dysplasia), with no Philadelphia chromosome or Platelet Derived Growth Factor Receptor (PDGFR) gene mutations—PDGFRA or PDGFRB.

Acute myeloid leukemia (AML) develops from premature white blood cells and starts in the bone marrow. Immature blasts quickly move into the blood and interfere with normal blood cell production. AML is divided into subtypes the classification proposed by French, American, and British leukemia experts (M0–M7 subtypes) (Bennett et al. 1976; Bloomfield and Brunning 1985; Lee et al. 1987). The classification proposed by The World Health Organization (WHO) system includes genes or chromosome changes that occur during AML formation (Vardiman et al. 2002; Falini et al. 2010).

1.2.1 Chronic Leukemia

A slow development is characteristic to this group of leukemia; it is a less aggressive form. Chronic leukemia is identified by the rapid increase of relatively mature, but still abnormal, white blood cells. It may take months or years to progress. Usually, in the case of such leukemia, new white blood cells form very rapidly and therefore in the blood appear abnormally many altered cells. Chronic leukemia is commonly found in the elderly population.

Chronic lymphoid leukemia (CLL) is the most common adult leukemia, it affects lymphocytes, and the primary cancer cells accumulate in bone marrow and lymphoid tissue. Around 95% of CLL cases are the B-cell type cases. And, about 1% of B-cell leukemia cases have a type called B-cell prolymphocytic leukemia. The T-cell type of CLL is now called T-cell prolymphocytic leukemia. In contrast to other mature B-cell neoplasms, CLL is not characterized by the presence of specific chromosomal translocations. Complex karyotypes are found in 10%; unrelated clones demonstrating the existence of cell subpopulations are frequent findings in this disease:

- trisomy12, found in 15–20% of CLL cases,
- del(13)(q14.3), found in 10–60% CLL cases,
- deletion 11q22-q23, involving ATM gene, is detected in 11–18% of CLL cases,
- deletion 17p13 (p53) occurs in 7–8% of CLL cases that are resistant to chemotherapy and have a short survival,
- 14q32:t(11;14)(q13;q32) with BCL1/IgH rearrangement and t(14;19)(q32;q13), and 14q32 rearrangements are found in about 4% of CLL cases, and
- del(6q) [0–6% of CLL], +8q24 [5% of CLL], +3, and +18.

Recent data suggest that the identification of CLL-specific DNA methylation, chromatin dynamics, and histone modifications as well as non-coding RNA could accurately explain the prognostic, prognostic, and therapeutic potential of CLL. Overall, the ability of CLL epigenomics to predict treatment response and resistance is increasing (Raval et al. 2006; Mansouri et al. 2018). Chemotherapy and targeted therapy or combination therapy are currently used to treat CLL. Drugs approved and/or in clinical trials for CLL are presented in Appendix A. Combination chemotherapy is more effective in both newly diagnosed and relapsed CLL.

Chronic myeloid leukemia (CML) is a rare type of blood cancer characterized by chromosomal changes—Philadelphia chromosomes, caused by a reciprocal translocation between parts of chromosomes 9 and 22, t(9;22)(q34;q11). This leads to the BCR-ABL1 fusion protein. CML is usually the case in the elderly and rarely in children. Other chromosomal alterations may also be detected by cytogenetic studies. But even when the Ph chromosome fails to identify, the BCR-ABL gene is found.

The successful treatment of most patients requires new therapeutic strategies focused on epigenetic processes—the development of this promising tool would improve the rate of successful therapy. Epigenetic changes occur in CML cases (Bugler et al. 2019). Therefore the epigenetic therapy in combination with currently approved therapy would significantly improve the chances of CML recovery. The following drugs are used to treat CML: targeted therapy (tyrosine kinase inhibitors: imatinib, dasatinib, nilotinib, bosutinib, and ponatinib), chemotherapy (e.g., hydroxyurea), and immunotherapy (e.g., interferon). In Appendix A the drugs and clinical trials for CML treatments are presented.

Chronic myelomonocytic leukemia (CMML) is a relatively rare type of blood cancer. In this form of leukemia, too many monocytes are detected in the blood. The WHO has included CMML in a group of blood cancers called myeloproliferative and myelodysplastic disorders. CMML is diagnosed by long-term (over 3 months) peripheral blood monocytosis (more than 10% of monocytes) and bone marrow dysplasia. CMML gene mutations are associated with the TET2 (~60%), SRSF2 (~50%), ASXL1 (~40%), and oncogenic RAS pathways (~30%) (Patnaik and Tefferi 2020). The ASXL1 protein is a member of the Polycomb group of proteins and is involved in a chromatin remodeling process (Asada et al. 2019). The TET family of proteins plays a role in DNA modification and therefore is involved in regulating the process of transcription. Therefore, epigenetic CMML therapy would be a promising step in the treatment of leukemia.

1.2.2 *Acute Leukemia*

Very fast development is characteristic to this form of leukemia. It is really aggressive form. Most often, there are many poorly differentiated cells in the blood, and cancer cells spread to the other organs. In acute leukemia, a rapid growth of abnormal cells, which accumulate in the bone marrow and blood, is inherent. Rapidly growing immature cells interfere with the production of normal mature blood cells. Acute lymphocytic leukemia (ALL) is the most common cancer in children, and it is the least common type of leukemia in adults.

Acute lymphoblastic leukemia (ALL) is a malignant transformation and proliferation of lymphoid stem cells in the bone marrow, blood, and extra medullary sites. ALL is divided into several types according to which lymphocytes are affected:

1. B-cell ALL—B lymphocytes are affected. These cells, which produce and mature in the bone marrow, play an important immune role in the production of antibodies; and
2. T-cell ALL— T lymphocytes are affected. T lymphocytes are known to be produced in the bone marrow, but these cells mature in the thymus gland. B-cell ALL accounts for about 85% of all cases and T-cell ALL for about 15%. ALL is the most common form of AML among childhood cancers.

The WHO subtypes related to ALL are:

- B-lymphoblastic leukemia/lymphoma:
  - not otherwise specified (NOS),
  - recurrent genetic abnormalities,
  - t(9;22)(q34.1;q11.2): BCR-ABL1,
  - t(v;11q23.3): KMT2A rearranged,
  - t(12;21)(p13.2;q22.1): ETV6-RUNX1,
  - t(5;14)(q31.1;q32.3): IL3-IGH,
  - t(1;19)(q23;p13.3): TCF3-PBX1,

  - hyperdiploidy,
  - hypodiploidy.

- T-lymphoblastic leukemia/lymphoma:
  - acute undifferentiated leukemia,
  - mixed phenotype acute leukemia (MPAL) with t(9;22)(q34.1;q11.2): BCR-ABL1,
  - MPAL with t(v;11q23.3): KMT2A rearranged,
  - MPAL, B/myeloid, NOS,
  - MPAL, T/myeloid, NOS.

The main treatment for ALL is usually chemotherapy. Stem cell therapy is also used in patients who have achieved remission. There are also other treatment options for ALL: radiation therapy, which can be used in combination with chemotherapy, and immunotherapy (see Appendix A). The studies on epigenetics in ALL cells have shown that DNA methylation and other factors of epigenetic regulation demonstrate the importance of epigenetic control in leukemogenesis, but the extended studies are required for epigenetic studies and their translation to epigenetic therapy of ALL (Nordlund and Syvaenen 2018).

Cases of acute myeloid leukemia (AML) are more common among the elderly, and the median age in the diagnosis is about 70 years. The prognosis for older patients with AML is generally considered poor due to other chronic diseases.

Acute myeloid leukemia (AML) is a very aggressive form of acute leukemia. In this disease, many immature, undifferentiated blasts are formed in the bone marrow. Many types of AML are distinguished from other types of blood cancers by their high blood blast content, reaching more than 20% (Seiter 2003). AML is a highly heterogeneous disease. Marked structural cellular changes under microscope and a large variety of chromosomal abnormalities are characteristic to it. AML is characterized by arrest of leukocyte precursor development at some early stages. The inhibition of cellular development takes place in two stages: (1) the production of normal blood cells is significantly reduced, resulting in anemia, thrombocytopenia, and neutropenia and (2) very rapid and uncontrolled proliferation of immature cells and associated reduced ability to kill by apoptosis (De Kouchkovsky and Abdul-Hay 2016). Such immature and intensely dividing cells pile up in blood and bone marrow and very often in the liver and spleen (Seiter 2013). In the case of AML, platelets and abnormal red blood cells may also form in the bone marrow.

Most cases of AML are classified by World Health Organization (WHO) and French–American–British (FAB). According to the FAB system, AML is subdivided into 8 subtypes according to the type of cells from which leukemia develops and the degree of cell maturation. Subtypes are specified based on the appearance of cells under the light microscope and on cytogenetic abnormalities.

Most updated genomic classifications were presented by the Cancer Genome Atlas Research Network and the other researchers (Network 2013; Papaemmanuil et al. 2016). In 2013, the 200 clinically annotated adult de novo AML genomes were analyzed using whole genome (50 cases) and whole exome (150 cases), and microRNA sequencing and DNA methylation analysis were performed. Twenty-three genes that were significantly mutated and 237 genes that were mutated in two or more samples were identified. In 2016, 5234 mutations were identified across 76 genes or genomic regions in 1540 AML patients. Studies of the AML genome reveal individual molecular subgroups, and many of the mutations inherent in the pathogenesis of AML are uncertain. The relationship between gene mutations and epigenetic changes is still unclear; therefore, further studies need to be continued, and personalized treatments should be used for AML patients.

It is very important to discuss the origin of acute myeloid leukemia cells—why and how they occur. During normal hematopoiesis, HSCs are known to differentiate into mature blood cells through intermediate cell populations. Each step of differentiation is strictly controlled by the involvement of both internal and external factors. The molecular pathogenesis of AML has not been fully proven. Prognostic and diagnostic markers have been identified, and this allows to suggest that specific genetic disorders play a key role in leukemogenesis (Betz and Hess 2010). These are structural changes in chromosomes such as t(15;17), t(8;21), inv(16), t(9;21), t(9;11), del5, del7, etc. Nevertheless, there is a normal karyotype and no detectable chromosomal abnormalities in nearly 50% of AML cases, even after complex assay analysis (high density comparative genomic hybridization or single nucleotide polymorphism) (Kohlmann et al. 2010; Suela et al. 2007; Walter et al. 2009). Several mutations carrying diagnostic and prognostic information have been identified during target sequencing analysis, including mutations in NPM1, FLT3, CEBPA, KIT, and TET2 (Bacher et al. 2010; Stirewalt et al. 2003; Stirewalt and Radich 2003; Stirewalt and Meshinchi 2010; Antar et al. 2020; Kiyoi et al. 2020). In further studies the recurrent mutations in the DNMT3A (Ley et al. 2010; Yasuda et al. 2014) and IDH1 genes (Mardis et al. 2009) were located. However, more than 25% of AML patients do not have known genes associated with leukemia (Dombret 2011; Shen et al. 2011). Moreover, it is difficult to define the molecular effects of recurrent mutations (e.g., whether the mutation is an initiating or a cooperating event).

Mutations in HSC stem cells may cumulate in long-living HSC (LL-HSC) cell populations which are able to recreate incessantly. Pre-LSCs (pre-leukemic stem cells) develop because of this continual cell division and regeneration. In other cases, mutations may take place in more mature cell populations that may not to regenerate incessantly as LL-HSC, but mutations provide this ability to regenerate—the so-called self-regenerating pre-LSC occurs. Last of all, pre-LSCs evolve into LSCs (leukemic stem cells) from which the blasts specific for leukemia emerge. LSCs are continually shared during the process of abnormal blast formation (Horton and Huntly 2012; Mesuraca et al. 2018). Nonetheless substantial progress was noticed in investigations of LSCs, due to it the development of numerous LSCs-targeted therapies accelerated (Wang et al. 2017; Pelosi et al. 2015).

It is hypothesized that AML is caused by at least two hematopoietic precursor mutations that confer survival and proliferative benefits and interrupt hematopoietic differentiation. These major oncogenic events are often classified by the two-hit model proposed by Gilliland (Dash and Gilliland 2001). Based on this model, there

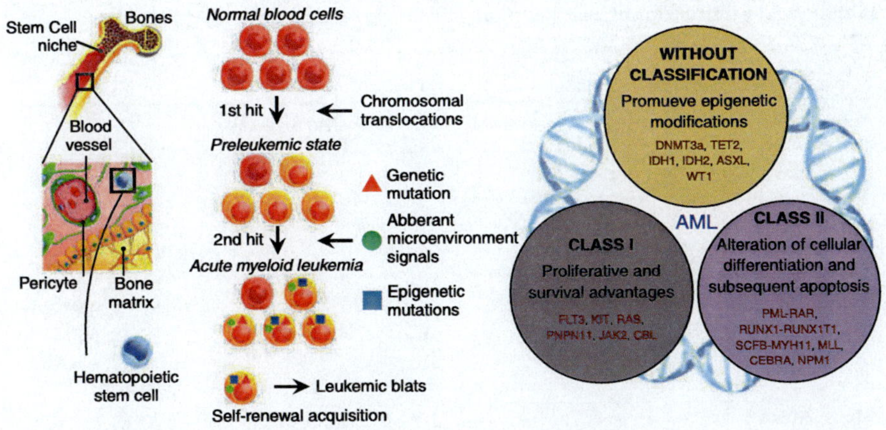

Fig. 1.2 Hematopoietic stem cell niche and mutations determinant in AML. The development of AML could be initiated by chromosomal translocations, genetic and epigenetic mutations as well as aberrant microenvironment signals that confer survival and rapid growth of immature cells and interrupt hematopoietic differentiation. Based on the observation, oncogenic lesions are often classified into the class I and class II as well as into the group with epigenetic alterations and no evident gene mutations and chromosomal translocations

are class I mutations conferring survival and proliferation benefits and class II mutations affecting cell differentiation and apoptosis (Fig. 1.2). Using modern technology, it has been found that there is a group of mutations that mainly promote epigenetic modifications.

1.3 Genetic Landscape of Leukemia

Chromosomal changes as translocations, deletions, insertions, inversions, monosomies, trisomies, polyploidy, and other aberrations usually are characteristic to nearly 55% of AML patients. In AML patients, the following predictions can be made based on the aberrations found - favorable, intermediate, and unfavorable. Based on that, patients with t(15;17)(q24;q21), inv(16)(p13;q22), and t(8;21)(q22;q22) have a favorable prognosis with good response to standard treatment and complete remissions. On the other hand, patients with t(9;11)(p22;q23) have intermediate prognosis, and patients with t(6;9)(p23;q34), inv(3)(q21;q26), and t(1;22)(p13;q13) have an unfavorable prognosis with poor response to treatment (Lagunas-Rangel et al. 2017). The distribution of gene mutations using NGS (next-generation sequencing) analysis was investigated in different subcategories of AML. The results are presented in Table 1.1. Approximately 13% of all AML patients have t(15;17) chromosomal alteration with additional gene mutations like FLT3-ITD (30% of all t(15;17)+ cases), FLT3-TKD (20%), and WT1 (10%).

Table 1.1 The distribution of gene mutations in different subcategories of AML

FAB type	Disease, cases, %	Chromosome alterations	Genes	Genes mutation by NGS[a]
M0	Undifferentiated acute myeloblastic leukemia (3%)	t(9;22)(q34;q11), del(5q), del(7q), +8, +13, t(12;13)(p13;q14)	ABL, BCR, EGR1, IRF1, CSF1, CDK6, ETV6, TTL	
M1	Acute myeloblastic leukemia with minimal maturation (15–20%)	+6(trisomia 6), +4		ASXL1, DNMT3a, ETV6, EZH2, FLT3, IDH1, IDH2, NPM1, TET2, NRAS, NUNX, SRF2
M2	Acute myeloblastic leukemia with maturation (25–30%)	+4, t(8;21)(q22;p22), t(6;9)(p23;q34), t(7;11)(p15;p15)	AML1, ETO, DEK, CAN(NUP214), HOXA9, NUP98	ASXL1, DNMT3a, CBL, ETV6, TET2, IDH1, IDH2, KIT, PHF6, RUNX
M3	Acute promyelocytic leukemia (5–10%)	t(15;17)(q22;q12), t(11;17)(q23;q12), t(11;17)(q13;q12), t(5;17)(q23;p12)	PML, RARA, PLZF, RARA, NuMa, RARA, NPM1, RARA	ASXL1
M4	Acute myelomonocytic leukemia (25–30%), M4 eos–acute myelomonocytic leukemia with eosinophilia	+22, +4, t(6;9)(p23;q34), inv(16)(p13;q22), t(10;11)(p11.2;q23), t(10;11)(p12;q23), t(3;7)(q26;q21)	DEK, CAN, MYTH11, CBFb, ABI1, MLL, AF10, MLL, EVI1, CDK6	ASXL1, CPS1, DNMT3a, FLT3, IDH1, IDH2, KIT, LRP1B, NPM1, NRAS, PHF6, RUNX
M5	Acute monocytic leukemia (2–9%)	t(9;11)(p22;q23), t(10;11)(p11.2;q23), t(10;11)(p12;q23)	AF9, MLL, ABI1, MLL, AF10, MLL	RUNX, ASXL1, DNMT3a, ETV6, FLT3, SRSF2, TET2, IDH1, NPM1, NRAS
M6	Acute erythroid leukemia (3–5%)	del(5q), del(7q)	EGR1, IRF1, CSF1R, ASNS, EPO, ACHE, MET	ASXL1, EZH2, IDH2, RUNX, SRSF2, TP53
M7	Acute megakaryoblastic leukemia (3–12%)	del(5q), del(7q), t(1;22)(p13;q13), t(11;12)(p15;p13)	EGR1, IRF1, CSF1R, ASNS, EPO, ACHE, MET, OTT, MAL, NUP98, JARID1A	ASXL1

[a]Next-generation sequencing (https://www.ncbi.nlm.nih.gov/pmc/articles/PMC5892606/)

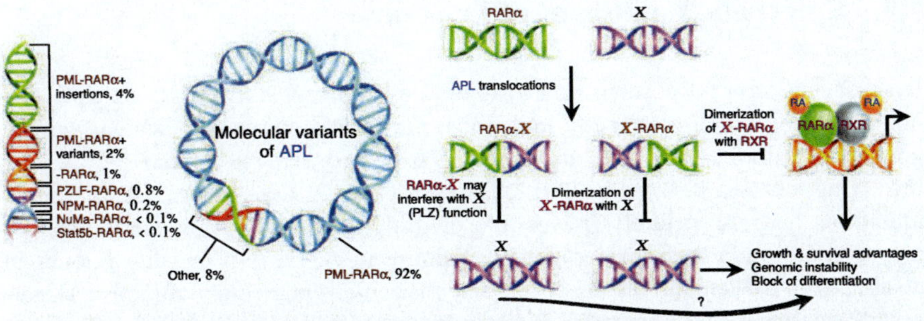

Fig. 1.3 The distribution of chromosomal alterations in APL. Various chromosomal translocations are detected in APL. Proteins (RARα-*X*) determined by chromosomal translocations can promote the development of leukemia by down-regulating target genes and block differentiation process

Another often alteration is NPM1 gene mutation (33% of all AML cases) with few gene alterations like DNMT3a (50% of all NPM1 gene mutation cases), IDH1 (6–8%), FLT3-ITD (40%), IDH2-R10 (15%), Cohesin (20%), and PTPN11 (15%). Nearly 13% of all AML cases belong to a group of patients with normal karyotype but with secondary gene alterations, like RUNX1 (40%), ASXL1 (30%), MLL-PTD (30%), EZH2 (5%), SRSF2 (20%), etc. (Table 1.1).

The most curable AML type is acute promyelocytic leukemia APL (M3). Human APL is related to five reciprocal translocations, in all cases associated with the retinoic acid receptor (RARα) gene, fused to promyelocytic leukemia (PML), nucleophosmin (NPM), promyelocytic leukemia zinc finger (or finger) protein (PLZ), NuMA, or STAT5b genes. The functioning of these fusion proteins appears to occur due to dysregulation of their respective partners. RARs are all-trans retinoic acid (ATRA)-dependent transcription factors connected to the management of hematopoietic cell differentiation. It should be noted that PLZF, PML, and STAT5b are related to the control of cellular mitotic and survival signaling (Rego and Pandolfi 2002; Hussain et al. 2019; Scaglioni and Pandolfi 2007). In more than 95% cases of APL, specific t(15;17) translocation forms PML-RARα and fusion proteins. PML-RARα results in oncogenic transcriptional suppression of ATRA gene targets (Fig. 1.3). APL patients are treated with all-trans retinoic acid (ATRA), which stimulates differentiation of PML-RARα-positive leukemic cells. Despite of crucial role of PML-RARα and RARα-PML fusion proteins in the leukemogenesis of APL, they do not confer a complete malignant phenotype to mouse promyelocytes (He et al. 1998).

Proliferation and maturation of hematopoietic stem cells are regulated in the bone marrow by external signals, involving cytokines fusion; proteins (RARα-X) determined by chromosomal translocations, insertions, and other alterations can promote the development of leukemia in these stem cells by down-regulating target genes, specifically disrupting the signaling pathways that lead to differentiation (Fig. 1.3). Due to interactions with other genetic dysregulations, the complete leukemic transformation may occur.

1.4 Epigenetic Landscape of Leukemia

It has been known for some time that the development and course of cancer is not determined only by genetic mutations. Epigenetic alterations, which regulate gene expression but do not alter the DNA strand sequence, also play a key role (Dawson and Kouzarides 2012; Dawson et al. 2012). As against to genetic mutations that are virtually irreversible or difficult to repair, epigenetic changes are reversible and therefore epigenetic therapy in AML is promising (Goldman et al. 2019; Bewersdorf et al. 2019) and make the epigenetics attractive targets for pharmacological intervention (Monaghan et al. 2019; Dhall et al. 2019). It has recently been argued that cancer can also be treated as a metabolic disease (Kreitz et al. 2019; Maher et al. 2018). Proponents of this theory argue that the genomic instability found in cells affected by cancer may be the result of cancer cell metabolism, not the other way around. However, the significance of epigenetic disorders and somatic mutations in the development of cancer has not yet been denied.

Epigenetic modifications (histone modification, DNA methylation, and chromatin remodeling) of healthy hematopoietic stem cells are significant in hematopoiesis and cell differentiation (Allis and Jenuwein 2016; Kramer and Challen 2017). The activity of epigenetic modifications is important in the pathophysiology of AML, and active research on application of epigenetic-directed therapies in AML is ongoing. Some treatment approaches addressing epigenetic mechanisms and AML heterogeneity are currently under investigation (Goldman et al. 2019).

1.4.1 Changes in Methylation in Carcinogenesis

Cancer initiation and progression are closely related to epimutations. The earliest observed change was DNA methylation. Global genome hypermethylation and hypomethylation of certain domains are specific to cancer. Epigenetic changes are thought to occur very early and may even contribute to cancer initiation (Feinberg et al. 2006; Yang et al. 2019).

Hypomethylation occurs in different parts of the genome, including areas important for genome stability (retrotransposons, introns, etc.). This can lead to chromosomal rearrangements and random gene activation. Oncogenes, which further accelerate the progression of cancer, may also be "on."

Hypermethylation is specific to certain sequences, usually promoter, which contain CpG islands. This epigenetic change contributes to the development of cancer by suppressing gene expression, thereby disrupting normal processes in cells. Some of these silenced genes may be cancer suppressors responsible for DNA repair, cell interactions, cell cycle, apoptosis, etc. The result is the initiation or progression of cancer (Bhattacharya and Patel 2018). DNA methylation alterations may not be directly included into this process. Other genes and proteins involved in reparation or transcription factors, are lost as the epigenome changes.

Although the importance of hypermethylation for gene expression is clear, it is not known exactly how targets are targeted for a seemingly random, methylation process. One possibility is that rapidly dividing cells gain growth advantage. Another is the interaction of DNA methyltransferases with transcription factors that target them (for example, the fusion PML-RARα protein in promyelocytic leukemia, which has more "targets" for interactions than normal PML and RARα proteins). Also, promoter hypermethylation is more likely to occur in areas of the genome where other epigenetic alterations occur intensively. Regions of DNA hypermethylation have been reported to be found where H3 histone is repressively methylated (Furuya et al. 2019).

1.4.2 Histone Modifications

The N-terminus of the histones that make up the nucleosome core of the protein undergoes various covalent modifications: methylation, phosphorylation, sumoylation, acetylation, ubiquitylation, and others (Fig. 1.4). After protein translation, certain amino acid residues are modified. These changes regulate the major cellular processes such as transcription, replication, and reparation. The totality of post-translational modifications of histones forms an "epigenetic memory" called histone code. This code designates the activity and structure of chromatin in different regions of the genome. Modifications occur due to chromatin structure alterations, either made by itself or through other proteins called "readers and editors" of the histone code.

Typical modifications include lysine acetylation, lysine (mono-, bi-, or tri-) and arginine (symmetric or asymmetric) methylation, and phosphorylation of serine, threonine, or tyrosine residues. Less common are modification of arginine by converting arginine to citrulline (believed to be a way to remove the methylation tag), binding of ADP (adenosine diphosphate) to glutamate or arginine, ubiquitylation and saponification of lysine, and others (Bannister and Kouzarides 2011).

The effect that histone modifications have on gene expression depends on their type and which amino acid residues are altered. Figure 1.4 represents DNA and main histone modifications conditioned the chromatin remodeling. Acetylation of lysine correlates with transcriptional activation, and methylation can show both expression activation and inhibition, depending on which amino acid is altered in the protein's histone sequence and to what degree (mono-, bi-, or trimethylated). For example, trimethylation of the lysine on position 4 of the histone H3 (H3K4me3) is common where the transcriptionally active gene promoter is located. The opposite result is obtained by trimethylation of histone H3 on K9 and K27 and histone H4 on K20 residues (Kouzarides 2007). Histone H3 modifications on K9 and K27 residues represent two major inhibitory mechanisms in mammalian cells: H3K9me3 acts in concert with DNA methylation and H3K27me3 acts alone (Bernstein et al. 2007).

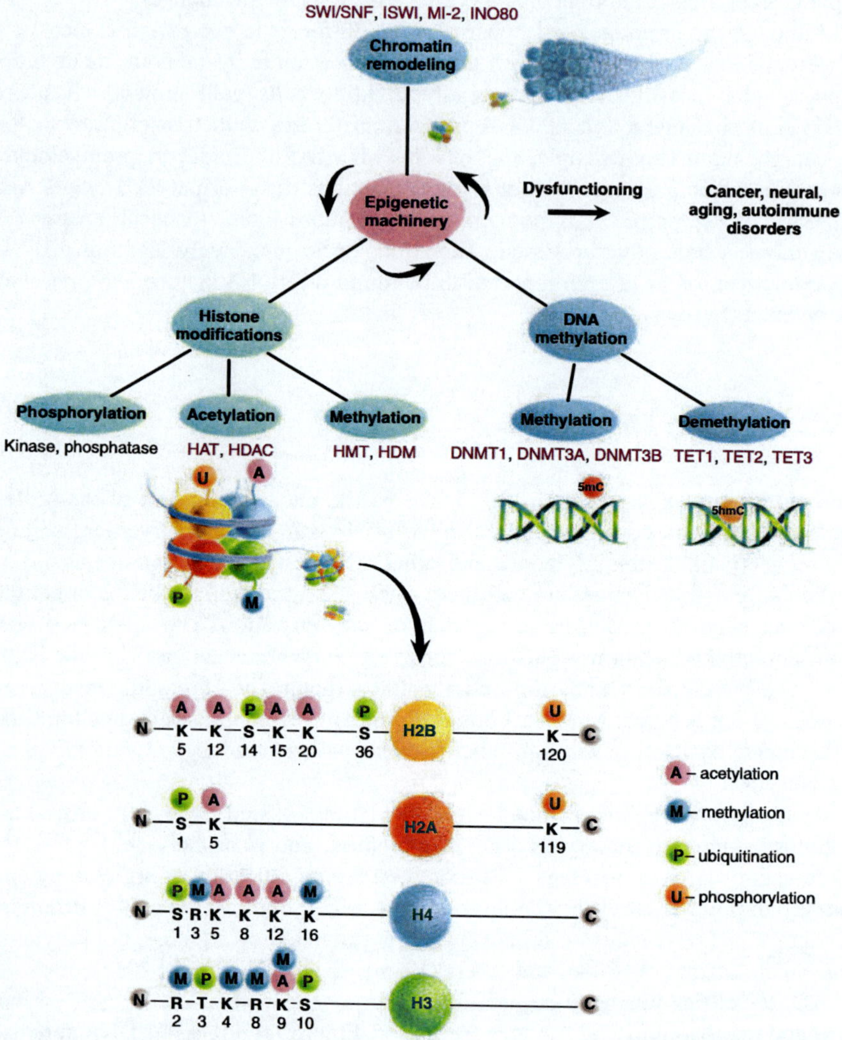

Fig. 1.4 Chromatin remodeling, DNA, and histone modifications. The modifications mostly on N-terminal histone tails act in diverse biological processes such as transcriptional activation/inactivation, chromosome packaging, and DNA damage/repair. DNA methylation can change the activity of gene expression—typically acts to repress gene transcription. The ten-eleven translocation (TET) family of 5mC hydroxylases may promote DNA demethylation by binding to CpG-rich regions to prevent unwanted DNMT activity and by converting 5mC to 5hmC

A "bivalent domain" was detected in stem cells of embryos. It is a combination of opposite modifications of H3K4me3 and H3K27me3, which represents a pause in gene expression. Their transcription depends on the direction in which the cells must evolve (Rodriguez-Paredes and Esteller 2011).

1.4.3 Interaction of DNA and Histone Modifications

As mentioned in previous chapters, DNA and histone modifications work together. Some HMTs (histone methyltransferases), such as G9a, SUV39H1, and PRMT5, can straightforwardly direct DNA methylation to specific regions of the genome, thereby suppressing gene expression. Moreover, HMTs and demethylases may not act directly on DNA methylation, but they regulate the stability of DNMT (DNA methyltransferases) proteins. An example is the SET7 histone methyltransferase, which methylates DNMT1 and degrades the latter into the proteasome (Pradhan et al. 2009). DNA methyltransferases can also cause gene silencing and chromatin condensation by activating histone deacetylases and methyl group binding proteins. DNA methylation via effector proteins induces repressive H3K9 methylation. Not all mechanisms of interactions are clear and examined. Even in the case of specific changes, it is not always known what happens earlier—DNA methylation or histone modification. Dependence on the epigenetic context is observed.

1.5 Treatment of Leukemia

1.5.1 Standard Treatment of AML

About two-third of AML patients achieve remission after standard chemotherapy, but patients experience multiple side effects such as fatigue, reduced resistance to infection, anemia, bleeding, nausea, hair loss, and changes in hormone levels in the body (www.cancerresearchuk.org). Between 25 and 50% of those who recover relapse within 5 years of stopping treatment. One-third of patients fail chemotherapy. New treatments are constantly being sought to improve the effectiveness of treatment and reduce side effects. Recent treatment methods rely on genetic testing and targeted therapy. Strategies of AML treatment are presented in Fig. 1.5.

Usually, after diagnosis the treatment for AML should begin at the earliest moment due to possible very fast progress of the disease. The acute promyelocytic leukemia (APL) subtype of AML is treated in a different way than other AML cases (Fig. 1.6). Two chemotherapy phases are more often characteristic to the treatment of patients with AML: induction or remission induction and consolidation (post-remission therapy), and for APL, the treatment has an additional maintenance phase.

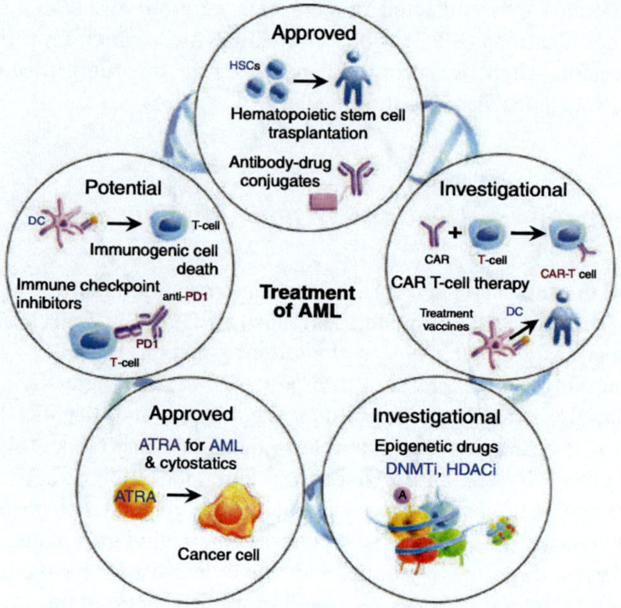

Fig. 1.5 Main strategies of AML treatment. Part of the treatment protocols indicated in the scheme has been already approved, but much remains to be investigated, including epigenetic drugs

During the remission induction phase, patients receive high doses of chemotherapy over a 5–7 day period. Currently, standard remission induction therapy includes 3 days of anthracycline and 7 days of cytarabine ("7 + 3 regimen"). In certain circumstances, it is possible to add the third drug in order to improve the chances of remission, e.g., midostaurin, gemtuzumab ozogamicin, cladribine, etoposide, fludarabine, etc. The drugs of chemotherapy kill leukemia cells as well as normal cells within bone marrow. Usually blood cells from bone marrow recover in approximately 2–3 weeks after induction. In this interval of time, in order to reduce the risk of infection, antibiotics and blood cell growth factors, which induce the bone marrow to produce normal white blood cell substances, are often prescribed to patients. Patients in remission continue to receive consolidation therapy. If patients do not respond, another induction treatment will be initiated immediately.

The essential drugs for APL treatment are non-chemo drugs, i.e., differentiating agents, for example, all-trans-retinoic acid (ATRA). Other treatments are chemotherapy (anthracycline drug like daunorubicin or idarubicin) or non-chemo agents (arsenic trioxide (ATO), gemtuzumab ozogamicin (Mylotarg)) alone or in combination with chemo, and transfusions of platelets or other blood products.

To prevent the relapse of leukemia, immediately after recovery, the consolidation therapy is initiated together with induction therapy, because more than 90% of patients without recurrent therapy are known to recur within weeks or months.

For patients who after induction treatment experience complete remission, consolidation therapy is recommended: chemotherapy with high doses in combination with autologous/allogeneic stem cell transplantation or chemotherapy with standard doses which does not include transplantation of stem cells. Patients with APL typically are treated as well as in remission, giving them some of the drugs they got during it, although the timing of treatment and the doses might differ. The best option for each person is chosen according to the prognosis of the risk of the leukemia coming back after treatment. The option depends as well on the other factors. Options for older people with AML or for those whose health is poor might include low-dose chemotherapy or targeted drug (like gemtuzumab ozogamicin, ivosidenib, enasidenib, etc.) alone (Fig. 1.6).

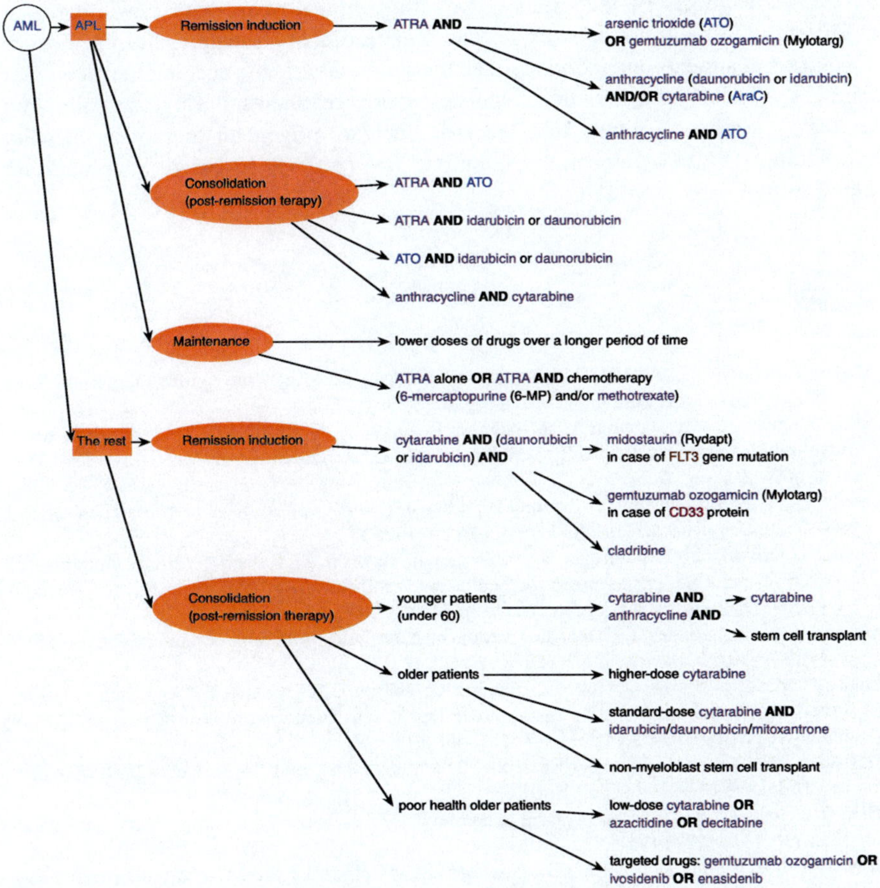

Fig. 1.6 AML treatment scheme representing the approved protocols

1.5.2 *Strategies to Improve Remission Induction*

Despite significant advances in AML therapy over the past 3 decades, two-thirds of young adults and 90% of older adults still die due to leukemia. Leukemia relapse can occur several months or years after primary remission; however, most relapses occur within 2 years of initial treatment. Patients who failed initial treatment can be divided into two categories: those who do not clear their primary cancer after several courses of induction chemotherapy and who return to cancer after remission has been achieved. There are currently several areas of active research aimed to improve the treatment of leukemia: invention of new drugs and trials of these drugs, the development of new treatment schemes with drugs which were already used to treat the other diseases, and the development of prognostic tests for better prognosis/diagnosis of the disease. Ongoing clinical trials with new drugs for AML are presented in Appendix A. The most promising therapies in this area are combined treatments and individualized therapies, taking into account the molecular changes of each case and the response to the treatments during the follow-up process. Despite all things that are still unclear, epigenetic therapy is steadily developing, and new targets and materials and methods for its development are being sought.

References

Allis CD, Jenuwein T (2016) The molecular hallmarks of epigenetic control. Nat Rev Genet 17(8):487–500. https://doi.org/10.1038/nrg.2016.59

Antar AI, Otrock ZK, Jabbour E, Mohty M, Bazarbachi A (2020) FLT3 inhibitors in acute myeloid leukemia: ten frequently asked questions. Leukemia. https://doi.org/10.1038/s41375-019-0694-3

Asada S, Fujino T, Goyama S, Kitamura T (2019) The role of ASXL1 in hematopoiesis and myeloid malignancies. Cell Mol Life Sci **76**:2511–2523

Bacher U, Haferlach C, Schnittger S, Kohlmann A, Kern W, Haferlach T (2010) Mutations of the TET2 and CBL genes: novel molecular markers in myeloid malignancies. Ann Hematol 89(7):643–652. https://doi.org/10.1007/s00277-010-0920-6

Bannister AJ, Kouzarides T (2011) Regulation of chromatin by histone modifications. Cell Res 21(3):381–395. https://doi.org/10.1038/cr.2011.22

Bennett JM, Catovsky D, Daniel MT, Flandrin G, Galton DAG, Gralnick HR, Sultan C (1976) Proposals for the classification of the acute leukaemias French-American-British (FAB) co-operative group. Br J Haematol 33(4):451–458

Bernstein BE, Meissner A, Lander ES (2007) The mammalian epigenome. Cell 128(4):669–681. https://doi.org/10.1016/j.cell.2007.01.033

Betz BL, Hess JL (2010) Acute myeloid leukemia diagnosis in the 21st century. Arch Pathol Lab Med 134(10):1427–1433

Bewersdorf JP, Shallis R, Stahl M, Zeidan AM (2019) Epigenetic therapy combinations in acute myeloid leukemia: what are the options? Ther Adv Hematol 10. https://doi.org/10.1177/2040620718816698

Bhattacharya P, Patel TN (2018) Microsatellite instability and promoter hypermethylation of DNA repair genes in hematologic malignancies: a forthcoming direction toward diagnostics. Hematology 23(2):77–82. https://doi.org/10.1080/10245332.2017.1354428

Bloomfield CD, Brunning RD (1985) FAB M7: acute megakaryoblastic leukemia–beyond morphology. Ann Intern Med 103(3):450–452

Borutinskaite VV, Savickiene J, Navakauskiene R (2008) Apoptotic effects of the novel histone deacetylase inhibitor BML-210 on HeLa cells. Biologija (3):217–220

Bugler J, Kinstrie R, Scott MT, Vetrie D (2019) Epigenetic reprogramming and emerging epigenetic therapies in CML. Front Cell Dev Biol 7:136

Dash A, Gilliland D (2001) Molecular genetics of acute myeloid leukaemia. Best Pract Res Clin Haematol 14(1):49–64. https://doi.org/10.1053/beha.2000.0115

Dawson MA, Kouzarides T (2012) Cancer epigenetics: From mechanism to therapy. Cell 150(1):12–27. https://doi.org/10.1016/j.cell.2012.06.013

Dawson MA, Kouzarides T, Huntly BJP (2012) Targeting epigenetic readers in cancer. N Engl J Med 367(7):647–657. https://doi.org/10.1056/NEJMra1112635

De Kouchkovsky I, Abdul-Hay M (2016) Acute myeloid leukemia: a comprehensive review and 2016 update. Blood Cancer J 6. https://doi.org/10.1038/bcj.2016.50

Dhall A, Zee BM, Yan F, Blanco MA (2019) Intersection of epigenetic and metabolic regulation of histone modifications in acute myeloid leukemia. Front Oncol 9. https://doi.org/10.3389/fonc.2019.00432

Dombret H (2011) Gene mutation and AML pathogenesis. Blood 118(20):5366–5367. https://doi.org/10.1182/blood-2011-09-379081

Falini B, Tiacci E, Martelli MP, Ascani S, Pileri SA (2010) New classification of acute myeloid leukemia and precursor-related neoplasms: changes and unsolved issues. Discov Med 10(53):281–292

Feinberg AP, Ohlsson R, Henikoff S (2006) The epigenetic progenitor origin of human cancer. Nat Rev Genet 7(1):21–33. https://doi.org/10.1038/nrg1748

Fiedler K, Brunner C (2012) The role of transcription factors in the guidance of granulopoiesis. Am J Blood Res 2(1):57–65

Furuya K, Ikura M, Ikura T (2019) Epigenetic interplays between DNA demethylation and histone methylation for protecting oncogenesis. J Biochem 165(4):297–299. https://doi.org/10.1093/jb/mvy124

Goldman SL, Hassan C, Khunte M, Soldatenko A, Jong Y, Afshinnekoo E, Mason CE (2019) Epigenetic modifications in acute myeloid leukemia: Prognosis, treatment, and heterogeneity. Front Genet 10. https://doi.org/10.3389/fgene.2019.00133

Grigoropoulos NF, Petter R, Van't Veer MB, Scott MA, Follows GA (2013a) Leukaemia update. part 1: diagnosis and management. BMJ-Br Med J 346. https://doi.org/10.1136/bmj.f1660

Grigoropoulos NF, Petter R, Van't Veer MB, Scott MA, Follows GA (2013b) Leukaemia update. part 2: managing patients with leukaemia in the community. BMJ-Br Med J 346. https://doi.org/10.1136/bmj.f1932

He L, Guidez F, Tribioli C, Peruzzi D, Ruthardt M, Zelent A, Pandolfi P (1998) Distinct interactions of PML-RAR alpha and PLZF-RAR alpha with co-repressors determine differential responses to RA in APL. Nat Genet 18(2):126–135. https://doi.org/10.1038/ng0298-126

Horton SJ, Huntly BJP (2012) Recent advances in acute myeloid leukemia stem cell biology. Haematologica Hematol J 97(7):966–974. https://doi.org/10.3324/haematol.2011.054734

Hussain L, Maimaitiyiming Y, Islam K, Naranmandura H (2019) Acute promyelocytic leukemia and variant fusion proteins: PLZF-RAR alpha fusion protein at a glance. Semin Oncol 46(2):133–144. https://doi.org/10.1053/j.seminoncol.2019.04.004

Kiyoi H, Kawashima N, Ishikawa Y (2020) FLT3 mutations in acute myeloid leukemia: Therapeutic paradigm beyond inhibitor development. Cancer Sci. https://doi.org/10.1111/cas.14274

Kohlmann A, Bullinger L, Thiede C, Schaich M, Schnittger S, Doehner K, Dugas M, Klein HU, Doehner H, Ehninger G, Haferlach T (2010) Gene expression profiling in AML with normal karyotype can predict mutations for molecular markers and allows novel insights into perturbed biological pathways. Leukemia 24(6):1216–1220. https://doi.org/10.1038/leu.2010.73

Kouzarides T (2007) Snapshot: Histone-modifying enzymes. Cell 131(4):802–e1. https://doi.org/10.1016/j.cell.2007.11.005

Kramer A, Challen GA (2017) The epigenetic basis of hematopoietic stem cell aging. Semin Hematol 54(1):19–24. https://doi.org/10.1053/j.seminhematol.2016.10.006

Kreitz J, Schoenfeld C, Seibert M, Stolp V, Alshamleh I, Oellerich T, Steffen B, Schwalbe H, Schnuetgen F, Kurrle N, Serve H (2019) Metabolic plasticity of acute myeloid leukemia. Cells 8(8). https://doi.org/10.3390/cells8080805

Lagunas-Rangel FA, Chavez-Valencia V, Gomez-Guijosa MA, Cortes-Penagos C (2017) Acute myeloid leukemia-genetic alterations and their clinical prognosis. Int J Hematol Oncol Stem Cell Res 11(4):328–339

Lee EJ, Pollak A, Leavitt RD, Testa JR, Schiffer CA (1987) Minimally differentiated acute nonlymphocytic leukemia: a distinct entity. Blood 70(5):1400–1406

Ley TJ, Ding L, Walter MJ, McLellan MD, Lamprecht T, Larson DE, Kandoth C, Payton JE, Baty J, Welch J, Harris CC, Lichti CF, Townsend RR, Fulton RS, Dooling DJ, Koboldt DC, Schmidt H, Zhang Q, Osborne JR, Lin L, O'Laughlin M, McMichael JF, Delehaunty KD, McGrath SD, Fulton LA, Magrini VJ, Vickery TL, Hundal J, Cook LL, Conyers JJ, Swift GW, Reed JP, Alldredge PA, Wylie T, Walker J, Kalicki J, Watson MA, Heath S, Shannon WD, Varghese N, Nagarajan R, Westervelt P, Tomasson MH, Link DC, Graubert TA, DiPersio JF, Mardis ER, Wilson RK (2010) DNMT3A mutations in acute myeloid leukemia. N Engl J Med 363(25):2424–2433. https://doi.org/10.1056/NEJMoa1005143

Maher M, Diesch J, Casquero R, Buschbeck M (2018) Epigenetic-transcriptional regulation of fatty acid metabolism and its alterations in leukaemia. Front Genet 9. https://doi.org/10.3389/fgene.2018.00405

Mansouri L, Wierzbinska JA, Plass C, Rosenquist R (2018) Epigenetic deregulation in chronic lymphocytic leukemia: clinical and biological impact. In: Seminars in Cancer Biology, vol 51. Elsevier, New York, pp 1–11

Mardis ER, Ding L, Dooling DJ, Larson DE, McLellan MD, Chen K, Koboldt DC, Fulton RS, Delehaunty KD, McGrath SD, Fulton LA, Locke DP, Magrini VJ, Abbott RM, Vickery TL, Reed JS, Robinson JS, Wylie T, Smith SM, Carmichael L, Eldred JM, Harris CC, Walker J, Peck JB, Du F, Dukes AF, Sanderson GE, Brummett AM, Clark E, McMichael JF, Meyer RJ, Schindler JK, Pohl CS, Wallis JW, Shi X, Lin L, Schmidt H, Tang Y, Haipek C, Wiechert ME, Ivy JV, Kalicki J, Elliott G, Ries RE, Payton JE, Westervelt P, Tomasson MH, Watson MA, Baty J, Heath S, Shannon WD, Nagarajan R, Link DC, Walter MJ, Graubert TA, DiPersio JF, Wilson RK, Ley TJ (2009) Recurring mutations found by sequencing an acute myeloid leukemia genome. N Engl J Med 361(11):1058–1066. https://doi.org/10.1056/NEJMoa0903840

Mesuraca M, Amodio N, Chiarella E, Scicchitano S, Aloisio A, Codispoti B, Lucchino V, Montalcini Y, Bond HM, Morrone G (2018) Turning stem cells bad: Generation of clinically relevant models of human acute myeloid leukemia through gene delivery- or genome editing-based approaches. Molecules 23(8). https://doi.org/10.3390/molecules23082060

Monaghan L, Massett ME, Bunschoten RP, Hoose A, Pirvan PA, Liskamp RMJ, Jorgensen HG, Huang X (2019) The emerging role of h3k9me3 as a potential therapeutic target in acute myeloid leukemia. Front Oncol 9. https://doi.org/10.3389/fonc.2019.00705

Morrison SJ, Scadden DT (2014) The bone marrow niche for haematopoietic stem cells. Nature 505(7483):327–334. https://doi.org/10.1038/nature12984

Network TCGAR (2013) Genomic and epigenomic landscapes of adult de novo acute myeloid leukemia. N Engl J Med 368(22):2059–2074

Nordlund J, Syvaenen AC (2018) Epigenetics in pediatric acute lymphoblastic leukemia. Semin Cancer Biol 51:129–138

Papaemmanuil E, Gerstung M, Bullinger L, Gaidzik VI, Paschka P, Roberts ND, Potter NE, Heuser M, Thol F, Bolli N et al (2016) Genomic classification and prognosis in acute myeloid leukemia. N Engl J Med 374(23):2209–2221

Patnaik MM, Tefferi A (2020) Chronic myelomonocytic leukemia: 2020 update on diagnosis, risk stratification and management. Am J Hematol 95(1):97–115

Pelosi E, Castelli G, Testa U (2015) Targeting LSCs through membrane antigens selectively or preferentially expressed on these cells. Blood Cells Mol Dis 55(4):336–346. https://doi.org/10.1016/j.bcmd.2015.07.015

Pradhan S, Chin HG, Esteve PO, Jacobsen SE (2009) Set7/9 mediated methylation of non-histone proteins in mammalian cells. Epigenetics 4(6):282–285

Raval A, Byrd JC, Plass C (2006) Epigenetics in chronic lymphocytic leukemia. Semin Oncol 33:157–166

Rego EM, Pandolfi PP (2002) Reciprocal products of chromosomal translocations in human cancer pathogenesis: key players or innocent bystanders? Trends Mol Med 8(8):396–405. https://doi.org/10.1016/S1471-4914(02)02384-5

Rodriguez-Paredes M, Esteller M (2011) A combined epigenetic therapy equals the efficacy of conventional chemotherapy in refractory advanced non-small cell lung cancer. Cancer Discov 1(7):557–559. https://doi.org/10.1158/2159-8290.CD-11-0271

Sawyers CL, Denny CT, Witte ON (1991) Leukemia and the disruption of normal hematopoiesis. Cell 64(2):337–350. https://doi.org/10.1016/0092-8674(91)90643-D

Scaglioni PP, Pandolfi PP (2007) The theory of APL revisited. In: Pandolfi P, Vogt P (eds) Acute promyelocytic leukemia: molecular genetics, mouse models and targeted therapy, Current topics in microbiology and immunology, vol 313, pp 85–100

Seiter K (2003) Diagnosis and management of core-binding factor leukemias. Curr Hematol Rep 2(1):78–85

Seiter K (2013) Myelodysplasia: New approaches. Curr Treat Options Oncol 14(2):156–169. https://doi.org/10.1007/s11864-013-0224-x

Shen Y, Zhu YM, Fan X, Shi JY, Wang QR, Yan XJ, Gu ZH, Wang YY, Chen B, Jiang CL, Yan H, Chen FF, Chen HM, Chen Z, Jin J, Chen SJ (2011) Gene mutation patterns and their prognostic impact in a cohort of 1185 patients with acute myeloid leukemia. Blood 118(20):5593–5603. https://doi.org/10.1182/blood-2011-03-343988

Stirewalt DL, Meshinchi S (2010) Receptor tyrosine kinase alterations in AML - biology and therapy. Cancer Treat Res 145:85–108

Stirewalt DL, Radich JP (2003) The role of flt3 in haematopoietic malignancies. Nat Rev Cancer 3(9):650–U1. https://doi.org/10.1038/nrc1169

Stirewalt DL, Meshinchi S, Radich JP (2003) Molecular targets in acute myelogenous leukemia. Blood Rev 17(1):15–23. https://doi.org/10.1016/S0268-960X(02)00057-7

Suela J, Alvarez S, Cigudosa JC (2007) DNA profiling by arrayCGH in acute myeloid leukemia and myelodysplastic syndromes. Cytogenet Genome Res 118(2–4):304–309. https://doi.org/10.1159/000108314

Vardiman JW, Harris NL, Brunning RD (2002) The World Health Organization (who) classification of the myeloid neoplasms. Blood 100(7):2292–2302

Walter MJ, Payton JE, Ries RE, Shannon WD, Deshmukh H, Zhao Y, Baty J, Heath S, Westervelt P, Watson MA, Tomasson MH, Nagarajan R, O'Gara BP, Bloomfield CD, Mrozek K, Selzer RR, Richmond TA, Kitzman J, Geoghegan J, Eis PS, Maupin R, Fulton RS, McLellan M, Wilson RK, Mardis ER, Link DC, Graubert TA, DiPersio JF, Ley TJ (2009) Acquired copy number alterations in adult acute myeloid leukemia genomes. Proc Natl Acad Sci U S A 106(31):12950–12955. https://doi.org/10.1073/pnas.0903091106

Wang X, Huang S, Chen JL (2017) Understanding of leukemic stem cells and their clinical implications. Mol Cancer 16. https://doi.org/10.1186/s12943-016-0574-7

Wilson A, Oser GM, Jaworski M, Blanco-Bose WE, Laurenti E, Adolphe C, Essers MA, Macdonald HR, Trumpp A (2007) Dormant and self-renewing hematopoietic stem cells and their niches. In: Kanz L, Weisel K, Dick J, Fibbe W (eds) Hematopoietic stem cells VI, Annals of the New York Academy of Sciences, vol 1106, pp 64–75. https://doi.org/10.1196/annals.1392.021. 6th Biennial International Symposium and Workshop on Hematopoietic Stem Cells, Univ Tubingen, Tubingen, 14–16, 2006

Yang X, Wong MPM, Ng RK (2019) Aberrant dna methylation in acute myeloid leukemia and its clinical implications. Int J Mol Sci 20(18). https://doi.org/10.3390/ijms20184576

Yasuda T, Ueno T, Fukumura K, Yamato A, Ando M, Yamaguchi H, Soda M, Kawazu M, Sai E, Yamashita Y, Murata M, Kiyoi H, Naoe T, Mano H (2014) Leukemic evolution of donor-derived cells harboring IDH2 and DNMT3A mutations after allogeneic stem cell transplantation. Leukemia 28(2):426–428. https://doi.org/10.1038/leu.2013.278

Chapter 2
Epigenetic Modifiers and Their Inhibitors in Leukemia Treatment

In this chapter the authors will attempt to describe how the inductor of differentiation, all-trans retinoic acid, together with agents used in standard therapy and inhibitors of epigenetic modifiers influences the cellular processes in acute promyelocytic leukemia cells. AML progresses rapidly, and therefore most patients have to get treatment soon after the disease was diagnosed. The initial goal of the treatment in most cases is to achieve remission of the patient. The long-term goal is to cure the disease. The mostly common therapy for AML patients is the chemotherapy, when one drug or a combination of different drugs is used to damage or kill the cancer cells. Sometimes different anticancer drugs (ATRA) are used to treat acute promyelocytic leukemia (subtype M3) and lead to cell differentiation and after it to apoptosis. So it is important to understand how chemicals (drugs) effectively contribute to leukemic cell differentiation, growth arrest, and apoptosis induction—processes important not only at the cellular level but also for identifying novel possible treatments for people with leukemia.

Today, there are few types of epigenetic modifiers like of DNA methyltransferase inhibitors, histone deacetylase inhibitors, lysine-specific demethylase 1 inhibitors, zeste homolog 2 inhibitors, and bromodomain and extra-terminal motif proteins inhibitors.

Very often epigenetic modifiers appear to have limited efficacy as single agents and often used in combination with other drugs such as targeted and/or as other epigenetic agents or standard therapy. However, such epigenetic modifiers provide a novel approach to epigenetic anticancer therapy.

Bromodomain and extra-terminal motif (BET) proteins can bind to acetylated lysine residues on histone tails and can regulate transcriptional (component of the Mediator complex (Denis et al. 2006)) and cell cycle (activates the promoters of few cell cycle regulatory genes, including Cyclin D11 and Cyclin E (Denis et al. 2000)). Few BET inhibitors in combination with other drugs were tested for potential therapeutic use. Currently in phase I of clinical trials are tested ABBV-075 (AML, multiple myeloma, and breast cancer), BI 894999 (hematologic malignancies), CPI-

R. Navakauskienė et al., *Epigenetics and Proteomics of Leukemia*,
https://doi.org/10.1007/978-3-030-68708-3_2

0610 (lymphoma, AML, ALL, and CML in blast crisis), and FT-1101 (AML and MDS), and in phase I/II of clinical trials are tested GSK525762/I-BET762 (hematologic malignancies), INCB054329 (hematologic malignancies, solid tumors), and OTX015/MK-8628 (glioblastoma multiforme).

Modifications of histone are essential for chromatin condensation/decondensation and further repression or activation of genes. It was showed that lysine-specific demethylase 1 (LSD1) overexpressed in breast cancer, prostate cancer, neuroblastoma, and acute myeloid leukemia cases. LSD1 demethylates methylation marks, leading to its activity as a transcription repressor (Maiques-Diaz et al. 2018). LSD1 inhibitors are investigated for cancer inhibition and also for the treatment of diseases other than cancer like viral infections and inflammation.

LSD1 inhibitors like GHK2879552, IMG-7289, SP-2577, INCB059872, and CC-90011 are currently in phase I of clinical trials. However, LSD1 inhibitors GHK2879552, IMG-7289 and SP-2577, and CC-90011 and INCB059872 on their website. However, all LSD1 inhibitors are under investigation, and no FDA-approved LSD1 inhibitor as a drug has been reported yet. Enhancer of zeste homolog 2 (EZH2) is a component of polycomb repressive complex 2 (PRC2) and can be involved in the regulation of cells proliferation, apoptosis, and senescence. Also a lot of studies showed EZH2 role in cancer initiation, progression, metastasis, metabolism, drug resistance, and immunity regulation (Duan et al. 2020).

Various types of EZH2 inhibitors have been developed. Both research and clinical trials are going on using drugs that are EZH2 inhibitors for different types of cancer. The first EZH2 inhibitor developed was 3-deazaneplanocin A (DZNep). However, DZNep is not specific to EZH2. DZNep globally inhibits histone methylation. Further scientists developed other specific EZH2 inhibitors like GSK126 (GSK2816126), EPZ005687, GSK343, GSK926, and tazemetostat (E7438/EPZ6438). The most promising tazemetostat is under evaluation in a series of clinical trials. Currently in phase I/III of clinical trials are EZH2 inhibitors such as tazemetostat and CPI-1205. However, many EXH2 inhibitors are experimental, and only tazemetostat was approved for the treatment of epithelioid sarcoma and preliminarily showed promising effects on follicular lymphoma.

Differentiation therapies are often described as those that stimulate malignant reversion, i.e., the malignant phenotype becomes curable. Clinically, these therapies have been most successful for the treatment of acute promyelocytic leukemia, while using ATRA (Fig. 2.1). Scientists do not know the exact differentiation mechanisms yet—it is unclear if differentiation occurs by stimulating terminal differentiation (arrest in G0 cell cycle phase), by triggering apoptosis, or by inducing differentiation "backward" to the non-malignant cell form. It is possible that all of these pathways are acting (Spira and Carducci 2003). As described previously (Minucci et al. 2001; Halsall and Turner 2016; San Jose-Eneriz et al. 2019), inhibition of histone deacetylase (HDAC) activity might represent a common strategy to treat myeloid leukemia as activation or repression of various cellular signaling pathways that could potentiate differentiation of cancer cell (Fig. 2.2).

Although there are probably mechanistic differences in how the various agents lead to differentiation, during this process malignant tumor cells should revert to a

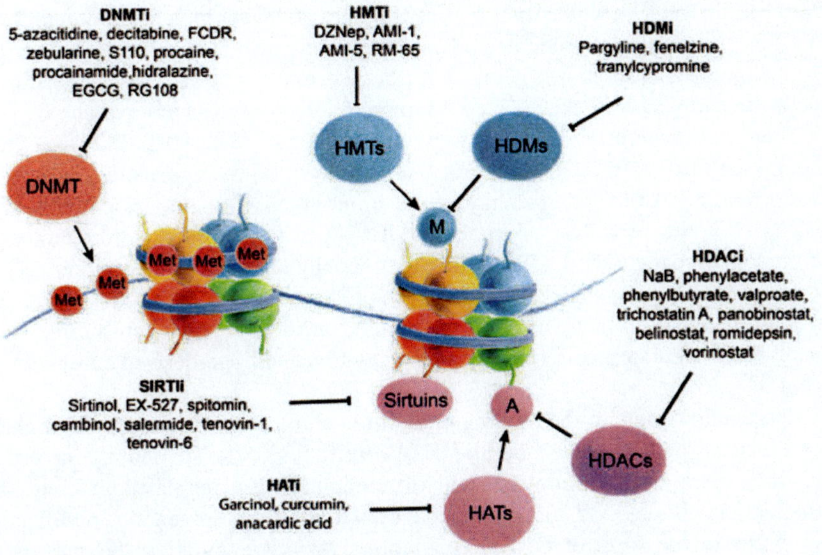

Fig. 2.1 Therapy options of AML via epigenetic modification

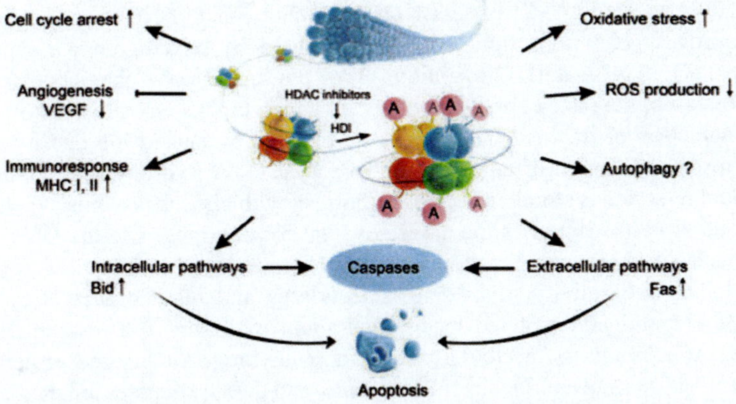

Fig. 2.2 Cellular processes induced by HDAC inhibitors. Schematic representation of HDAC inhibitors' anticancer activities. The antitumor potential of HDACis relies on their ability to influence many cellular processes that are deregulated in tumor cells. On the whole, inhibition of cell cycle, activation of differentiation, and apoptosis are important HDAC inhibitors' antitumor activities

more benign form, in which their replication rates are lower than in malignant forms, and therefore a tumor burden decreases. They might also have a lower tendency for distant metastatic spread to occur, and the process might also restore traditional apoptotic pathways, all of which could improve the prognosis of a patient.

There are several drugs that were approved to treat AML like azacitidine and cytarabine (DB00928 and DB00987, a pyrimidine nucleoside analogues), daunorubicin (DB00694, an anthracycline aminoglycoside), etoposide (DB00773, a podophyllotoxin derivative), decitabine (DB01262, a small intercalating molecule), gemtuzumab ozogamicin (DB00056, a monoclonal anti-CD33 antibody), glasdegib (DB11978, a sonic hedgehog receptor inhibitor), idarubicin (DB01177, an anthracycline antineoplastic agent), midostaurin (DB06595, an antineoplastic agent), and mitoxantrone (DB01204, an anthracenedione-derived antineoplastic agent).

Epigenetics refers to the modification of the chromatin structure without changing the base pair sequence of the DNA itself, which is an underlying process in which gene transcription and cell differentiation are regulated in both physiological conditions and malignant cell transformation. Epigenetic modifications can occur in the form of DNA methylation/hydroxymethylation, histone protein modifications (methylation and acetylation), and changes to higher order chromatin structures. Methylation of CpG islands in the DNA generally suppresses gene transcription and is mediated by DNA methyltransferases (DNMT). Changes in DNA methylation have been linked to AML development, and the treatment with hypomethylating agents (e.g., decitabine, azacitidine) that inhibit DNMT has been successfully used in AML patients. On the other hand, DNA hydroxymethylation improves gene transcription and is mediated by α-ketoglutarate-dependent enzymes, e.g., TET2. IDH1/2 mutations allow the formation of the oncometabolite 2-hydroxyglutarate instead of α-ketoglutarate which blocks DNA hydroxymethylation. The action of mutated IDH1/2 can be blocked by enasidenib and ivosidenib which restores function of enzymes orchestrating DNA hydroxymethylation. The DNA double-strand is stored in cells as a complex with histone proteins. Acetylation of histone proteins decreases the access of transcription factors to the DNA strand and in such a way prevents gene transcription. Histone acetylation status is regulated by balancing the activity of histone deacetylases and histone acetylases which can be therapeutically targeted by bromodomain inhibitors and histone deacetylase (HDAC) inhibitors. Methylation and demethylation of histone proteins can happen at various sites of the histone molecule, and these processes are mediated by histone methyltransferases and histone demethylases. DOT1L is a histone H3K79 methyltransferase, while EZH1/EZH2 methylates histone H3K27, and both are implicated in leukemogenesis and can be targeted by specific inhibitors. LSD1 inhibitors might block histone demethylation.

Some of them we used in combination with differentiation agent ATRA for *in vitro* studies to investigate anti-proliferative and differential effect on leukemia cell lines (Appendix B, Appendix C).

Transcriptional activation of genes requires local remodeling of chromatin. Dynamic alterations in the nucleosomal packaging of DNA must happen to allow transcriptional proteins to contact with the DNA template. The realization that these proteins that regulate the modification of chromatin participate in leukemia chromosomal rearrangements gives a spur to study the chromatin structure. Previous reports brought hope that pharmacological manipulation of chromatin remodeling might develop into an effective and specific strategy that will help to treat leukemia (Advani et al. 1999; Slack and Rusiniak 2000).

2.1 *In vitro* Studies of ATRA Effects on APL Cells

APL has become a curable subtype of acute myeloid leukemia because of successful use of ATRA treatment (Zhou et al. 2007). The combination of chemotherapy with ATRA further improved the 5-year overall survival of patients. Complete remission can be reached after ATRA treatment, and howeverrelapse can occur very rapidly, and to prevent this event, ATRA was used in combination with other chemotherapeutic agents. To get over the limitations of ATRA in relapsed or refractory patients, the scientists all over the world struggled to find the effective alternative therapies by using epigenetic modifiers (He et al. 1998; Greenblatt and Nimer 2014; Valiuliene et al. 2017; Valiulienė et al. 2016; Valiuliene et al. 2015; Noguera et al. 2019), because pharmacological concentrations of ATRA not in all APL cases are able to disrupt the polycomb repressor complex (Fig. 2.3).

In our studies we used ATRA treatment as positive control for effective differentiation induction. We used different combinations of chemical agents with or without ATRA to improve AML/APL treatment outcomes. HL-60, NB4, KG1, and other human leukemia cell lines were used as *in vitro* model systems (see Appendix B for characteristics of cell lines used in the studies).

HL-60 cell line is an attractive model to study differentiation of human myeloid cells. HL-60 cells are extremely sensitive to the various apoptotic stimuli including damage of DNA, because of lack of p53 protein, and high expression of Myc (Kim et al. 2007). HL-60 proliferates continuously in suspension culture (RPMI media) with nutrients (10% FBS) and antibiotics. The doubling time of HL-60 cells is approximately 24–48 h, and the cell cycle distribution during proliferation is presented in Table 2.1. We showed that nearly 2–7% of HL-60 control cells have CD11b on their surface after 48 h of proliferation in untreated (control) condition (Table 2.1). Spontaneous differentiation to mature granulocytes can be stimulated by compounds such as dimethyl sulfoxide (DMSO) or all-trans retinoic acid (Tables 2.1 and 2.2). Other compounds, e.g., Vit. D3-1,25-dihydroxyvitamin D3, 12-O-tetradecanoylphorbol-13-acetate (TPA), and GM-CSF may stimulate HL-60 to differentiate to macrophage-like, monocytic, and eosinophil phenotypes, respectively (Mangelsdorf et al. 1984; Olsson et al. 1983). We showed that only higher concentration of DMSO can induce differentiation of HL-60 cells after 96 h of treatment. The concentration below 1.1% DMSO does not influence differentiation process in cells.

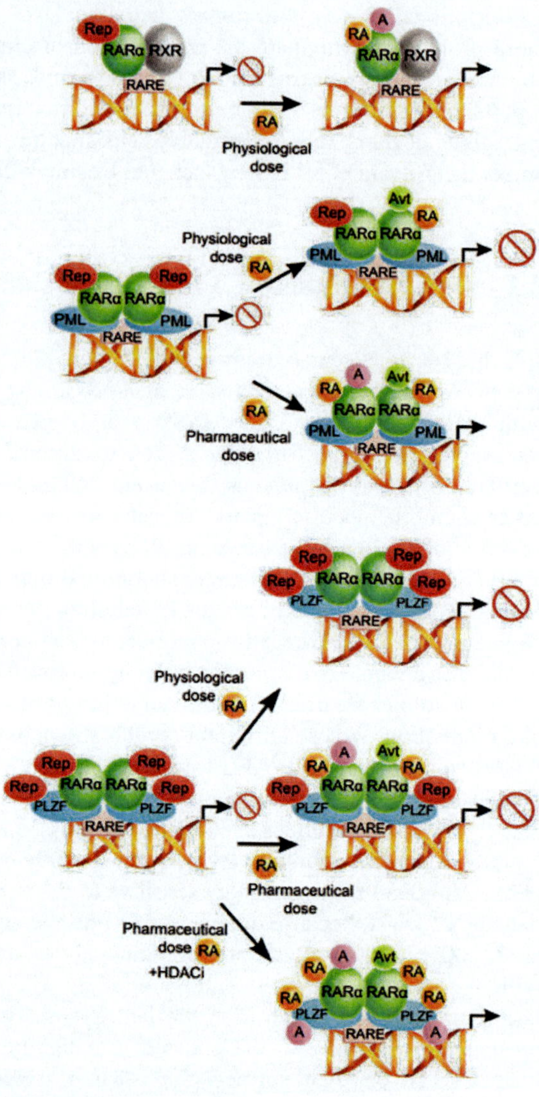

Fig. 2.3 All-trans retinoic acid treatment and chromatin remodeling. Binding of physiological dose of all-trans retinoic acid (ATRA or RA) to RXR-RARα or pharmacological dose to PML-RARα leads to the chromatin remodeling—repression complex (Rep) disruption and specific histone modifications. Transcription of ATRA target genes can also be initiated by PLZF-RARα after treatment with combination of pharmacological dozes of ATRA and HDACi

In our studies we used different concentrations of ATRA (0.3–1 μM) and measured growth of HL-60 cells, viability, differentiation, and apoptosis (Valiulienė et al. 2016; Vitkeviciene et al. 2019). Afterexposure to the differentiation inducer ATRA (0.5–1 μM), the viability of cells after 96 h of treatment was 70–82%, the cells were accumulated in G1 cell cycle phase, DNA fragmentation was increased in a time-dependent manner (Savickiene et al. 1999), and the number of differentiated (CD11b positive) cell counts was observed up to 80% higher after 1 μM ATRA treatment (Table 2.1). We identified that by using different concentrations of

Table 2.1 Characteristics of HL-60 control treated with 1 μM ATRA: growth, cell cycle distribution, viability, differentiation, and apoptosis (results represented as HL-60 cells control/treated with 1 μM ATRA). Results are given as mean ± SEM ($n = 3$)

	HL-60 cells control ⇝ treated with 1 μM ATRA				
	0 h	24 h	48 h	72 h	96 h
Cell count, 10^6/mL	0.5 ⇝ 0.5	0.7 ⇝ 0.65	0.98 ⇝ 0.9	1.2 ⇝ 0.95	1.1 ⇝ 0.93
Cell viability, %	98 ⇝ 98±3	97 ⇝ 95±5	92 ⇝ 85±4	85 ⇝ 82 ±6	78 ⇝ 75±10
Cell cycle distribution, %					
G0/G1	42	39 ⇝ 38	42 ⇝ 56	45 ⇝ 65	n/a
S	49	52 ⇝ 43	48 ⇝ 31	45 ⇝ 23	n/a
G2/M	9	9 ⇝ 19	10 ⇝ 13	10 ⇝ 12	n/a
subG1	–			12 ⇝ 28	n/a
DNA fragmentation, %	– ⇝ –	– ⇝ 3	– ⇝ 5	– ⇝ 10	– ⇝ 15
Trypan blue+ cells, 10^4/mL	2	12 ⇝ 15±3	12 ⇝ 15±4	25 ⇝ 12±5	72 ⇝ 75±6
CD11b, %	–	– ⇝ 48±3	3 ⇝ 65±7	7 ⇝62±6	7.3 ⇝ 74±6
NBT+ cells, %	–	1 ⇝ 20±3	2 ⇝ 30±3	2 ⇝ 45±6	2 ⇝ 58±8

Table 2.2 Viability and differentiation of HL-60 cells after treatment of ATRA and DMSO (96 h). Results are given as mean ± SEM ($n = 3$)

ATRA, nM	300	500	1000	DMSO, %	1	1.1	1.2
Viability, %	84 ± 3	81 ± 5	81 ± 6	Viability, %	92 ± 3	84 ± 5	81 ± 4
NBT+ cells, %	25 ± 2	40 ± 3	80 ± 5	NBT+ cells, %	0	0	50 ± 3

Table 2.3 NB4 control cells and cells treated with 1 μM ATRA: cell growth, viability, and distribution via cell cycle phases, differentiation, and apoptosis analysis (results represented as NB4 cells control/treated 1 μM ATRA). Results are given as mean ± SEM ($n = 3$)

	NB4 cells control ⇝ treated with 1 μM ATRA				
	0 h	24 h	48 h	72 h	96 h
Cell count, 10^6/mL	0.5 ⇝ 0.5	1.0 ⇝ 0.65	1.4 ⇝ 0.9	1.45 ⇝ 0.95	1.7 ⇝ 0.66
Cell viability, %	98 ⇝ 98 ± 2	95 ⇝ 77 ± 5	90 ⇝ 72 ± 4	87 ⇝ 71 ± 8	84 ⇝ 70 ± 10
CD11b, %	–	5 ⇝ 48 ± 2	12 ⇝ 65 ± 10	12 ⇝ 62 ± 24	13 ⇝ 76 ± 7
NBT+ cells, %	–	5 ⇝ 20 ± 2	12 ⇝ 30 ± 3	12 ⇝ 45 ± 4	13 ⇝ 58 ± 4

ATRA, the viability of HL-60 cells slightly changed, but maximum differentiation was achieved with 1 μM ATRA concentration.

It was showed in the literature that ATRA was metabolized and its concentration decreased by 50% in the medium after 2 days (Parthasarathy and Mehta 1998). In our study we compared single dose of ATRA and ATRA added every other day from the start of induction or daily, replacing the medium with a new portion of ATRA. HL-60 cells were cultivated for 4 days and differentiation induction was measured. Results showed that one additional introduction of ATRA increased cell differentiation at day 3 by 16%, but at the fourth day of treatment, results were the same (63% and 65%) as after single dose of ATRA. After daily addition of ATRA, the amount of differentiated cells was 20% higher compared to the single use of ATRA (data not shown). To conclude, the optimal concentration that allowed to induce the maximum differentiation of HL-60 cells is 1 μM ATRA (Table 2.2), and this concentration induces slight growth inhibition and cell accumulation in the G0/G1 cell cycle phase with further differentiation (up to 70–80%) and apoptosis. That concentration widely will be used in further studies alone or combined with other chemical agents to obtain anti-proliferative effects of chemical agents.

In our studies we widely used another APL cell line, NB4, as typical cell line that carries t(15;17)(q24.1;q21.2) chromosomal translocation (see Appendix B for characteristics of cell lines used in the studies). NB4 cell line was treated with 1 μM ATRA and anti-proliferative effects were observed (Table 2.3). Growth conditions were the same as for HL-60 cells. It was showed that doubling time is 25–45 h. It was observed that 1 μM ATRA slightly inhibited cell growth and cell viability. This concentration induced differentiation up to 75% after 96 h of treatment and apoptosis up to 50% (96 h) compared to control cells.

Another acute myeloid leukemia cell line used in our studies is KG1 (M1 type) cell line. KG1 immature myeloblastic cell line has t(11;17) (q23;q21) translocation PLZF-RARα fusion protein that can recruit histone deacetylase and that lead to

Table 2.4 KG1 cells treated with 1–10 μM ATRA: cell growth, viability, differentiation, and apoptosis analysis. Results are given as mean ± SEM ($n = 3$)

Duration	Treatment	Cell count, 10^4/mL	Viability, %	NBT+ cells	Dead cells, %
3 days	*Control cells*	279	77	0	23
	ATRA, 1 μM	164	80	28	20
	ATRA, 2 μM	163	79	54	21
	ATRA, 3 μM	167	73	54	27
	ATRA, 4 μM	184	76	39	24
	ATRA, 5 μM	201	82	60	18
	ATRA, 6 μM	190	76	50	24
	ATRA, 7 μM	202	71	56	29
	ATRA, 10 μM	153	71	52	29
4 days	*Control cells*	227	59	0	41
	ATRA, 1 μM	149	64	42	36
	ATRA, 2 μM	146	61	55	39
	ATRA, 3 μM	149	54	33	46
	ATRA, 4 μM	266	66	44	34
	ATRA, 5 μM	145	78	55	22
	ATRA, 6 μM	178	59	48	41
	ATRA, 7 μM	201	50	43	50
	ATRA, 10 μM	153	56	55	44

poorly response to the differentiation inducer all-trans retinoic acid (see Appendix B for characteristics of cell lines used in the studies). We treated KG1 cell line with different concentrations of ATRA (1–10 μM), and cell growth, viability, differentiation potential, and apoptosis were detected. As presented in Table 2.4, based on results (Savickiene et al. 2009) after 72–96 h of treatment, the optimal concentration of ATRA for differentiation induction is 2–3 μM. After treatment of KG1 cells with 2 μM ATRA, we observed cell growth inhibition up to 40%, the viability of the cells was similar to that of control cells (79%), and the differentiation rate was 54% compared to control cells. Higher concentrations of ATRA did not affect differentiation rate compared with 2 μM concentration. 1 μM ATRA slightly induced granulocytic differentiation (up to 28%) and that is 2 times less than 2 μM ATRA concentration. In conclusion, KG1 cells are more resistant for ATRA treatment, and we chose 2–3 μM concentration of ATRA for further experiments.

The results of these studies are presented in Fig. 2.4 and showed that ATRA was efficient in inducing time- and dose-dependent differentiation of leukemic cells (HL-60, NB4, and KG1) and the combination of ATRA with other therapy could be very promising in treatment of APL.

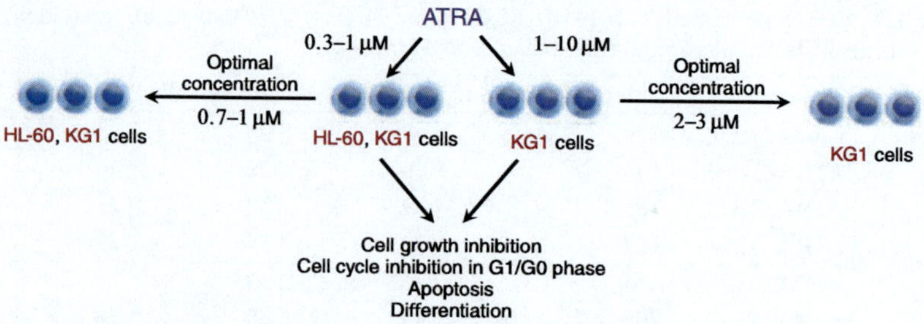

Fig. 2.4 Resumptive scheme of the effect of ATRA treatment on AML cell lines

2.2 Effects of ATRA Combination with Nucleotide Analogues and Ionizing Radiation

Recently differentiation therapy, when tumor cells are induced to differentiate by various differentiation agents and reenter into pathway of normal cells, has been applied in oncology. The purpose of our studies was to elucidate the possibilities of differentiation therapy in the cell culture system as the model *in vitro* using different inducers, nucleotide analogues and their combinations, and radiation hormesis (the effect of small doses of gamma radiation). It was shown that monoblast lines, for example, U937 might develop locomotor capacity only after addition of agents such as dibutyryl cyclic AMP (dbcAMP) or cytokines that drive differentiation into monocyte morphology.

In our studies (Savickiene and Gineitis 2003; Savickiene et al. 2010a) the human acute promyelocytic leukemia HL-60 cells were induced to differentiate by the following inducers: dimethylsulphoxide (DMSO), all-trans retinoic acid (ATRA), and 2′-O-dibutyryladenosine 3′,5′-cyclic monophosphate (dbcAMP). Optimal concentration of these agents was determined, by which cells differentiated well without any toxic effects. The optimal concentration of dbcAMP was 350 μM, and this concentration induces differentiation of HL-60 cells up to 68% after 96 h of treatment. After 120 h of treatment, both dbcAMP (35 μM) and ATRA (700 nM) induced similar differentiation level in HL-60 cells (68% and 62%) to granulocytes. These inducers suppressed cell growth up to 25–27% and viability (36% after ATRA and 13% after dbcAMP treatment) compared with untreated HL-60 cells. dbcAMP, which was added to the cell culture before induction with ATRA, increased ATRA-induced differentiation, while the same short pretreatment with ATRA 2 times decreased dbcAMP-mediated differentiation. Our results are generalized in Fig. 2.5. The data obtained by us are in agreement with the literature results that combination of ATRA with cAMP analogs can upregulate the expression of leukocyte alkaline phosphatase (LAP, a marker for the differentiation of the granulocyte) (Gianni et al. 1995).

Also it was shown that the effects of ATRA and dbcAMP on promyelocytic cells are closely related and cellular events are regulated at a different level within a

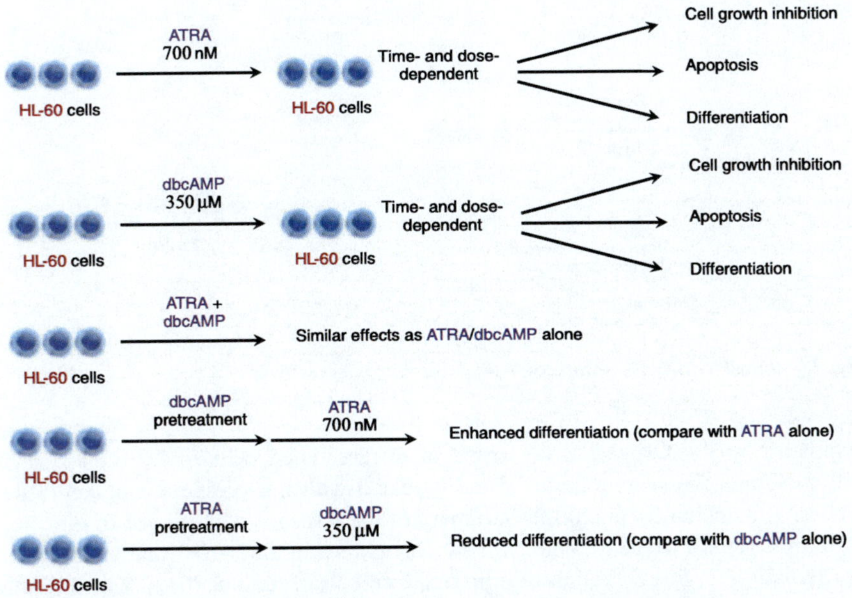

Fig. 2.5 Effects of dbcAMP alone and in combination with ATRA on AML cell lines

common signal transduction pathway (PLC/IP/PKC and the cAMP/PKA) (Lopez-Pedrera et al. 2001).

In the studies (Savickiene et al. 1999; Savickiene and Gineitis 2003; Savickiene et al. 2002) we used nucleotide analogues like 6-methylmercaptopurine riboside (6-MMPR), 6-thioguanine (6-TG), cytarabine (AraC), and 3-deazauridine (3-DU), which can incorporate into both the DNA and/or the RNA. We determined that these agents reduced HL-60 cell growth (6-MMPR—~31%, 6-TG— ~87%, AraC—~68%, and 3-DU—56%) at the fourth day of treatment but did not induce cell differentiation (6-TG and 3-DU) or induced slightly up to 15–25% (6-MMPR, AraC). These analogues in combination with ATRA increased granulocytic differentiation with different efficiency. The best stimulating effect was showed by 6-MMPR and 3-DU in combination with ATRA. Some nucleoside analogues (6-TG, 3-DU) in combination with dbcAMP suppressed cell differentiation. Pretreatment on uninduced cells with all analogues revealed the same effect of suppression of dbcAMP-mediated cell differentiation. 3-DU was the most efficient in all treatments. HL-60, NB4, and K562 cells treated with small doses of 3-DU (2.5–15 µM) for 24 h or full time (96 h) of differentiation with ATRA stimulated differentiation quite early (Savickiene et al. 2002). Moreover, after 4 days there were 2 times more mature cells than in cell culture treated only with ATRA. Doses of 3-DU higher than 10 µM had a cytotoxic effect and caused the appearance of apoptotic cells. Pretreatment with 3-DU for 24 h and the following treatment with other nucleoside analogues significantly increased the level of mature cells up to 50–80%. Flow

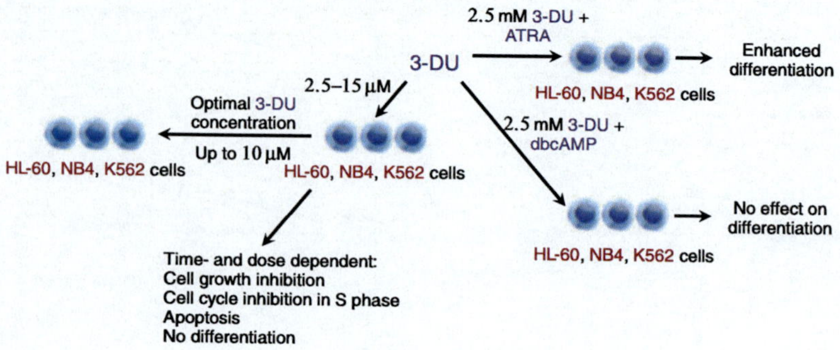

Fig. 2.6 Summary of 3-DU effects on AML cells

cytometric analysis revealed the arrest of differentiated cells in G0/G1 phase of cell cycle and the accumulation of cells treated with nucleoside analogues in the S phase (Savickiene et al. 2002). Summary of these results is presented in Fig. 2.6.

In other study we also demonstrated (Savickiene et al. 2010a) that small doses of irradiation (1–50 cGy) slightly suppressed cell proliferation. 20–30% of the cells were differentiated after irradiation with 50 cGy. Radiation hormesis did not have a positive effect on ATRA-induced differentiation. The results of these studies showed that nucleoside analogues, especially 3-DU, in combination with ATRA were efficient in inducing differentiation of leukemic cells and they could be considered as potential agents for therapy of APL.

2.3 Role of Protein Kinases and Their Inhibitors During Leukemic Cell Differentiation

Aiming to improve the strategy for therapy of AML, it is important to understand better the mechanisms that regulate the differentiation processes. In our studies (Savickiene et al. 1999, 2010b; Pivoriunas et al. 2007) we investigated effects of PKC/PKA inhibitors (Calphostin C and isoquinoline (H-89) and bisindolylmaleimide I (GF109203X)) on differentiation of AML cells.

It was shown in the literature that H-89 is a potent PKA inhibitor, which also inhibits several other kinases such as S6K1, ROCKII, MSK1, PKBα, PKA, and MAPKAP-K1b, and it also has effect on some others, including Kv1.3 K+ channels, β1AR, and β2AR (Lochner and Moolman 2006; Chijiwa et al. 1990). Bisindolylmaleimide I (GF109203X) as an ATP-competitive PKC inhibitor might be a tool to study the involvement of PKC in signal transduction pathways (Toullec et al. 1991). GF 109203X produces reversal activity on P-glycoprotein and MRP-mediated multidrug resistance. PKC inhibition by GF109203X distinctly lowers carbachol-stimulated ERK1/ERK2 activation and the subsequent proliferation of

SNU-407 colon cancer cells (Park and Cho 2012). Calphostin C as a powerful and selective inhibitor of the protein kinase C (PKC) interacts with phorbol ester binding site and the regulatory DAG binding site. It was found that inhibition of PKC is light dependent. At higher concentrations it inhibits protein kinase G, cAMP-dependent protein kinase, myosin light chain kinase, pp60v-src protein tyrosine kinase, and DAG kinase. It also inhibits phospholipases D1 and D2 (Mamoon et al. 2004). It stimulates apoptotic fragmentation of DNA and cell death. Calphostin C inhibits protein kinase C (PKC) isoenzymes by covalent modification of the lipid-binding regulatory domain. Exposure of cells to Calphostin C causes PKC-independent effects including growth inhibition, disruption of intracellular transport, and stimulation of apoptosis suggesting actions at additional targets (Zhu et al. 1998). Phospholipase D (PLD) enzymes are PKC activation targets. *In vitro*, Calphostin C inhibits activity of PLD1 and PLD2 with the half-maximal inhibitory concentration (IC50) of approximately 100 nM. Lipid and protein activators of these enzymes do not overcome inhibition, and it does not involve blockade of phosphatidylinositol 4,5-bisphosphate (PIP2)-dependent PLD binding to liposomes in substrate. According to research in which a series of deletion and point mutants of the enzymes were investigated, it is possible to argue that Calphostin C targets the PLD catalytic domain. Inhibition of PLD by Calphostin C *in vitro* involves stable and presumably irreversible modification of the enzyme. Activity of both PLD1 and PLD2 can be inhibited by Calphostin C treatment of intact cells in a manner that is separate from upstream actions of PKC.

The human promyelocytic leukemia cell line HL-60 was prompted to granulocytic differentiation by using dbcAMP or ATRA. H8, genistein, and staurosporine were used as inhibitors of PKA, PTK, and PKC, and during different stages of differentiation of granulocytic cells, the influence of distinct protein kinases on differentiation process was investigated (Savickiene et al. 1995). The influence of protein kinase inhibition on granulocytic maturation is in the order of PTK > PKC > PKA for the dbcAMP-mediated differentiation and PKC > PKA > PTK for the ATRA-mediated differentiation at the terminal stage of differentiation. The outcome shows that during leukemia cell differentiation the protein kinase activities are stage specific and the used inducer influences this particularity.

In the research we investigated the effects of protein kinase C (PKC) and protein kinase A (PKA) on leukemia cells by using their specific inhibitors (Calphostin C and GF 109203X and H-89). For this study HL-60 cells were induced by differentiating agents, ATRA and dbcAMP, and protein kinase inhibitors additionally were used. Studies have shown that PKC inhibition activates a non-differentiated apoptosis program in HL-60 cells, whereas PKA inhibition reduces differentiation by not affecting the apoptosis. We have found that leukemic cell differentiation and apoptosis involving PKC and PKA are differentially regulated (Savickiene et al. 1999). When HL-60 cells were exposed to specific inhibitors of protein kinases and protein phosphatases (lavendustine, Go6976, Go6983, and sodium vanadate), histone deacetylase inhibitor (sodium butyrate), or O- and N-glycosylation inhibitors (IPTG and tunicamycin), alterations of the NFκB transcription factor associated with cell differentiation process were detected. This is a significant factor in the downstream stage of HL-60 cell differentiation (Savickiene et al. 2002). Investigating the combined effect of a PKC inhibitor Calphostin C with differentiation inducers ATRA and dbcAMP, we found a causal relationship between signal transduction involving PKC and the transcription factor NFκB and an intracellular balance of reactive oxygen molecules that was critical for leukemic cell differentiation or apoptosis (Savickiene et al. 2010a, 1999).

Terminal leukemic cell differentiation involves cell cycle arrest and termination of a lineage specific genetic program. The phorbol ester PMA activates many signaling pathways (including the phosphoinositide-3-kinase (PI3K) pathway) that lead to the leukemic cell differentiation into monocytes. During the research using THP-1 monocytic leukemia and NB4 cells of promyelocytic leukemia, we first detected the abnormal PI3K signaling pathway blockage effects in these cells and the association with cell regulatory proteins and transcription factors, which ultimately determine PMA-induced monocytic differentiation (Pivoriunas et al. 2004; Treigyte et al. 2006; Pivoriunas et al. 2007).

2.4 Etoposide and Z-VAD(OH)-FMK Effects on AML Cells

Due to resistance to apoptosis of cells with prolonged survival, they can become a primary oncogenic reason, since normally a balance between apoptosis and proliferation is maintained in separate tissues. Apoptosis molecularly implies a cascade of caspase proteases that are released from the mitochondria and is characterized by chromatin margination, nucleosome ladder formation, condensation, and early nuclear condensation. In the apoptosis pathway, caspases are the mediators of cellular destruction which can schematically be assigned to two groups: "initiator" caspases (including caspase-2, caspase-8, caspase-9, and caspase-10) and effectors or executors (caspase-3, caspase-6, and caspase-7). Previous investigations have shown that inhibitors, based on the cleavage specificity of known substrates of caspases, are able to inhibit apoptosis in various systems. Benzyloxycarbonyl-Val-Ala-Asp (OMe) fluoromethylketone (Z-VAD(OH)-FMK), an ICE-like protease inhibitor, inhibits apoptosis in various cell types (Navakauskiene et al. 2004a). Etoposide (VP16), the topoisomerase II inhibitor, as well as the other drugs, is used to treat various tumors, e.g., lung and breast cancer or leukemia.

The objectives of the present study were to evaluate apoptosis-associated protein patterns in HL-60 cells as induced with etoposide and affected by the broad caspase and apoptosis inhibitor Z-VAD(OH)-FMK. HL-60 cells were treated for different times in culture with either only 68 μM etoposide or plus 25 mM Z-VAD(OH)-FMK or pretreated for 1 h with 25 μM Z-VAD(OH)-FMK. Also the cell was treated with all-trans retinoic acid and etoposide. We detected the time-dependent effect on apoptosis in HL-60 cell populations as evaluated by morphology. ATRA-treated cells after 96 h showed 20% of apoptotic cells. Etoposide stimulated apoptotic cell appearance after 3 h of treatment, and at 18 h this fraction of cells increased up to 100% (Navakauskiene et al. 2002, 2004a). The constant presence of Z-VAD(OH)-

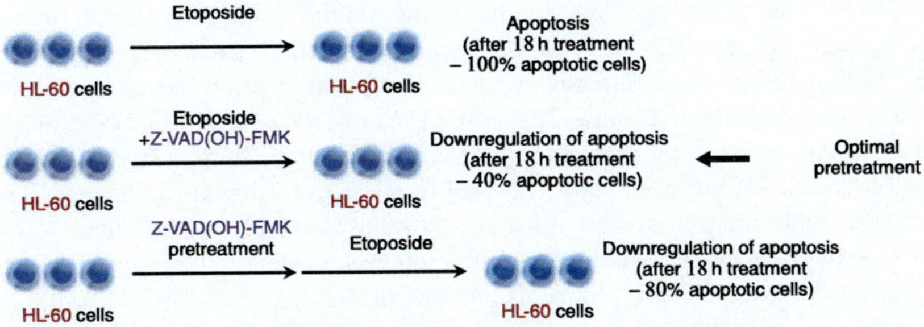

Fig. 2.7 Effects of etoposide and Z-VAD(OH)-FMK on HL-60 cells

FMK in the cell population significantly decreased the level of apoptotic cells to a great extent. Additionally, the number of apoptotic cells pretreated with Z-VAD(OH)-FMK started to increase after 3 h, and at 18 h it increased up to 80%. The proportion of apoptotic cells treated with etoposide and Z-VAD(OH)-FMK was only 40% after 8 h (Navakauskiene et al. 2004b) (Fig. 2.7). Cells treated with 68 μM etoposide for 0.5–18 h or differentiated for 120 h with 1 μM of ATRA have the percentages of apoptotic cells after 1, 2, 3, 8, and 18 h of etoposide treatment 7 ± 1.2, 35 ± 2.3, 60 ± 2.5, 85 ± 3.2, and 96 ± 2.3%, respectively. After 120 h of induction, there were 50 ± 2.7% apoptotic cells in the differentiating population. DNA fragmentation was investigated in control proliferating HL-60 cells and cells grown in the presence of 1 μM ATRA or 68 μM etoposide. In ATRA-treated cells DNA fragmentation started at 72 h of differentiation, and at 120 h the smallest fragments of DNA reached the nucleosomal level (145 bp), while in etoposide-treated cells after 3 h of treatment DNA was degraded at the nucleosomal level (Navakauskiene et al. 2004a). We suggest that Z-VAD(OH)-FMK can prevent leukemia cells from apoptotic stimuli caused by etoposide. Comprehension of signal transduction in apoptosis initiation by screening for biomarkers related with apoptosis may create new leukemia treatment targets.

2.5 Enhanced Effect of Epigenetic Modifiers on APL Cell Proliferation and Death

The standard therapeutic approaches for AML continue to be based on anthracyclines and cytarabine and for APL cases combination of anthracyclines with ATRA. In our study (Vitkeviciene et al. 2019), we analyzed the possibility of using conventional therapy (idarubicin + all-trans retinoic acid) plus epigenetic agents to treat APL. *In vitro*, we used NB4 and HL-60 cell lines that were exposed to the HMT inhibitor 3-deazaneplanocin A and the HDAC inhibitor belinostat together with idarubicin and all-trans retinoic acid. It was shown in the literature that 3-deazaneplanocin A can inhibit cell proliferation and induce apoptosis in different cancer cells. Likewise, belinostat inhibits cell growth and induces apoptosis in different human cells, APL cell lines NB4, and HL-60 (Savickiene et al. 2014). Belinostat is already used in therapy as approved drug and is well tolerated (Campbell and Thomas 2017). In our study, we detected that combination of belinostat, 3-deazaneplanocin A, and conventional therapy (idarubicin + all-trans retinoic acid) inhibited APL cell proliferation and survival in greater manner compared with idarubicin and all-trans retinoic acid (Fig. 2.8). There was no critical growth of cytotoxicity. New combination activated cell apoptosis and cell cycle arrest during G0/G1 phase. Treatment of HL-60 cells for 24 h with new combination (with a higher dose of idarubicin (8 nM)) resulted in cell cycle arrest in G2 phase. Idarubicin is known to inhibit DNA topoisomerase II, thereby disrupting DNA synthesis and arresting cells in G2 phase (Hollingshead and Faulds 1991). Meanwhile, ATRA and belinostat caused cell cycle arrest of NB4 and HL-60 (Savickiene et al. 2014), whereas 3-deazaneplanocin A induced gastric cancer cell clusters in G0/G1 cell cycle phase (Pan et al. 2016). In this study, the combination of belinostat, 3-deazaneplanocin A, and ATRA with a smaller dose of idarubicin (2 nM) induced NB4 and HL-60 cell accumulation in G0/G1 phase as well. We found that the newly proposed combination hasted granulocytic cell differentiation more than conventional treatment. This is consistent with preceding findings that 3-deazaneplanocin A and belinostat intensified all-trans retinoic acid-induced granulocytic differentiation (Valiuliene et al. 2017).

In conclusion (Fig. 2.9), our proposed combination of 3-deazaneplanocin A, belinostat, all-trans retinoic acid, and idarubicin had a greater effect on the inhibition of cell proliferation and survival and induction of apoptosis than treatment with all-trans retinoic acid and idarubicin alone.

2.6 Effects of Vitamin B3 and Phenylbutyrate on APL Cells

Niacin, or vitamin B3, helps to improve general good health. It is known that higher amounts of niacin can decrease cholesterol levels and lower cardiovascular risks. On the other hand, niacin (vitamin B3), a class III HDAC inhibitor, and its related compounds induce differentiation in human leukemia cells (e.g., HL-60) (Iwata et al. 2003).

In our study (Merzvinskyte et al. 2006) (Fig. 2.10) HL-60 cells were exposed to vitamin B3 at different concentrations: 2.5, 5, 7.5, and 10 mM. Cell growth and differentiation of HL-60 cells were examined was during 5 days of treatment. Control, untreated with HDAC inhibitor HL-60 cells, showed faster growth than those induced with various concentrations of vitamin B3. Control cell growth increased significantly after 48–72 h. Control cell amount after 96–120 h was decreased up to 10^4 cells/mL. We observed time- and dose-dependent cell growth inhibition after treatment with different vitamin B3 concentrations. But at 120 h point of all treatments, the growth inhibition effect was more or less the same.

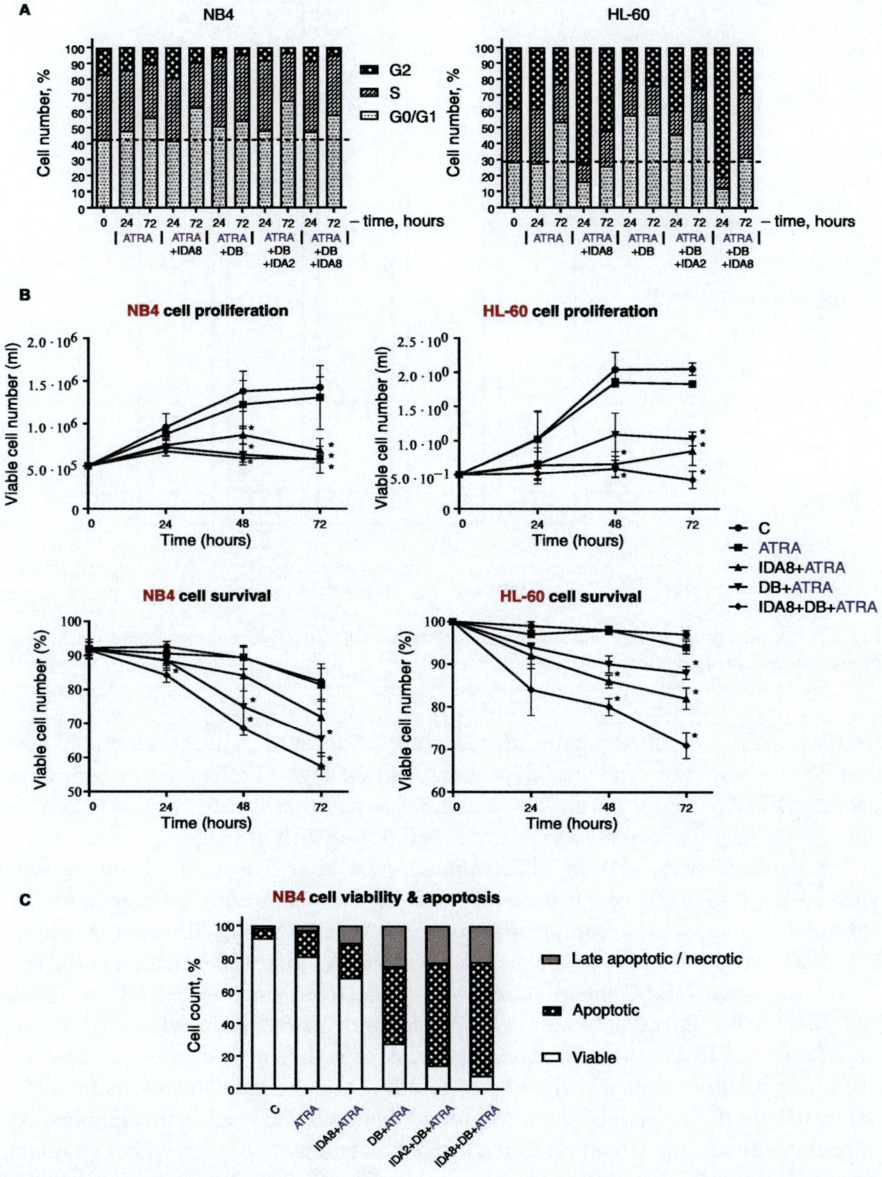

Fig. 2.8 NB4 and HL-60 cell lines' cell cycle distribution, proliferation, survival, and apoptosis after treatment with epigenetic agents plus conventional treatment: ATRA—all-trans retinoic acid, IDA2—idarubicin 2 nM, IDA8— idarubicin 8 nM, and DB—3-deazaneplanocin A with belinostat. According to Vitkeviciene et al. (2019)

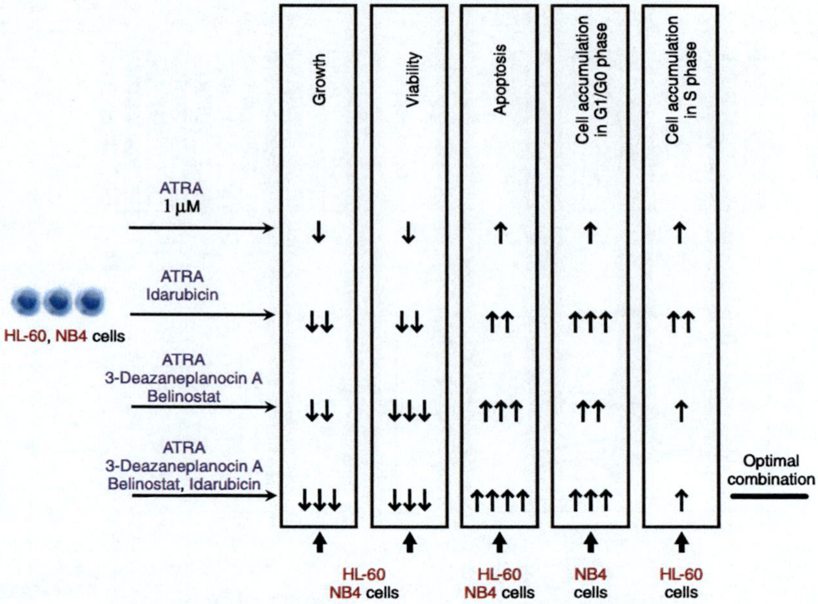

Fig. 2.9 Summary of ATRA and combined treatment with idarubicin, 3-deazaneplanocin and belinostat treatment on AML cell lines

The viability of HL-60 cells treated with 2.5–10 mM concentrations of vitamin B3 was also time- and dose-dependent, and the highest (10 mM) concentration decreased cell viability up to 72% compared with control cells. 2.5 mM dose of vitamin B3 slightly decreased viability of cell during 72 h of treatment.

We detected only 10% of differentiated cells after 5 days of treatment with vitamin B3 (2.5 mM), which was very similar to spontaneous differentiation of control cells (6%). The combination of ATRA and vitamin B3 after 120 h of treatment was 60%, and this value is less than ATRA alone 120-h treatment (80%).

We also used HDAC inhibitor phenylbutyrate to enhance the effect of ATRA and vitamin B3. Phenylbutyrate reversibly inhibits histone deacetylases (HDACs) of classes I and II, and this may cause an overall growth in gene expression, boosted cell differentiation, reduced cellular proliferation, and the induction of apoptosis in susceptible tumor cell populations. We found that pretreated cells with combination of phenylbutyrate and vitamin B3 for 6 h and then treatment of cells with ATRA and vitamin B3 was the most effective on early HL-60 differentiation induction (at 24 h of treatment up to 86%) with no apoptosis (less than 10%).

In conclusion, the higher was the concentration of vitamin B3 during treatment of HL-60 cells, the more the growth of the cells was inhibited, and only the lowest concentration slightly induced differentiation, which was close to the spontaneous differentiation. ATRA inhibits cell growth and viability better than combination with vitamin B3, and also ATRA is the best inducer of HL-60 cell differentiation (Fig. 2.11).

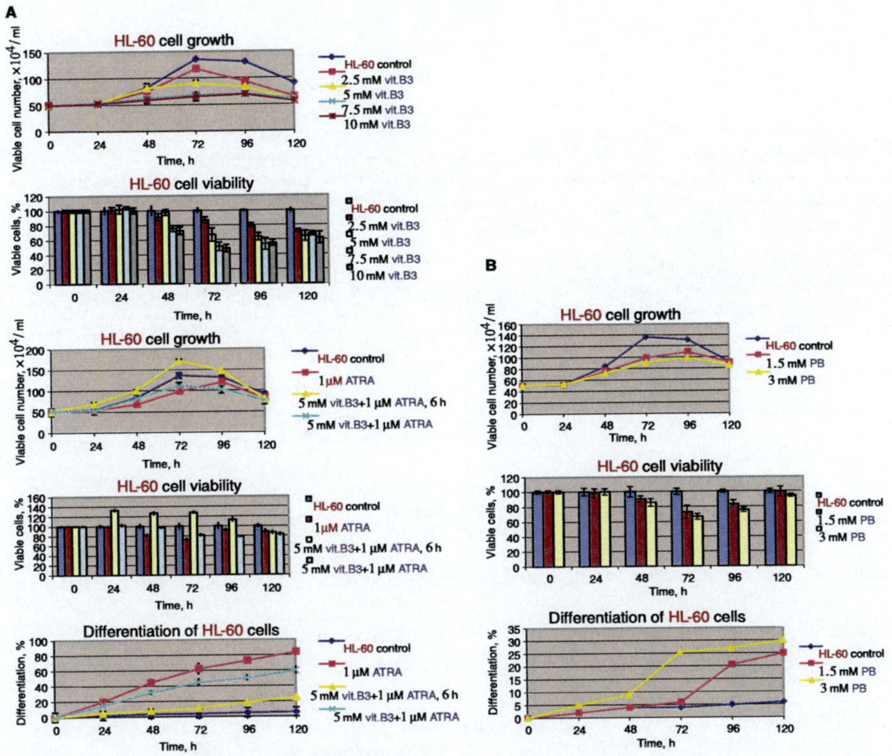

Fig. 2.10 The effects of HDAC inhibitors vitamin B3 and phenylbutyrate (PB) on acute promyelocytic leukemia cell line HL-60 proliferation viability and granulocytic differentiation. Results are given as mean ± SEM ($n = 3$)

2.7 Sodium Butyrate and Trichostatin A Impact on Granulocytic Differentiation of APL Cells

It was shown previously (Weston et al. 2003; Peng et al. 2010) that trichostatin A (TSA) is a histone deacetylases inhibitor which regulates cancer cell proliferation. We detected that NB4 cells exposed to TSA with 500 nM concentration died much more rapidly and after only 4 days had a viability of just over 10%. TSA alone induced differentiation only after 3 days (28%), and cell distribution via cell cycle phases was different than control cells (at 72 h control cells G1/G0—62%, S—21%, and G2/M—16%; TSA treated cells G1/G0—41.3%, S—35.3%, and G2/M—23.4%). In the cells exposed to 500 nM TSA, the death process also dramatically increased, and nearly 45% of the apoptotic cells were reached after 3 days of treatment. After treatment with combination of ATRA and TSA, we observed similar to TSA alone treatment results (Table 2.5).

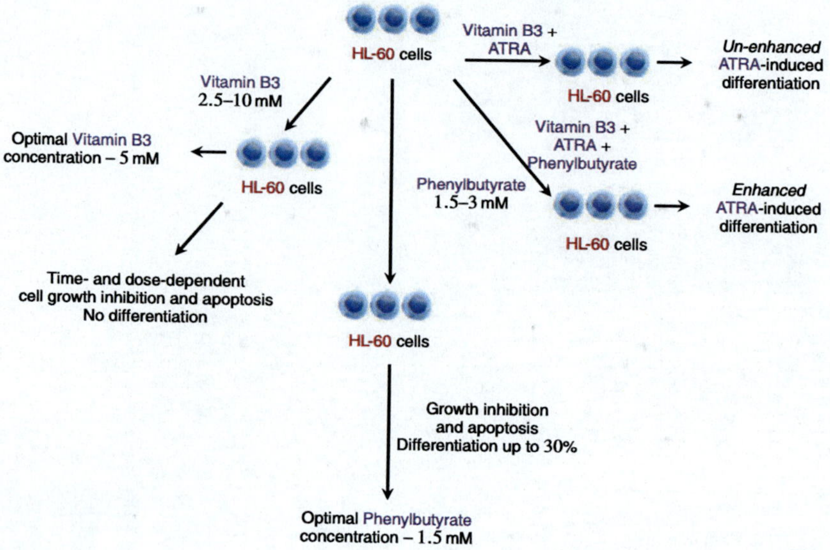

Fig. 2.11 Effects of vitamin B3 alone and in combination with ATRA on AML cell line HL-60

Table 2.5 NB4 cells treated with 500 nM TSA alone or in combination with 1 μM ATRA: cell growth, viability, differentiation, and apoptosis analysis. Results are given as mean ± SEM ($n = 3$)

	Days long treatment	Cell count, 10^4/mL	Viability, %	G1/G0 | S | G2/M distribution, %	CD11b+ cells	Dead cells, %
3	*Control NB4 cells*	140	83	62 | 21 | 17	15	35
	500 nM TSA	40	20	41 | 36 | 24	35	47
	500 nM TSA + 1 μM ATRA	40	32	46 | 41 | 12	30	37
4	*Control NB4 cells*	124	83	n/a	n/a	n/a
	500 nM TSA	15	12	n/a	n/a	n/a
	500 nM TSA + 1 μM ATRA	26	27	n/a	n/a	n/a

Very similar effect was observed after treatment of HL-60 cell with sodium butyrate. In the literature, it was shown that sodium butyrate stimulates differentiation of malignant cells *in vitro*. Also a clinical research was done, and the conclusions were made that there were no important changes between groups of patients treated with standard protocol versus standard treatment with sodium butyrate addition (Miller et al. 1987). It can be due to a short half-life of sodium butyrate *in vivo* or lower doses that can be used in clinics versus *in vitro* studies.

We treated HL-60 cells with sodium butyrate with 3 mM concentration and detected its effect on proliferation and differentiation of cell. As early as the second day after exposure, the viability of cell population began to decrease. The third day

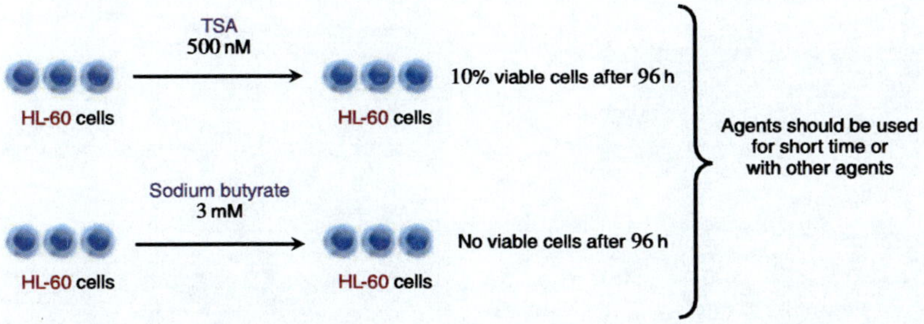

Fig. 2.12 Effects of HDAC inhibitors trichostatin A and sodium butyrate on HL-60 cells

showed a particularly strong decrease in viability compared to the control culture. By the fourth day, all treated cells had already died. We detected early induction of differentiation (up to 25%) in sodium butyrate-treated population after 48 h, but after 72 h of treatment, the amount of differentiated cell began to decrease, probably due to cell viability and proliferation inhibition. After 48 h of treatment, we observed strong proliferation inhibition.

In conclusion (Fig. 2.12), sodium butyrate impact upon proliferation and viability of HL-60 cells is very significant, and it is enough only short (up to 6 h) treatment for induction of differentiation and other processes like chromatin remodeling. The TSA should also be used in combination with other agents for best effect.

2.8 Combination of BML-210 and ATRA Induce Granulocytic Differentiation of APL Cells

It is known that histone deacetylase inhibitors affect chromatin remodeling, influence gene expression, and provoke cancer cells apoptosis. Histone deacetylase inhibitors are now used to treat different diseases like myeloma, sarcoma, lymphoma with Zolina (varinostat, SAHA), myeloma, breast cancer with Istodax (romidepsin, FK228), AML, ovarian cancer with Beleodaq (belinostat, PXD101), prostate cancer, myeloma with Farydak (panobinostat, LBH-589), ovary cancer, and AML with Depacon (valproic acid) (Yoon and Eom 2016).

In these studies (Savickiene et al. 2006a; Borutinskaite and Navakauskiene 2015), we examined *in vitro* effects of, at that time, new histone deacetylase inhibitor BML-210 (N-(2-Aminophenyl)-N/ phenyloctanol diamine). Since its mechanism of activity has not been described, we treated human leukemia cell lines (NB4, HL-60, THP-1, and K562) with different BML-210 doses alone or in combination with ATRA/hemin (Fig. 2.13). It was estimated that BML-210 suppresses the growth of all cell lines and incentivizes apoptosis in a dose- and time-dependent manner. Optimal BML-210 concentration with low cytotoxic effect on HL-60 was 5–10 μM. The higher concentration (20 μM) affected inhibition of viability up to 75% and induced

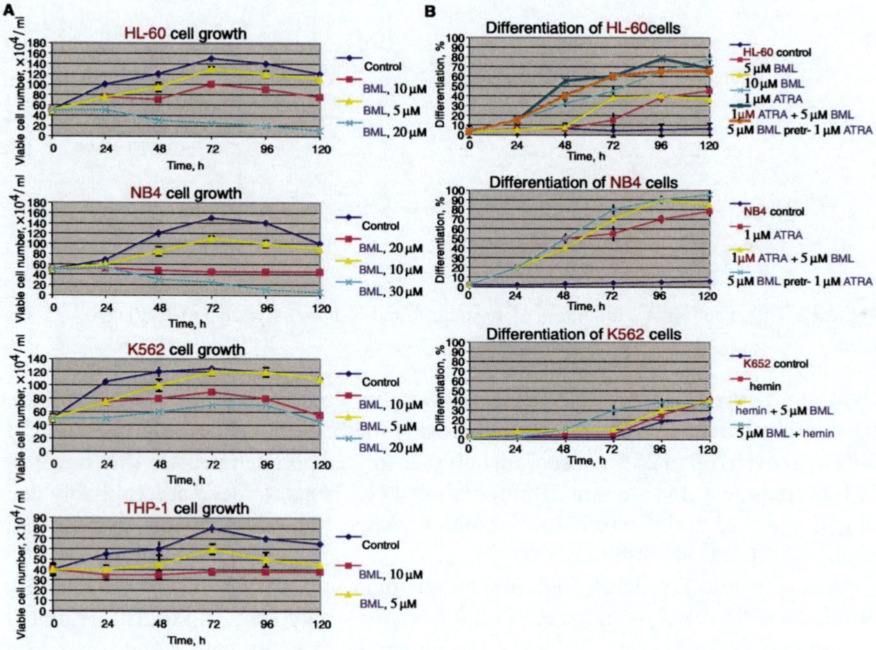

Fig. 2.13 The effects of HDAC inhibitor BML-210 alone or plus ATRA and hemin on HL-60, NB4, K562, and THP-1 cells

apoptosis up to 50% after 48 h of treatment. In accordance with comparable results gained by trypan blue, acridine orange/ethidium bromide staining, and flow cytometry, the authors argue that BML-210 provokes cytotoxicity through the pathway of apoptosis. BML-210 used alone induces HL-60 and K562 cell differentiation (up to 30%) to erythrocytes and granulocytes. We observed CD11b (early myeloid differentiation marker) expression and NBT-reduction in HL-60, but not NB4 cells. This variety can be due to more noticeable histone acetylation in HL-60 cells or to direct/indirect variations of the transcriptional profile. Combination with differentiation agents—all-trans retinoic acid and hemin—significantly intensifies differentiation (up to 80–90%). BML-210 alone or plus ATRA/hemin caused cell cycle arrest in G1 phase, which led to changes of gene transcription and influenced gene expression.

To conclude, for the first time we demonstrated class I HDAC inhibitor BML-210's effects on APL cell proliferation, differentiation, and apoptosis (Fig. 2.14). Our results argue that BML-210 could be a high-potential antileukemic agent that could provoke apoptosis and modulate differentiation through the modulation of gene expression and histone acetylation.

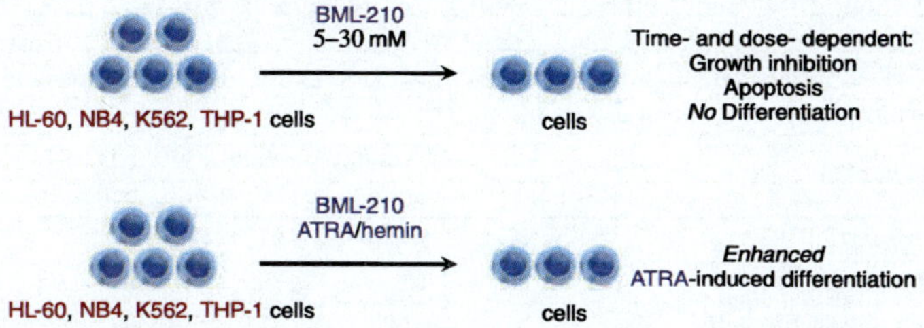

Fig. 2.14 Effects of histone deacetylase inhibitor BML-210 alone and plus ATRA and hemin on AML cells

2.9 Potential Histone Deacetylase Inhibitor FK228

FK228 (Depsipeptide or Romidepsin) is a prodrug and a potent histone deacetylase 1 and 2 (HDAC1 and HDAC2) inhibitor. It has a synergistic effect with ATRA *in vitro* and is able to stimulate differentiation in acute promyelocytic leukemia. Previously it was found as synthetic peptides that may reverse or inhibit diverse epigenetic modifications of histones and the DNA. Peptides may have an influence on the expression of non-coding RNAs, for example, lncRNAs and the maturation of miRNAs as well (Janssens et al. 2019).

We investigated *in vitro* activities of FK228 in AML cell lines (Savickiene et al. 2006b, 2009, 2011). We used various concentrations of FK228 (0.25–1 ng/ml) to the treatment of NB4 and HL-60 leukemia cells during 120 h. We detected that in both cell lines, FK228 induced cell growth inhibition up to 20–40% compared to control cells at the third day and induced cell death. We observed that FK228 treatment effects were dose- and time-dependent. Also we found that growth inhibition in HL-60 cells was more intensive in a shorter period of time. We also treated cell with higher concentration of FK228, but exposure of cells to FK228 at a concentration of 10 ng/ml provoked irreversible death of cells after 8 h (data not shown). Combined treatment with ATRA (1 μM) caused the cumulative anti-proliferative and apoptotic effects, which were more pronounced in HL-60 cells. Pretreatment of FK228 for 6 h before treatment with ATRA or ATRA with vitamin B3 has similar effect as ATRA alone treatment (Table 2.6).

As a result, the date we have collected shows facts about myeloid cell line-specific, differential activity of FK228 in the enhancement of ATRA-mediated differentiation that is related with the regulation of gene expression mediated through chromatin remodeling (Fig. 2.15).

Table 2.6 HL-60, NB4, and KG1 cells treated for 3 days with 0.2–1 ng/mL concentrations of FK228(FK), in combination with 1 μM ATRA, in pretreatment before treatment with ATRA (1 μM) or vitamin B3 (5 mM, B3): cell growth, viability, differentiation, and apoptosis analysis. Results are given as mean ± SEM ($n = 3$)

Cells	Treatment	Growth, %	Viability, %	NBT+ cells	Dead cells, %
HL-60	*Control*	100	91	0	9
	0.25 ng/mL FK	46	39	0	40–61
	0.5 ng/mL FK	62	42	0	58–70
	1 ng/mL FK	57	36	0	64–80
	0.5 ng/mL FK + 1 μM ATRA	36	18	–	82
	ATRA	65	83	56	17
	FK pretreat., 1 μM ATRA	91	83	73	17
	1 μM ATRA + 5 mM B3	87	79	59	21
	FK pretreat., 1 μM ATRA + 5 mM B3	100	88	80	12
NB4	*Control*	100	93	0	7
	0.25 ng/mL FK	38	44	0	56
	0.5 ng/mL FK	30	30	0	70
	1 ng/mL FK	25	16	0	84
	0.5 ng/mL FK + 1 μM ATRA	50	61	42	39
	ATRA	72	90	57	10
	FK pretreat., 1 μM ATRA	72	94	56	6
	1 μM ATRA + 5 mM B3	53	89	61	11
	FK pretreat., 1 μM ATRA + 5 mM B3	85	97	72	3
KG1	*Control*	100	91	–	9
	2 ng/mL FK + 3 μM ATRA	62	62	–	30
	3 μM ATRA	75	83	–	17
	2 ng/mL FK pretreat., 3 μM ATRA	65	86	–	14

2.10 Combined Effect of Belinostat, ATRA, and 3-deazaneplanocin A on AML Cells

Belinostat (Beleodaq and PXD101) is an approved drug to treat the patients with Peripheral T-Cell Lymphoma (PTCL). It was showed that anthracycline-based therapies (vincristine, doxorubicin, cyclophosphamide, and prednisone (CHOP)) are the first-line treatment for patients with PTCL. However, prognosis remains unsatisfactory with most patients relapsing within 5 years. The study with combination of belinostat and CHOP showed promising results (Johnston et al. 2015). Since belinostat is not approved for AML/APL treatment and its mechanism of action in AML cells is unclear, we studied belinostat alone or in combination with other agents (ATRA and 3-deazaneplanocin A (DZNep)) and antileukemic effects on APL cells (Savickiene et al. 2014; Valiuliene et al. 2015; Valiulienė et al. 2016; Valiuliene et al. 2017; Vitkeviciene et al. 2019). We used different concentrations (0.2–2 μM) of chemical agent belinostat alone or plus differentiation

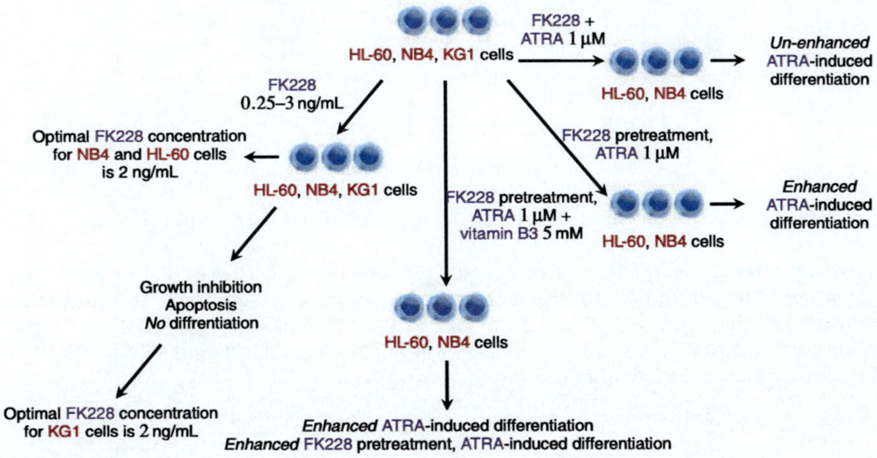

Fig. 2.15 Summary of histone deacetylase inhibitor FK228's effects on AML cells

Table 2.7 The effect of belinostat on HL-60 and NB4 cell proliferation, viability, and apoptosis. Results are given as mean ± SEM ($n = 3$)

Duration	Treatment	Growth, %	Viability, %	NBT+ cells	Dead cells, %
3 days	*Control HL-60 cells*	100	91	–	9
	0.2 µM belinostat	77	82	–	18
	0.5 µM belinostat	62	67	–	33
	1 µM belinostat	52	21	–	79
4 days	*Control NB4 cells*	100	92	n/a	15
	0.2 µM belinostat	53	78	n/a	22
	0.5 µM belinostat	33	13	n/a	87
	1 µM belinostat	31	0	n/a	100

inducer ATRA. We detected that high doses of belinostat alone inhibited cell growth up to 100%, and more intensity was observed in NB4 cells after 72 h of treatment with 2 µM (Table 2.7). Cells accumulated in the G1/G0 cell cycle phase and apoptosis rate were high (up to 90–100%).

We determined that optimal concentration and good anti-proliferative response with low cytotoxicity can be achieved with 0.2–0.5 µM concentration of belinostat. We showed that belinostat accelerates ATRA-mediated differentiation upon sequential treatments after 2 days, but eventually the level of differentiation was analogous to the ATRA-treated cells after longer treatment. This impact was more evident in HL-60 cells compared to NB4 cells (Fig. 2.16).

After treatment of cells with ATRA, belinostat, and DZNep, the most significant decrease in viability of NB4 and HL-60 cell lines was observed with combination of 1 µM ATRA, 0.2 µM belinostat, and 0.5 µM DZNep. After 72 h of combinational treatment, the number of NB4 and HL-60 viable cells was 0.29 million/ml and

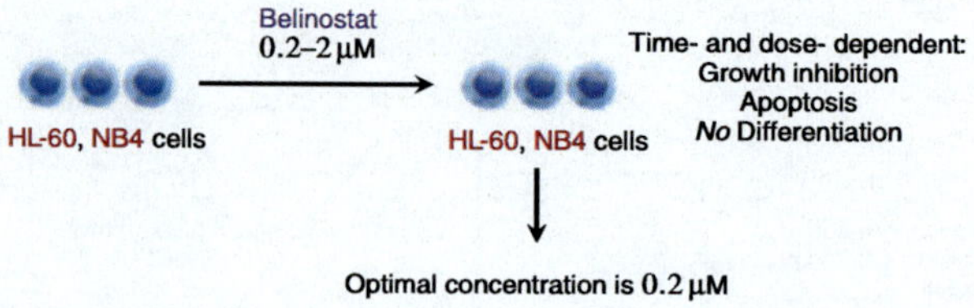

Fig. 2.16 Effects of histone deacetylase inhibitor belinostat on HL-60 and NB4 cells

Table 2.8 The inhibitory effect of belinostat (Bel) alone or in combination with ATRA and DZNep on NB4 and HL-60 cells. The inhibitory effect (%) was evaluated after 24, 48, and 72 h of treatment with 1 μM of ATRA, 0.2 μM belinostat, 0.5 μM DZNep, 1 μM ATRA + 0.2 μM belinostat, 1 μM ATRA + 0.5 μM DZNep, and 1 μM ATRA + 0.2 μM belinostat + 0.5 μM DZNep. Results are given as mean ± SEM ($n = 3$)

Cells	Treatment	24 h	48 h	72 h
NB4	*Control*	0	0	0
	ATRA	37.8	49.9	44.8
	Belinostat	49.9	48.7	30.4
	DZNep	64.8	44.8	14.8
	ATRA + Belinostat	52.1	72.4	73.6
	ATRA + DZNep	37.4	56.6	55.4
	ATRA + Belinostat + DZNep	67.4	87.2	87.5
HL-60	*Control*	0	0	0
	ATRA	9.8	20.9	24.6
	Belinostat	29.5	26.9	28.2
	DZNep	13.7	29.0	22.4
	ATRA + Belinostat	36.8	48.9	58.0
	ATRA + DZNep	37.6	43.1	34.0
	ATRA + Belinostat + DZNep	36.8	70.5	87.7

0.49 million/ml, respectively, and viability was 45.8% and 66.3%; control cells were 1.22 million/ml and 1.55 million/ml, respectively. Viability was more significantly reduced by the effects of ATRA and ATRA + DZNep than individual ATRA, belinostat, or DZNep. In addition, it can be seen that belinostat and belinostat cotreatment with ATRA showed more growth inhibitory effect compared with ATRA alone, only DZNep, or ATRA + DZNep. After combined exposure with ATRA + belinostat + DZNep NB4 cells after 72 h, amount of dead cells was close to 55% and HL-60—30%. In NB4 cells, ATRA + belinostat amount of dead cells after 72 h exposure was approximately 28% and 14% in HL-60 cells. After exposure to belinostat NB4 cells, the amount of dead cells at 72 h reached 22% and HL-60—15%. Calculation of cell growth inhibition showed that both NB4 and HL-60 cells were inhibited most by combination of ATRA + belinostat + DZNep (approximately 87% inhibition after 72 h in both cases) and ATRA + belinostat (growth inhibition after 72 h was about 73% and 58%, respectively) (Table 2.8).

Comparing the viability of HL-60 and NB4 cells after treatment with ATRA + belinostat + DZNep, we observed that the cell viability of the HL-60 line was less pronounced than that of NB4 cells. Student's t-test showed a statistically significant difference ($p \leq 0.05$) in the viability of NB4 and HL-60 cells exposed to ATRA + belinostat + DZNep at 48 and 72 h. The results show that NB4 and HL-60 leukemic cell lines respond differently to ATRA + belinostat + DZNep—NB4 cells more sensitive than HL-60 cells.

Evaluation of differentiation of NB4 and HL-60 leukemic cells showed similar levels of differentiation after ATRA and combination of ATRA + belinostat + DZNep treatments. The influence of belinostat and DZNep on differentiation was not significant—these compounds alone did not induce differentiation. After 72 h the most differentiated cells were found in ATRA + belinostat + DZNep-treated culture. In NB4 cells exposed to all three compounds, differentiation after 24 h reached 7.2%, after 48 h—22%, and after 72 h—86%, control cell differentiation after 24 h—0.3%, after 48 h— 2.2%, and after 72 h—3.2%. In HL-60 cells exposed to all 3 compounds, differentiation after 24 h was almost 13%, after 48 h—24%, and after 72 h—72–83%, in control HL-60 cells after 24 h—1.8%, after 48 h—3.9%, and after 72 h—4% of differentiated cells. Cell differentiation data showed that HL-60 cells undergo differentiation earlier than NB4 cells. After 72 h of treatment, the levels of differentiation are similar in both lines.

We observed that the maximum number of apoptotic (early apoptosis—annexin V-FITC (+)/PI2 (-)) and already dead (late apoptosis and/or necrosis—annexin V-FITC (+)/PI2 (+)) cells are reached after treatment of cells with ATRA + belinostat + DZNep. This confirms the apoptotic/dead cell analysis results from cells staining with trypan blue. The majority of cells in the early apoptosis stage were detected after exposure to ATRA + belinostat + DZNep. In conclusion, belinostat and DZNep alone inhibited growth of leukemic cells and improved differentiation of APL when ATRA was used for treatment.

2.11 Histone Methylation Inhibition Benefits to APL Differentiation

Histone methylation is epigenetic modification essential when regulating the gene expression. Histone methylation affects the compactness of chromatin, and usually amino acid residues of lysine (Lys), arginine (Arg), or glutamine (Gln) are methylated. Recently, phase I clinical trials of the DOT1 suppressor EPZ-5676 have been conducted, and its effects on acute lymphoid and myeloid leukemia and myelodysplastic syndrome were evaluated (Clinical Trial Identification in ClinicalTrials.gov Database No. NCT02141828; Study Completed 2016). The HMT inhibitor, the EZH2 suppressor EPZ-6438, is presently being investigated in B-cell lymphoma phase I clinical trials (Clinical Trial Identifications ClinicalTrials.gov Database No. NCT01897571).

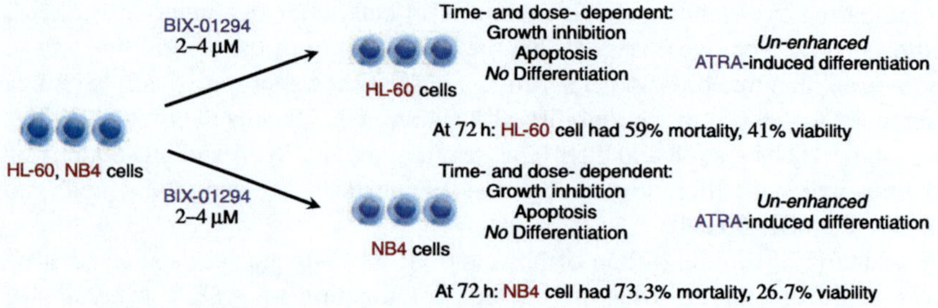

Fig. 2.17 Effects of histone methylase inhibitor BIX-01294 on HL-60 and NB4 cells

In our studies we have used BIX-01294 as a HMT inhibitor (Savickiene et al. 2014; Vitkeviciene et al. 2018). It was shown that by regulating H3K9 methylation and cell cycle, G9a inhibitor BIX-01294 inhibits the proliferation and provokes apoptosis of acute T lymphoblastic leukemia cells (Huang et al. 2017). G9a is a histone methyltransferase which induces histone H3 lysine 9 (H3K9) mono- and dimethylation, and it has also been found to methylate other proteins like p53, MyoD, CDYL1, WIZ, and ACINUS (Casciello et al. 2015).

The antileukemic activity of BIX-01294 was investigated on two human acute promyelocytic leukemia cell lines HL-60 and NB4 (Savickiene et al. 2014). BIX-01294 was shown to inhibit the growth of both cell lines concentration dependently. Also it notably decreased HL-60 and NB4 cell viability, when used at higher (4–5 μM) concentrations. The analysis of cell cycle upon 48 h of treatment demonstrated the shift in the proportion of cells in G0/G1 phase and a concentration-dependent reduction (about 1.6–2-fold) in G2/M phase. Also the treatment of NB4 and HL-60 cells with 1 μM ATRA, 2 μM BIX-01294, 1 μM ATRA + 2 μM BIX, and 4 μM BIX-01294 was performed in this study, and cell growth, viability, and death were evaluated. It was found that the most growth-reducing and anti-proliferative effects were observed in NB4 and HL-60 cell lines after treatment with 4 μM BIX-01294 after 72 h: NB4 cells had 73.3% apoptosis and 26.7% viability and HL-60 cells had 59% and 41%, respectively. The effect of the 4 μM BIX-01294 was stronger and induced greater apoptotic effect than the 2 μM BIX-01294 (Fig. 2.17). Exposure to 1 μM ATRA + 2 μM BIX-01294 also had a significant effect on cell viability and death: in NB4 cells, 72% mortality was observed after 72 h, and in HL-60, almost 70% mortality and 18% viability. These data also show that HL-60 cell death is lower after exposure to 4 μM BIX-01294, ATRA + 2 μM BIX-01294, and 2 μM BIX-01294. Cell differentiation was also assessed using the effects of 2 μM BIX-01294, 1 μM ATRA, and 1 μM ATRA + 2 μM BIX-01294. As in the case discussed above, the differentiation inducer ATRA induced a marked change in differentiation: in NB4 cells after 24 h nearly 13% and HL-60 about 21%. After 72 h the levels of differentiation between the two cell lines are similar (effect 1 μM ATRA: NB4 75%, HL-60 77% and effect 1 μM ATRA + 2 μM BIX-01294: NB4 almost 61%, HL-60 75%). The effect of 2 μM BIX-01294 differentiation was virtually inducible. Thus, the previous HL-60 cellular response and faster induction of differentiation of these cells are observed after all-trans retinoic acid exposure with BIX-01294 as well as with belinostat and DZNep. This could be explained by the fact that in HL-60 cells, RARα is immediately activated upon binding to ATRA (no fusion protein PML-RARα) and may be involved in activation of differentiation genes, whereas NB4 cells are predominantly degraded by PML-RARα, and therefore the effect of ATRA is delayed.

2.12 DNMT Inhibitors Contribution Toward Leukemia Cell Granulocytic Differentiation

2.12.1 Effects of DNA Methyltransferase Inhibitor Zebularine on APL Cells

Inhibition of DNA methyltransferase (DNMT) activity was also found to have an inhibitory effect on tumor formation, i.e., inhibitors of DNA methyltransferases can demethylate tumor suppressor genes and thereby suppress the cancer cell-specific phenotype (Brueckner and Lyko 2004). DNA methyltransferase inhibitors can be divided into 2 groups: nucleosides and non-nucleosides. The first group includes inhibitors such as 5-azacitidine (5-Aza-CR), 5-azadeoxycytidine (5-Aza-CdR), and zebularine. Non-nucleoside inhibitors may include compounds such as procaine and RG108. In this study (Savickiene et al. 2012a,b,c), each of the cell line (HL-60, NB4 and KG1) was exposed to zebularine (ZEB), all-trans retinoic acid (ATRA), and a combination of the two chemical agents, or the cells were pretreated with zebularine for 48 h and then exposed to all-trans retinoic acid (ZEB → ATRA). ZEB concentrations in all cell lines were 20 μM, and ATRA concentration for the treatment of NB4 and HL-60 cells was 1 μM and for KG1 treatment—3 μM. Effects' evaluations were performed every 24 h. For ZEB → ATRA, calculations were started after adding ATRA. We can state that in the human leukemic cell lines tested, the DNMT inhibitor zebularine and its combined treatment with ATRA inhibited the growth of NB4 cells the most.

Changes in the viability of each cell line were determined every 24 h by comparing the number of live and dead cells. We determined that the best inhibition of viability was affected by the effects of ZEB and ZEB → ATRA in all cell lines (up to 30%). Also we showed that zebularine alone did not induce differentiation of cells. We showed that CD11b expression is significantly increased in all three cell lines after exposure to ATRA and ZEB → ATRA. This is most prominent in KG1 cells (over 90%). The apoptosis analysis revealed that zebularine appears to have the greatest influence on the number of apoptotic cells in all lines (summarized in Fig. 2.18). This is especially significant in KG1 cells—61.7%. The effects of ATRA and ZEB → ATRA in NB4 and KG1 cells are circumscribed and only slightly above the control level. However, ATRA-exposed HL-60 cells show a

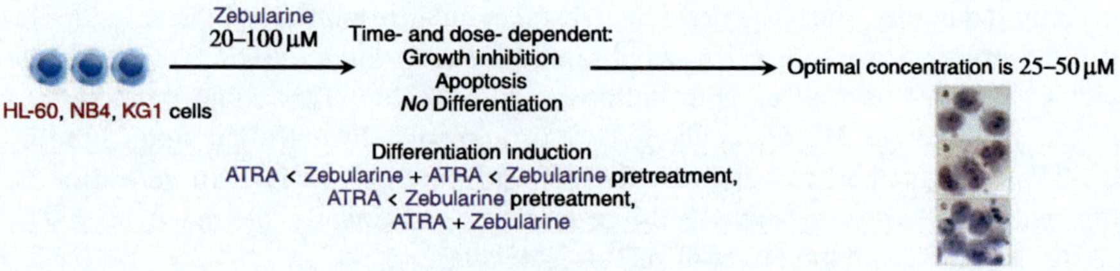

Fig. 2.18 Effects of DNA methyltransferase inhibitor zebularine on AML cells

presence of 31.4% of apoptotic cells, i.e., almost twice as much as ZEB → ATRA. In all three cell lines after evaluation of cell cycle, the number of cells in the cell cycle phase G1/G0 exposed to ATRA and ZEB → ATRA is greater than 10% for NB4 and KG1 and 20% for HL-60. The number of cells in the cell cycle phase S is reduced after exposure to ATRA and ZEB → ATRA. Zebularine has no effect on the distribution of NB4 and HL-60 cells between the cycle phases, but the proportion of KG1 cells in the cell cycle phase G1/G0 is higher compared to control cells. As the cell cycle phase subG1 assay showed that KG1 cells are particularly due to apoptosis, it can be assumed that the phase G1/G0 is higher than the other cells due to the apoptotic state rather than to the differentiated one.

After cell growth, viability, and differentiation analysis, we can summarize that ZEB → ATRA affected all cell lines the most compared with zebularine or ATRA treatment. The combination of these agents resulted in the slowest growth of cells, the lowest level of viability on the last day (fourth), and a marked increase in differentiation. Thus, it can be stated that ATRA-induced differentiation leaded to inhibition of cell growth and loss of viability and is improved by exposure of cells to the DNMT inhibitor zebularine.

2.12.2 *Procaine, Decitabine, and All-Trans Retinoic Acid Treatment Cause APL Cell Differentiation*

This section presents the results obtained with the NB4 and HL-60 leukemic cells exposed to DNA methyltransferase (DNMT) inhibitor procaine and differentiation-inducing all-trans retinoic acid (ATRA). For comparison, NB4 and HL-60 cells were exposed to another DNMT inhibitor decitabine. Comparative data on growth, viability, and differentiation of control and treated cells were obtained (Borutinskaite et al. 2016). The cell cycle of the NB4 line and its changes, as well as the results of the expression of the surface marker CD11b, were also examined. Intrinsic changes in the expression of DNMT, cell surface protein E-cadherin, and other molecules at the protein and gene levels are presented in the other chapter of the monograph.

We used such concentrations of DNMT inhibitor procaine: 0.5, 1, 3, and 5 mM. All-trans retinoic acid was added to HL-60 and NB4 cells at 1 μM concentration.

The cells were additionally exposed to another popular DNA methyltransferase inhibitor, decitabine, at concentrations of 0.5 μM and 1 μM. A few more experiments were also performed: procaine (3 and 5 mM) was exposed for 48 h, then the medium was changed, and the same combinations were again applied. Cells were cultured with chemicals for 96 h and counted with a hemocytometer every 24 h. We detected that the highest concentration of procaine has the greatest effect on both cell lines. Procaine (3 mM) maintains approximately constant influence over time (growth inhibition—NB4 25–35% and HL-60 40–50%). This concentration is effective from the onset of action and creates stable conditions for growth inhibition. For this reason, such a concentration of procaine is chosen for further experiments. It was also observed that procaine arrests growth of human leukemia cells and together with ATRA provokes differentiation. The majority of differentiated CD11b positive cells were reached within 72 h. Cells were most differentiated by ATRA and its combination with procaine. Despite fluctuations (48 h), the DNMT inhibitor procaine contributes to differentiation and results in increased differentiation. This combination arrests cell cycle in G0/G1 phase. Also it was showed that the highest rate of apoptosis was reached after 5 mM procaine treatment, already after 24 h (18%) and 48 h (30%). Other treatments with 3 mM procaine or 0.5 μM decitabine induce apoptosis at similar levels (48 h—~15% and ~19%, respectively). Cell accumulation in the G0/G1 phase is most noticeable in the ATRA-exposed sample (20% more than in controls). The cell amount decreases in S and G2/M phases, indicating cell cycle arrest in G0/G1 phase. The combined treatment of procaine plus ATRA already after 24 h' starts to arrest cell cycle. The fact that a single procaine does not arrest the cell cycle (in G0/G1 phase) but acts only in combination with ATRA indicates that ATRA is the protagonist here. Cells treated with decitabine accumulate in S phase. In this case, cell cycle arrest occurs by maintaining normal levels of DNA (G0/G1) and preventing it from doubling (arrest in S phase). Procaine inhibits the cell cycle in S and G2/M phases. The results obtained confirm the findings of previous studies (Villar-Garea et al. 2003). In conclusion, the DNMT inhibitor procaine was found to inhibit HL-60 and NB4 cell growth and to induce differentiation in combination with ATRA. Also combination of procaine and ATRA inhibit cell cycle in the G0/G1 cycle phase. Earlier differentiation marker CD11b expression is induced as early as 24 h after treatment to a combination of ATRA and ATRA with 3 mM procaine.

2.12.3 DNA Methyltransferase Inhibitor RG108 Has Anti-proliferative Effect on APL Cells

In these studies (Savickiene et al. 2012a,b), NB4, KG1, and HL-60 cells were exposed to RG108, ATRA, and PB (phenylbutyrate) alone and in various combinations, and cell differentiation, growth arrest, and apoptosis induction processes were evaluated. In tested NB4 and HL-60 cell lines, RG108 inhibited cell proliferation

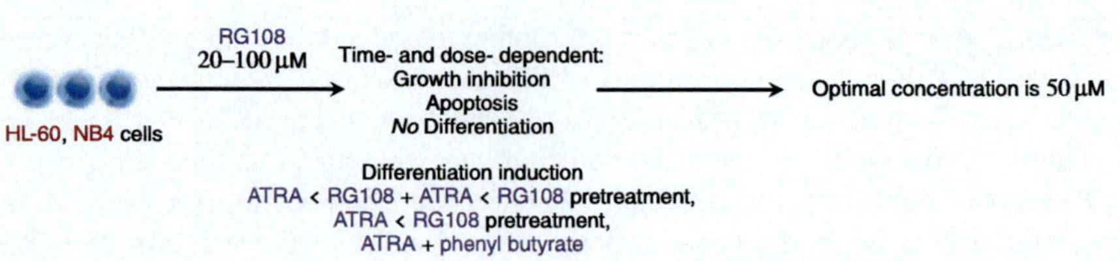

Fig. 2.19 Summary of DNA methyltransferase inhibitor RG108's effects on HL-60 and NB4 cells

in a time-dependent but not a dose-dependent manner and did not induce myeloid differentiation. We determined that 24 h exposure to RG108 was sufficient to reveal growth inhibitory impact on cell lines. Treatment of the cells with 50 μM RG replicates the growth of control cells, and only the number of cells is reduced by about 8% during the whole culture time. After 24 h of incubation with RG108 with the following addition of ATRA, cell proliferation was decreased in comparison to control leukemic cells. In combination of RG108 with ATRA and phenylbutyrate, the cells almost stop to proliferate after 24 h of treatment. Control cell differentiation reflects spontaneous NB4 cell differentiation (up to 1%). After treatment with 50 μM RG108, differentiated cells were present only up to 1%, so we observed that RG108 alone did not influence the induction of NB4 cell differentiation (Fig. 2.19). We compare different combinations of RG108 with all-trans retinoic acid or phenylbutyrate and can conclude that temporal pretreatment with RG108 before ATRA provoked remarkable (up to 62%) acceleration of granulocytic differentiation (Table 2.9).

The anti-proliferative RG108 impact on HL-60 cells was analyzed during 4 days of treatment with various concentrations of RG108 (25–300 μM). We determined that RG108 had only minor anti-proliferative and growth inhibitory (9–21%) impact at day 4 at concentrations of 25–100 μM. Only huge dose of 300 μM was allowed to reach the maximal effect (78%) after 96 h of treatment (Fig. 2.19).

2.13 Promising Benefits on Antileukemic Effect Demonstrate EGCG

Epigallocatechin—3—gallate (EGCG) is a plant compound that acts both as HDACi and as DNMTi. It reduces global DNA methylation level, changes gene expression, and induces apoptosis. EGCG could be a potential agent used to treat cancer. In the research the effect of EGCG on leukemic cells was investigated (Vitkeviciene et al. 2018; Borutinskaitė et al. 2018).

We treated NB4 cells with EGCG (10–40 μM) and differentiation inductor all-trans retinoic acid (1 μM). We determined that all concentrations of EGCG are shown to inhibit growth compared to control cells unaffected by EGCG. The

Table 2.9 HL-60 cells treated with RG108 alone and various combinations with sodium phenylbutyrate (PB) and all-trans retinoic acid (ATRA) characteristics. Results are given as mean ± SEM ($n = 3$)

Time	Cellular response	Control	RG108	RG108 + ATRA	RG108 → ATRA	RG108 → ATRA + PB	1 μM ATRA
0 h	Cell count, 10^6/mL	0.5 ± 0.1	0.5 ± 0.1	0.5 ± 0.1	0.5 ± 0.1	0.5 ± 0.1	0.5 ± 0.1
	Viability, %	100 ± 3	100 ± 2	100 ± 1	100 ± 3	100 ± 2	100 ± 1
	NBT+ cells, %	0	0	0	0	0	0
24 h	Cell count, 10^6/mL	0.7 ± 0.2	0.8 ± 0.2	0.7 ± 0.15	0.7 ± 0.14	0.6 ± 0.13	0.5 ± 0.2
	Viability, %	96 ± 2	88 ± 2	94 ± 2	82 ± 2	95 ± 2	95 ± 2
	NBT+ cells, %	1	0	20 ± 2	28 ± 2	29 ± 3	21 ± 4
48 h	Cell count, 10^6/mL	1.1 ± 0.3	1.2 ± 0.2	0.94 ± 0.2	0.83 ± 0.2	0.7 ± 0.2	0.82 ± 0.2
	Viability, %	92	81	82	82	87	87
	NBT+ cells, %	2 ± 2	0	30 ± 2	42 ± 2	51 ± 2	35 ± 2
72 h	Cell count, 10^6/mL	1.6 ± 0.5	1.3 ± 0.4	1.0 ± 0.2	0.9 ± 0.14	0.7 ± 0.2	0.8 ± 0.2
	Viability, %	90 ± 2	60 ± 2	74 ± 5	80 ± 4	82 ± 4	82 ± 5
	NBT+ cells, %	2	2	35 ± 2	55 ± 4	62 ± 5	48 ± 2

most pronounced differences in exposure to the concentrations used occurred at day 3, i.e., in 72 h from the start of the experiment, 40 μM EGCG inhibited most growth. After 96 h growth inhibition dependence on EGCG concentration remained unchanged: 10 μM < 20 μM < 30 μM < 40 μM. All-trans retinoic acid, like EGCG, also inhibited cell growth compared to control cells, but the effect of one ATRA was less than that of ATRA + EGCG. Differences in growth of ATRA and EGCG + ATRA-affected cells occurred as early as 48 h. We also see that at day 3 and day 4, there was almost no difference in the growth of cells exposed to EGCG 20 μM, 30 μM, and 40 μM with ATRA, whereas EGCG alone was more pronounced. A concentration of 10 μM of EGCG had the minimal impact on viability of cells and was therefore no longer used in subsequent experiments. Further efforts were made to determine the effect of EGCG and ATRA combinations on pretreated cells for 24 h with EGCG alone (cells were exposed to three different concentrations of EGCG for 24 h), and then 1 μM ATRA was added after media was changed (EGCG → ATRA). With these reagent combinations, differences in growth begin to appear at day 2 and were most pronounced after 96 h. The trend remains the same as in previous experiments, with most growth inhibited by cells exposed to the highest concentration of EGCG and at least 20 μM EGCG. However, the comparison of the results of this experiment with the effects of EGCG and EGCG + ATRA without rinsing shows that cells proliferate more in this case. A sharp decrease in the number of live control and all-trans retinoic acid treated cells observed at day 4 may be due to a significant increase in cell counts, decrease in nutrient medium, and acidification of the medium.

A concentration of 40 μM of EGCG can promote up to 20% differentiation. ATRA and EGCG + ATRA induce differentiation significantly—up to 80% in differentiated cell populations (summarized results presented in Table 2.10). The greatest differentiation effect was observed after 4 days with 40 μM EGCG + ATRA. This combination works more strongly than others on day 3. No substantial difference was noticed between the remaining combinations of EGCG + ATRA concentrations in inducing differentiation for all the 4 days. However, differentiation is induced by effects of 40 μM EGCG + ATRA and 40 μM EGCG → ATRA. Cell distribution in different cell cycle phases revealed that after EGCG and EGCG + ATRA treatment for 72 h, NB4 cells cycle was stopped in the G0/G1 cell cycle phase.

After the analysis of apoptotic cells, we showed that EGCG exposure was most prominent after day 1 and was almost 18% ATRA after 24 h (the number of apoptotic cells was less than 12%). It can be claimed that the combination effects of EGCG → ATRA and EGCG + ATRA were more influenced by all-trans retinoic acid than EGCG because the percentage of apoptotic cells was less than that of EGCG-exposed cells. The percentage of EGCG → ATRA apoptosis was approximately 9% and that of EGCG + ATRA was approximately 14%. After 3 days, the trend for apoptosis remained the same, with the highest number of cells killed by EGCG, as much as 41%, while all-trans retinoic acid was the lowest. Effect

Table 2.10 Cellular response of NB4 cells to the treatment with EGCG alone and various combinations with all-trans retinoic acid (ATRA). Results are given as mean ± SEM ($n = 3$)

Time	Cellular response	Control	40 μM EGCG	40 μM EGCG + ATRA	40 μM EGCG → ATRA	1 μM ATRA
0 h	Cell count, 10^6/mL	0.5	0.5	0.5	0.5	0.5
	Viability, %	100	100	86	100	100
	NBT+ cells, %	0	0	0	0	0
24 h	Cell count, 10^6/mL	0.780	0.51	0.5	0.53	0.56
	Viability, %	96	94	82	95	88
	Apoptosis, %	10.30	8.7	14.1	12	17.7
	G1/G0 | S | G2/M distribution, %	60 | 21 | 19	72 | 18 | 10	76 | 14 | 10	76 | 13 | 11	53 | 27 | 20
	NBT+ cells, %	1	17	30	21	0
48 h	Cell count, 10^6/mL	1.1	0.4	0.56	0.82	0.6
	Viability, %	92	82	82	87	81
	NBT+ cells, %	2	35	48	35	6
72 h	Cell count, 10^6/mL	1.6	0.56	0.6	0.8	0.56
	Viability, %	90	74	80	82	60
	Apoptosis, %	11	14	28.1	12.5	41.2
	G1/G0 | S | G2/M distribution, %	–	85 | 10 | 5	62 | 16 | 22	68 | 14 | 18	78 | 12 | 10
	NBT+ cells, %	2	73	50	48	11
96 h	Cell count, 10^6/mL	2.0	0.4	0.46	0.64	0.32
	Viability, %	88	57	66	80	44
	NBT+ cells, %	3	78	91	64	20

of ATRA after EGCG flush on apoptosis after 72 h was almost the same as ATRA alone, so probably after some time all-trans retinoic acid had the main impact on cell viability.

The impact of EGCG on growth, viability, and differentiation of HL-60 cells was investigated (summarized results presented in Table 2.11). We found similar effects on HL-60 leukemic cells like in NB4 cells. EGCG reduced HL-60 cells viability, slightly inhibited the leukemic cell cycle in the G0/G1 phase, and did not induce differentiation. In addition, we tested EGCG effect on non-malignant WI38 (a diploid human cell line composed of fibroblasts derived from lung tissue of a 3-month-gestation aborted female fetus) cell and on amniotic fluid stem cell proliferation by MTT. Treatment with EGCG did not have any effect on tested cell proliferation for 48 h, and only after 72 h, the highest used concentration of EGCG (140 μM) caused slightly decreased cell proliferation (to 85% and 87% compared to untreated cell proliferation in WI38 and amniotic fluid stem cells, respectively).

In conclusion, NB4 cell growth is most inhibited by the DNMT/HDAC inhibitor EGCG treatment and EGCG + ATRA combinations without rinsing. It was also found that in all the cases studied, the number of living cells was most substantially reduced by the highest amount used, i.e., 40 μM, EGCG concentration. From the viability studies, we can state that the viability of NB4 cells is mainly reduced by 40 μM EGCG and to a lesser extent by 40 μM EGCG + ATRA, while 40 μM EGCG + ATRA is affected even less by these two effects. Growth and viability are most strongly inhibited by 40 μM EGCG and 40 μM EGCG + ATRA, while differentiation is most induced by 40 μM EGCG → ATRA. Therefore, these EGCG and all-trans retinoic acid combinations have been selected for further studies of cellular processes.

Complex effects of different structure and activity of chromatin modifiers—HDAC inhibitors (BML-210, belinostat, FK228, sodium phenylbutyrate, vitamin B3, and EGCG), DNA methyltransferase inhibitors (RG108, zebularine), histone methyltransferase inhibitors (BIX-01294), and other agents— nucleoside analogues, inhibitors of cell signaling molecules—protein kinases and phosphatases, and low dose ionizing in combination with differentiation inducer ATRA generate effective cellular processes leading to leukemia cells' induction to differentiation, proliferation inhibition and apoptosis initiation.

Table 2.11 Cellular response of HL-60 cells to the treatment with EGCG alone and various combinations with all-trans retinoic acid (ATRA). Results are given as mean ± SEM ($n = 3$)

Time	Cellular response	Control	20 μM EGCG	30 μM EGCG + ATRA	20 μM EGCG → ATRA	30 μM EGCG → ATRA
0 h	Cell count, 10^6/mL	0.5	0.5	0.5	0.5	0.5
	Viability, %	100	100	100	100	100
	NBT+ cells, %	0	0	0	0	0
24 h	Cell count, 10^6/mL	0.76	0.65	0.67	0.66	0.56
	Viability, %	96	94	95	92	90
	NBT+ cells, %	1	28	17	17	25
48 h	Cell count, 10^6/mL	0.166	0.87	0.85	1.0	0.81
	Viability, %	92	86	86	89	88
	NBT+ cells, %	2	36	34	27	38
72 h	Cell count, 10^6/mL	0.154	0.53	0.55	0.92	0.74
	Viability, %	90	79	67	85	84
	NBT+ cells, %	2	50	48	48	50
96 h	Cell count, 10^6/mL	1.3	40	36	0.82	0.5
	Viability, %	88	58	50	72	64
	NBT+ cells, %	3	70	75	57	84

References

Advani R, Warnke R, Rosenberg S (1999) Treatment of multicentric Castleman's disease complicated by the development of non-Hodgkin's lymphoma with high-dose chemotherapy and autologous peripheral stem-cell support. Ann Oncol 10(10):1207–1209. https://doi.org/10.1023/A:1008366721816

Borutinskaitė V, Virkšaitė A, Gudelytė G, Navakauskienė R (2018) Green tea polyphenol EGCG causes anti-cancerous epigenetic modulations in acute promyelocytic leukemia cells. Leukemia Lymphoma 59(2):469–478. https://doi.org/10.1080/10428194.2017.1339881

Borutinskaite V, Navakauskiene R (2015) The histone deacetylase inhibitor BML-210 influences gene and protein expression in human promyelocytic leukemia NB4 cells via epigenetic reprogramming. Int J Mol Sci 16(8):18252–18269. https://doi.org/10.3390/ijms160818252

Borutinskaite V, Bauraite-Akatova J, Navakauskiene R (2016) Anti-leukemic activity of DNA methyltransferase inhibitor procaine targeted on human leukaemia cells. Open Life Sci 11(1):322–330. https://doi.org/10.1515/biol-2016-0044

Brueckner B, Lyko F (2004) DNA methyltransferase inhibitors: old and new drugs for an epigenetic cancer therapy. Trends Pharmacol Sci 25(11):551–554. https://doi.org/10.1016/j.tips.2004.09.004

Campbell P, Thomas CM (2017) Belinostat for the treatment of relapsed or refractory peripheral T-cell lymphoma. J Oncol Pharm Pract 23(2):143–147

Casciello F, Windloch K, Gannon F, Lee JS (2015) Functional role of G9a histone methyltransferase in cancer. Front Immunol 6:487. https://doi.org/10.3389/fimmu.2015.00487

Chijiwa T, Mishima A, Hagiwara M, Sano M, Hayashi K, Inoue T, Naito K, Toshioka T, Hidaka H (1990) Inhibition of forskolin-induced neurite outgrowth and protein phosphorylation by a newly synthesized selective inhibitor of cyclic AMP-dependent protein kinase, N-[2-(p-bromocinnamylamino)ethyl]-5-isoquinolinesulfonamide (H-89), of PC12D pheochromocytoma cells. J Biol Chem 265(9):5267–5272

Denis GV, Vaziri C, Guo N, Faller DV (2000) RING3 kinase transactivates promoters of cell cycle regulatory genes through E2F. Cell Growth Differ Mol Biol J Am Assoc Cancer Res 11(8):417

Denis GV, McComb ME, Faller DV, Sinha A, Romesser PB, Costello CE (2006) Identification of transcription complexes that contain the double bromodomain protein Brd2 and chromatin remodeling machines. J Proteome Res 5(3):502–511

Duan R, Du W, Guo W (2020) EZH2: a novel target for cancer treatment. J Hematol Oncol 13:1–12

Gianni M, Terao M, Norio P, Barbui T, Rambaldi A, Garattini E (1995) All-trans-retinoic acid and cyclic adenosine-monophosphate cooperate in the expression of leukocyte alkaline-phosphatase in acute promyelocytic leukemia-cells. Blood 85(12):3619–3635. https://doi.org/10.1182/blood.V85.12.3619.bloodjournal85123619

Greenblatt SM, Nimer SD (2014) Chromatin modifiers and the promise of epigenetic therapy in acute leukemia. Leukemia 28(7):1396–1406. https://doi.org/10.1038/leu.2014.94

Halsall JA, Turner BM (2016) Histone deacetylase inhibitors for cancer therapy: An evolutionarily ancient resistance response may explain their limited success. Bioessays 38(11):1102–1110. https://doi.org/10.1002/bies.201600070

He L, Guidez F, Tribioli C, Peruzzi D, Ruthardt M, Zelent A, Pandolfi P (1998) Distinct interactions of PML-RAR alpha and PLZF-RAR alpha with co-repressors determine differential responses to RA in APL. Nat Genet 18(2):126–135. https://doi.org/10.1038/ng0298-126

Hollingshead LM, Faulds D (1991) Idarubicin – a review of its pharmacodynamic and pharmacokinetic properties, and therapeutic potential in the chemotherapy of cancer. Drugs 42(4):690–719. https://doi.org/10.2165/00003495-199142040-00010

Huang Y, Zou Y, Lin L, Ma X, Huang X (2017) Effect of BIX-01294 on proliferation, apoptosis and histone methylation of acute T lymphoblastic leukemia cells. Leukemia Res 62:34–39. https://doi.org/10.1016/j.leukres.2017.09.015

Iwata K, Ogata S, Okumura K, Taguchi H (2003) Induction of differentiation in human promyelocytic leukemia HL-60 cell line by niacin-related compounds. Biosci Biotechnol Biochem 67(5):1132–1135. https://doi.org/10.1271/bbb.67.1132

Janssens Y, Wynendaele E, Vanden Berghe W, De Spiegeleer B (2019) Peptides as epigenetic modulators: therapeutic implications. Clin Epigenet 11. https://doi.org/10.1186/s13148-019-0700-7

Johnston PB, Cashen AF, Nikolinakos PG, Beaven AW, Barta SK, Bhat G, Song T, Choi MR, Allen LF, de Vos S, Oki Y, Deng C, Foss FM (2015) Safe and effective treatment of patients with peripheral T-Cell lymphoma (PTCL) with the novel HDAC inhibitor, belinostat, in combination with CHOP: Results of the Bel-CHOP phase 1 trial. Blood 126(23): 253–253

Kim EK, Kwon KB, Han MJ, Song MY, Lee JH, Ko YS, Shin BC, Yu J, Lee YR, Ryu DG, Park JW, Park BH (2007) Induction of G1 arrest and apoptosis by Scutellaria barbata in the human promyelocytic leukemia HL-60 cell line. Int J Mol Med 20(1):123–128

Lochner A, Moolman JA (2006) The many faces of H89: A review. Cardiovasc Drug Rev 24(3–4):261–74. https://doi.org/10.1111/j.1527-3466.2006.00261.x

Lopez-Pedrera C, Dobado-Berrios P, Ros R, Torres A, Garcia-Navarro S, Jardi M, Felez J, Velasco F (2001) Signal transduction pathways underlying the expression of tissue factor and thrombomodulin in promyelocytic cells induced to differentiate by retinoid acid and dibutyryl camp. Thromb Haemost 85(6):1031–1036

Maiques-Diaz A, Spencer GJ, Lynch JT, Ciceri F, Williams EL, Amaral FMR, Wiseman DH, Harris WJ, Li Y, Sahoo S, et al. (2018) Enhancer activation by pharmacologic displacement of LSD1 from GFI1 induces differentiation in acute myeloid leukemia. Cell Rep 22(13):3641–3659

Mamoon AM, Baker RC, Farley JM (2004) Regulation of acetylcholine-induced phosphorylation of PLD1 in porcine tracheal smooth muscle. J Biomed Sci 11(6):810–817

Mangelsdorf DJ, Koeffler HP, Donaldson CA, Pike JW, Haussler MR (1984) 1,25-Dihydroxyvitamin D3-induced differentiation in a human promyelocytic leukemia cell line (HL-60): receptor-mediated maturation to macrophage-like cells. J Cell Biol 98(2):391–398

Merzvinskyte R, Treigyte G, Savickiene J, Magnusson KE, Navakauskiene R (2006) Effects of histone deacetylase inhibitors, sodium phenyl butyrate and vitamin B3, in combination with retinoic acid on granulocytic differentiation of human promyelocytic leukemia HL-60 cells. In: Diederich M (ed) Signal transduction pathways, PT B: stress signaling and transcriptional control, Annals of the New York Academy of Sciences, vol 1091, pp 356–367. https://doi.org/10.1196/annals.1378.080, cell Signaling World 2006 Conference, Luxembourg, Luxembourg, Jan 25–28, 2006

Miller AA, Kurschel E, Osieka R, Schmidt CG (1987) Clinical pharmacology of sodium butyrate in patients with acute leukemia. Eur J Cancer Clin Oncol 23(9):1283–1290. https://doi.org/10.1016/0277-5379(87)90109-x

Minucci S, Nervi C, Lo Coco F, Pelicci P (2001) Histone deacetylases: a common molecular target for differentiation treatment of acute myeloid leukemias? Oncogene 20(24):3110–3115. https://doi.org/10.1038/sj.onc.1204336

Navakauskiene R, Treigyte G, Pivoriunas A, Savickiene J (2002) Cell cycle inhibitors in retinoic acid- and etoposide-mediated biological responses. Biologija 2:64–67

Navakauskiene R, Treigyte G, Gineitis A, Magnusson KE (2004a) Identification of apoptotic tyrosine-phosphorylated proteins after etoposide or retinoic acid treatment of HL-60 cells. Proteomics 4(4):1029–1041. https://doi.org/10.1002/pmic.200300671

Navakauskiene R, Treigyte G, Savickiene J, Gineitis A, Magnusson KE (2004b) Alterations in protein expression in HL-60 cells during etoposide-induced apoptosis modulated by the caspase inhibitor ZVAD.fmk. In: Diederich M (ed) Signal transduction pathways, chromatin structure, and gene expression mechanisms as therapeutic targets, Fdn Rech Canc & Sang; Novartis Luxembourg; Q8 Petr, Annals of the New York Academy of Sciences, vol 1030, pp 393–402. https://doi.org/10.1196/annals.1329.0049

Noguera IN, Catalano G, Banella C, Divona M, Faraoni I, Ottone T, Arcese W, Voso MT (2019) Acute promyelocytic leukemia: Update on the mechanisms of leukemogenesis, resistance and on innovative treatment strategies. Cancers 11(10). https://doi.org/10.3390/cancers11101591

Olsson IL, Breitman TR, Sarngadharan MG, C GR (1983) Mechanisms for induction of differentiation in the human promyelocytic cell line HL-60. Haematol Blood Transfus 28:384–389

Pan YM, Wang CG, Zhu M, Xing R, Cui JT, Li WM, Yu DD, Wang SB, Zhu W, Ye YJ, Wu Y, Wang S, Lu YY (2016) STAT3 signaling drives EZH2 transcriptional activation and mediates poor prognosis in gastric cancer. Mol Cancer 15. https://doi.org/10.1186/s12943-016-0561-z

Park YS, Cho NJ (2012) EGFR and PKC are involved in the activation of ERK1/2 and p90 RSK and the subsequent proliferation of SNU-407 colon cancer cells by muscarinic acetylcholine receptors. Mol Cell Biochem 370(1–2):191–198. https://doi.org/10.1007/s11010-012-1410-z

Parthasarathy R, Mehta K (1998) Altered metabolism of all-trans-retinoic acid in liposome-encapsulated form. Cancer Lett 134(2):121–128

Peng CY, Jiang J, Zheng HT, Liu XS (2010) Growth-inhibiting effects of arsenic trioxide plus epigenetic therapeutic agents on leukemia cell lines. Leuk Lymph 51(2):297–303. https://doi.org/10.3109/10428190903486212

Pivoriunas A, Navakauskiene R, Gineitis A (2004) Inhibition of phosphatidylinositol 3-kinase activity blocks nuclear accumulation of protein kinase zetta during granulocytic differentiation of HL-60 cells. Biologija 2:46–48

Pivoriunas A, Savickiene J, Treigyte G, Tunaitis V, Navakauskiene R, Magnusson KE (2007) PI3-K signaling pathway suppresses PMA-induced expression of p21WAF1/Cip1 in human leukemia cells. Mol Cell Biochem 302(1–2):9–18. https://doi.org/10.1007/s11010-007-9419-4

San Jose-Eneriz E, Gimenez-Camino N, Agirre X, Prosper F (2019) HDAC inhibitors in acute myeloid leukemia. Cancers 11(11). https://doi.org/10.3390/cancers11111794

Savickiene J, Gineitis A (2003) 3-Deazauridine triggers dose-dependent apoptosis in myeloid leukemia cells and enhances retinoic acid-induced granulocytic differentiation of HL-60 cells. Int J Biochem Cell Biol 35(10):1482–1494. https://doi.org/10.1016/S1357-2725(03)00130-4

Savickiene J, Gineitis A, Shanbhag VP, Stigbrand T (1995) Protein kinase inhibitors exert stage specific and inducer dependent effects on HL-60 cell differentiation. Anticancer Res 15(3):687–692

Savickiene J, Gineitis A, Stigbrand T (1999) Modulation of apoptosis of proliferating and differentiating HL-60 cells by protein kinase inhibitors: suppression of PKC or PKA differently affects cell differentiation and apoptosis. Cell Death Differ 6(7):698–709. https://doi.org/10.1038/sj.cdd.4400541

Savickiene J, Kazlauskaite N, Characiejus D, Treigyte G, Navakauskiene R (2002) Combined differentiating effects of retinoic acid and nucleoside analogues on acute promyelocytic leukemia. Acta Med Lituanica 9:22–27

Savickiene J, Borutinskaite VV, Treigyte G, Magnusson KE, Navakauskiene R (2006a) The novel histone deacetylase inhibitor BML-210 exerts growth inhibitory, proapoptotic and differentiation stimulating effects on the human leukemia cell lines. Eur J Pharmacol 549(1–3):9–18. https://doi.org/10.1016/j.ejphar.2006.08.010

Savickiene J, Treigyte G, Borutinskaite V, Navakauskiene R, Magnusson KE (2006b) The histone deacetylase inhibitor FK228 distinctly sensitizes the human leukemia cells to retinoic acid-induced differentiation. In: Diederich M (ed) Signal transduction pathways, PT B: Stress signaling and transcriptional control, Annals of the New York academy of sciences, vol 1091, pp 368–384. https://doi.org/10.1196/annals.1378.081, cell Signaling World 2006 Conference, Luxembourg, Luxembourg, Jan 25–28, 2006

Savickiene J, Treigyte G, Magnusson KE, Navakauskiene R (2009) Response of retinoic acid-resistant KG1 cells to combination of retinoic acid with diverse histone deacetylase inhibitors. In: Diederich M (ed) Natural compounds and their role in apoptotic cell signaling pathways. Annals of the New York academy of sciences, vol 1171, pp 321–333. https://doi.org/10.1111/j.1749-6632.2009.04718.x

Savickiene J, Treigyte G, Aleksandraviciene C, Navakauskiene R (2010a) Low-dose ionizing radiation effects on differentiation of HL-60 cells. Central Eur J Biol 5(5):600–612. https://doi.org/10.2478/s11535-010-0085-2

Savickiene J, Treigyte G, Gineitis A, Navakauskiene R (2010b) A critical role of redox state in determining HL-60 cell granulocytic differentiation and apoptosis via involvement of PKC and NF-kappa B. In Vitro Cell Dev Biol Animal 46(6):547–559. https://doi.org/10.1007/s11626-010-9296-0

Savickiene J, Treigyte G, Vistartaite G, Tunaitis V, Magnusson KE, Navakauskiene R (2011) C/EBPalpha and PU.1 are involved in distinct differentiation responses of acute promyelocytic leukemia HL-60 and NB4 cells via chromatin remodeling. Differentiation 81(1):57–67. https://doi.org/10.1016/j.diff.2010.08.003

Savickiene J, Treigyte G, Borutinskaite VV, Navakauskiene R (2012a) Antileukemic activity of combined epigenetic agents, DNMT inhibitors zebularine and RG108 with HDAC inhibitors, against promyelocytic leukemia HL-60 cells. Cell Mol Biol Lett 17(4):501–525. https://doi.org/10.2478/s11658-012-0024-5

Savickiene J, Treigyte G, Jazdauskaite A, Borutinskaite VV, Navakauskiene R (2012b) DNA methyltransferase inhibitor RG108 and histone deacetylase inhibitors cooperate to enhance NB4 cell differentiation and E-cadherin re-expression by chromatin remodeling. Cell Biol Int 36(11):1067–1078

Savickiene J, Treigyte G, Jonusiene V, Bruzaite R, Borutinskaite VV, Navakauskiene R (2012c) Epigenetic changes by zebularine leading to enhanced differentiation of human promyelocytic leukemia NB4 and KG1 cells. Mol Cell Biochem 359(1–2):245–261. https://doi.org/10.1007/s11010-011-1019-7

Savickiene J, Treigyte G, Valiuliene G, Stirblyte I, Navakauskiene R (2014) Epigenetic and molecular mechanisms underlying the antileukemic activity of the histone deacetylase inhibitor belinostat in human acute promyelocytic leukemia cells. Anti-Cancer Drugs 25(8):938–949. https://doi.org/10.1097/CAD.0000000000000122

Slack J, Rusiniak M (2000) Current issues in the management of acute promyelocytic leukemia. Ann Hematol 79(5):227–238. https://doi.org/10.1007/s002770050585

Spira AI, Carducci MA (2003) Differentiation therapy. Curr Opin Pharmacol 3(4):338–343. https://doi.org/10.1016/S1471-4892(03)00081-X

Toullec D, Pianetti P, Coste H, Bellevergue P, Grand-Perret T, Ajakane M, Baudet V, Boissin P, Boursier E, Loriolle F, other (1991) The bisindolylmaleimide GF 109203X is a potent and selective inhibitor of protein kinase C. J Biol Chem 266(24):15771–15781

Treigyte G, Savickiene J, Pivoriunas A, Navakauskiene R (2006) Effects of inhibition of phosphoinositide 3-kinase and p53 on monocytic differentiation driven leukaemia cells with different p53 status. Biologija 2:93–98

Valiulienė G, Treigytė G, Savickienė J, Matuzevičius D, Alksnė M, Jarašienė-Burinskaja R, Bukelskienė V, Navakauskas D, Navakauskienė R (2016) Histone modifications patterns in tissues and tumours from acute promyelocytic leukemia xenograft model in response to combined epigenetic therapy. Biomed Pharmacother 79:62–70. https://doi.org/10.1016/j.biopha.2016.01.044

Valiuliene G, Stirblyte I, Cicenaite D, Kaupinis A, Valius M, Navakauskiene R (2015) Belinostat, a potent HDACi, exerts antileukaemic effect in human acute promyelocytic leukaemia cells via chromatin remodelling. J Cell Mol Med 19(7):1742–1755. https://doi.org/10.1111/jcmm.12550

Valiuliene G, Stirblyte I, Jasnauskaite M, Borutinskaite V, Navakauskiene R (2017) Anti-leukemic effects of HDACi belinostat and HMTi 3-Deazaneplanocin A on human acute promyelocytic leukemia cells. Eur J Pharmacol 799:143–153. https://doi.org/10.1016/j.ejphar.2017.02.014

Villar-Garea A, Fraga MF, Espada J, Esteller M (2003) Procaine is a DNA-demethylating agent with growth-inhibitory effects in human cancer cells. Cancer Res 63(16):4984–4989

Vitkeviciene A, Baksiene S, Borutinskaite V, Navakauskiene R (2018) Epigallocatechin-3-gallate and BIX-01294 have different impact on epigenetics and senescence modulation in acute and chronic myeloid leukemia cells. Eur J Pharmacol 838:32–40. https://doi.org/10.1016/j.ejphar.2018.09.005

Vitkeviciene A, Skiauteryte G, Zucenka A, Stoskus M, Gineikiene E, Borutinskaite V, Griskevicius L, Navakauskiene R (2019) HDAC and HMT inhibitors in combination with conventional therapy: A novel treatment option for acute promyelocytic leukemia. J Oncol 2019:11. https://doi.org/10.1155/2019/6179573

Weston AD, Blumberg B, Underhill TM (2003) Active repression by unliganded retinoid receptors in development: less is sometimes more. J Cell Biol 161(2):223–228. https://doi.org/10.1083/jcb.200211117

Yoon S, Eom GH (2016) HDAC and HDAC inhibitor: From cancer to cardiovascular diseases. Chonnam Med J 52(1):1–11. https://doi.org/10.4068/cmj.2016.52.1.1

Zhou GB, Zhang J, Wang ZY, Chen SJ, Chen Z (2007) Treatment of acute promyelocytic leukaemia with all-trans retinoic acid and arsenic trioxide: a paradigm of synergistic molecular targeting therapy. Philos Trans R Soc B Biol Sci 362(1482):959–971. https://doi.org/10.1098/rstb.2007.2026

Zhu DM, Narla RK, Fang WH, Chia NC, Uckun FM (1998) Calphostin C triggers calcium-dependent apoptosis in human acute lymphoblastic leukemia cells. Clin Cancer Res 4(12):2967–2976

Chapter 3
Chromatin Remodeling During Leukemia Cell Proliferation and Differentiation

During the development of cells of the hematopoietic system, errors in process regulation are possible, which influence the development of blood cancer—leukemia. Epigenetic changes influence DNA methylation and histone modifications (acetylation, methylation, phosphorylation, etc.). Acute myeloid leukemia can be treated using a differentiation inducer ATRA in combination with inhibitors of histone deacetylation, methylation, and DNA methylation. This increases the ability of leukemia cells to differentiate into granulocytes, induces cell death by apoptosis, and inhibits uncontrolled cell division. During all these processes there is an intensive transformation of the level of gene expression of cells, protein synthesis, and the physiological state of cells.

In recent years, many different epigenetic modifiers were developed. Great amount of clinical trials have shown that in order to treat some types of cancer possibly DNMT/HDAC/HMT inhibitors will be applied effectively. Actually, to develop efficient strategies of cancer therapy one of the greatest challenges in the future will be the approximation of epigenetic therapy and conventional chemotherapy.

The changes in gene and protein expression affected by induced chromatin remodeling in leukemia were analyzed in our research.

3.1 Chromatin Functional Regulation by Histone Modifications

Chromatin modifications are modifications that affect the regulation of gene expression and do not alter the DNA sequence. Such modifications include post-translational modifications (PTMs) of chromatin-containing proteins, changes in nucleosome composition, DNA methylation, and the action of antisense micro-RNA. Chemical modifications of DNA and histones have a profound effect on

R. Navakauskienė et al., *Epigenetics and Proteomics of Leukemia*,
https://doi.org/10.1007/978-3-030-68708-3_3

the regulatory processes of gene expression. Most histone PTMs are dynamic and require specific enzymes for each PTM to catalyze the attachment and removal of appropriate chemical groups. Many post-translational histone modifications affect chromatin activity by attracting large protein factors with specific reader domains that recognize different modifications.

Hyperacetylation of histones can be detected in active chromatin areas. In contrast, deacetylation of histones and methylation of DNA are detected in heterochromatin. Improper and malfunctioning of the histones and DNA modifications can cause the cell malignancy and the development of different forms of cancer. Modeling of histone acetylation has a significant role in the overall chromatin state and, most importantly, in the overall cell phenotype. It has been found that acetylation of histones can change both nucleosome position and gene activity.

HDACi is widely studied and used in the treatment of cancer and inflammatory diseases. Importantly, inhibitors of histone deacetylases can remove or at least reduce the abnormal inhibition of gene expression that occurs during cancer development. HDACi is a very promising compound because of its ability to inhibit growth and induce differentiation and apoptosis in cells of cancer (Dawson and Kouzarides 2012). HDACi contributes to a decrease in the expression of anti-apoptotic proteins, such as some members of the anti-apoptotic Bcl-2 family. In human myeloma cell lines, it was found that HDACi is the reason of the decrease in Bcl-xL, Mcl-1, and XIAP. Similar results were found in studies of the impact of HDACi on human leukemia cells. HDAC inhibition was empirically detected as a new type of cancer therapy (Ceccacci and Minucci 2016). The biology of different HDAC isoforms and their associations with tumorigenesis is just beginning to be explained. New knowledge is mostly obtained from the perceived clinical potential of HDAC inhibitors (Ammerpohl et al. 2004; Paris et al. 2008; Ceccacci and Minucci 2016). It is not yet clear if a more precise understanding of the certain roles played by different HDAC isoforms during human tumorigenesis will contribute not only to development of isoform-specific inhibitors, but also to more efficient or less toxic antitumor therapeutics, in comparison to the multiclass HDAC inhibitors that are presently being evaluated clinically. Rationally designed combinations of HDAC inhibitors with different types of approved or investigational anticancer agents look promising in tumor cell culture systems, but they still need to be based on clinical trials (Budillon et al. 2007; Ceccacci and Minucci 2016). Various HDAC inhibitors are able to alter transcription, both positively and negatively, for about 2% of the total human genes. While this number appears to be small, common genes affected by these inhibitors are known to play a crucial role in controlling cell cycle and differentiation or apoptosis process (Budillon et al. 2007; Sekhavat et al. 2007; Ceccacci and Minucci 2016).

The other important histone modification for chromatin remodeling is that histone methylation can have both activating and inhibitory effects on transcription depending on the location of the methylated amino acid residue in the histone. The enzymes that catalyze the methylation of histone Lys residues are called histone lysine methyltransferases (HMT). In contrast to histone lysine acetyltransferases (HAT), HMTs are highly selective enzymes that catalyze only the methylation of the corresponding Lys residues, for example, G9a methylate H3K9, EZH2—H3K27. HMT usually forms large multiprotein complexes, which as a rule include other histone-modifying proteins (HAT, HDAC, DNMT, histone demethylases). HMT can act as co-activators or co-inhibitors of transcription factors and participate in gene-target promoters by attracting complexes involved in epigenetic regulation. Thus, it can be argued that histone methylation and related processes are not only responsible for changes in gene expression but also have a significant role in the development of cancers.

In our study we applied different HDAC and HMT inhibitors for treatment of leukemia cells. We used at that time novel HDAC inhibitors BML-210, FK228, and other frequently used chemicals like belinostat, phenylbutyrate, vitamin B3 (Borutinskaite and Navakauskiene 2015; Savickiene et al. 2006b,a, 2009, 2011, 2012a, 2014a; Merzvinskyte et al. 2006; Valiulienė et al. 2016; Vitkeviciene et al. 2019). Also we demonstrated the potential use of HMT inhibitors' BIX-01294, 3-deazaneplanocin A (DZNep) for leukemia treatment (Valiulienė et al. 2016; Valiuliene et al. 2017; Vitkeviciene et al. 2019). We have demonstrated that EGCG could be a promising antileukemic agent that act as HDACi, HMTi, and DNMTi (Borutinskaitė et al. 2018; Vitkeviciene et al. 2019).

3.1.1 HDAC Inhibitor BML-210 Affects Leukemia Cells via Chromatin Remodeling

BML-210 was found to be one of the compounds HDAC inhibitors that selectively induced histone but not tubulin acetylation for which HDAC6 is responsible (Wong et al. 2003; Haggarty et al. 2003). Presently the precise molecular mechanisms by which the benzamide derivatives including BML-210 inhibit HDAC function remain unclear. There is provided evidence of direct binding to transcription factor MEF2, which is believed to indicate a previously unknown mechanism of action of these molecules, which is not linked to the active HDAC center but to the protein–protein interaction between HDAC and its functional partners (Jayathilaka et al. 2012).

In our studies, we have examined the novel HDAC inhibitor, BML-210, and its biological impact on protein expression, transcriptional regulation of cell cycle, and apoptosis regulating genes by transcription factors Sp1, NFκB, and p53, which causes the inhibition of leukemia cell proliferation, induction of differentiation, and apoptosis (Borutinskaite and Navakauskiene 2015; Savickiene et al. 2006a).

We determined that different doses of BML-210 from 5 to 30 μM inhibited leukemia cells (NB4, HL-60, THP-1, and K562) growth in a time- and dose-dependent manner. BML-210 plus differentiating inducer (ATRA for NB4 and HL-60; hemin for K562 cells) enhanced ATRA and hemin induced differentiation (Savickiene et al. 2006a), for more detailed prospects see Chap. 2. The data is based on the morphological and immunochemical observations and demonstrated stronger PU.1 binding to the HNE promoter (Savickiene et al. 2006a); the expression of the last onset is usually up-regulated during the initial phases of granulocytic/monocytic differentiation. Also we observed that BML-210 used alone can induce differentiation only in HL-60 and K562 cells, but not in NB4 cells. This effect correlated with changes in histone H4 acetylation and direct/indirect changes of the transcription profile.

In our study (Borutinskaite and Navakauskiene 2015), we detected that BML-210 at a concentration up to 20 μM created cytotoxic effects on NB4 cells and stimulated apoptotic cell death (up to 60%). We identified impact of BML-210 on HDAC expression level in NB4 cells. We revealed that BML-210 down-regulated HDAC1 mRNA level in dose-dependent manner and the best effect was observed during first hours of treatment: BML-210 at 10 μM dose suppressed gene expression up to 36% after 8 h of treatment; the 20 μM concentration of BML-210 suppressed HDAC1 expression up to 74% after 8 h of treatment. The alterations in expression of HDAC2 and HDAC3 were very low and not substantial. The HDAC1 protein expression level was minimal after 48 h of treatment with 20 μM of BML-210. Also we investigated HDAC activity changes in NB4 cells after treatment with BML-210 was applied (Borutinskaite and Navakauskiene 2015). It was noticed that after BML-210 treatment the activity of HDAC lowers. In NB4 cell line the greatest decrease (85%) of the HDAC activity was observed after 48 h of treatment with 20 μM BML-210. In the study we also performed mass spectrometry analysis to identify few proteins the expression levels of which changed after BML-210 treatment (for details see Chap. 4). Further we examined expression level of some genes after BLM-210 treatment: endoplasmin, ENPL; 14-3-3 protein eta, 1433F; calreticulin, CALR; proliferating cell nuclear antigen, PCNA; lactoylglutathione lyase, LGUL; and chloride intracellular channel protein 1, CLIC1. We observed that examined gene expression level correlated to the proteomic data. Our analysis demonstrated that after treatment with BML-210 (both 10 and 20 μM) the expression of calreticulin, endoplasmin, 14-3-3 protein eta, and PCNA decreased. The products of these genes are significant for cell proliferation, differentiation, and apoptosis processes. The CLIC1 and LGUL gene expressions lowered only after treatment with 20 μM BML-210, but not after treatment with 10 μM BML-210. Since the chloride channel proteins important for stabilization of cell membranes and LGUL play some role in the cell growth, the down-regulation of their expression correlates with apoptotic processes induced by BML-210 treatment.

We detected that BML-210 alone very effectively influenced histone H4 hyperacetylation in HL-60 and NB4 cells (Savickiene et al. 2006a). In control NB4 cells, histone H4 as a rule exhibits in Ac 0–Ac 2 forms, but after treatment with BML-210 hyperacetylation of histone H4 occurs. We compared the intensity of hyperacetylation of histone H4 caused by BML-210 and phenylbutyrate (PB). We demonstrated that phenylbutyrate caused histone H4 hyperacetylation during first hours of treatment (Merzvinskyte et al. 2006). We found that BML-210 in dose of 10 μM affected hyperacetylation of histone H4 and this effect was time-dependent, but at the 48 h after treatment the hyperacetylation effect started to reduce. On the other hand, HDAC inhibitor phenylbutyrate at 1.5 mM concentration induced much stronger effect on hyperacetylation of histone H4 (Fig. 3.1). During the comparison

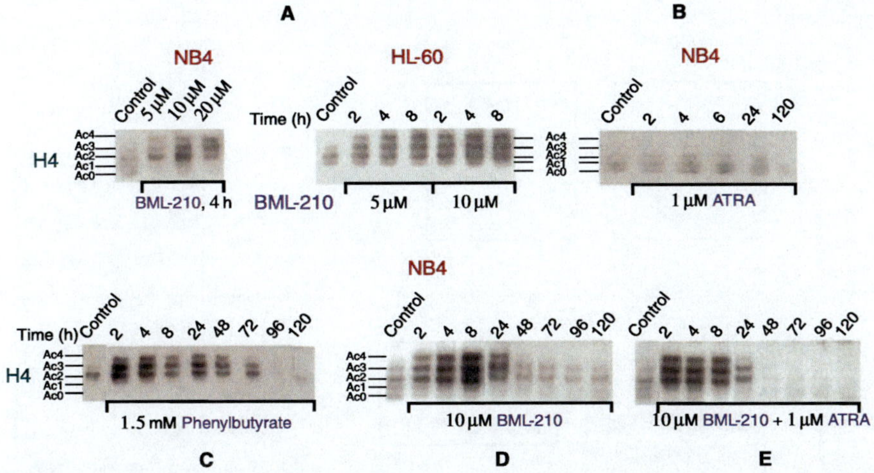

Fig. 3.1 Time- and dose-dependent hyperacetylation of histone H4 in the response to HDAC inhibitors. (**a**) NB4 and HL-60 cells were treated with different concentrations of BML-210 for 4–8 h. NB4 cells were treated with 1 μM ATRA alone (**b**), 1.5 mM phenylbutyrate alone (**c**), 10 μM BML-210 alone (**d**), and 10 μM BML-210 in combination with 1 μM ATRA (**e**) for indicated time. Histones were isolated from untreated or HDAC inhibitor-treated cells resolved on 15% polyacrylamide-acetic acid-urea gels and examined by Western blot analysis using antibodies specific to hyperacetylated histone H4. The five acetylation states of histone H4 are indicated representing non-acetylated (Ac 0), mono- (Ac 1), di- (Ac 2), tri- (Ac 3), and tetra- (Ac 4) acetylated forms. According to Savickiene et al. (2006a), License No 4832400121586

of NB4 and HL-60 cells, more pronounced effect was obtained in HL-60 cells. ATRA alone did not stimulate histone hyperacetylation, but the combined treatment with BML-210 caused the intense hyperacetylation of histone H4 even after 2 h. These results demonstrated that BML-210 and ATRA can have a synergetic impact on histone H4 hyperacetylation (Fig. 3.1).

We employed the electrophoretic mobility shift assay (EMSA) method to study protein–DNA interactions in leukemic cells after treating cells with epigenetic modifiers (BML-210, phenylbutyrate, FK228) alone or in combination with differentiation inducer ATRA (Figs. 3.2, 3.6). Even the EMSA method is not direct evaluator of epigenetic changes; nevertheless, this method enables to determine if transcription factors analyzed by us are capable of binding to a specific DNA sequence, thereby activating gene expression.

In order to examine the binding activity of transcription factors Sp1, NFκB, p53, and PU.1, we conducted EMSA with radiolabelled oligonucleotides and nuclear extracts from untreated NB4 cells, and from cells treated with BML-210 alone or plus ATRA.

We demonstrated a certain Sp1 binding to the consensus sequence and the binding site 3 (Sp1–3) of the p21 promoter, p53 binding affinity to the p21 promoter or NFκB to the consensus sequence and to the FasL promoter, or PU.1 to the human

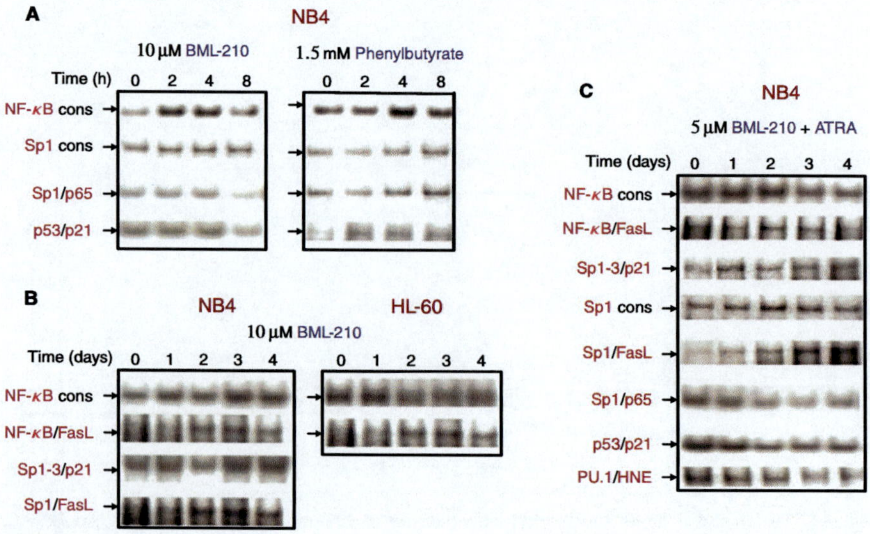

Fig. 3.2 Transcription factors binding activity in response to BML-210 alone or in combination with ATRA. (**a**) NB4 cells where treated with 1.5 mM phenylbutyrate or 10 µM BML-210 for 2–8 h. (**b**) NB4 and HL-60 cells were treated with 10 µM BML-210 for 4 days. (**c**) NB4 cells were treated with 5 µM BML-210 in combination with 1 µM ATRA during 4 days. EMSA was performed using a total 10 µg protein from each nuclear extract and oligonucleotides containing binding sites in the indicated promoters. Arrows indicate DNA complexes with Sp1, NFκB, PU.1, and p53 proteins. According to Savickiene et al. (2006a), License No 4832400121586

neutrophil elastase (HNE) promoter (Savickiene et al. 2006a). The binding of those transcription factors to the promoters was effectively obstructed by a 50 or 100-fold excess of the unlabeled oligonucleotide (cold), or by a mutated oligonucleotide.

NFκB stands for a family of inducible transcription factors involved in the controlling of individual cellular processes, such as cell growth, differentiation, apoptosis, immunity, and inflammation (Grants et al. 2020). HDAC inhibitors, e.g., sodium butyrate or TSA, improved NFκB-mediated gene induction (Adam et al. 2003; Quivy and Van Lint 2004). The most abundant form of NFκB is a p50/p65 heterodimer, in which p65 contains the transcriptional activation domain. Both the subunits are acetylated upon activation, and p300/CBP (CREB-binding protein) is involved (Chen and Greene 2003; Mukherjee et al. 2013). p65 binds to CBP and its homolog p300 (Roth et al. 2001; Mukherjee et al. 2013), while p50 fails to recruit the transcriptional co-activators. This caused the transcriptional activation, supposedly via related histone-directed acetylase activity, which contains localized chromatin (Kiernan et al. 2003). Our findings show that BML-210 improves NFκB binding to its consensus sequence during the first hours of treatment. This initial effect most manifested in HL-60 cells and coincided with the induction of cell differentiation.

Later we determined whether treatment with various HDAC inhibitors, BML-210 and phenylbutyrate, influenced the binding activity of transcription factors in NB4 and HL-60 cells. By EMSA, both HDAC inhibitors promoted NFκB or Sp1 binding to their consensus sequences and Sp1 to the p65 promoter, or p53 to the p21 promoter during the first 8 h (Fig. 3.2). We have shown different NFκB binding activity to consensus sequence in HL-60 and NB4 cells during 4 days of treatment with 10 μM BML-210. This was related to differentiation induction in HL-60 cells only causing to improved complex formation the first two days. NFκB binding to the FasL promoter elevated in both cell lines. Besides, it was shown that a significant functional role of NFκB in p21 induction, G1 arrest, and cell maturation by HDAC inhibitor, sodium butyrate, and the disruption of NFκB pathway was a reason of the cells engagement into an alternative, apoptotic program (Dai et al. 2003). In NB4 cells, binding activity of Sp1 to the FasL or Sp1–3 to the p21 promoter was manifested on days 3–4. We detected increased Sp1 binding activity to the p65 promoter and NFκB to the consensus sequence during 2–4 h of BML-210 treatment. This correlated temporally with the elevated p65 protein level in nucleus of NB4 cell, which could justify the transcriptional activation of NFκB-activable genes. There are two classes of such genes: those comprising constitutively and immediately accessible NFκB binding sites in their promoter and those that require the hyperacetylation to become accessible to NFκB (Saccani et al. 2001). HDAC inhibitors would thus increase NFκB-dependent transcription by obstructing the HDAC activity related to p65, consequently causing histone hyperacetylation and thereby a higher level of gene expression.

In HL-60 cells, we did not notice Sp1 binding, due to Sp1 protein existence in a truncated form, which cannot bind to the GC-rich elements in DNA (Savickiene et al. 2005; Deniaud et al. 2009). p53 protein conformation for DNA binding to the promoters may influence the modulating of the rate of its acetylation and further recruitment of chromatin remodeling factors (Dornan et al. 2003; Vogelstein et al. 2000). Lagger et al. (2003) have shown that p53 cooperates with Sp1 in the activation of the p21 promoter. It is known that HDAC inhibitors, such as sodium butyrate, apicidin, TSA, SAHA, and others, are able to stimulate p21 (Joseph et al. 2005). Our findings verify increased binding activity of both Sp1–3 and p53 to the p21 promoter during 24 h exposure to BML-210 alone or plus ATRA in NB4 cells. After such treatment, p21 protein level elevated in both the cytoplasm and the nucleus instantly.

Combined treatment with BML-210 and ATRA led to increased binding capacity of many transcription factors (NFκB, PU.1, Sp1–3, p53) to the promoters examined during first two days and following decline on days 3–4 (Fig. 3.2). On the other hand, Sp1 binding to the FasL promoter markedly elevated on days 3–4, which is possibly related to the induction of apoptosis at this time-point (Savickiene et al. 2006a). The summarized scheme of BML-210 effect on leukemia cells is presented in Fig. 3.3.

These results allow to argue that HDAC inhibitor BML-210 has an antileukemic effect—it is able to induce apoptosis and modulate differentiation during the modulation of histone acetylation and gene expression.

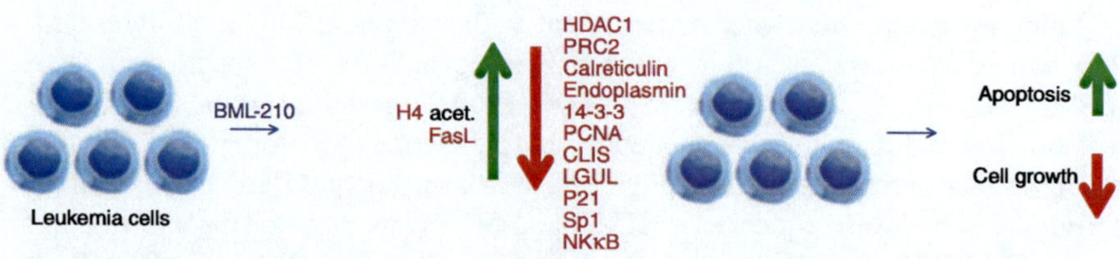

Fig. 3.3 Summarized BML-210 effect on leukemia cells

3.1.2 Histone Modifications Initiated by HDACi Phenylbutyrate and Vitamin B3

Short-chain fatty acid compounds, like butyrates, belong to a class that is presently confirmed for the clinical application (Magrin et al. 2020), inducing cell differentiation and/or apoptosis programs in a broad range of neoplastic cells. The boundaries of its clinical use due to rapid metabolism and a very short half-life in the blood may be overcome by butyrate derivates or prodrugs of butyric acid (Gozzini et al. 2003). In leukemia cells, HDAC inhibitors, such as butyrate derivate and suberoylanilide hydroxamic acid (SAHA), promote differentiation or apoptosis (Vrana et al. 1999). Several HDAC inhibitors are now being examined clinically: SAHA is presently in a phase II clinical study against T-cell lymphomas and metastatic squamous cell carcinomas of the head and neck; MS-275 undergoes tests if it will be possible to use in treatment of advanced solid tumors or lymphomas and poor-risk hematological malignancy; CI-994 is applied to treat leukemia (Kelly et al. 2003).

In our study we used HDACi phenylbutyrate in combination with HDAC Sir2 inhibitor vitamin B3 (Merzvinskyte et al. 2006; Savickiene et al. 2011). Earlier it was demonstrated that some niacin-related agents, like vitamin B3 and others, which belong to Sir2 family of HDAC, can be used to induce differentiation in leukemia cells (Iwata et al. 2003).

We demonstrated that HDAC inhibitors—phenylbutyrate and vitamin B3—caused rapid histone H3 and H4 modifications. HDACi combinations have the various impacts on cell cycle arrest, differentiation, and apoptosis. The most promising treatment for differentiation therapy is detected applying 6 h pretreatment with phenylbutyrate and vitamin B3 before the combined exposition to ATRA with vitamin B3, which considerably accelerates and increases cell differentiation (up to 95%) during 48 h of treatment.

We determined that the HL-60 control cells have few histone H4 acetylated forms: Ac0, Ac1, and Ac2. This suggests that one of the causes of leukemia is HDAC dysfunction. The effect of HDAC inhibitor phenylbutyrate at different times and concentrations on acetylation was evaluated. We found that 3 mM phenylbutyrate has the greatest effect on HDAC inhibition, after 6 h of treatment when histone H4 becomes completely acetylated: we detected Ac3 and Ac4 forms. Meanwhile, after 96 and 120 h treatment with PB histone H4 is completely

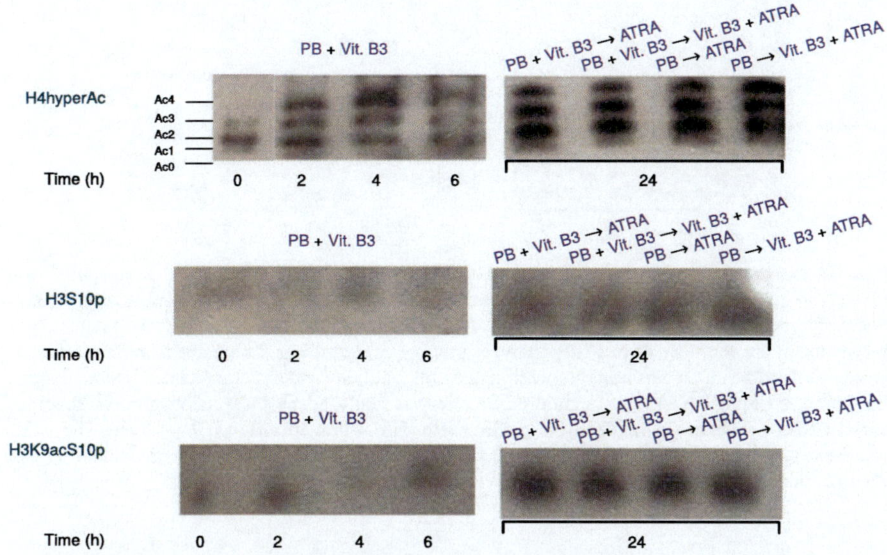

Fig. 3.4 Modifications of histones H4 and H3 in the response to HDACi and ATRA. HL-60 cells were treated with 5 mM vitamin B3 together with 3 mM phenylbutyrate (PB) for indicated time or 6 h pretreated with 5 mM vitamin B3 together with 3 mM phenylbutyrate (PB) with following 24 h treatment of 1 μM ATRA alone or together with 5 mM vitamin B3. Histones from control or treated HL-60 cells' nucleus were isolated, resolved on 15% polyacrylamide-acetic acid-urea (AU) gels, and examined by Western blot analysis using specific antibodies. According to Merzvinskyte et al. (2006), License No 4830251467958

deacetylated. In the case of vitamin B3, the H4 acetylation started at 2 h of treatment but with low intensity compared with PB, and after 24 h of treatment with vitamin B3 histone H4 was completely deacetylated. Also we have found changes in phosphorylation and acetylation of histone H3 after combined treatments with phenylbutyrate, vitamin B3, and ATRA, but not phenylbutyrate or vitamin B3 alone (Fig. 3.4) (Merzvinskyte et al. 2006). The elevated level of histone H3 and H4 modifications is related to initiation and maturation stages of HL-60 cell differentiation. Our findings allow arguing that the chromatin remodeling applying HDAC inhibitors provides a rationale for the treatment of acute promyelocytic leukemia. Our other study (Savickiene et al. 2011) was dedicated to find the effects of phenylbutyrate, vitamin B3, ATRA on human leukemia cells (HL-60, NB4) and the role of selected transcription factors. The binding of hyperacetylated histone H4 to the promoter of the G-CSFR gene containing the C/EBPα binding site in control and various combinations of agents in leukemia cells were investigated (Savickiene et al. 2011). Hyperacetylated histone H4 was found to bind to the G-CSFR gene after exposure to 3 mM phenylbutyrate (PB) for 6 h and 3 mM PB ⇝ in 1 μM ATRA + 5 mM vitB3 (48 h). When exposed to these combinations of agents, histone H4 is modified in the promoter region of the G-CSFR gene to which C/EBPα binds, which is likely to activate G-CSFR gene expression (Fig. 3.5).

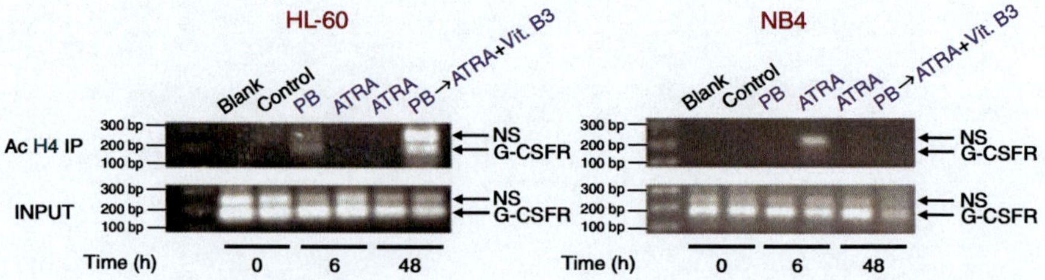

Fig. 3.5 Association of histone H4 acetylation with the G-CSFR promoter in response to HDACi in HL-60 and NB4 cells. Cells were treated with 1 μM ATRA for 48 h, or with 3 mM phenylbutyrate (PB) for 6 h, or with 1 μM ATRA and 5 mM vitamin B3 (vit. B3) after PB-pretreatment for the next 48 h. ChIP analysis was performed using chromatin from control and treated cells. Chromatin was immunoprecipitated with antibodies to hyperacetylated histone H4 or no antibody (input DNA). Precipitated chromatin was analyzed by PCR using primers specific for G-CSFR promoter at the C/EBPα binding site. NS— non-specific PCR product. The data represent two independent experiments showing similar results. According to Savickiene et al. (2011), License No 4832391034680

To conclude, it was demonstrated that treatment with 3 mM PB ⇝ 1 μM ATRA + 5 mM vit. B3 causes histone H4 hyperacetylation at specific sites of RARβ, G-CSF, and G-CSFR genes.

Another study was performed with KG1 cells (Savickiene et al. 2009). We demonstrated that HDAC inhibitors, like PB, vitamin B3 in various combinations with ATRA, can reduce growth of cells and induce differentiation. Also we have showed that PB effected histone H4 hyperacetylation process the most and this effect was strongest at first 8 h of treatment. Recent investigations suggested the better result of combination HDAC inhibitors with DNMT inhibitors on antileukemic activity of KG1 cells (Shaker et al. 2003). Chromatin immunoprecipitation analysis has demonstrated that phenylbutyrate and its combinations with ATRA and vitamin B3 result in histone H4 acetylation in the p21 promoter regions corresponding to p53 and/or Sp1 sites. This concurred with the activation of the transcription factor p53 binding activity to the p21 promoter in electrophoretic mobility shift assay. The findings justify the possibility of applying the combination of agents for therapeutic strategy in ATRA-resistant acute myeloid leukemia to induce both differentiation and apoptosis.

Our findings allow us to offer an opinion that the chromatin remodeling applying HDAC inhibitors and differentiation agent all-trans retinoic acid could be a rational way for acute promyelocytic leukemia treatment.

3.1.3 HDAC Inhibitor FK228 Induces Antileukemic Effects via Changes in Histone Modifications

FK228 (or FR901228, depsipeptide) contains a cyclic structure and exhibits powerful antitumor activity against mouse and human tumor cells (Panicker et al.

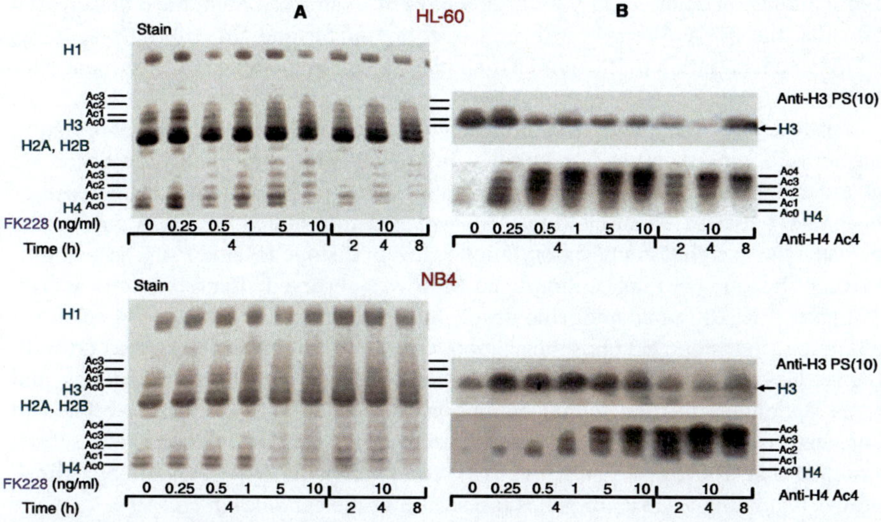

Fig. 3.6 FK228 induces a dose- and time-dependent histone H4 and H3 modifications. Histones from nuclear proteins of untreated or HDACi-treated NB4 and HL-60 cells were subjected to AU (15% polyacrylamide, acetic acid, and urea) electrophoresis, gel staining with Brilliant Blue G-Colloidal (**a**) and immunoblotting (**b**) with anti-acetylated histone H4 and anti-phosphorylated histone H3 at serine 10 antibodies. The five acetylation states of histone are indicated, representing un-acetylated (Ac 0), mono- (Ac 1), di- (Ac 2), tri- (Ac 3), and tetra- (Ac 4) acetylated forms. According to Savickiene et al. (2006b), License No 4833110834912

2010). FK228 has been studied in preclinical studies using models of acute myeloid leukemia (AML) and chronic lymphocytic leukemia (CLL). These studies justified efficient inhibition of HDAC and antitumor activity (Mori et al. 2004; Piekarz et al. 2001). However, its clinical application has been limited by progressive constitutional symptoms and requires a different schedule. Its unique effects have also been investigated in clinical studies of acute lymphocytic leukemia, small lymphocytic lymphoma, T cell lymphoma, or advanced small cell or non-small cell lung cancer.

In this investigation we studied antileukemic impact of FK228 on APL cell. We used NB4 cells, which have specific chromosomal translocation (15;17) and HL-60 cells with negative functional p53 status. Various concentrations of HDAC inhibitor FK228 (0.25, 0.5, 1 ng/mL) were applied in order to treat the NB4 and HL-60 leukemia cells. In HL-60 cells, HDACi FK2208 induced a better cell growth inhibition and caused cell death during shorter exposure time. In control HL-60 and NB4 cells histone H4 was predominantly detected in un-acetylated (Ac 0) and mono-acetylated (Ac 1) forms (Fig. 3.6). We showed that FK228 alone stimulated a dose- and time-dependent boost in histone H4 acetylation in NB4 and HL-60 cells. The greatest histone H4 hyperacetylation at FK228 concentrations of 5–10 ng/mL was noticed after 4 h; despite this fact, such doses resulted in high cytotoxic effect and marked cell death up to 90% after 4 days of treatment. We noticed that in HL-60 cells, the tri- and tetra- acetylated isoforms of histone H4 exhibited at lower HDACi concentration (0.25 ng/mL) than in NB4 cell line after 4 h of treatment with FK228.

Further, after treatment with FK228 we detected histone H3 phosphorylation on Serine 10 in HL-60 and NB4 cells. In HL-60 cells, raise of FK228 doses led to a decrease of histone H3 phosphorylation simultaneously with an increase of histone H4 hyperacetylation (Fig. 3.6), which demonstrates the close connection between the acetylation/phosphorylation status of histone H3 and H4. The accumulation of histone H3 in phosphorylated form was observed after 8 h treatment with 10 ng/mL FK228 (an apoptogenic dose). In less apoptosis-sensitive NB4 cells, the alterations in histone H3 phosphorylation were alike, but happened slower. Overall, higher histone acetylation (hyperacetylation) initiates chromatin remodeling and transcriptional activation, while histone phosphorylation is frequently related to chromatin condensation that incorporates mitosis and apoptosis (Cheung et al. 2000; Lorenzo et al. 2002; Park and Kim 2012). Treatment with HDACi and trichostatin A allows regulating a subset of genes (about 9% of the genome) in HL-60 cells. Over 50% of these genes were transcription augmenters or transcription factors. Certain number of genes was changed in expression, while others had opposite regulatory transcription profiles (Hake et al. 2004).

In our study we investigated the NFκB binding activity in reaction to FK228 treatment of NB4 and HL-60 cells. We detected that various doses of FK228 resulted in cell line-specific boost in the intensity of the NFκB and DNA complex formation at concentrations of 0.5–5 ng/mL and decline in both cell lines at the apoptotic dose of 10 ng/mL (Fig. 3.7). HDACi FK228 also enhances the NFκB binding to the FasL promotor region in a dose-dependent manner. In HL-60 and NB4 cells, DNA binding to NFκB was changed in reaction to 5 ng/mL FK228 (Fig. 3.7) and more distinguished at 8 h exposure (Fig. 3.7), which correlates with the histone H4 hyperacetylation during first 8 h (Fig. 3.7). The treatment of ATRA plus FK228 (0.25 ng/mL) led to maintenance of increased NFκB binding activity during cell differentiation and suppression throughout differentiation leading apoptosis on day 3 or 4 in HL-60 and NB4 cells, accordingly. Nevertheless, in this instance, no up-regulation of NFκB binding to the FasL promotor was observed in cells undergoing apoptosis. However, this activity declines due to treatment with ATRA, justifying the assumption that the Fas/FasL pathway possibly does not participate in the apoptosis of terminally differentiated leukemia cells (Savickiene et al. 2005; Kwon et al. 2002; Kikuchi et al. 1996).

FK228 was efficient in provoking apoptosis in cell lines that are various according to p53 status. Lately it was stated that a synthetic compound pifithrin (PFT) is a specific inhibitor of p53 transactivation and there was a proposal to treat cancer with it (Kaiser et al. 2011; Kanno et al. 2015). Nevertheless, PFT demonstrates various effects; for example, the restraint of cell growth and stimulation or improvement of p53-dependent apoptosis in mouse epidermal cells JB637 or, conversely, the protection of neurons and cardiomyocytes against apoptosis provoked by DNA-damaging agents and doxorubicin, accordingly (Kaji et al. 2003; Ma et al. 2018). Our

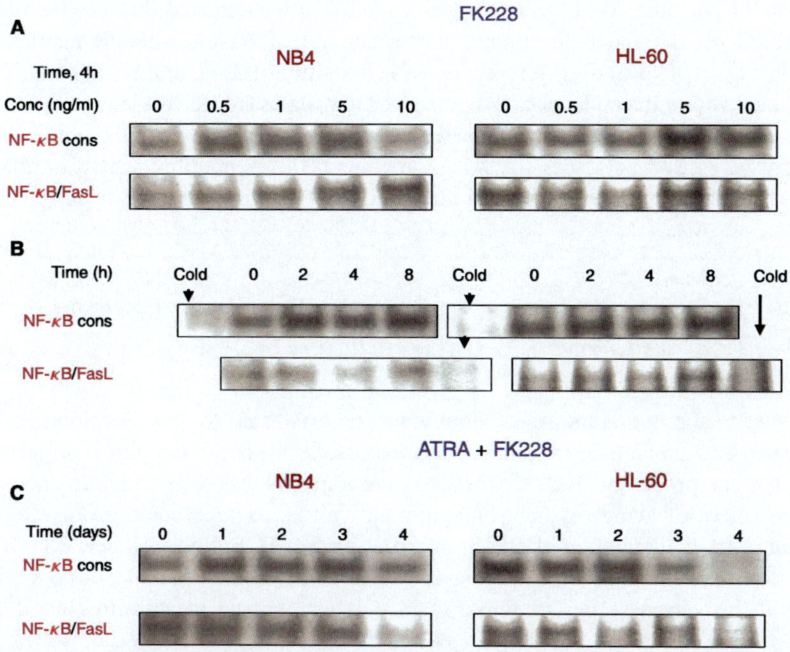

Fig. 3.7 Alterations in NFκB binding activity in response to FK228 alone or in combination with ATRA. Nuclear extracts were prepared from control cells and treated with different doses of FK228 during 4 h (**a**) or treated with 5 ng/ml FK228 for different time (**b**), and in combination with ATRA (1 μM) and FK228 (0.25 ng/ml) during 4 days (**c**). EMSA was performed using a total 15 μg protein from each nuclear extract and oligonucleotides, containing NFκB consensus motif or NFκB binding site in the FasL promoter. Specific DNA complexes with NFκB were eliminated competitively by addition of 100-fold molar excess of unlabeled oligonucleotide (cold). According to Savickiene et al. (2006b), License No 4833110834912

investigations show that p53-defective HL-60 cells exhibited improved apoptotic potential in contrast to NB4 upon treatment with 30 μM PFT alone. PFT-pretreated NB4 cells with functional p53 displayed improved cell survival after following treatment with FK228. One of the presumed mechanisms of PFT activity is the restraint of p53 translocation from the cytoplasm to the nuclei and prohibition of its binding to certain DNA sites (Kaji et al. 2003). The latter suggestion is justified in our study by the declined p53 binding activity to the p21 promoter in NB4 cells treated with PFT alone or together with FK228. PFT as well lowered the activation of p53-regulated genes, including bax, cyclin G, and mdm2 that control the degradation rate of p53 (Salih et al. 2002; Xie et al. 2018). Besides, PFT-induced apoptosis can be mediated through a p53-independent pathway by the participation of p38 and extracellular kinase activities of ERKs, or by the up-regulation of proapoptotic Bax and activation of Caspases (Kim et al. 2010). In our investigations, PFT stimulates apoptosis in p53-negative HL-60 cells and does not obstruct the anti-proliferative action of FK228.

Also in our other study (Savickiene et al. 2011) we detected that pre-treatment of FK228 for 6 h and then combination treatment of ATRA with PB resulted in high level (up to 90%) of differentiated cells in culture (HL-60 and NB4) cells. That correlated with induced level of hyperacetylated histone H4 (Savickiene et al. 2011).

In conclusion, our data confirmed that myeloid cell line-specific, differential activity of FK228 enhances the ATRA-mediated differentiation, which is related to the regulation of gene expression mediated through chromatin remodeling.

3.1.4 Belinostat Effects on Epigenetic Changes in Leukemia Cells and Promotes Differentiation

Epigenetic alterations make a significant contribution to the development of leukemia and epigenetic modifiers such as histone deacetylases play a significant role in these processes. HDAC inhibitors are an attractive scheme while choosing the treatment of acute myeloid leukemia as well as its type acute promyelocytic leukemia (APL) (Rajak et al. 2013; Wang et al. 2020). Belinostat (Bel, PXD101), a hydroxamate-like HDAC inhibitor, is known to be promising and can be applied to treat the various types of tumors, since it was shown to restrain class I and class II HDACs enzymatic activity (Khan et al. 2008; Northfelt et al. 2007); for detailed information on chemicals used in the study, see Appendix C. Different studies revealed its activity resulting in cell cycle arrest, apoptosis, and restraint of cell proliferation (Gravina et al. 2012; Qian et al. 2006). However, information is still lacking on the molecular effects of belinostat on leukemia cells. In our study (Savickiene et al. 2014b; Valiuliene et al. 2017, 2015), we tested the impact of belinostat and its combination with differentiating agent all-trans retinoic acid (ATRA) and/or epigenetic modifier 3-deazaneplanocin A (DZNep) on human promyelocytic leukemia NB4 (FAB-M3, bearing typical APL translocation t(15;17)) and HL-60 cells (FAB-M2, do not have typical APL translocation t(15;17); for detailed characterization of cells used in the study, see Appendix B).

We demonstrated that exposure to belinostat alone in a dose-dependent manner inhibited cell proliferation but did not induce cell differentiation (see Chap. 2). Also we found that apoptosis correlated with activation of caspase 3, degradation of PARP-1, and alterations in the expression of survivin, cyclin E1, and cyclin A2 proteins that lead to the cell cycle arrest (Savickiene et al. 2014b). The combination of belinostat with ATRA was found to decrease the proliferation of NB4 and HL-60 cells and arrest the cell cycle in the G0/G1 phase. In addition, combined effects of belinostat with ATRA or belinostat and DZNep with ATRA were found to hasten and stimulate greater granulocytic differentiation in comparison with ATRA alone. These findings are alike to those of other HDAC inhibitors applied in leukemia granulocytic differentiation therapy (Valiuliene et al. 2017, 2015) (Chap. 2). For the following experiments we applied selected optimal concentrations for treatments: belinostat alone (0.2 μM), ATRA alone (1 μM), combined treatments (0.2 μM belinostat +1 μM ATRA or 0.2 μM belinostat +1 μM ATRA + 0.5 μM DZNep), pretreatment for 4 h with combination of 0.2 μM belinostat and 0.5 μM DZNep with the following treatment with 1 μM ATRA. We assessed how all these treatments effect gene expression and protein level modulation of specific epigenetic modifiers (HDAC1, HDAC2, PCAF, Policomb complex) along with the cell cycle and differentiation regulators (p27, C/EBPα, C/EBPϵ, PPARγ, Cyclin E1, Cyclin E2, etc.) (Savickiene et al. 2014b; Valiuliene et al. 2017, 2015).

RT-qPCR analysis revealed a decline in HDAC1 gene expression after NB4 cells exposure with belinostat, ATRA alone, and their combination during first 24 h. Nevertheless, after 72 h of treatment the levels of HDAC1 elevated in comparison with the control cells (Table 3.1). Action of belinostat applied alone resulted in the greatest inhibition of HDAC1. The combination of belinostat with ATRA and DZNep or pretreatment of belinostat and DZNep, which was followed by ATRA treatment, elevated HDAC1 level during the first 24 h in comparison to control cells or belinostat alone treatment. Effects alike were observed in HL-60 cells (Table 3.1) (Valiuliene et al. 2017, 2015). The HDAC1 protein expression level declined both in HL-60 and NB4 cells after treatment with ATRA or belinostat alone, or by applying their combination (Table 3.1).

The expression level of HDAC2 mRNA and its protein was also analyzed in our investigations. The HDAC2 mRNA level in NB4 treated cells had resembling regulation as HDAC1—during first 24 h they were down-regulated with the following up-regulation. Protein level was down-regulated only after combination of ATRA, belinostat, and DZNep (Table 3.1). We found that HL-60 cells exhibited different patterns—up-regulation of HDAC2 mRNA level during first day of treatment with ATRA, belinostat or their combination with down-regulation after 72 h of treatment (Table 3.1) (Valiuliene et al. 2015). The HDAC2 protein level was down-regulated at 72 h point after all used treatments (ATRA alone, belinostat alone, their combination, Table 3.1).

Since belinostat in combination with ATRA and/or DZNep stimulated differentiation in HL-60 and NB4 cells, and we know from literature that PPARγ and C/EBPα are essential for granulocytic differentiation initiation (Paz-Priel and Friedman 2011), mRNA levels of these genes were investigated in our research (Savickiene et al. 2014b; Valiuliene et al. 2017, 2015). Our study justified that expression of transcription factor C/EBPα is down-regulated during ATRA-mediated differentiation processes in both cell lines (NB4 and HL-60). The decline of C/EBPα mRNA level was noticed in HL-60 cells treated with DZNep + belinostat ⇝ ATRA combination (Table 3.1). We determined no effect in NB4 cells and up-regulation in HL-60 cells of C/EBPα mRNA level after treatment with belinostat alone or in combination with ATRA/±DZNep (Table 3.1). The C/EBPγ mRNA was up-regulated after all treatments in both cell lines (Table 3.1), especially when leukemia cells were provoked to differentiate with various agents (ATRA, its combination with belinostat/±DZNep). Findings alike were noticed by other authors on C/EBPα, C/EBPγ and PU.1 expression levels after treatment of HDAC inhibitor, valproic acid, and ATRA.

We also evaluated histone acetyltransferase PCAF gene expression and determined that belinostat alone had no impact on PCAF gene expression; while ATRA

Table 3.1 Effect of HDAC inhibitor belinostat on NB4 and HL-60 cells' gene and protein expression after single and combined treatment (C—untreated cells)

		Treatment/NB4 cells												
		ATRA			Belinostat (Bel)			ATRA + Bel			ATRA + Bel + DZNep		DZNep + Bel → ATRA	
Gene/protein of interest	C	6 h	24 h	72 h	6 h	24 h	72 h	6 h	24 h	72 h	24 h	72 h	24 h	72 h
HDAC1 mRNA	1	0.7	1–1.1	1.2	0.5	0.5–0.7	1.25	0.75	1.2	1.6	1.5	1.4	1.2	1
HDAC2 mRNA	1	0.9	2.1	0.8	0.46	0.75	1.5	0.65	0.6	0.8	0.8	0.9	0.8	0.5
HDAC3 mRNA	1	0.8	1.1	1.5	0.9	1	2	-	3	2	2.5	2.5		
PCAF mRNA	1	0.6	7.8	3.8	0.5	1	0.7	0.5	1	4	0.8	3.3	–	-
p27 mRNA	1	0.3	0.25	2.1	0.5	1.3	4.1	0.5	1.1	2.3	–	–	–	-
C/EBPα mRNA	1	0.5	0.4	0.5	0.3	2	1.4	0.6	0.6	0.3	1	1.5	1.6	0.9
C/EBPε mRNA	1	–	13	26	–	2	3	–	–	–	7	15	17	29
PPARγ mRNA	1	–	1	7	–	0.8	1	–	–	–	13	30	0.4	0.3
EZH2 mRNA	1	4	1.1	3	–	0.5	0.6	–	–	–	0.3–1.1	0.9–1.7	1.6	1.2
SUZ12 mRNA	1	0.7	1	1	–	0.6	0.5	–	–	–	1.15	1.5	0.7	0.3
EED mRNA	1	0.7	0.4	2.1	–	0.8	1.2	–	–	–	1.6	2.7	0.7	0.6
EZH2 protein	1	0.3	–	0	0.4	–	0.3	–	–	–	0.6	0	–	–
SUZ12 protein	1	1	–	1	1	–	0.5	–	–	–	0.8	0.1	–	–
EED protein	1	0.9	–	0.5	1.4	–	2.6	–	–	–	1.5	1.2	–	–
HDAC1 protein	1	0.9	–	0.7	0.6	–	0.4	0.5	–	0.2	0.4	0.3	–	-
HDAC2 protein	1	2.12	–	1	2	–	2.1	1.3	–	0.8	0.9	0.2	–	–
H4hyperAc protein	1	2.2	–	8.3	17.5	–	9.6	20.8	–	9.5	17.3	14	–	-
H3K27me3 protein	1	1.5	–	2.7	2	–	2.4	–	–		1.7	3.3	–	–
miR-223	1	–	2.6	7	–	1.6	0.8	–	4	9	2.7	5.5	–	–
H3K9me2 protein	1	0.6	–	2.4	0.8	–	0.8	–	–		0.6	1.1	–	–

		Treatment/HL-60 cells												
HDAC1 mRNA	1	–	1.1	1.2	–	0.7	1	–	–	–	1.5	3.2	1.3	1.1
HDAC2 mRNA	1	–	1.8	0.8	–	1	0.4	–	–	–	1.4	0.4	0.9	0.8
HDAC3 mRNA	1	1.4	1.1	1	2.5	1.5	1.6	–	1.1	2	4	4.5	–	–
PCAF mRNA	1	14.7	1.5	3.3	1.7	1.2	1	1.5	–	3.6	1.5	1.6	–	–
p27 mRNA	1	–	–	–	–	–	–	–	–	–	–	–	–	–
C/EBPα mRNA	1	2	0.2	0.5	1	1	0.85	0.3	0.1	0.15	0.8	0.9	0.5	0.4
C/EBPϵ mRNA	1	–	10	12	–	2	3	–	–	–	9	15	40	13
PPARγ mRNA	1	–	0.7	0.8	–	0.9	0.8	–	–	–	10	12	1	0.3
EZH2 mRNA	1	1.5	1.1	0.5	–	1	2.5	–	–	–	1.2	0.9	1.15	0.6
SUZ12 mRNA	1	2	2.2	1	–	1.5	0.8	–	–	–	2.1	0.9	2.7	0.7
EED mRNA	1	1.2	0.7	1	–	0.9	0.5	–	–	–	1.9	1.6	1.2	0.9
EZH2 protein	1	0.4	–	0	1.4	–	0.3	–	–	–	1.4	0	–	–
SUZ12 protein	1	0.6	–	0.5	0.6	–	0.5	–	–	–	0.5	0.05	–	–
EED protein	1	0.2	–	1	0.7	–	0.4	–	–	–	0.4	0.01	–	–
HDAC1 protein	1	0.6	–	0.7	0.5	–	0.4	0.5	–	0.5	0.7	0.8	–	–
HDAC2 protein	1	0.9	–	0.4	1.1	–	0.9	1.1	–	0.6	0.7	0.02	–	–
H4hyperAc protein	1	2.6	–	24.4	13.7	–	9.4	–	–	–	22	22	–	–
H3K27me3 protein	1	0.7	–	2.1	–	–	0.9	–	–	–	1	1.7	–	–
H3K9me2 protein	1	–	–	–	–	–	–	–	–	–	–	–	–	–
miR-223	1	–	1	3	–	1	1.5	–	2	6	1	2.8	–	–

or ATRA combined with belinostat treatment dramatically increased PCAF mRNA levels in NB4 cells after 72 h treatment (Table 3.1). It was demonstrated by others that ATRA raises the expression levels of p300/CREB-binding protein-associated factor (PCAF) in the other tissues, i.e., in mouse kidney tissues (Kumar et al. 2014; Blanco et al. 1998). In this research we also showed that in NB4 cells 6 h treatment with belinostat down-regulated PCAF gene expression up to 2 times in comparison to control NB4 cells, and later PCAF gene expression was restored (Table 3.1). However, we did not observe such effect in HL-60 cells (Table 3.1). It is known that PCAF is a histone acetyltransferase that is related to p300/CBP and competes with E1A that is related to the oncogenic impacts in different cancers. As the histone acetyltransferase, it could modulate p53 and p21 transcription via stress-stimulated modification of histones (acetylation of H3) and other proteins (Love et al. 2012). We determined in our study (Valiuliene et al. 2015) that belinostat alone and, at a lower level, co-administration with ATRA compared with ATRA applied alone were more effective in inducing p27 gene expression (Table 3.1) (Valiuliene et al. 2015, 2017). It has been demonstrated earlier by other researchers that ATRA stimulates PCAF expression and accumulation in nuclei of P19 carcinoma cells (Zhang et al. 2010). It has also been assumed that rise in PCAF mRNA levels enhances retinoid-dependent gene expression and the overexpression of PCAF caused the expression of the granulocytic differentiation marker CD11b at the mRNA level in leukemia cells (Sunami et al. 2017). It can only be speculated on the basis of Hirano and co-authors research that down-regulation of PCAF gene and protein expression may be related to belinostat's proapoptotic impact, as it was shown that down-regulation of PCAF sensitized human prostate cancer PC3 cells to chemotherapeutic therapy, prompted G1 arrest and apoptosis (Hirano et al. 2010).

In order to evaluate the mechanisms by which belinostat acts to change chromatin structure, two histone modifications (H4 acetylation and H3K9 acetylation), representing the active chromatin state and leading to opening of the chromatin structure, were tested. As presented in Fig. 3.8, when 0.5–2 μM of belinostat was applied to treat HL-60 and NB4 cells for short time (1–3–6 h), the cells contained the isoforms of highly acetylated histone H4 up to tetra-acetylated. H4 hyperacetylation was more notable after 3 h of treatment with all doses of belinostat in both cell lines, but tri- and tetra-acetylated forms emerged soon (at 1 h) in HL-60 cells. In NB4 cells, hyperacetylation of histone H4 was kept for 3–6 h, but lowered after the drug exposure for 24 h (Fig. 3.8). In order to carry out the more detailed study of histone H3 modifications, histones were fractionated on TAU gel and stained. Due to this histone H3 resolved from histones H2A and H2B. The positions of H3 variants (H3.1, H3.2, and H3.3) were identified, applying antibodies against the total histone H3, and immunoblot study was accomplished for histone H3K9 acetylation (Fig. 3.8). Thus, belinostat prompted lysine hyperacetylation of both histones H3 and H4 with some time-related variations between both cell lines.

Next we demonstrated that belinostat dose-dependently depletes HDACs, EZH2, SUZ12 and leads to histone H4 and H3 modifications in leukemia cells (Fig. 3.9). Reduction in HDAC1 protein level correlated with a rise in histone H4 hyperacetylation in NB4 and HL-60 cell lines up to tetra-acetylated form (Fig. 3.9c, Table 3.1).

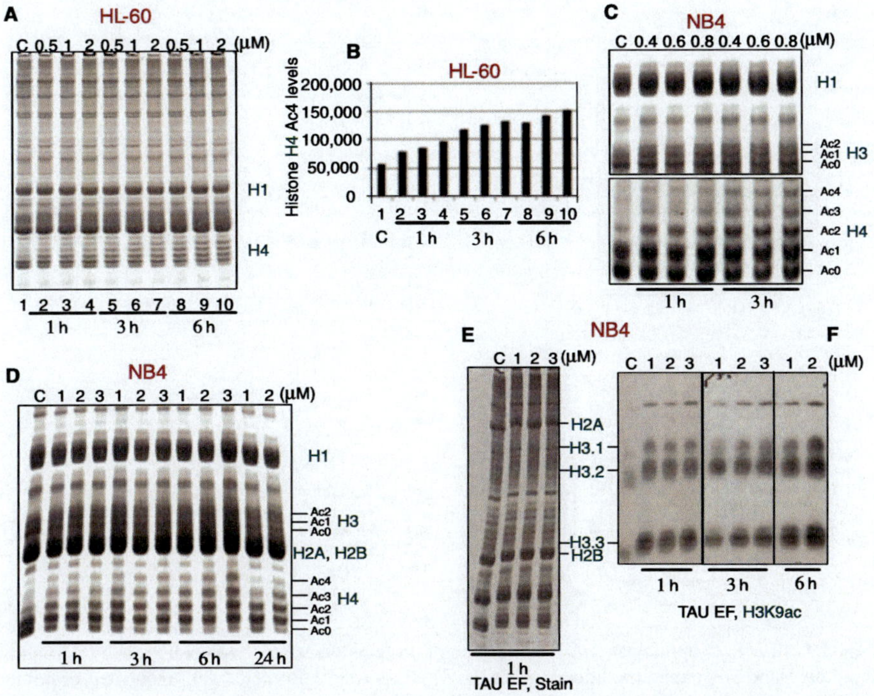

Fig. 3.8 Belinostat induces a dose- and time-dependent accumulation of histones H3 and H4 acetylation in HL-60 and NB4 cells. (**a**) Acid-urea (AU) gel after staining with Brilliant Blue G-Colloidal represents acetyl-isoforms of histone H4 (Ac0-Ac4) in HL-60 cells after treatment with the indicated doses of belinostat for 1, 3, and 6 h. Histone H1 band served as a control of histone extraction and protein loading. (**b**) Histone H4 hyperacetylation levels in the gel determined by scanning densitometry. (**c**) Acetyl-isoforms of histone H3 and H4 in stained AU gel after NB4 cell treatment with intermediate doses (0.4–0.8 μM) of belinostat for the indicated times. (**d**) Western blot analysis of acetylated H4 isoforms after NB4 cell treatment with high doses (1–3 μM) of belinostat for the indicated times. (**e**) Triton-acid-urea (TAU) gel after staining with Brilliant Blue G-Colloidal. (**f**) Western blot analysis of the gel using antibodies against acetylated histone H3K9 after NB4 cell treatment with the indicated doses of belinostat for 1–6 h. The data is representative of three independent experiments showing similar results. According to Savickiene et al. (2014b), License No 4833581426843

Our group has shown that belinostat provokes accumulation of H4K16ac mark, which at the same time is related to the transcriptional activation. It was also demonstrated that belinostat up-regulates H3K9 acetylation level depending on treatment period and applied dose including histone H4 hyperacetylation (Table 3.1) (Savickiene et al. 2014b). We observed no significant changes in H3K14ac mark after treatment with ATRA or its combination with HDAC inhibitor belinostat and HMT inhibitor DZNep (Vitkeviciene et al. 2019).

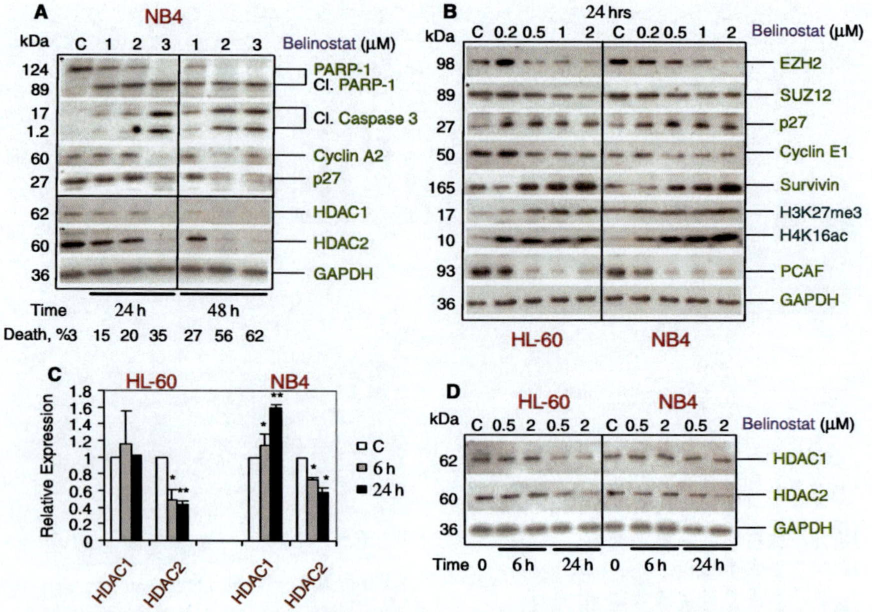

Fig. 3.9 Effects of belinostat on the expression of proteins associated with cell growth inhibition, apoptosis, and epigenetic modifications. (**a**, **b**, **d**) HL-60 and NB4 cells were incubated or not in the presence of increasing doses (0.2–3 μM) of belinostat for the indicated times. Western blot analysis was performed with the indicated antibodies on the total NB4 cell lysates. GAPDH served as the loading control. (**a**) The percentages of dead cells in cultures where cells were lyzed at the end of incubation period and subjected to Western blot analysis. (**c**) In HL-60 and NB4 cells treated with 2 μM belinostat for 6 and 24 h, HDAC1 and HDAC2 gene expression relative to control at each time-point was determined by Q-PCR using GAPDH for normalization. Results are mean ± SD ($n = 3$), $P \leq 0.05$ (∗), $P \geq 0.01$ (∗∗) indicate significant differences from untreated control. According to Savickiene et al. (2014b), License No 4833581426843

Due to HDACs linking with histone methyl transferases and DNA methyl transferases (DNMTs), belinostat has an effect not only on HDACs or HATs, but also on other chromatin remodeling enzymes. So, we demonstrated in our study that therapy with belinostat effects polycomb repressive complex 2 (PRC2) subunits' EZH2, SUZ12, EED mRNA, and proteins: SUZ12 mRNA up-regulation at 24 h point with further reduction and protein expression down-regulation after 24 h (Table 3.1). In our other (Savickiene et al. 2014b) study we showed that belinostat depleted EZH2 and SUZ12 protein levels in a dose-dependent manner in one cell line as well as the other, and the depletion was more noticeable while applying the higher doses of belinostat (Savickiene et al. 2014b). Our results correlate with other authors' study where Fiskus et al. (2009) showed that another HDACi panabinostat (LBH589) reduces the amount of PRC2 subunits in primary AML cells. However, treatment of belinostat alone had no significant effect on

H3K27 trimethylation in NB4 and HL-60 cells but was slightly down-regulated in HL-60 cells during first 6 h of ATRA treatment (Table 3.1) (Savickiene et al. 2014b).

In general, we demonstrated that belinostat was able to affect HDAC1 and HDAC2 at protein and/or mRNA levels in a dose-dependent and cell type-specific manner, which led to almost complete disappearance of HDAC1 and HDAC2 protein in dying cells (Fig. 3.9). It may seem that the actions of HDAC inhibitors are dependent on cell context and vary based on the differential expression profile of HDACs, which in all likelihood regulate the sensitivity to different types and dosages of HDAC inhibitors. Our findings demonstrate that incessant exposure to ATRA with a low dose of belinostat or temporal pre-treatment substantially accelerated and improved ATRA-induced HL-60 and NB4 cell differentiation. Undoubtedly, the acetylation state of lysine residues controls local and higher order chromatin structure, including acetylation of histone H4K16, which results in chromatin decondensation, the binding of chromatin-related proteins, and following growth of transcriptional activity. The data obtained from our experiments allow us to conclude that the rise of histone acetylation may be involved in apoptosis by facilitating chromatin accessibility cooperatively with few other histone modifications, resulting in apoptotic-associated chromatin alterations, such as chromatin condensation or mid-nucleosomal DNA fragmentation.

To establish the modified histones that include acetylated histone H4, trimethylated H3K27/K4, and association with promoter regions of certain genes, we performed chromatin immunoprecipitation (ChIP) assay (Valiuliene et al. 2015). Immunoprecipitation of chromatin was performed and changes of histone H3 (H3K4me3 and H3K27me3) methylation and H4 hyperacetylation in promoter sequences of PRC2 components, C/EBPα, C/EBPϵ coding genes were evaluated after treatment with combination of ATRA, belinostat, DZNep, or belinostat alone (Table 3.2). We revealed that after treatment with belinostat alone, basal hyperacetylation levels in C/EBPα, C/EBPϵ promoter regions did not change a lot compared with control cells. These data correlate with NBT test data, showing that belinostat alone is incapable to prompt granulocytic differentiation (Valiuliene et al. 2015). It was also detected that belinostat alone induced association of histone hyperacetylated form with p27 promoter region up to 1.5 times (Valiuliene et al. 2015). Also we found no significant changes in association of histone H3 (H3K27) with promoter sequences of the PRC2 genes (except EZH2 after 72 h) after ATRA + belinostat + DZNep treatment. Also we detected some changes in association of H3K4me3 with C/EBPϵ and CSFR promoter regions. Such ChIP results also partially confirm the results of gene expression, where the levels of mRNA in NB4 cells encoding PRC2 components were not significantly altered (Table 3.1). We found that H3K4me3 and H3K27me3 methylation in C/EBPα promoter sequences decreased after combined treatment of DZNep + belinostat + ATRA for 72 h (Table 3.2). The association level of histone H3K4me3 and the main target of C/EBPα, the CSF3R promoter and exon 1 of the granulocyte colony stimulating receptor (G-CSFR) gene, were evaluated. The effect of ATRA + belinostat + DZNep treatment on H3K27me3 decreases in CSF3R promoter (1.5 times), and H3K4me3 increases (2 times). Because the G-CSF receptor is important for granulocyte

Table 3.2 Treatment effect on acetylated histone H4, trimethylated H3K27/K4 association with promoter regions of certain genes. ChIP with antibody against selected histone was performed with control NB4 and HL-60 cells and NB4 and HL-60 cells treated with 1 μM ATRA for 12 and 48 h, 2 μM belinostat 6 h, in combination with 0.5 μM DZNep + 0.8 μM belinostat +1 μM ATRA (72 h), or pretreatment of 0.5 μM DZNep +0.8 μM belinostat (4 h) ⇝ 1 μM ATRA (12 and 48 h). Specimens were further tested using qPCR analysis. Data is represented as a fold-change in percent input (compared to untreated control). "↑" denotes a fold-change increase, "↓" denotes a fold-change decrease, whereas "–" stands for having not changed or a change that was not higher than 1.5 fold. Represented data is mean ($n = 2$). S.D. (not shown) was less than 10%

Histone	H4ac							H3K27me3	H3K4me3
Treat-ments	ATRA (12 h)	ATRA (48 h)		Bel (6 h)	DZNep + Bel ⇝ ATRA (12 h)		DZNep + Bel ⇝ ATRA (48 h)	DZNep + Bel + ATRA (72 h)	DZNep + Bel + ATRA (72 h)
Gene	HL-60	NB4	HL-60	NB4	NB4	HL-60	HL-60	NB4	NB4
EZH2	–	–	–	–	–	–	–	1.6 ↓	1.5 ↓
SUZ12	–	–	–	–	–	–	–	2 ↓	–
EED	–	–	–	–	–	–	–	-	1.5 ↓
C/EBPα	2.5–3 ↑	–	2–2.5 ↑	–	–	3–3.5 ↑	6.5– ↑	8 ↓	6 ↓
C/EBPϵ	2.5–3 ↑	–	–	–	–	14–14.5 ↑	10–10.5 ↑	–	2.3 ↑
PPARγ	12–12.5 ↓	1.5–2 ↑	–	–	1.5–2 ↑	7–7.5 ↓	–	–	–
p27	–	–	–	1.5 ↑	–	–	–	–	–
CSFR	–	–	–	–	–	–	–	1.5 ↓	2 ↑

maturation—regulating neutrophil proliferation, differentiation, and survival—it could be argued that exposure to chemical compounds (ATRA + belinostat + DZNep) influenced changes in H3K27 trimethylation in promoter sequences—chromatin activity increases (increases H3K4me3) in this gene, which influenced the course of leukemic cell differentiation.

We also performed immunoprecipitation of chromatin, to evaluate the composition of protein complexes interacting with hyperacetylated histone H4 in NB4 cells after treatment with belinostat alone. For this, we used mass spectrometry analysis after ChIP experiments. These methods helped us to find out which proteins interact with hyperacetylated histone H4 in control NB4 cells and those cells to which belinostat therapy was applied. Quantitative alterations in determined proteins in control and belinostat treated cells are described in Chap. 4. In summary, we could suggest that treatment of AML cells with belinostat prompts the association of some proteins like calprotectin (S100A8/S100A9), TMPRSS11A protease with hyperacetylated histone H4, and that this can affect the formation of NETs (neutrophil extracellular traps) or similar structures.

3.1.5 Combined Epigenetic Treatment Promotes Conventional Leukemia Cell Differentiation

In our next study we evaluated ATRA and its combination with belinostat and DZNep effects on different genes participating in the regulation cell cycle, apoptosis, and proliferation (Vitkeviciene et al. 2019). Also such therapies were combined idarubicin used in conventional therapy. Gene expression alterations of cell cycle inhibitors ATM, p53, p21, p27, and Rb and cell cycle activator CCNE2 (cyclin E2) were investigated in NB4 and HL-60 cells after 6 and 72 h of treatment with 3-deazaneplanocin A, belinostat with ATRA: we found that combination of used drugs mostly influenced the CCNE, p21, and p53 expression levels in NB4 cells and ATM, p21, and p27 in HL-60 cells; ATRA treatment alone affected mostly p27, Rb in NB4 cells and p21 in HL-60 cells (Figs. 3.10, 3.11). Cell cycle activator CCNE2 (cyclin E2) gene expression level corresponded with the findings of cell cycle distribution analysis; at the beginning CCNE2 expression raised, but, later, after 72 h, it lowered. We know from the literature that the proteins encoded by p53, p21, Rb, and CCNA2 genes are involved in signaling pathways in response to DNA damage that can trigger cellular senescence, interrupting the cell cycle. The ATM (ataxia telangiectasia mutated) gene-encoded kinase is a protein involved in the DNA damage checkpoint. Also, ATM functions as a tumor suppressor. It inhibits cancer-related metabolism and thereby inhibits cell proliferation and may induce senescence (Aird et al. 2015).

It is known that the p53 protein is a transcription factor that directly binds to DNA. The activation of the p53 gene by DNA damage, hypoxia, or other factors induces the expression of the cyclin-dependent kinase inhibitor p21 (Fig. 3.12).

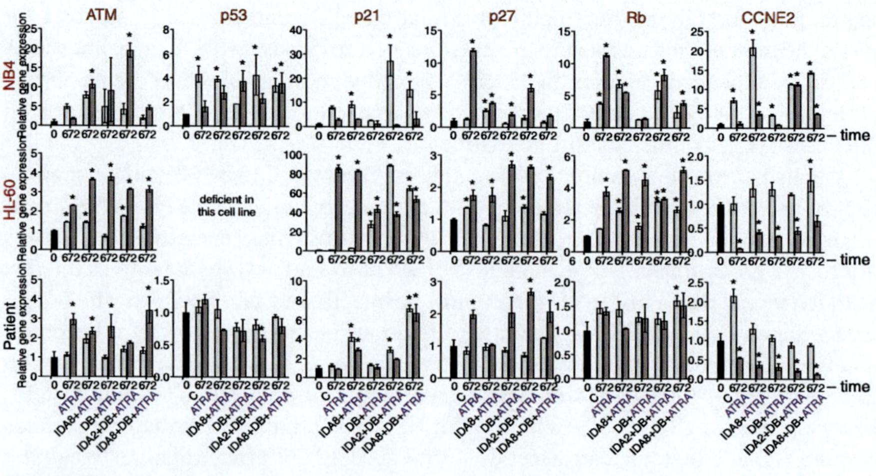

Fig. 3.10 Gene expression analysis after treatment with epigenetic agents in combination with conventional treatment. NB4 and HL-60 cells and APL patient promyelocytes were treated with 1 μM all-trans retinoic acid (ATRA), 2 nM or 8 nM idarubicin (IDA2 or IDA8), 0.5 μM 3-deazaneplanocin A (**d**), and 0.2 μM belinostat (**b**) in different combinations; C—untreated cells; treatment time—6 and 72 h. Cell cycle inhibition related gene expression changes after treatment were measured using RT-qPCR ΔΔCt method. GAPDH was used as a "housekeeping" gene; results are presented as changes in comparison to untreated cells; results are mean ± S.D. ($n = 3$); $P \leq 0.05$ (∗), calculated by one-way ANOVA statistical test. According to Vitkeviciene et al. (2019), Creative Commons Attribution License

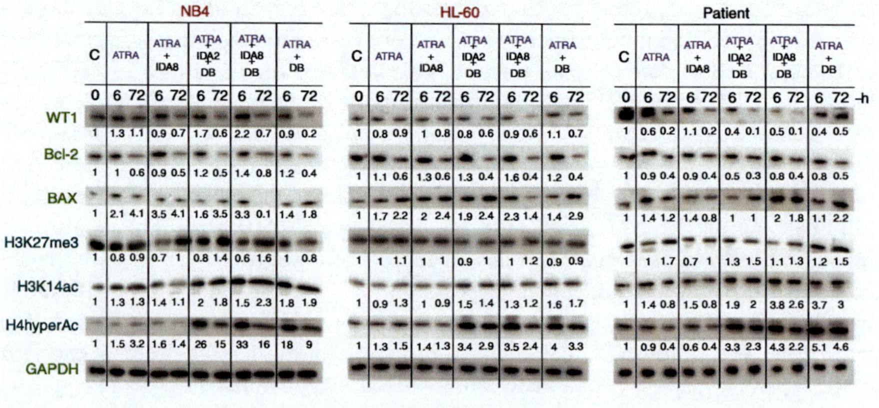

Fig. 3.11 Protein expression changes after NB4, HL-60, and APL patient's cell treatment with epigenetic agents in combination with conventional treatment. Cell samples were treated with 1 μM all-trans retinoic acid (ATRA), 2 nM or 8 nM idarubicin (IDA2 or IDA8), 0.5 μM 3-deazaneplanocin A (D), and 0.2 μM belinostat (B) in different combinations for 6 and 72 h; C—untreated cells. (**a**) Protein level changes were assessed by immunoblot analysis. Intensity of protein bands was measured using `ImageJ` software and normalized to the GAPDH loading control; results are presented as changes in comparison to untreated cells. (**b**) Graphical visualization of relative band intensity of detected protein levels as measured using `ImageJ` software and normalized to the GAPDH loading control; results are mean ± S.D. ($n = 2$). According to Vitkeviciene et al. (2019), Creative Commons Attribution License

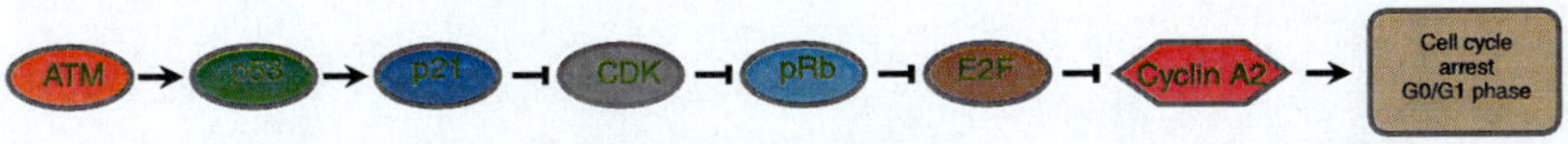

Fig. 3.12 Signaling pathway of protein associated with cell cycle arrest. Activation of the tumor suppressor gene p53 by the ATM protein promotes the expression of the cyclin-dependent kinase inhibitor p21 and thus inhibits the ability of cyclin-dependent kinases (CDKs) to phosphorylate the retinoblastoma protein (pRb). Unphosphorylated Rb locks the transcription factor E2F and thus inhibits the expression of the CCNA2 gene encoding Cyclin A2

p21 is known to inhibit cyclin-dependent kinase (CDK) phosphorylation of the retinoblastoma (Rb) protein-encoding gene Rb (Campisi and Di Fagagna 2007; Al Bitar and Gali-Muhtasib 2019).

Unphosphorylated Rb binds to the transcription factor E2F and inhibits its activation by inhibiting the expression of the CCNA2 gene encoding cyclin A2 protein. This stops the cell cycle transition from the G1 cycle phase to S (Chandler and Peters 2013). Normally, the cyclin A2 protein binds and stimulates CDC2 and CDK2 kinases, thereby initiating cell transition from G1/S or G2/M cycle phases (Yam et al. 2002; Dachineni et al. 2016).

As epigenetic factor in leukemia cell differentiation and functioning we investigated some important miRNA. It is known that miR-223 (micro-RNA) plays a significant role in myeloid cell differentiation (Gilicze et al. 2014; Liao et al. 2017). The expression is practically undetectable in non-hematopoietic cells. This micro-RNA is closely related to the transcription factor C/EBPα already discussed in our study: C/EBPα acts as transcriptional activator of the miR-223 gene. Up-regulation of miR-223 expression level promotes granulocytic differentiation and also contributes to inhibition of cell division (Minucci et al. 2001). MiR-223 is known to block E2F1 translation, thereby causing cell cycle arrest accompanying myeloid differentiation. In AML cells, the expression of this micro-RNA is suppressed, leading to an increase in E2F1 expression. Expression of miR-223 (gene MIR223) in NB4 and HL-60 cells after ATRA, belinostat, and DZNep (single and combined effects) was investigated after 24 and 72 h of treatments in our study. In both cell lines, expression levels were found to increase with leukemia cell differentiation (Table 3.1). In NB4 and HL-60 cells, maximal increases were observed after 72 h of exposure to ATRA and belinostat. Significant increases in expression level were also observed after exposure to ATRA alone and its combination with belinostat and DZNep. NB4 cells show higher expression than HL-60 cells. Effects of belinostat alone had little impact on miR-223 expression (expression level similar to control cells). Exposure to ATRA induces the expression of miR-223, which contributes to the stimulation of differentiation. Concluding, we have shown that HMT inhibitor 3-deazaneplanocin A and HDAC inhibitor belinostat together with ATRA promote differentiation of leukemia cells. Also such epigenetic combination improved conventional treatment (all-trans retinoic acid + idarubicin) for acute promyelocytic leukemia *in vitro* and *ex vivo*. Improvement of treatment with epigenetic agents resulted in chromatin remodeling that is related to with chromatin

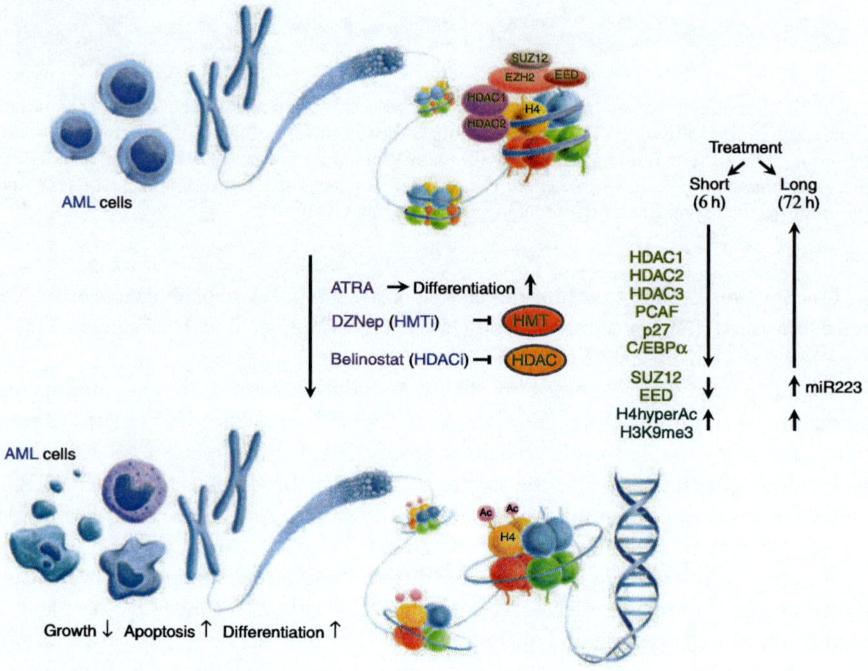

Fig. 3.13 Scheme of chromatin remodeling after combined epigenetic treatment

relaxation and improved transcription. Such alterations might upregulate genes significant for leukemia treatment. Our findings show to an increased inhibited cell proliferation and survival, induced apoptosis, accelerated granulocytic differentiation, and changes in gene and protein expression including modifications (Fig. 3.13).

3.1.6 *Combined Epigenetic Treatment in Leukemia Xenograft Model*

Previous studies in other groups have shown that AML cells (HL-60, KG1, and KG1a) injected into BALB/c nude mouse have formed proliferating tumors such as myelosarcoma, whereas these cells were not detected in hematopoietic tissues. Notably, inoculated leukemia cells maintained their ability to proliferate *in vivo*. However, the total number of peripheral blood cells and the percentage of different cell types remained unchanged in AML xenograft BALB/C nude mice (Kohnken et al. 2017).

In our *in vivo* study we developed a mouse xenograft APL model where NB4 cells with typical APL translocation t(15;17) were injected into immunodeficient NOG mice (Valiulienė et al. 2016). We further characterized the xenograft mouse model of APL in several aspects: (1) peripheral blood composition data of APL xenograft mice and xenograft mice treated with combination of DZNep, belinostat, and ATRA (DBR); (2) leukemia cell infiltration; (3) tumor development; and (4) gene expression evaluation.

In our studies, the presence of NB4 cells injected into immunodeficient NOG mice showed a marked increase in the cell population containing human CD33+, CD45+ in mice peripheral blood samples (Table 3.3). These results were also correlated with an increase in white blood cells (WBCs) (Table 3.3). This may be attributed to the occurrence of hyperleukocytosis, which is quite common in some APLs (Testi et al. 2005). We demonstrated that treatment with ATRA and epigenetically active compound (belinostat, DZNep) reduced CD33+ and CD45+ cells and increased CD15+ cells in the peripheral blood of APL xenograft mice. An increase in CSF3 gene expression was also observed (Table 3.3). In addition, epigenetic therapy with ATRA, belinostat, and DZNep prolonged the life span of APL xenograft mice and protected them from tumor formation: AML-DBR mice survived for 46 days ± 5 days, whereas untreated APL mice survived for 41 days ± 1 day (Valiulienė et al. 2016).

WT1 expression in peripheral blood of healthy individuals is known to be very low or even undetectable, but it has been shown that acute lymphoid or acute myeloid leukemia increases WT1 mRNA levels and can serve as a prognostic marker (Lindstedt et al. 2014; Yoon et al. 2017). We expanded our research and evaluated WT1 gene and protein expression. We sought to determine whether WT1 expression could serve as a marker for assessing the effectiveness of applied therapies. The obtained results showed that WT1 protein expression was lower in tumors of APL xenograft mice than in control NB4 cells (Valiulienė et al. 2016). WT1 mRNA levels were increased in peripheral blood of APL xenograft mice in contrast to control mice. Notably, treatment with epigenetic compounds (ATRA, belinostat, DZNep) reduced WT1 gene expression by 2.5-fold (Valiulienė et al. 2016). Following the uptake of NB4 cells with high WT1 expression in NOG mice, WT1 protein expression was increased correspondingly in the lungs, brain, and liver of these mice (no increase in WT1 protein was observed in the kidney and spleen). Treatment with epigenetic therapy (ATRA, belinostat, DZNep) partially suppressed WT1 protein expression in the tissues of these organs (Valiulienė et al. 2016).

We also evaluated the expression of PML-RARA (a marker related to APL pathogenesis) in the blood of APL xenograft mice. PML-RARA mRNA was found in the blood of APL xenograft mice, as expected, but not in APL xenograft mice to the therapy of which the epigenetic (ATRA, belinostat, DZNep) compounds were used (Table 3.3). These results are in agreement with other published works indicating that treatment with single ATRA is capable of significantly reducing PML-RARA gene expression in APL xenograft mice (Patel et al. 2012). Pharmacological doses of ATRA are known to prompt both proteolytic degradation of PML-RARα and ability corepressor / HDAC complexes release of PML-RARα, which in turn silence target genes in ATRA (Melnick and Licht 1999; Wu et al. 2018). As a result, treatment with HDAC or DNMT inhibitors in APL blasts is

Table 3.3 Disease response of APL xenografted mice to combined epigenetic treatment. Peripheral blood counts are presented relatively to BALB/C (N), untreated control (NOG mice, C) on days 25–35 after APL xenografted mice (APL) treatment with DZNep, belinostat, and ATRA (APL-DBR). According to Valiulienė et al. (2016)

Variables		Treatment group			
		N	C	APL	APL – DBR
Survival data (days post exposure)		–	–	41.0 ±1	46.0 ±5
Tumor formation		-	-	Tumors in neck, bladder, and abdominal areas; gut ulcered	Gut ulcered
Peripheral blood counts (relative to control)	WBC	0.7	1.0	4.1 ±0.3	2.3 ±0.1
	LYM	0.7	1.0	0.6 ±0.2	0.8 ±0.1
	MON	0.6	1.0	1.3 ±0.2	1.1 ±0.2
	NEU	1.2	1.0	1.1 ±0.1	1.1 ±0.1
	EOS	0.8	1.0	0.2 ±0.0	0.8 ±0.2
	HGB	0.8	1.0	0.7 ±0.0	0.9 ±0.1
	HCT	0.9	1.0	0.8 ±0.0	0.9 ±0.1
	RBC	0.8	1.0	0.7 ±0.1	0.9 ±0.1
	MCV	1.1	1.0	1.0 ±0.1	1.0 ±0.0
	PLT	0.4	1.0	1.3 ±0.3	1.3 ±0.2
Gene expression (relative to APL)	WT1	-	-	1.0	0.4 ±0.1
	PML-RARα	-	-	1.0	-
	CSF3	-	-	1.0	5.3 ±0.6
	G-CSFR	-	-	1.0	0.5 ±0.2
Flow cytometric analysis, %	CD15	2.2	3.1	4.0 ±0.0	8.3 ±5.4
	CD33	1.6	1.2	12.2 ±4.6	3.8 ±2.8
	CD45	2.8	2.9	16.8 ±6.7	5.1 ±2.9

capable of restoring ATRA differentiation, both *in vitro* and *in vivo* (Fazi et al. 2005; Ho et al. 2013). Our investigations demonstrated that our chosen xenograft model can be successfully applied to evaluate the therapeutic efficacy of epigenetic inhibitors (Table 3.3).

Different histone epigenetic markers are found in cells and tissues of cancer patients and healthy organisms. As it is known (Bojang Jr and Ramos 2014), in healthy cells, promoter sequences of tumor suppressor genes are rich in active transcription markers, such as H4ac and H3K4me3, and in satellite regions, many inhibitory markers are found, like H3K9me3 and H3K27me3. Cancer cells undergo epigenetic changes that result in the loss of inhibitory tags in the satellite regions and drastic reduction of histone acetylation levels in the promoter regions of the tumor suppressor genes and the inhibitory tags H3K9me3 and H3K27me3. Previous studies (Grigoryev et al. 2006) have shown that the H3K9me3 modification is most frequently found in somatic tissues of adult mice. It has also been demonstrated that histone deacetylation plays a significant role in the compacting of chromatin to heterochromatin (this chromatin structure is typical of somatic tissues).

Notably, control APL xenograft mice developed tumors (neck, bladder, and abdomen) as well as intestinal lesions (ulcers). In this study, we evaluated the influence of epigenetic therapy (ATRA, belinostat, DZNep) on changes in histone modifications in tumors and APL xenograft mice tissues. We have shown that the levels of H3K4me3 and H3K9me3 modifications are higher in APL xenograft mouse tumors than in cells NB4, whereas the levels of H4ac and H3K27me3 are lower (Valiulienė et al. 2016). In comparison with the control NOG mice, levels of epigenetic modifications, both activating and inhibiting gene expression, in APL xenograft NOG mice decreased in the lung, spleen (except H3K4me3), and kidney (except H3K9me3), and more reduced levels of these modifications after treatment of xenograft mice with epigenetic (ATRA, belinostat, DZNep) compounds (Valiulienė et al. 2016). The most significant lesions were found in the lungs, which not only showed the highest levels of H4ac, H3K4me3, H3K9me3, and H3K27me3 modifications, but also significantly increased WT1 protein expression. The levels of all studied modifications were lower in the brains of AML xenograft mice than in the control one, but both activating tags and the highly inhibitory H3K9me3 tag were significantly increased following epigenetic therapy of AML xenograft mice. Interestingly, the levels of the investigated epigenetic markers (except H4ac) increased most in liver of APL xenograft mice. Notably, the liver of APL xenograft mice also exhibited extremely strong WT1 protein expression, which was slightly suppressed after treatment with DZNep, belinostat, and ATRA (Valiulienė et al. 2016). It should be added that not only the expression of WT1 protein was altered in the liver of APL xenograft mice following exposure to epigenetic compounds, but an increase in histone H4ac, H3K4me3, H3K9me3, and H3K27me3 modifications was also observed. This may be related to liver function, their response to APL cell infiltration, or treatment.

In conclusion, our data related to belinostat impact on cell growth, differentiation, gene and protein expression, including epigenetic modifications, justified potential value belinostat has in APL therapy. In this research some new insights in possible molecular mechanisms of belinostat were also demonstrated. Also, our *in vivo* studies have shown for the first time that tumors and tissues of APL xenograft mice exhibit a variety of epigenetic modifications, reflecting changes in the epigenetics of adult cancerous tissue and which epigenetic changes in tissues occur.

3.1.7 HMT Inhibitor BIX-01294 Effects on Epigenetic Changes of DNA and Proteins

We evaluated the anti-proliferative activity of BIX-01294 in two different human APL cell lines, HL-60 and NB4 (Savickiene et al. 2014a); for detailed information on chemicals and cell lines used in the study see Appendices B and C). Also it was investigated how BIX-01294 can affect chronic leukemia cells K562. It was showed in other studies that BIX-01294 (diazepin–quinazolin–amine derivative) selectively regulates the G9a HMTase that leads to dimethylation of histone H3 at K9 *in vitro* and this effect can be restored upon removal of the inhibitor (Kubicek et al. 2007).

It should be noted that BIX-01294 itself did not provoke cell differentiation between HL-60 and NB4 but effectively prompts ATRA-induced cell differentiation of promyelocytic leukemia cells (Savickiene et al. 2014a); Chap. 2). Also we tried to investigate BIX-01294 in combination with other epigenetic modifiers, such as HDAC inhibitors (belinostat, BML-210) and DNMT inhibitors (zebularine, RG108). Compared with ATRA-treated cells, we noticed that combination of BIX-01294, ATRA, and one of the HDACi/DNMTi influenced differentiation in NB4 and HL-60 cells. This combination affected DNMT1 and H3K9me2 levels in different ways, but the strongest H3K9me2 inhibition was detected after BIX-01294 treatment alone in both cell lines (Fig. 3.14).

In our study we determined the effect of BIX-01294 as HMT inhibitor EHMT2 (G9a) gene expression level. We have shown that the 48-h treatment with BIX-01294 does not reduce the transcription level of EHMT2. A pronounced increase in expression was observed in NB4 cells treated with ATRA plus BIX (24 h point— 5; 48 h—9.5; 72 h—8.6 times). There was no such significant change in HL-60 cells (Table 3.4).

Our study also assessed alterations in the expression of CEBPα and CEBPϵ genes related to granulocytic APL cell differentiation. It is known from the literature that during granulocytic differentiation, CEBPα expression decreases, and CEBPϵ increases. This tendency towards expression of these genes has also been observed in this study. We determined CEBPϵ expression after treatment with ATRA, BIX-01294, and its combination and noticed that expression is highest in the samples where cells were most differentiated (Table 3.4). CEBPα expression was also studied after the cell was exposed to BIX-01294 and its combination with ATRA. In this case, we detected slight down-regulation of CEBPα expression (Savickiene et al. 2014a), Table 3.4).

Expression level of histone acetyltransferase PCAF (KAT2B-K (lysine) acetyltransferase 2B) was also assessed after treatment with BIX-01294 alone or its combination with ATRA. We detected very high expression of PCAF miRNA after therapy of NB4 cells with ATRA and BIX-01294 combination (Table 3.4). We did not detect the same effect in HL-60 cells. This increase in NB4 cells could be associated with p53 activation by direct binding of p53 to the PCAF gene promoter (Watts et al. 2004). Since the p53 deficiency is detected in HL-60 cells, it could be argued that this is the reason for the differences in the expression of the PCAF gene between the two cell lines: the expression in NB4 cells increases by more than 100 times, and HL-60 only by approximately 4 times (Table 3.4).

We detected that BIX-01294 treatment induces dose- and time-dependent cell death, but however the Propidium Iodite and Anexin-V test revealed that cell death was largely due not to apoptosis, but to potentially other mechanisms of cell death. Also these findings were correlated with the absence of PARP-1 and caspase-3 cleavage and further activation (Savickiene et al. 2014a), Fig. 3.15. To clarify the molecular mechanisms by which BIX-01294 contributes to the death of non-apoptotic cells we performed analysis of autophagy- related gene

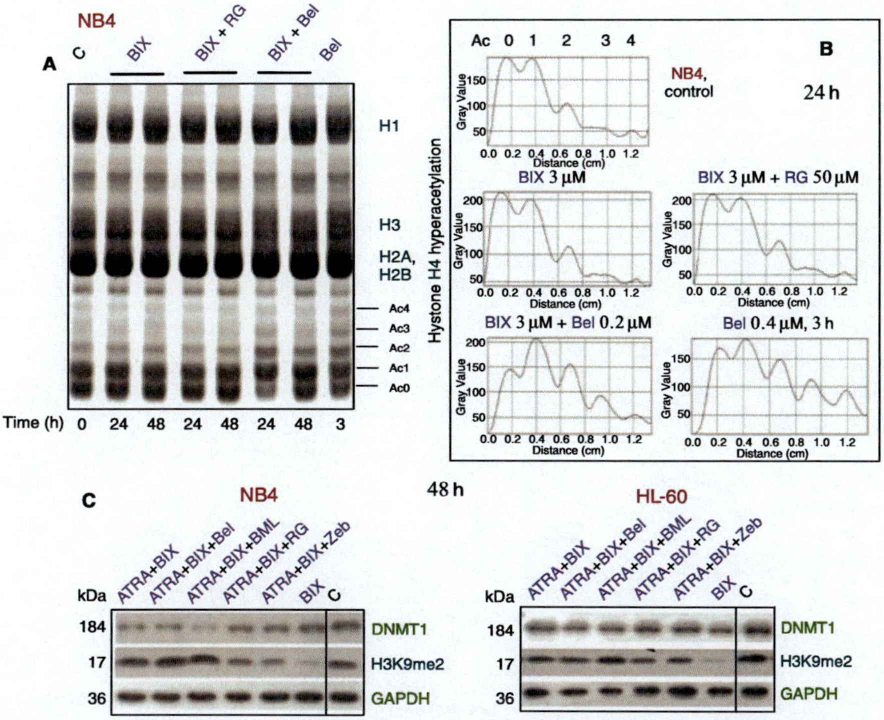

Fig. 3.14 Analysis of epigenetic changes in response to treatments of HL-60 and NB4 cells with BIX-01294, ATRA, and HDAC or DNMT inhibitors. (**a**) Acid-urea (AU) gel after staining with Brilliant Blue G-Colloidal represents acetyl-isoforms of histone H4 (Ac0-Ac4) in NB4 cells treated with 3 μM BIX-01294 (BIX) alone or together with 0.2 μM belinostat (Bel) or 50 μM RG108 (RG) for the indicated time. Histone H1 band served as a control of histone extraction and protein loading. (**b**) Histone H4 hyperacetylation levels in the gel determined by scanning densitometry. (**c**) HL-60 and NB4 cells were treated with 2 μM BIX-01249 as a single drug or cotreated with 1 μM ATRA, HDACi (0.2 μM belinostat, 5 μM BML-210 (BML)) and DNMTi (20 μM zebularine (Zeb) or 50 μM RG108) for 48 h. Western blot analysis was performed with the indicated antibodies on the total cell lysates. C—untreated cells. GAPDH served as the loading control. The data represent one of three independent experiments showing similar results. According to Savickiene et al. (2014a), License No 5070770686871

expression level. We evaluated Beclin1, a marker that indicates cell death by autophagy (Maskey et al. 2013), gene expression changes after treatment with BIX-01294 and revealed no significant changes (Table 3.4). The ATG5 as a key protein participating in the extension of the phagophore membrane in autophagy vesicles was investigated in this study. A small significant change was observed only in HL-60 cells exposed to 2 μM BIX-01294, after 48 h—expression increased by about 30% (Table 3.4). The effect of BIX-01294 on different cancer cells has been shown to be different. In adenocarcinoma cells, BIX-01294 induces apoptosis processes by initiating the internal (mitochondrial) pathway (Wang et al. 2017). The

Table 3.4 Summary of gene expression changes in leukemia cells treated with BIX-01294 alone and in combination with ATRA. C— untreated cells

Gene	C	Treatment/ NB4 cells											
		BIX 2 μM			BIX 4 μM			ATRA 1 μM			ATRA 1 μM + BIX 2 μM		
		24 h	48 h	72 h	24 h	48 h	72 h	24 h	48 h	72 h	24 h	48 h	72 h
PPARγ mRNA	1	0.51	0.7	0.7	1	1.5	2.2						
C/EBPϵ mRNA	1	1.3	1.1	1.2				9.3	20	22	10	10	23
C/EBPα mRNA	1							0.2	0.4	0.2	0.6		
PCAF mRNA	1	3.8	2.2	5				4	6	2.1	12	46	112
G9a mRNA	1	1.7	1.5	1.7	1.1	1.2	3				0.9	1.7	1.5
Beclin1 mRNA	1	1	1	1.1	1.4	1.5	2.2						
ATG5 mRNA	1		1.2			1.7							
		Treatment/HL-60 cells											
PPARγ mRNA	1	1.1	0.6	1.1	0.3	0.4	2						
C/EBPϵ mRNA	1	4	2.3	3				11	10	19	8	7	17
C/EBPα mRNA	1							0.2	0.2	0.3	0.4		
PCAF mRNA	1							1.3	1.4	3	0.5	1	4.5
G9a mRNA	1	1.2	1.1	1.2	0.7	0.7	1.7				5	9.5	8.6
Beclin1 mRNA	1	2.5	1.2	1.5	1.2	1.1	1.5						
ATG5 mRNA	1		1.2			1.3							

effect of BIX-01294 on breast and rectal cancer cells leads to autophagy-dependent cell death (Kim et al. 2013). Previous studies have shown that the repression of EHMT2 due to RNA interference causes significant changes in the morphology of cancer cells (prostate, lung, and breast cancer), up-regulation of the expression of galactosidase, and inhibition of telomerase activity. Because of the decrease in H3K9me2 in the centromeres' part of chromatin, the structure of the centromeres disrupts themselves. Chromosome instability, in turn, can continue to lead to cell aging (Andriani et al. 2016). The results of our ATG5 and PPARγ gene expression studies—a clear increase in the expression of these genes after exposure to BIX-01294 (Table 3.4)—suggest that BIX-01294 therapy for NB4 and HL-60 cells may induce cellular aging processes. PPAR activation is known to have a significant role in the regulation of signaling pathways (Altucci et al. 2001; Lu et al. 2005). It is particularly important to note that PPAR, combined with the growth of the survivin, inhibits TRAIL induced apoptosis, and it is due to this pathway that NB4 cells killed (Castedo et al. 2004). ATG5, in turn, is required for an autophagy independent mitotic catastrophe that occurs in cancer affected cells after exposure to non-lethal doses of DNA-damaging drugs or antimitotic chemical compounds (Kim et al. 2010). As it is known, survivin not only protects cells from apoptosis in the internal (mitochondrial) pathway, but also regulates cell division and participates in mitotic spindle verification processes (Dohi et al. 2004; Vernier et al. 2011). And our studies have shown that the survivin protein content increases slightly (Fig. 3.15) after the treatment with BIX-01294 in NB4 cells and the cells are no longer able to move to the phase of cycle G2/M (Savickiene et al. 2014a). These results were also positively correlated with an increase of the H3K9me3—one of the heterochromatin marks (Fig. 3.15).

It is important to note that the effects with BIX-01294 caused cellular-specific and dose-dependent alterations in the methylation level of H3K9, surviving, and cyclin E1 (Fig. 3.15). This, in turn, may also have an impact on the regulation of the G1/S phase. Surprisingly, the 48 h effect of the 4–5 μM BIX-01294 increased cyclin E1, H3K9me2 and H3K9me3 levels exclusively in NB4 cells. Also, PPARγ and ATG5 gene expression (Table 3.4) also increased in NB4 cells. These data also correlated with the stop of the NB4 cell cycle in cell cycle phase S (Savickiene et al. 2014a). It should be noted that the 3 μM BIX-01294 concentrations decreased H3K9me2 levels in both cell lines after 48 h of exposure. Similar alterations in the protein content of p27, surviving, and chromatin modifying enzymes HDAC1 and DNMT1 were also observed in both HL-60 and NB4 cells (Fig. 3.15).

Also in our other study (Vitkeviciene et al. 2018) we investigated the effects of BIX-01294 on acute (NB4) and chronic (K562) leukemia cells. We used larger doses of BIX-01294 on K562 cells compared to NB4 cells (7–8 μM instead of 3–4 μM, respectively). It was found that chronic myeloid leukemia cells are highly resistant to different chemotherapeutic agents due to the expression of Bcr-Abl, which prevents apoptosis (Cilloni and Saglio 2012). However, K562 cells have also been shown to be resistant to interferon and imatinib treatment. By others it was shown that after exposure to BIX-01294 K562 cell proliferation was explicitly inhibited and apoptosis was prompted (Loh et al. 2014).

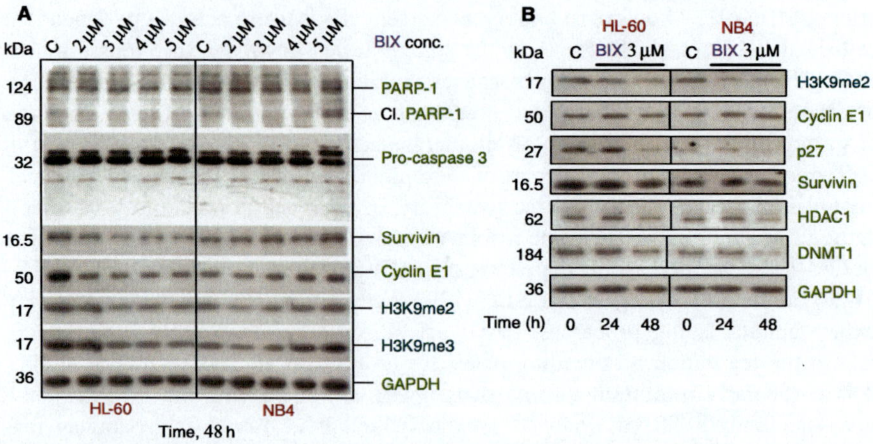

Fig. 3.15 BIX-01294 induced changes in cell cycle or survival regulating protein expression and histone H3K9 modification state. (**a, b**) HL-60 and NB4 cells were treated with the indicated doses of BIX-01294 for 24 or 48 h. Western blot analysis was performed with the indicated antibodies on the total cell lysates. C—untreated cells. GAPDH served as the loading control. The data represent one of the three independent experiments showing similar results. According to Savickiene et al. (2014a), License No 5070770686871

We demonstrated that BIX-01294 up-regulated cell cycle inhibitors p53, p21, and Rb and down-regulated cell cycle activator CCNA2 in both K562 and NB4 cell lines. Therefore, we assessed if BIX-01294 can stimulate cellular senescence. We have found that BIX-01294 did not stimulate senescence (no SA-β-gal+ cell staining). BIX-01294 prompted cell cycle arrest and up-regulation of HMGA2, but neither up-regulation nor phosphorylation of ATM was observed (Vitkeviciene et al. 2018).

In some cases of cancer p53 gene promoter tends to be methylated, for example, it has been shown to be methylated in about half of all ovarian cancer cases (Chmelarova et al. 2013), and substantial hypermethylation was noticed among cervical cancer patients (Jha et al. 2012). It has also been shown that p53 methylation correlates with decreased gene expression in a subgroup of patients with ALL (Agirre et al. 2003). Still, in our study (Vitkeviciene et al. 2018) we have found that p53 promoter is unmethylated in K562 and NB4 cells. Thus, increased p53 expression after BIX-01294 therapy could not be justified by gene promoter demethylation. The same could be argued about p21 gene as its promoter was determined to be unmethylated as well. This correlates with earlier findings that p21 promoter is unmethylated in most types of human cancer including different myeloid leukemias (Scott et al. 2006).

In our work, we examined changes in the expression of SPI1 (PU.1) and CBFB after treatment with BIX-01294 alone. HMT inhibitor BIX-01294 showed a beneficial impact on these hematopoietic cell differentiation-related genes in both cell lines. Thus, our findings showed that BIX-01294 has the ability to improve the differentiation of myeloid cells when used in combination with cell differentiation

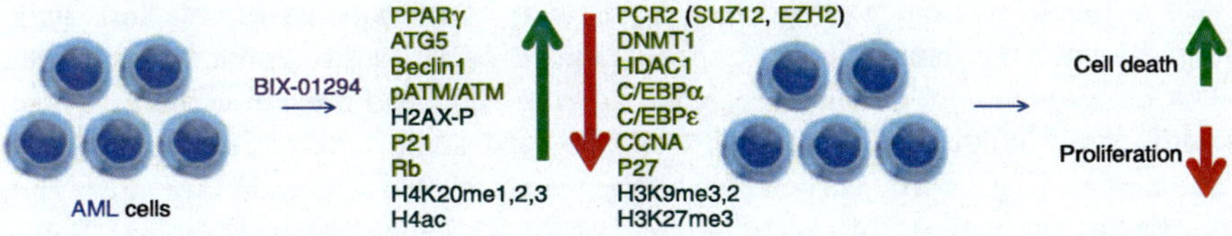

Fig. 3.16 Summarized BIX-01294 effect on leukemia cells

inducers such as ATRA. Since one of the known causes of cancer is epigenome changes we analyze the effects of BIX-01294 on the epigenetic modulation of myeloid leukemia cells. Chromatin enriched with histone modification H3K9me3 acts as an HP1α interaction protein that promotes the formation of a DNMT1 repressor complex that leads to tumor suppression (Pradhan et al. 2009). Changes in PRC2 complex are also associated with the development of various forms of cancer. Other authors have shown that BIX-01294 regulates the PRC2 component EHZ2 (Balasubramanian et al. 2010) and DNMT1 (Singh et al. 2013). In our research, we analyze the expression of H3K9me3, HP1α, DNMT1, PRC2 complex proteins (EZH2 and SUZ12), and changes in methylation of H4K20 after therapy. Acute promyelocytic leukemia cells demonstrated susceptibility to epigenetic modulation BIX-01294. However, BIX-01294 had less effect on K562 cells in contrast to NB4 cells (Vitkeviciene et al. 2018).

In conclusion, we for the first time demonstrated that the single epigenetic modifier inhibitor BIX-01294 and its combination with differentiation inducer ATRA reduced expression level of genes and their coded proteins associated with chromatin transformation (SUZ12, EED, EZH2, HDAC1), and increased the expression level of genes and proteins associated with cell cycle (CDKN1B) and differentiation (CEBPE, PPARG) processes. After treatment, chromatin-activating histone modifications (H4ac) have also been observed and inhibition of H3K9me2, H3K9me3, and H3K27me3 has been detected. The summarized scheme is presented in Fig. 3.16. These results are even enhanced by the combined treatments.

3.1.8 Multifunctional Epigallocatechin Gallate Reduces Leukemia Cell Growth and Induces Apoptosis

Epigallocatechin gallate (EGCG) is an exceptional plant compound that gets a lot of attention for its potential positive impact on health. Formally known as epigallocatechin gallate, EGCG is a type of plant-based compound called catechin. Catechins may be further categorized into a larger group of plant compounds known as polyphenols. EGCG and other related catechins function as potent antioxidants that may protect cells against damage caused by free radicals. Free radicals are highly reactive particles formed in human body that can damage cells when there are too much of them. Eating foods high in such antioxidants as catechins may help diminish the damage made by free radicals. Moreover, research suggests that such catechins as EGCG may decrease inflammation and prevent certain chronic conditions, including diabetes, heart disease, and some types of cancers. As an antioxidant, EGCG protects cells from damage associated with oxidative stress and suppresses the activity of pro-inflammatory chemicals produced in human body, such as tumor necrosis factor-alpha (TNF-α).

It has been shown that EGCG can act as an anticancer agent for such diseases as breast, colorectal, gastric, prostate, esophageal, and lung cancer. This has been tested in *in vitro* and *in vivo* systems. EGCG obstructed cell growth and induced apoptosis (Pan et al. 2016). In our studies (Borutinskaitė et al. 2018; Vitkeviciene et al. 2018) we detected that growth of leukemia cells (NB4, HL-60, K562) was reduced by EGCG in a dose- and time-dependent manner, as well as apoptosis was induced. Our results allow us to argue that fully efficient and non-toxic EGCG concentrations are 30–40 μM for NB4, HL-60 cells, and 120–140 μM for K562 cells. EGCG concentrations used in our studies are very similar to those identified for therapy of other cancer cells (Pan et al. 2016). Because EGCG is not highly biostable, oxidation reactions occur at neutral-alkaline pH and in the presence of dissolved oxygen; different EGCG delivery systems in order to improve its stability and bioavailability are being studied (Granja et al. 2016). Transcription factors like C/EBPα and C/EBPϵ are important for cell growth and differentiation (Park et al. 1999; Gery et al. 2005; Suh et al. 2006). Reduced activity of these factors plays a role in development of hematopoietic disorders, e.g., acute myeloid leukemia (AML). It was shown that C/EBPα and C/EBPϵ transcription factors expression can be induced after cell treatment with all-trans retinoic acid (ATRA) and leads to suppression of cell growth and partial cell differentiation *in vitro* (Truong et al. 2003).

In our investigations, we demonstrated up-regulation of C/EBPα and C/EBPϵ in cells (HL-60, NB4) treated with EGCG (Borutinskaitė et al. 2018). Also, we prepared chromatin immunoprecipitation experiments to find if EGCG treatment can change the binding of acetylated H3K14 histone and H4 hyperacetylated histone to the promoter regions. We have found that EGCG treatment improved the binding of acetylated H3K14 histone and H4 hyperacetylated histone to the promoter regions of C/EBPα and C/EBPϵ genes. However, we did not detect cell differentiation after EGCG treatment (see Chap. 2).

Further, we showed that EGCG up-regulated p53, p21, and Rb in NB4 and K562 cell lines and down-regulated CCNA2 in both cell lines with a lower effect on K562 cell (Vitkeviciene et al. 2018). We also detected up-regulation of p27 mRNA level after NB4 and HL-60 cells treatment with EGCG (Borutinskaitė et al. 2018). Our findings correlated with studies of other authors where EGCG induces cell cycle arrest and alters different cell cycle activators and inhibitors expression like cell cycle activators CCNA2, CCNB1, CCND1, E2F1, cyclin D1, CDK4/6, phosphorylated Rb (inhibition), and p21 gene expression (activation) in cancer cells (Gan et al. 2018; Mayr et al. 2015; Chen et al. 2004; Zhang et al. 2012).

Also we demonstrated that EGCG up-regulated histone H4 acetylation in time- and dose-dependent manner, also down-regulation of H3K9me2, had no significant effect on the H3K14ac in NB4 and HL-60 cells. Histone H2AX was also investigated in our study (Borutinskaitė et al. 2018; Vitkeviciene et al. 2018). From the literature we know that ATM can phosphorylate H2AX and that leads to altered chromatin that recruits few downstream DDR proteins (Campisi and Di Fagagna 2007). ATM and H2AX protein phosphorylation alterations after treatment with EGCG were investigated in our research. We demonstrated that ATM phosphorylation was up-regulated in both cell lines (NB4, K562); however, phosho-H2AX level only raised in NB4 cells after EGCG treatment. Total H2AX amount after EGCG treatment stayed unaltered in K562 and NB4 cells. It was showed that ATM phosphorylates other proteins such as p53, which can result in the cell cycle arrest, senescence, or apoptosis. We detected that after EGCG therapy, level of ATM mRNA expression almost doubled. We also analyzed methylation sites in ATM promoter region and revealed that ATM promoters are unmethylated in NB4 and K562 cells that did not receive treatment (Vitkeviciene et al. 2018). We observed unchanged methylation status of ATM promoters after EGCG treatment. We can conclude that methylation changes do not result in ATM expression alterations. It is common knowledge that DNA damage results in protein kinase ATM phosphorylation.

We found that the up-regulation of p27 gene showed stronger relation of hyper-acetylated H4 and acetylated H3K14 with p27 promoter after treatment with EGCG. Consequently, EGCG has a beneficial impact on p27 activity, which is known to result in the cell accumulation during the G0/G1 cycle phase. Moreover, we noticed up-regulated PCAF gene expression level and strong binding impact of H3K14ac on the PCAF promoter after EGCG therapy for the cell (Borutinskaitė et al. 2018). PCAF (P300/CBP Associated Factor) gene encodes lysine acetyltransferase, which acetylates histone lysines and, besides that, directly controls p53 by acetylation of Lys320 in the C-terminal portion of p53. The latter gene expression is suppressed in almost all types of cancer (Nishimori et al. 2000). As a result, elevated PCAF expression may activate p53 and thus can result in the cell cycle arrest and tumor suppression. Moreover, analysis of cell cycle distribution showed that NB4 and HL-60 cells accumulated in G0/G1 phase, and thus p27 and PCAF gene expression results were justified. These data confirm showed EGCG capacity to prevent cell proliferation. Among many anticancerous activities, EGCG has been shown to function as DNMT and HDAC inhibitor (Khan et al. 2015). Hypermethylation in CpG islands in promoter regions that occurs with the help of DNMT is a significant mechanism to affect genes that inhibit cancer. Because of that DNMT inhibitors are used to induce demethylation of CpG and silence gene expression (Fang et al. 2003). The level of acetylation of histones is crucial for the transcription of inhibitory genes, which determines their use in cancer therapy (Nandakumar et al. 2011). EGCG has been shown to interact with DNMT3B and HDAC1, thereby reducing the activity of DNMT3B and HDAC1 in human HeLa cells (Khan et al. 2015). In addition, it was found that EGCG reduces the general level of DNA methylation by inhibiting the expression of DNMT3a, DNMT3b mRNA, and proteins in

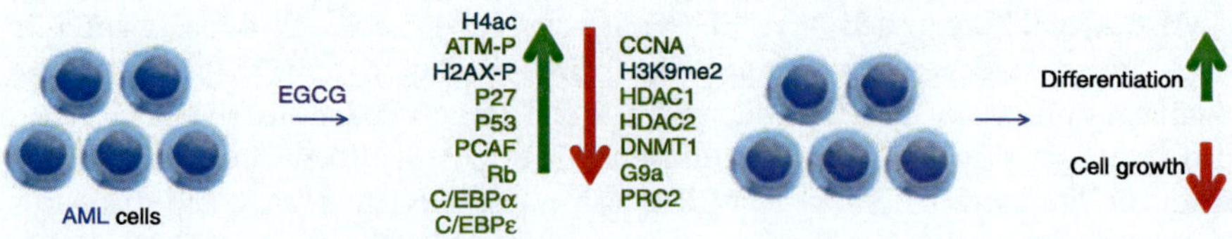

Fig. 3.17 Summarized EGCG effect on leukemia cells

A431 cells (Nandakumar et al. 2011). In our research we demonstrated that EGCG is able to impede DNMT and HDAC in NB4 and HL-60 cells (Borutinskaitė et al. 2018). Substantially down-regulated DNMT1, HDAC1, and HDAC2 gene expression was presented.

We have shown down-regulation of epigenetic modifier G9a gene expression and lowered levels of histone H3K9me2 modification, which is catalyzed by G9a. It was demonstrated that HMT G9a was up-regulated in different tumor types (Zhang et al. 2015). Due to our demonstration that EGCG obstructs DNMT, HDAC, and G9a, it could be applied as a possible inhibitor of epigenetic modifiers in APL cells. PRC2 is a chromatin modifier in charge of chromatin silencing through trimethylation of lysine 27 on H3. The enzyme EZH2 is the catalytic component of the PRC2 because it has HMT activity. Nevertheless, EZH2 is active only in combination with at least two other nonenzymatic partners: SUZ12 and EED. In this research, we have found that EGCG results in the down-regulated gene expression of PRC2 core components in NB4 and HL-60 cells. Analysis of protein expression analysis justified considerably lowered expression of EZH2 and SUZ12. Moreover, ChIP experiments demonstrated that EGCG therapy diminished the binding effect of acetylated H3K14 and hyperacetylated H4 to the PCR2 complex genes' (EZH2, SUZ12, EED) promoter regions (Borutinskaitė et al. 2018). It was demonstrated earlier that EGCG treatment diminishes BMI-1 (main component of PRC1) and EZH2 levels in skin tumor SCC-13 cells (Balasubramanian et al. 2010).

Thus, our data support EGCG down-regulating effect on polycomb group chromatin modifiers. Overall, we argue that EGCG is an efficient growth inhibition agent for APL cells. It regulates expression of cell cycle/differentiation regulatory genes and modifies chromatin structure via inhibiting epigenetic modifiers (Fig. 3.17).

3.2 DNA Methylation and Epigenetic Regulation

Numerous genetic and epigenetic alterations may lead to the development of leukemic cells carrying unlimited self-renewal and drug-resistant characteristics that are related with the initiation of leukemia. Novel therapies created to attack leukemia cells could remove cancer without the risks and side effects of current treatments that destroy not only cancer cells, but healthy cells in the body as well. The epigenetic therapy is targeted on new tendencies for cancer stem cells therapy—combined epigenetic therapy when the DNA methylation is inhibited. DNA methylation is related to the regulation of gene expression; therefore, the DNA methyltransferase inhibitors are under investigations by establishing their impact on cancer cell growth, apoptosis, and differentiation mechanisms by applying new drugs altogether with the existing therapies. New treatments may be created by using DNMT inhibitors that attack the leukemic cell population.

3.2.1 Procaine Effects on Epigenetic Changes of DNA and Proteins

Epigenetic changes in DNA influence important intracellular processes such as gene expression and DNA replication. Methylation of promoter regions of tumor suppressor genes and deacetylation of histones lead to gene silencing and condensed chromatin formation.

The aim of our study was to analyze the impacts of DNA methyltransferase (DNMT) inhibitor procaine on growth, growth inhibition, viability, and differentiation of human leukemia cells. Human leukemia NB4 cell line was treated with procaine alone or together with differentiation inducer all-trans retinoic acid (ATRA) (Borutinskaite et al. 2016). It is known that procaine and procainamide (a derivative of procaine) can restore gene expression by interacting with areas of DNA that are rich in CpG cells, for instance, the RARβ 2 gene (for detailed information on chemicals used in the study see Appendix C). Therefore, procaine, which binds to DNA but is not incorporated therein, can be very useful in cancer therapy (Sheikh and Dua 2019). Therapy with procaine can have less side effects than decitabine and others presently used to treat AML and MDS. The long-term and safe application of procaine as a local anesthetic with prominent properties may result into its transition to preclinical and early clinical epigenetic-based cancer treatment research (Villar-Garea et al. 2003; Gao et al. 2009; Esteller 2005; Lv et al. 2010).

In our study we have found that procaine inhibited growth of human leukemia cell line NB4 in a time-/dose-dependent manner and used together with ATRA induced differentiation very similar to differentiation level when ATRA is applied alone (see Chap. 2, for detailed information on cell lines used in the study see Appendix B). The expression of molecules associated with differentiation (CD11b, E-cadherin, G-CSF), apoptosis (PPARγ), and DNA methylation (DNMT1, DNMT3A, DNMT3B) alters in human leukemia cells after the procaine therapy.

To identify demethylating impacts of procaine or its combination with ATRA on DNMT expression level in NB4 cells, we conducted experiments on gene (DNMT1, DNMT3A, and DNMT3B) expression and Western blotting of DNMT1 protein (Table 3.5). The human cells are known to have a family of DNA methyltransferases, whose members DNMT1, DNMT3a, and DNMT3b are responsible for adding a methyl (CH_3) group to the CG dinucleotide cytosine and thus alter gene expression. Due to DNMT1 during replication process the exact pattern of the DNA methylation on newly synthesized DNA is being repeated. DNMT1 expression during the cell cycle is tightly regulated by few mechanisms, and highest DNMT1 expression can be observed during the S phase (Chen and Li 2004; Delpu et al. 2013). Once again, DNA methyltransferases DNMT3a and DNMT3b are involved in determining DNA methylation patterns during development (Chen and Li 2004; Delpu et al. 2013). In this research, we showed that after 24 h of treatment procaine suppressed DNMT1 gene expression. This impact was exacerbated by ATRA and procaine combination therapy. Also the similar down-regulation of the DNMT3b gene expression was detected. Nevertheless, the expression of the DNMT3a gene was down-regulated by procaine alone, whereas the combination with ATRA during therapy had no influence on this process (Table 3.5).

Further we investigated the expression of cadherin after procaine treatment of NB4 cells, because cadherin is important in cancer progression. It is known that cadherin promoter region in many cancer cells is hypermethylated and that leads to gene suppression. The E-cadherin can influence tumor cells by adhesion, invasion, and activation of signaling pathways. The role and the mechanism of E-cadherin silencing in hematopoietic cells is not well investigated. We found some evidence in the literature that one of the possible mechanisms is that acetylation of histones in the region of the E-cadherin promoter appears after treatment of leukemia cells with some epigenetic modifiers (Jordaan et al. 2013). Few studies on the reactivation of functional E-cadherin in some cancer cell lines applying the demethylating agent 5-aza-2′-deoxycytidine (5AzaC) have been published in the literature (Cheishvili et al. 2015). We have shown in our work that procaine can recover cadherin expression as early as after 24 h (Table 3.5). We also found that treating cells with procaine and its combination with ATRA or ATRA alone resulted in increased acetylation of histone H3 lysine K9 and serine phosphorylation (S10). H3 phosphorylation (specifically serine 10) was found to be directly involved in the induction of immediate–early genes, for instance, c-jun, c-fos, and c-myc. Also it was found that MSK1 kinase can be triggered by growth factor and stress phosphorylates H3 *in vitro* as well. H3 phosphorylation at serine 10 and 28 is also important for proper chromosome condensation during mitosis and meiosis (Park and Kim 2012). Also we detected increase in the histone H3K4 trimethylation after treatment with procaine (Table 3.5) and this correlated with growth inhibitory and cell death activation.

We also investigated the G-CSF mRNA expression level after procaine therapy and after treatment with both procaine and ATRA. Since procaine alone did not affect differentiation of NB4 cells, we also did not detect G-CSF mRNA in our samples (Table 3.5). Only combined treatment of procaine with ATRA as well as ATRA alone up-regulated G-CSF mRNA level, and that correlated with differentiation induction in NB4 cells up to 76% after 72 h. Other important gene for differentiation process like PPARγ was investigated. We observed induction of PPARγ mRNA level after treatment with procaine alone (Table 3.5).

Our findings allow us to argue that procaine can be further investigated with its regard to epigenetic and differentiation therapy especially when applied in

Table 3.5 Gene and protein expression levels in leukemia cells treated with ATRA, procaine alone with different concentrations, and in combination with ATRA, and decitabine. Intensity of gene and protein bands was measured using ImageJ software: every sample amount was calculated according to GAPDH amount; results are presented as changes in comparison to control untreated cells (C). The data are presented as mean ± SD ($n = 3$)

Gene/Protein	C	ATRA 1 μM		Procaine 3 mM		Procaine 5 mM		Procaine 5 mM +ATRA 1 μM		Decitabine 0.5 μM	
		24 h	96 h	24 h	96 h	24 h	96 h	24 h	96 h	24 h	96 h
G-CSF mRNA	1	1.7	1.7	1	1	1	1	1.7	1.7	1.3	1
DNMT1 mRNA	1	1	1	0.8	1	1	1	1	1	1	1
DNMT3A mRNA	1	0.8	0.5	0.7	0.9	0.7	0.7	1	1.1	1	1
DNMT3B mRNA	1	0.8	0.7	0.88	0.85	0.84	0.84	0.6	0.6	0.8	0.7
PPARγ mRNA	1	0.8	0.8	1	1	0.85	0.9	0.9	0.8	0.9	0.7
DNMT1 protein	1	0.7	0.1	0.7	0.2	0.7	0.10	0.6	0	0	0
Cadherin protein	1	1.3	1.4	2	1.9	1.7	2	1.3	1.7	1.7	1.6
H3K4me3 protein	1	0.6	0.7	0.8	0.8	0.7	0.7	0.6	0.7	0.7	0.6
H3K9ac/S10p protein	1	2.5	2.5	1	2	2.2	2.5	2	2.2	2.3	2.3

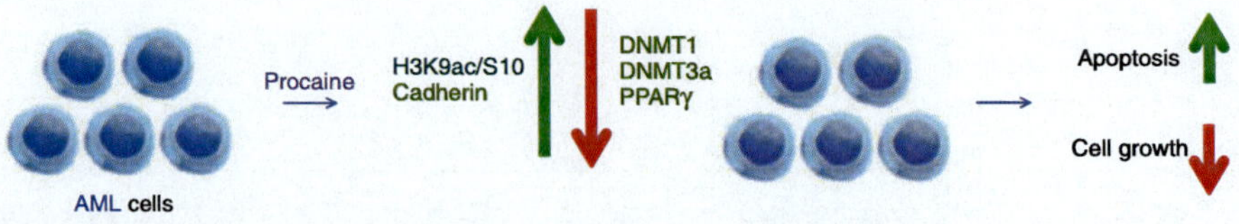

Fig. 3.18 Summarized scheme of procaine effect on leukemia cells

combination with all-trans retinoic acid or similar agents. Procaine arrests growth of human leukemia cells and in combination with ATRA prompts differentiation. This combination arrests cell cycle in G0/G1 phase. Procaine leads to the DNA methyltransferases expression reduction. Human leukemia cells under the treatment with procaine improve the expression of molecules related to the differentiation (CD11b, E-cadherin, G-CSF) and apoptosis (PPARγ). Furthermore, this DNMT inhibitor improves transcription activating histone modifications, i.e., H3K4me3 and H3K9ac/S10p (Fig. 3.18). Overall, procaine as DNMT inhibitor can be applied in further thorough studies on epigenetic and differentiation therapy especially when used in combination with all-trans retinoic acid or similar agent.

3.2.2 DNMT Inhibitors, RG108 and Zebularine Effects on Epigenetic Changes of DNA and Proteins

DNA methylation is indisputably important during organism development. In advanced organisms, the need for DNA methylation is weakened/decreasing and only the maintenance of the non-coding part of the genome becomes important. Methylation of CpG islands requires only a few classes of genes, but these genes are very important and crucial. It has been demonstrated in mice that homozygous mutations in DNMT genes are fatal (Takebayashi et al. 2007). The methylation pattern in cancer cell lines is the exact opposite: methylation of CpG islands is elevated, but the genome-wide methylation level is lower (Cheung et al. 2000). We know from the different studies that promoter hypermethylation often inactivates tumor suppressor genes. Inhibition of DNA methyltransferase activity was also found to have an inhibitory effect on tumor formation, i.e., inhibitors of DNA methyltransferases can demethylate tumor suppressor genes and thus suppress the cancer cell-specific phenotype (Brueckner and Lyko 2004; Cheishvili et al. 2015).

Because DNA methylation is important in leukemogenesis, the development of new therapeutic strategies by applying several DNMT inhibitors is possible. In the literature there are some data on the biological impacts of various hypomethylating agents (Chuang et al. 2005; Flotho et al. 2009; Duchmann and Itzykson 2019). However, the exact mechanism of action is not entirely clear. In our few studies (Savickiene et al. 2012a,b,c), we performed analysis of two DNMT inhibitors, the non-nucleoside agent RG108 and nucleoside agent zebularine in order to determine their antileukemic function in human promyelocytic leukemia NB4 cells, HL-60 cells, and ATRA-resistant early-myeloblastic KG1 cells (for detailed characterization of chemicals and cells used in the study see Appendices B and C).

It was determined that zebularine functions by the formation of covalent complexes between DNMT and zebularine-incorporated DNA (Zhou et al. 2002), leading to depletion of DNMTs (Cheng et al. 2004). Presently, this compound has been suggested as a good fit in *in vivo* studies and some clinical trials due to its stability, minimal toxicity, oral bioavailability, and selectivity to attack tumor cells (Marquez et al. 2005; Scott et al. 2007; Yoo et al. 2008). Nevertheless, zebularine needs prolonged exposure and high doses due to its limited DNA incorporation and low metabolism (Ben-Kasus et al. 2005). In our study we found that zebularine (20–60 μM) inhibited leukemia (NB4, HL-60) cell growth, suppressed cell viability, and activated apoptosis in a time- and dose-dependent manner. For KG1 cells we used higher doses of zebularine (20–100 μM), but effect was very similar, such as in NB4 or HL-60 cells. We found that zebularine induced apoptosis in APL cells (HL-60) and was toxic to the cells at higher doses than 100 μM (100–200 μM) (Savickiene et al. 2012a). Activation of apoptosis at low doses of zebularine correlated with up-regulation of caspase 3 activity and PAR-1 cleavage (Savickiene et al. 2012c). We revealed for the first time that zebularine treatment (continuous or temporal) can accelerate and then enhance ATRA-induced granulocytic differentiation in leukemia cells (NB4, HL-60, KG1). It was showed in the literature that methylated genes are transcriptionally silenced and demethylation by DNMT inhibitors can release the methyl-CpG binding protein (MeCP2) from the promoter and prepare chromatin for further histone acetylation by HDAC (El-Osta et al. 2002). We decided to evaluate if HDAC inhibitors (phenylbutyrate (PB) or BML-210) can impact the gene expression level and facilitate differentiation effects observed after treatment with zebularine and ATRA. We found that HDAC inhibitors, for instance, sodium phenylbutyrate (1 mM) and BML-210 (50 μM) in combination with ATRA after a temporal (48 h) pre-treatment with 20 μM zebularine without the additive cytotoxicity improved ATRA-mediated differentiation. The side effects of HDAC inhibitors were more clearly observed in KG1 cells (1.8–2-fold increase) than in NB4 cells (1.3–1.4-fold increase). In conclusion, zebularine in combination with HDAC inhibitors assisted in overcoming the differentiation block both in ATRA-sensitive NB4 and ATRA-resistant KG1 cells.

Also we investigated effects of novel small-molecule inhibitor, RG108, on proliferation of leukemia cells. It is known that RG108 acts without being involved into the DNA, effectively inhibits DNMTs at the active site, and results in the demethylation of genomic DNA and reactivation of tumor suppressor genes, having little toxicity in cell lines of human cancer (Brueckner et al. 2005; Stresemann et al. 2006). We detected that RG108 alone in different doses (20–100 μM) can induce time-dependent, but not a dose-dependent leukemia cells (NB4) growth inhibition and was non-toxic at a wide spectrum of concentrations (Savickiene et al. 2012b). The decreased toxicity demonstrated by RG108 can be characterized by its incapacity to influence the methylation of repeated centromeres' sequences that are able to contribute to chromosome stability (Brueckner et al. 2005). Like in the case of zebularine, we showed that RG108 in combination with ATRA or ATRA +PB enhanced ATRA-mediated differentiation into granulocytes in NB4 cells (Savickiene et al. 2012b).

Hypomethylating agents with antitumor potential have already demonstrated their advantage in treatment of different human cancers (Tsai and Baylin 2011). In our study we compare the zebularine and RG108 effects on other APL cell line, HL-60 (Savickiene et al. 2012a). For the first time, we demonstrated that both zebularine and RG108 are effective in enhancing ATRA-induced differentiation into granulocytes in promyelocytic HL-60 leukemia cells. Comparing the effects of both DNMT inhibitors, we found that zebularine has a more pronounced effect on cell growth and apoptosis than RG108. Previous work by other authors has shown that zebularine can effectively inhibit AML cell line proliferation and accumulate cells during the G2/M cell cycle phase (Scott et al. 2007; Lemaire et al. 2005). Several groups of scientists have argued that the anti-proliferative effect of zebularine on many human cancer lines is related to the increase in inhibitors of p21 WAF1, p16 INK4A, or p15 INK4B, cyclinCDK complexes that inhibit G1 and S phase (Scott et al. 2007; Cheng et al. 2003). Cancer cells exposed to zebularine lack DNMT1, and down-regulation of DNMT3b (both regulated by the cell cycle) and DNMT3a (occurs only during S phase) occurs (Cheng et al. 2003; Velicescu et al. 2002). DNMT inhibitors such as zebularine disrupt DNMT regulation, decrease cellular DNMT levels, induce DNA replication arrest during S phase, inhibit histone synthesis, reduce histone acetylation during S phase, or induce cell cycle arrest in G2/M, resulting in the cell growth and apoptosis activation (Milutinovic et al. 2003; Billam et al. 2010). Other work has shown that zebularine induces apoptosis in AML cell lines (Scott et al. 2007) by proapoptotic degradation of Bax, anti-apoptotic Bcl-2, PARP, or caspase 3 (Billam et al. 2010; Neureiter et al. 2007).

Here, in our studies we detected that zebularine and RG108 alone or plus ATRA down-regulated DNMT1 mRNA and then protein level in a time- and dose-dependent manner (Savickiene et al. 2012a,c). But more remarkable inhibition effect was induced by zebularine alone at 100 μM concentration (DNMT1 protein was down-regulated 95% (Savickiene et al. 2012a). As we know from the literature, in leukemia cell blasts the DNMTs expression level is high and DNA methylation is related to the loss of E-cadherin (Corn et al. 2000; Melki et al. 2000).

In our studies we revealed changes in E-cadherin expression and methylation status in leukemia HL-60, NB4, KG1 cells after treatment with RG108 or zebularine. We showed that adherin protein expression level increased after treatment with RG108, zebularine alone, or its combination with ATRA. That correlated with up-regulated cadherin mRNA level. It was shown by other authors that zebularine can affect genomic DNA demethylation process alongside with re-expression of methylation-silenced genes, for instance, p16, p15, or E-cadherin (Brueckner et al. 2005; Stresemann et al. 2006). Other authors demonstrated that RG108 affected p16, SERP-1 and TIMP-3 expression level (Brueckner et al. 2005). Similarly, in our studies (Savickiene et al. 2012a,b) we revealed E-cadherin promoter methylation in control (untreated) HL-60 cells, KG1, and NB4 cells by methylatios-specific PCR analysis. Our findings allow us to argue that zebularine and RG108 attacked DNMT1 and caused incomplete demethylation of E-cadherin promoter and the gene reactivation as demonstrated by E-cadherin mRNA and protein expression observed after therapy with zebularine or RG108 even at cytostatic doses of 50 μM. The increase in E-cadherin was detected and after leukemia cell treatment with HDAC inhibitors phenylbutyrate, BML-210 and even differentiation agent ATRA.

In our study (Savickiene et al. 2012c) we investigated p15 methylation status and protein level after KG1 and N cells therapy with zebularine and found that there were no changes in some methylation sites of p15 promoter regions. But p15 protein level was up-regulated after 72 h treatment with zebularine and ATRA (Savickiene et al. 2012c). Our findings justify recent data that zebularine therapy for AML cells with hypermethylated p15 promoter reactivates p15 expression and stimulates apoptosis at demethylating dosages (Scott et al. 2007). Moreover, improved p15 induction was noticed after AML193 cells' co-treatment with zebularine and HDAC inhibitor trichostatin A. In our research, we have not shown any detectable p15 transcript and protein in NB4 and KG1 cells and demonstrated low potency of zebularine that was used either alone or plus ATRA to reactivate p15 until 48 h of treatments. Nevertheless, p15 re-expression at 72 h was related to the higher degree of granulocytic differentiation during combined treatment.

In our studies we used the differentiation-inducing agent ATRA. We detected that treatment with ATRA alone also can reactivate E-cadherin. Other authors noticed that ATRA can down-regulate the DNMT enzyme activity that can lead to DNMT-HDAC interaction reduction in ATRA target genes (Fazi et al. 2005). In our research we observed the correlation between E-cadherin re-expression (mRNA, protein) and level of differentiation in leukemia cells (KG1, HL-60, NB4) after treatment with zebularine and RG108 (Savickiene et al. 2012a,b,c). Therefore E-cadherin was suggested as an indicator beneficial in monitoring the efficiency of treatment based on DNA demethylation in promyelocytic leukemia. It should be pointed out that the impacts of distinct DNMT inhibitors on gene transcription can be un-associated with the direct promoter DNA hypomethylation, but they may take place through regional enrichment of histone acetylation at the gene promoter (Flotho et al. 2009).

We also investigated how zebularine and RG108 can alter chromatin structure. The summarized scheme is presented in Fig. 3.19. We detected changes in methylation of histone H3K4 and acetylation of histone H4, which represent the active state of chromatin and caused the opening chromatin structure and gene expression (Strahl and Allis 2000). We found that zebularine or RG108 alone did not induce changes in H4 acetylation and H3K4me3 methylation. Only pretreatment with RG108 with further treatment with combination of PB or PB+ATRA can induce very high H3K4me3 methylation and H4 acetylation degree in leukemia cells, causing a relaxed and active chromatin state for transcriptional activation.

We also performed chromatin immunoprecipitation (ChIP) analysis and we can conclude that zebularine treatment leads to acetylation of histone H4 and methylation of histone H3 at K4 or acetylation of H3 at K9. These modifications represent the transcriptionally active promoters of genes such as p21 and cadherin.

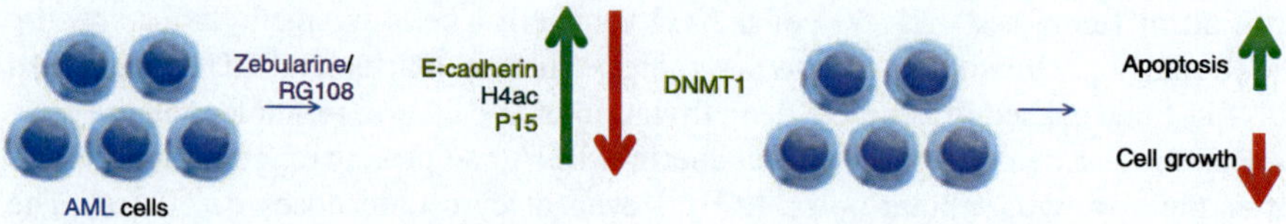

Fig. 3.19 Summarized zebularine/RG108 effect on leukemia cells

Our results revealed that binding of the inactive E-cadherin promoter due to its methylation with hyperacetylated histone H4 and methylated histone H3 at K4 was enhanced by zebularine therapy for leukemic KG1 and NB4 cells, which resulted in transcriptional changes of E-cadherin and formation of chromatin open structure.

In our research we studied the impact of zebularine on activation of other negative cell cycle regulator p21, which has a significant role in the growth arrest and control of transition of G1 to S phase. By comparison with the methylation of p15 in acute leukemia, genes encoding other CDKIs, for instance, p21, were observed as non-methylated (Dexheimer et al. 2017). To identify the impact of zebularine on the pattern of acetylation/methylation of histones related to the active p21 gene, we applied primers to p21 promoter regions (P2, P3) and exon 2 (E2). Primer pairs (P2) and (P3) correlated with the places in the proximal promoter region and in the distal promoter region close to p53 consensus sequence, respectively. Employing ChIP assay, we discovered that zebularine prompted association of acetylated H4 and methylated H3 on K4 with p21 promoter P2 site in KG1 cells and with P2 and P3 sites in NB4 cells. Elevated modifications of histones H4 and H3 related to the P3 region corresponding to p53 site concur with its transcriptional activation and growth inhibition in p53-positive NB4 cells. The induction of p53 may cause its binding with transcription factor Sp1, dissociation of HDAC1 from complex, and acetylation of the p21 promoter (Lagger et al. 2003). The potential mechanism of p21 induction by DNMT inhibitors may include the release of HDAC1 without the need of promoter demethylation that was shown in recent research where decitabine activates unmethylated p21 in KG1 and KG1a cells in association with the release of HDAC1 and elevated acetylated histone H3 at the unmethylated p21 promoter (Scott et al. 2006). Our analysis with AML cell lines justifies the clinical potential of zebularine that has been demonstrated to exhibit an *in vivo* antitumor impact via oral or intraperitoneal administration in tumor-bearing mouse (Cheng et al. 2003).

We also assessed the impact of therapy with ATRA and belinostat, as single agents, and their combined treatment belinostat + ATRA on NB4 cells' global DNA methylation patterns. ATRA applied alone did not exert any important activity toward global DNA methylation modulation, while the augmentation in global DNA methylation was observed upon therapy with belinostat (global DNA methylation elevated by 15–38% after 6–24 h of treatment with 0.2 μM belinostat, while in later time-points the earlier methylation level was restored). In contrast, the rise in DNA methylation degree after combined therapy consisting of belinostat + ATRA was more sudden and more noticeable in comparison with therapy consisting of belinostat applied alone (DNA methylation increased by 40% after 6 h treatment). It should be emphasized that prolonged treatment with belinostat + ATRA (72 h) down-regulated the global DNA methylation more than 14%.

The findings of our research justify the opinion that the usefulness of DNMT inhibitor relies not so much on its own DNMT-inhibiting characteristics but rather on its potential to improve the activity of other drugs applied in combination. Moreover, the combined strategy can help to reduce the dosages of each drug and offers the most successful approaches. Here, we suggest low toxic culture conditions with short-term exposure (24–48 h) to zebularine and RG108 in sequential combinations with ATRA and HDAC inhibitor PB that may be promising in the improvement of differentiation therapy in acute leukemia cases.

References

Adam E, Quivy V, Bex F, Chariot A, Collette Y, Vanhulle C, Schoonbroodt S, Goffin V, Nguyen TLA, Gloire G, et al. (2003) Potentiation of tumor necrosis factor-induced NF-κB activation by deacetylase inhibitors is associated with a delayed cytoplasmic reappearance of IκBα. Mol Cell Biol 23(17):6200–6209

Agirre X, Vizmanos JL, Calasanz MJ, Garcia-Delgado M, Larrayoz MJ, Novo FJ (2003) Methylation of CpG dinucleotides and/or CCWGG motifs at the promoter of TP53 correlates with decreased gene expression in a subset of acute lymphoblastic leukemia patients. Oncogene 22(7):1070–1072

Aird KM, Worth AJ, Snyder NW, Lee JV, Sivanand S, Liu Q, Blair IA, Wellen KE, Zhang R (2015) ATM couples replication stress and metabolic reprogramming during cellular senescence. Cell Rep 11(6):893–901

Al Bitar S, Gali-Muhtasib H (2019) The role of the cyclin dependent kinase inhibitor p21cip1/waf1 in targeting cancer: molecular mechanisms and novel therapeutics. Cancers 11(10):1475

Altucci L, Rossin A, Raffelsberger W, Reitmair A, Chomienne C, Gronemeyer H (2001) Retinoic acid-induced apoptosis in leukemia cells is mediated by paracrine action of tumor-selective death ligand trail. Nat Med 7(6):680–686

Ammerpohl O, Thormeyer D, Khan Z, Appelskog IB, Gojkovic Z, Almqvist PM, Ekström TJ (2004) HDACi phenylbutyrate increases bystander killing of HSV-tk transfected glioma cells. Biochem Biophys Res Commun 324(1):8–14

Andriani GA, Almeida VP, Faggioli F, Mauro M, Tsai WL, Santambrogio L, Maslov A, Gadina M, Campisi J, Vijg J, et al. (2016) Whole chromosome instability induces senescence and promotes SASP. Sci Rep 6:35218

Balasubramanian S, Adhikary G, Eckert RL (2010) The Bmi-1 polycomb protein antagonizes the (-)-epigallocatechin-3-gallate-dependent suppression of skin cancer cell survival. Carcinogenesis 31(3):496–503

Ben-Kasus T, Ben-Zvi Z, Marquez VE, Kelley JA, Agbaria R (2005) Metabolic activation of zebularine, a novel DNA methylation inhibitor, in human bladder carcinoma cells. Biochem Pharmacol 70(1):121–133

Billam M, Sobolewski MD, Davidson NE (2010) Effects of a novel DNA methyltransferase inhibitor zebularine on human breast cancer cells. Breast Cancer Res Treat 120(3):581–592

Blanco JCG, Minucci S, Lu J, Yang XJ, Walker KK, Chen H, Evans RM, Nakatani Y, Ozato K (1998) The histone acetylase PCAF is a nuclear receptor coactivator. Genes Develop 12(11):1638–1651

Bojang Jr P, Ramos KS (2014) The promise and failures of epigenetic therapies for cancer treatment. Cancer Treat Rev 40(1):153–169

Borutinskaite V, Navakauskiene R (2015) The histone deacetylase inhibitor BML-210 influences gene and protein expression in human promyelocytic leukemia NB4 cells via epigenetic reprogramming. Int J Mol Sci 16(8):18252–18269. https://doi.org/10.3390/ijms160818252

Borutinskaite V, Bauraite-Akatova J, Navakauskiene R (2016) Anti-leukemic activity of DNA methyltransferase inhibitor procaine targeted on human leukaemia cells. Open Life Sci 11(1):322–330. https://doi.org/10.1515/biol-2016-0044

Borutinskaitė V, Virkšaitė A, Gudelytė G, Navakauskienė R (2018) Green tea polyphenol EGCG causes anti-cancerous epigenetic modulations in acute promyelocytic leukemia cells. Leukemia Lymphom 59(2):469–478. https://doi.org/10.1080/10428194.2017.1339881

Brueckner B, Lyko F (2004) DNA methyltransferase inhibitors: old and new drugs for an epigenetic cancer therapy. Trends Pharmacol Sci 25(11):551–554. https://doi.org/10.1016/j.tips.2004.09.004

Brueckner B, Boy RG, Siedlecki P, Musch T, Kliem HC, Zielenkiewicz P, Suhai S, Wiessler M, Lyko F (2005) Epigenetic reactivation of tumor suppressor genes by a novel small-molecule inhibitor of human DNA methyltransferases. Cancer Res 65(14):6305–6311

Budillon A, Di Gennaro E, Bruzzese F, Rocco M, Manzo G, Caraglia M (2007) Histone deacetylase inhibitors: a new wave of molecular targeted anticancer agents. Rec Pat Anti-Cancer Drug Discov 2(2):119–134

Campisi J, Di Fagagna FD (2007) Cellular senescence: when bad things happen to good cells. Nat Rev Mol Cell Biol 8(9):729–740

Castedo M, Perfettini JL, Roumier T, Andreau K, Medema R, Kroemer G (2004) Cell death by mitotic catastrophe: a molecular definition. Oncogene 23(16):2825–2837

Ceccacci E, Minucci S (2016) Inhibition of histone deacetylases in cancer therapy: lessons from leukaemia. Br J Cancer 114(6):605–611

Chandler H, Peters G (2013) Stressing the cell cycle in senescence and aging. Curr Opin Cell Biol 25(6):765–771

Cheishvili D, Boureau L, Szyf M (2015) DNA demethylation and invasive cancer: implications for therapeutics. Br J Pharmacol 172(11):2705–2715

Chen LF, Greene WC (2003) Regulation of distinct biological activities of the NF-κB transcription factor complex by acetylation. J Mol Med 81(9):549–557

Chen T, Li E (2004) Structure and function of eukaryotic DNA methyltransferases. In: Current topics in developmental biology, vol 60. Elsevier, Amsterdam, pp 55–89

Chen JJ, Ye ZQ, Koo MWL (2004) Growth inhibition and cell cycle arrest effects of epigallocatechin gallate in the NBT-II bladder tumour cell line. BJU Int 93(7):1082–1086

Cheng JC, Matsen CB, Gonzales FA, Ye W, Greer S, Marquez VE, Jones PA, Selker EU (2003) Inhibition of DNA methylation and reactivation of silenced genes by zebularine. J Natl Cancer Inst 95(5):399–409

Cheng JC, Weisenberger DJ, Gonzales FA, Liang G, Xu GL, Hu YG, Marquez VE, Jones PA (2004) Continuous zebularine treatment effectively sustains demethylation in human bladder cancer cells. Mol Cell Biol 24(3):1270–1278

Cheung P, Tanner KG, Cheung WL, Sassone-Corsi P, Denu JM, Allis CD (2000) Synergistic coupling of histone H3 phosphorylation and acetylation in response to epidermal growth factor stimulation. Mol Cell 5(6):905–915

Chmelarova M, Krepinska E, Spacek J, Laco J, Beranek M, Palicka V (2013) Methylation in the p53 promoter in epithelial ovarian cancer. Clin Transl Oncol 15(2):160–163

Chuang JC, Yoo CB, Kwan JM, Li TW, Liang G, Yang AS, Jones PA (2005) Comparison of biological effects of non-nucleoside DNA methylation inhibitors versus 5-aza-2′-deoxycytidine. Mol Cancer Therap 4(10):1515–1520

Cilloni D, Saglio G (2012) Molecular pathways: Bcr-abl. Clin Cancer Res 18(4):930–937

Corn PG, Smith BD, Ruckdeschel ES, Douglas D, Baylin SB, Herman JG (2000) E-cadherin expression is silenced by 5′ CpG island methylation in acute leukemia. Clin Cancer Res 6(11):4243–4248

Dachineni R, Ai G, Kumar DR, Sadhu SS, Tummala H, Bhat GJ (2016) Cyclin A2 and CDK2 as novel targets of aspirin and salicylic acid: a potential role in cancer prevention. Mol Cancer Res 14(3):241–252

Dai Y, Rahmani M, Grant S (2003) An intact NF-κB pathway is required for histone deacetylase inhibitor induced G1 arrest and maturation in U937 human myeloid leukemia cells. Cell Cycle 2(5):465–470

Dawson M, Kouzarides T (2012) Cancer epigenetics: from mechanism to therapy. Cell 150(1):12–27

Delpu Y, Cordelier P, Cho WC, Torrisani J (2013) DNA methylation and cancer diagnosis. Int J Mol Sci 14(7):15029–15058

Deniaud E, Baguet J, Chalard R, Blanquier B, Brinza L, Meunier J, Michallet MC, Laugraud A, Ah-Soon C, Wierinckx A, et al. (2009) Overexpression of transcription factor Sp1 leads to gene expression perturbations and cell cycle inhibition. PloS One 4(9):e7035. https://doi.org/10.1371/journal.pone.0007035

Dexheimer GM, Alves J, Reckziegel L, Lazzaretti G, Abujamra AL (2017) DNA methylation events as markers for diagnosis and management of acute myeloid leukemia and myelodysplastic syndrome. Dis Markers 2017:5472893

Dohi T, Beltrami E, Wall NR, Plescia J, Altieri DC (2004) Mitochondrial survivin inhibits apoptosis and promotes tumorigenesis. J Clin Investig 114(8):1117–1127

Dornan D, Shimizu H, Burch L, Smith AJ, Hupp TR (2003) The proline repeat domain of p53 binds directly to the transcriptional coactivator p300 and allosterically controls DNA-dependent acetylation of p53. Mol Cell Biol 23(23):8846–8861

Duchmann M, Itzykson R (2019) Clinical update on hypomethylating agents. Int J Hematol 110(2):161–169

El-Osta A, Kantharidis P, Zalcberg JR, Wolffe AP (2002) Precipitous release of methyl-CpG binding protein 2 and histone deacetylase 1 from the methylated human multidrug resistance gene (MDR1) on activation. Mol Cell Biol 22(6):1844–1857

Esteller M (2005) DNA methylation and cancer therapy: new developments and expectations. Curr Opin Oncol 17(1):55–60

Fang MZ, Wang Y, Ai N, Hou Z, Sun Y, Lu H, Welsh W, Yang CS (2003) Tea polyphenol (−)-epigallocatechin-3-gallate inhibits DNA methyltransferase and reactivates methylation-silenced genes in cancer cell lines. Cancer Res 63(22):7563–7570

Fazi F, Travaglini L, Carotti D, Palitti F, Diverio D, Alcalay M, McNamara S, Miller WH, Coco FL, Pelicci PG, et al (2005) Retinoic acid targets DNA-methyltransferases and histone deacetylases during APL blast differentiation in vitro and in vivo. Oncogene 24(11):1820–1830

Fiskus W, Wang Y, Sreekumar A, Buckley KM, Shi H, Jillella A, Ustun C, Rao R, Fernandez P, Chen J, et al (2009) Combined epigenetic therapy with the histone methyltransferase EZH2 inhibitor 3-deazaneplanocin A and the histone deacetylase inhibitor panobinostat against human AML cells. Blood J Am Soc Hematol 114(13):2733–2743

Flotho C, Claus R, Batz C, Schneider M, Sandrock I, Ihde S, Plass C, Niemeyer C, Lübbert M (2009) The DNA methyltransferase inhibitors azacitidine, decitabine and zebularine exert differential effects on cancer gene expression in acute myeloid leukemia cells. Leukemia 23(6):1019–1028

Gan RY, Li HB, Sui ZQ, Corke H (2018) Absorption, metabolism, anti-cancer effect and molecular targets of epigallocatechin gallate (EGCG): an updated review. Crit Rev Food Sci Nutr 58(6):924–941

Gao Z, Xu Z, Hung MS, Lin YC, Wang T, Gong M, Zhi X, Jablons DM, You L (2009) Procaine and procainamide inhibit the Wnt canonical pathway by promoter demethylation of WIF-1 in lung cancer cells. Oncol Rep 22(6):1479–1484

Gery S, Tanosaki S, Bose S, Bose N, Vadgama J, Koeffler HP (2005) Down-regulation and growth inhibitory role of C/EBPα in breast cancer. Clin Cancer Res 11(9):3184–3190

Gilicze AB, Wiener Z, Tóth S, Buzás E, Pállinger É, Falcone FH, Falus A (2014) Myeloid-derived microRNAs, miR-223, miR27a, and miR-652, are dominant players in myeloid regulation. BioMed Res Int 2014:870267

Gozzini A, Rovida E, Sbarba PD, Galimbert S, Santini V (2003) Butyrates, as a single drug, induce histone acetylation and granulocytic maturation: possible selectivity on core binding factor-acute myeloid leukemia blasts. Cancer Res 63(24):8955–8961

Granja A, Pinheiro M, Reis S (2016) Epigallocatechin gallate nanodelivery systems for cancer therapy. Nutrients 8(5):307

Grants JM, Wegrzyn-Woltosz J, Hui T, O'Neill K, Shadbolt M, Knapp DJHF, Parker JDK, Deng D, Gopal A, Docking TR, et al. (2020) Altered microRNA expression links IL6 and TNF-induced inflammaging with myeloid malignancy. Blood 135(25):2235–2251

Gravina GL, Marampon F, Giusti I, Carosa E, Di Sante S, Ricevuto E, Dolo V, Tombolini V, Jannini EA, Festuccia C (2012) Differential effects of PXD101 (belinostat) on androgen-dependent and androgen-independent prostate cancer models. Int J Oncol 40(3):711–720. https://doi.org/10.3892/ijo.2011.1270

Grigoryev SA, Bulynko YA, Popova EY (2006) The end adjusts the means: heterochromatin remodelling during terminal cell differentiation. Chromosome Res 14(1):53–69

Haggarty SJ, Koeller KM, Wong JC, Butcher RA, Schreiber SL (2003) Multidimensional chemical genetic analysis of diversity-oriented synthesis-derived deacetylase inhibitors using cell-based assays. Chem Biol 10(5):383–396

Hake SB, Xiao A, Allis CD (2004) Linking the epigenetic 'language' of covalent histone modifications to cancer. Br J Cancer 90(4):761–769

Hirano G, Izumi H, Kidani A, Yasuniwa Y, Han B, Kusaba H, Akashi K, Kuwano M, Kohno K (2010) Enhanced expression of PCAF endows apoptosis resistance in cisplatin-resistant cells. Mol Cancer Res 8(6):864–872

Ho AS, Turcan S, Chan TA (2013) Epigenetic therapy: use of agents targeting deacetylation and methylation in cancer management. OncoTargets Ther 6:223

Iwata K, Ogata S, Okumura K, Taguchi H (2003) Induction of differentiation in human promyelocytic leukemia HL-60 cell line by niacin-related compounds. Biosci Biotechnol Biochem 67(5):1132–1135. https://doi.org/10.1271/bbb.67.1132

Jayathilaka N, Han A, Gaffney KJ, Dey R, Jarusiewicz JA, Noridomi K, Philips MA, Lei X, He J, Ye J, et al (2012) Inhibition of the function of class IIa HDACs by blocking their interaction with MEF2. Nucleic Acids Res 40(12):5378–5388

Jha AK, Nikbakht M, Jain V, Sehgal A, Capalash N, Kaur J (2012) Promoter hypermethylation of p73 and p53 genes in cervical cancer patients among north Indian population. Mol Biol Rep 39(9):9145–9157

Jordaan G, Liao W, Sharma S (2013) E-cadherin gene re-expression in chronic lymphocytic leukemia cells by HDAC inhibitors. BMC Cancer 13(1):88

Joseph J, Wajapeyee N, Somasundaram K (2005) Role of p53 status in chemosensitivity determination of cancer cells against histone deacetylase inhibitor sodium butyrate. Int J Cancer 115(1):11–18

Kaiser M, Kühnl A, Reins J, Fischer S, Ortiz-Tanchez J, Schlee C, Mochmann LH, Heesch S, Benlasfer O, Hofmann WK, et al. (2011) Antileukemic activity of the HSP70 inhibitor pifithrin-μ in acute leukemia. Blood Cancer J 1(7):e28–e28

Kaji A, Zhang Y, Nomura M, Bode AM, Ma WY, She QB, Dong Z (2003) Pifithrin-α promotes p53-mediated apoptosis in JB6 cells. Molecular Carcinogenesis: Published in Cooperation with the University of Texas MD Anderson Cancer Center 37(3):138–148

Kanno Si, Kurauchi K, Tomizawa A, Yomogida S, Ishikawa M (2015) Pifithrin-alpha has a p53-independent cytoprotective effect on docosahexaenoic acid-induced cytotoxicity in human hepatocellular carcinoma HepG2 cells. Toxicol Lett 232(2):393–402

Kelly WK, Richon VM, O'Connor O, Curley T, MacGregor-Curtelli B, Tong W, Klang M, Schwartz L, Richardson S, Rosa E, et al. (2003) Phase I clinical trial of histone deacetylase inhibitor: suberoylanilide hydroxamic acid administered intravenously. Clin Cancer Res 9(10):3578–3588

Khan N, Jeffers M, Kumar S, Hackett C, Boldog F, Khramtsov N, Qian X, Mills E, Berghs SC, Carey N, et al (2008) Determination of the class and isoform selectivity of small-molecule histone deacetylase inhibitors. Biochem J 409(2):581–589

Khan MA, Hussain A, Sundaram MK, Alalami U, Gunasekera D, Ramesh L, Hamza A, Quraishi U (2015) (-)-Epigallocatechin-3-gallate reverses the expression of various tumor-suppressor genes by inhibiting DNA methyltransferases and histone deacetylases in human cervical cancer cells. Oncol Rep 33(4):1976–1984

Kiernan R, Brès V, Ng RWM, Coudart MP, El Messaoudi S, Sardet C, Jin DY, Emiliani S, Benkirane M (2003) Post-activation turn-off of NF-κB-dependent transcription is regulated by acetylation of p65. J Biol Chem 278(4):2758–2766

Kikuchi H, Iizuka R, Sugiyama S, Gon G, Mori H, Arai Ma, Mizumoto K, Imajoh-Ohmi S (1996) Monocytic differentiation modulates apoptotic response to cytotoxic anti-Fas antibody and tumor necrosis factor α in human monoblast U937 cells. J Leukocyte Biol 60(6):778–783

Kim S, Han J, Lee SK, Hur SM, Koo M, Cho DH, Bae SY, Choi MY, Shin I, Yang JH, et al. (2010) Pifithrin-α, an inhibitor of P53 transactivation, up-regulates COX-2 expression through an MAPK-dependent pathway. Pharmacology 86(5–6):313–319

Kim Y, Kim YS, Kim DE, Lee JS, Song JH, Kim HG, Cho DH, Jeong SY, Jin DH, Jang SJ, et al (2013) BIX-01294 induces autophagy-associated cell death via EHMT2/G9a dysfunction and intracellular reactive oxygen species production. Autophagy 9(12):2126–2139

Kohnken R, Porcu P, Mishra A (2017) Overview of the use of murine models in leukemia and lymphoma research. Front Oncol 7:22

Kubicek S, O'Sullivan RJ, August EM, Hickey ER, Zhang Q, Teodoro ML, Rea S, Mechtler K, Kowalski JA, Homon CA, et al. (2007) Reversal of H3K9me2 by a small-molecule inhibitor for the g9a histone methyltransferase. Mol Cell 25(3):473–481

Kumar P, Periyasamy R, Das S, Neerukonda S, Mani I, Pandey KN (2014) All-trans retinoic acid and sodium butyrate enhance natriuretic peptide receptor a gene transcription: role of histone modification. Mol Pharmacol 85(6):946–957

Kwon SH, Ahn SH, Kim YK, Bae GU, Yoon JW, Hong S, Lee HY, Lee YW, Lee HW, Han JW (2002) Apicidin, a histone deacetylase inhibitor, induces apoptosis and Fas/Fas ligand expression in human acute promyelocytic leukemia cells. J Biol Chem 277(3):2073–2080

Lagger G, Doetzlhofer A, Schuettengruber B, Haidweger E, Simboeck E, Tischler J, Chiocca S, Suske G, Rotheneder H, Wintersberger E, et al (2003) The tumor suppressor p53 and histone deacetylase 1 are antagonistic regulators of the cyclin-dependent kinase inhibitor p21/WAF1/CIP1 gene. Mol Cell Biol 23(8):2669–2679

Lemaire M, Momparler LF, Bernstein ML, Marquez VE, Momparler RL (2005) Enhancement of antineoplastic action of 5-aza-2′-deoxycytidine by zebularine on L1210 leukemia. Anti-Cancer Drugs 16(3):301–308

Liao Q, Wang B, Li X, Jiang G (2017) miRNAs in acute myeloid leukemia. Oncotarget 8(2):3666

Lindstedt I, Lindgren MA, Andersson E, Engström W (2014) The WT1 gene–its role in tumourigenesis and prospects for immunotherapeutic advances. in vivo 28(5):675–681

Loh SW, Ng WL, Yeo KS, Lim YY, Ea CK (2014) Inhibition of euchromatic histone methyltransferase 1 and 2 sensitizes chronic myeloid leukemia cells to interferon treatment. PLoS One 9(7):e103915

Lorenzo E, Ruiz-Ruiz C, Quesada AJ, Hernández G, Rodriguez A, López-Rivas A, Redondo JM (2002) Doxorubicin induces apoptosis and CD95 gene expression in human primary endothelial cells through a p53-dependent mechanism. J Biol Chem 277(13):10883–10892

Love IM, Sekaric P, Shi D, Grossman SR, Androphy EJ (2012) The histone acetyltransferase PCAF regulates p21 transcription through stress-induced acetylation of histone H3. Cell Cycle 11(13):2458–2466

Lu M, Kwan T, Yu C, Chen F, Freedman B, Schafer JM, Lee EJ, Jameson JL, Jordan VC, Cryns VL (2005) Peroxisome proliferator-activated receptor γ agonists promote TRAIL-induced apoptosis by reducing survivin levels via cyclin D3 repression and cell cycle arrest. J Biol Chem 280(8):6742–6751

Lv G, Chen Z, Zheng J, Wei F, Jiang H, Zhang R, Wang X (2010) Theoretical study of the interaction pattern and the binding affinity between procaine and DNA bases. J Mol Struct: Theochem 939(1–3):44–52

Ma J, Kavelaars A, Dougherty PM, Heijnen CJ (2018) Beyond symptomatic relief for chemotherapy-induced peripheral neuropathy: targeting the source. Cancer 124(11):2289–2298

Magrin GL, Di Summa F, Strauss FJ, Panahipour L, Mildner M, Magalhães Benfatti CA, Gruber R (2020) Butyrate decreases ICAM-1 expression in human oral squamous cell carcinoma cells. Int J Mol Sci 21(5):1679

Marquez VE, Kelley JA, AGbaria R, Ben-Kasus T, Cheng JC, Yoo CB, Jones PA (2005) Zebularine: a unique molecule for an epigenetically based strategy in cancer chemotherapy. Ann NY Acad Sci 1058(1):246–254

Maskey D, Yousefi S, Schmid I, Zlobec I, Perren A, Friis R, Simon HU (2013) ATG5 is induced by DNA-damaging agents and promotes mitotic catastrophe independent of autophagy. Nat Commun 4(1):1–14

Mayr C, Wagner A, Neureiter D, Pichler M, Jakab M, Illig R, Berr F, Kiesslich T (2015) The green tea catechin epigallocatechin gallate induces cell cycle arrest and shows potential synergism with cisplatin in biliary tract cancer cells. BMC Complement Altern Med 15(1):194

Melki JR, Vincent PC, Brown RD, Clark SJ (2000) Hypermethylation of e-cadherin in leukemia. Blood J Am Soc Hematol 95(10):3208–3213

Melnick A, Licht JD (1999) Deconstructing a disease: RARα, its fusion partners, and their roles in the pathogenesis of acute promyelocytic leukemia. Blood J Am Soc Hematol 93(10):3167–3215

Merzvinskyte R, Treigyte G, Savickiene J, Magnusson KE, Navakauskiene R (2006) Effects of histone deacetylase inhibitors, sodium phenyl butyrate and vitamin B3, in combination with retinoic acid on granulocytic differentiation of human promyelocytic leukemia HL-60 cells. In: Diederich M (ed) Signal transduction pathways, PT B: stress signaling and transcriptional control. Annals of the New York academy of sciences, Cell signaling world 2006 conference, Luxembourg, Luxembourg, Jan 25–28, 2006, vol 1091, pp 356–367. https://doi.org/10.1196/annals.1378.080

Milutinovic S, Zhuang Q, Niveleau A, Szyf M (2003) Epigenomic stress response knockdown of DNA methyltransferase 1 triggers an intra-S-phase arrest of DNA replication and induction of stress response genes. J Biol Chem 278(17):14985–14995

Minucci S, Nervi C, Lo Coco F, Pelicci P (2001) Histone deacetylases: a common molecular target for differentiation treatment of acute myeloid leukemias? Oncogene 20(24):3110–3115. https://doi.org/10.1038/sj.onc.1204336

Mori N, Matsuda T, Tadano M, Kinjo T, Yamada Y, Tsukasaki K, Ikeda S, Yamasaki Y, Tanaka Y, Ohta T, et al (2004) Apoptosis induced by the histone deacetylase inhibitor FR901228 in human T-cell leukemia virus type 1-infected T-cell lines and primary adult T-cell leukemia cells. J Virol 78(9):4582–4590

Mukherjee SP, Behar M, Birnbaum HA, Hoffmann A, Wright PE, Ghosh G (2013) Analysis of the RelA: CBP/p300 interaction reveals its involvement in NF-κB-driven transcription. PLoS Biol 11(9)

Nandakumar V, Vaid M, Katiyar SK (2011) (-)-epigallocatechin-3-gallate reactivates silenced tumor suppressor genes, Cip1/p21 and p 16 INK4a, by reducing DNA methylation and increasing histones acetylation in human skin cancer cells. Carcinogenesis 32(4):537–544

Neureiter D, Zopf S, Leu T, Dietze O, Hauser-Kronberger C, Hahn EG, Herold C, Ocker M (2007) Apoptosis, proliferation and differentiation patterns are influenced by zebularine and SAHA in pancreatic cancer models. Scand J Gastroenterol 42(1):103–116

Nishimori H, Nishikawa R, Fujimaki T, Nakagomi T, Matsutani M, Huang HJS, Cavenee WK (2000) Analysis of the p300/CBP-associated factor (PCAF) gene in astrocytic tumors. J Neuro-Oncol 46(1):17–22

Northfelt DW, Marschke Jr RF, Bonnem E, Ooi CE, Gerwien R, Hawthorne T, Clarke A, Buhl-Jensen P, Von Hoff D, Grem J (2007) A phase Ib/II study of PXD101 alone and in combination with 5-fluorouracil in patients with advanced solid tumors. J Clin Oncol 25(18_suppl):3501–3501

Pan YM, Wang CG, Zhu M, Xing R, Cui JT, Li WM, Yu DD, Wang SB, Zhu W, Ye YJ, Wu Y, Wang S, Lu YY (2016) STAT3 signaling drives EZH2 transcriptional activation and mediates poor prognosis in gastric cancer. Mol Cancer 15. https://doi.org/10.1186/s12943-016-0561-z

Panicker J, Li Z, McMahon C, Sizer C, Steadman K, Piekarz R, Bates SE, Thiele CJ (2010) Romidepsin (fk228/depsipeptide) controls growth and induces apoptosis in neuroblastoma tumor cells. Cell Cycle 9(9):1830–1838

Paris M, Porcelloni M, Binaschi M, Fattori D (2008) Histone deacetylase inhibitors: from bench to clinic. J Med Chem 51(6):1505–1529

Park CH, Kim KT (2012) Apoptotic phosphorylation of histone H3 on Ser-10 by protein kinase cδ. PLoS One 7(9):e44307

Park DJ, Chumakov AM, Vuong PT, Chih DY, Gombart AF, Miller WH, Koeffler HP (1999) CCAAT/enhancer binding protein ε is a potential retinoid target gene in acute promyelocytic leukemia treatment. J Clin Investig 103(10):1399–1408

Patel S, Zhang Y, Cassinat B, Zassadowski F, Ferre N, Cuccuini W, Cayuela JM, Fenaux P, Bonnet D, Chomienne C, et al (2012) Successful xenografts of AML3 samples in immunodeficient NOD/shi-SCID IL2Rγ-/- mice. Leukemia 26(11):2432–2435

Paz-Priel I, Friedman A (2011) C/EBPα dysregulation in AML and ALL. Crit Rev Oncog 16(1–2):93–102

Piekarz RL, Robey R, Sandor V, Bakke S, Wilson WH, Dahmoush L, Kingma DM, Turner ML, Altemus R, Bates SE (2001) Inhibitor of histone deacetylation, depsipeptide (fr901228), in the treatment of peripheral and cutaneous T-cell lymphoma: a case report. Blood J Am Soc Hematol 98(9):2865–2868

Pradhan S, Chin HG, Esteve PO, Jacobsen SE (2009) Set7/9 mediated methylation of non-histone proteins in mammalian cells. Epigenetics 4(6):282–285

Qian X, LaRochelle WJ, Ara G, Wu F, Petersen KD, Thougaard A, Sehested M, Lichenstein HS, Jeffers M (2006) Activity of PXD101, a histone deacetylase inhibitor, in preclinical ovarian cancer studies. Mol Cancer Therapeut 5(8):2086–2095

Quivy V, Van Lint C (2004) Regulation at multiple levels of NF-κB-mediated transactivation by protein acetylation. Biochem Pharmacol 68(6):1221–1229

Rajak H, Singh A, Dewangan PK, Patel V, Jain DK, Tiwari SK, Veerasamy R, Sharma PC (2013) Peptide based macrocycles: selective histone deacetylase inhibitors with antiproliferative activity. Curr Med Chem 20(14):1887–1903

Roth SY, Denu JM, Allis CD (2001) Histone acetyltransferases. Ann Rev Biochem 70(1):81–120

Saccani S, Pantano S, Natoli G (2001) Two waves of nuclear factor κB recruitment to target promoters. J Exp Med 193(12):1351–1360

Salih HR, Starling GC, Brandl SF, Pelka-Fleischer R, Haferlach T, Hiddemann W, Kiener PA, Nuessler V (2002) Differentiation of promyelocytic leukaemia: alterations in Fas (CD95/Apo-1) and Fas Ligand (CD178) expression. Br J Haematol 117(1):76–85

Savickiene J, Treigyte G, Magnusson KE, Navakauskiene R (2005) p21 (Waf1/Cip1) and fasl gene activation via Sp1 and NFκb is required for leukemia cell survival but not for cell death induced by diverse stimuli. Int J Biochem Cell Biol 37(4):784–796

Savickiene J, Borutinskaite VV, Treigyte G, Magnusson KE, Navakauskiene R (2006a) The novel histone deacetylase inhibitor BML-210 exerts growth inhibitory, proapoptotic and differentiation stimulating effects on the human leukemia cell lines. Eur J Pharmacol 549(1–3):9–18. https://doi.org/10.1016/j.ejphar.2006.08.010

Savickiene J, Treigyte G, Borutinskaite V, Navakauskiene R, Magnusson KE (2006b) The histone deacetylase inhibitor FK228 distinctly sensitizes the human leukemia cells to retinoic acid-induced differentiation. In: Diederich M (ed) Signal transduction pathways, PT B: stress signaling and transcriptional control. Annals of the New York academy of sciences. Cell signaling world 2006 conference, Luxembourg, Jan 25–28, 2006, vol 1091, pp 368–384. https://doi.org/10.1196/annals.1378.081

Savickiene J, Treigyte G, Magnusson KE, Navakauskiene R (2009) Response of retinoic acid-resistant KG1 cells to combination of retinoic acid with diverse histone deacetylase inhibitors. In: Diederich M (ed) Natural compounds and their role in apoptotic cell signaling pathways.

Annals of the New York academy of sciences, vol 1171, pp 321–333. https://doi.org/10.1111/j.1749-6632.2009.04718.x

Savickiene J, Treigyte G, Vistartaite G, Tunaitis V, Magnusson KE, Navakauskiene R (2011) C/EBPalpha and PU.1 are involved in distinct differentiation responses of acute promyelocytic leukemia HL-60 and NB4 cells via chromatin remodeling. Differentiation 81(1):57–67. https://doi.org/10.1016/j.diff.2010.08.003

Savickiene J, Treigyte G, Borutinskaite VV, Navakauskiene R (2012a) Antileukemic activity of combined epigenetic agents, DNMT inhibitors zebularine and RG108 with HDAC inhibitors, against promyelocytic leukemia HL-60 cells. Cell Mol Biol Lett 17(4):501–525. https://doi.org/10.2478/s11658-012-0024-5

Savickiene J, Treigyte G, Jazdauskaite A, Borutinskaite VV, Navakauskiene R (2012b) DNA methyltransferase inhibitor RG108 and histone deacetylase inhibitors cooperate to enhance NB4 cell differentiation and E-cadherin re-expression by chromatin remodeling. Cell Biol Int 36(11):1067–1078

Savickiene J, Treigyte G, Jonusiene V, Bruzaite R, Borutinskaite VV, Navakauskiene R (2012c) Epigenetic changes by zebularine leading to enhanced differentiation of human promyelocytic leukemia NB4 and KG1 cells. Mol Cell Biochem 359(1–2):245–261. https://doi.org/10.1007/s11010-011-1019-7

Savickiene J, Treigyte G, Stirblyte I, Valiuliene G, Navakauskiene R (2014a) Euchromatic histone methyltransferase 2 inhibitor, BIX-01294, sensitizes human promyelocytic leukemia HL-60 and NB4 cells to growth inhibition and differentiation. Leukemia Res 38(7):822–829

Savickiene J, Treigyte G, Valiuliene G, Stirblyte I, Navakauskiene R (2014b) Epigenetic and molecular mechanisms underlying the antileukemic activity of the histone deacetylase inhibitor belinostat in human acute promyelocytic leukemia cells. Anti-Cancer Drugs 25(8):938–949. https://doi.org/10.1097/CAD.0000000000000122

Scott SA, Dong WF, Ichinohasama R, Hirsch C, Sheridan D, Sanche SE, Geyer CR, DeCoteau JF (2006) 5-Aza-2′-deoxycytidine (decitabine) can relieve p21WAF1 repression in human acute myeloid leukemia by a mechanism involving release of histone deacetylase 1 (HDAC1) without requiring p21WAF1 promoter demethylation. Leukemia Res 30(1):69–76

Scott SA, Lakshimikuttysamma A, Sheridan DP, Sanche SE, Geyer CR, DeCoteau JF (2007) Zebularine inhibits human acute myeloid leukemia cell growth in vitro in association with p15ink4b demethylation and reexpression. Exp Hematol 35(2):263–273

Sekhavat A, Sun JM, Davie JR (2007) Competitive inhibition of histone deacetylase activity by trichostatin A and butyrate. Biochem Cell Biol 85(6):751–758

Shaker S, Bernstein M, Momparler LF, Momparler RL (2003) Preclinical evaluation of antineoplastic activity of inhibitors of DNA methylation (5-aza-2′-deoxycytidine) and histone deacetylation (trichostatin A, depsipeptide) in combination against myeloid leukemic cells. Leukemia Res 27(5):437–444

Sheikh NK, Dua A (2019) Procaine. In: StatPearls [Internet]. StatPearls Publishing, Treasure Island

Singh V, Sharma P, Capalash N (2013) DNA methyltransferase-1 inhibitors as epigenetic therapy for cancer. Current Cancer Drug Targets 13(4):379–399

Strahl BD, Allis CD (2000) The language of covalent histone modifications. Nature 403(6765):41–45

Stresemann C, Brueckner B, Musch T, Stopper H, Lyko F (2006) Functional diversity of DNA methyltransferase inhibitors in human cancer cell lines. Cancer Res 66(5):2794–2800

Suh HC, Gooya J, Renn K, Friedman AD, Johnson PF, Keller JR (2006) C/EBPα determines hematopoietic cell fate in multipotential progenitor cells by inhibiting erythroid differentiation and inducing myeloid differentiation. Blood 107(11):4308–4316

Sunami Y, Araki M, Kan S, Ito A, Hironaka Y, Imai M, Morishita S, Ohsaka A, Komatsu N (2017) Histone acetyltransferase p300/creb-binding protein-associated factor (PCAF) is required for all-trans-retinoic acid-induced granulocytic differentiation in leukemia cells. J Biol Chem 292(7):2815–2829

Takebayashi Si, Tamura T, Matsuoka C, Okano M (2007) Major and essential role for the DNA methylation mark in mouse embryogenesis and stable association of DNMT1 with newly replicated regions. Mol Cell Biol 27(23):8243–8258

Testi AM, Biondi A, Coco FL, Moleti ML, Giona F, Vignetti M, Menna G, Locatelli F, Pession A, Barisone E, et al. (2005) GIMEMA-AIEOPAIDA protocol for the treatment of newly diagnosed acute promyelocytic leukemia (APL) in children. Blood 106(2):447–453

Truong BTH, Lee YJ, Lodie TA, Park DJ, Perrotti D, Watanabe N, Koeffler HP, Nakajima H, Tenen DG, Kogan SC (2003) CCAAT/Enhancer binding proteins repress the leukemic phenotype of acute myeloid leukemia. Blood J Am Soc Hematol 101(3):1141–1148

Tsai HC, Baylin SB (2011) Cancer epigenetics: linking basic biology to clinical medicine. Cell Res 21(3):502–517

Valiulienė G, Treigytė G, Savickienė J, Matuzevičius D, Alksnė M, Jarašienė-Burinskaja R, Bukelskienė V, Navakauskas D, Navakauskienė R (2016) Histone modifications patterns in tissues and tumours from acute promyelocytic leukemia xenograft model in response to combined epigenetic therapy. Biomed Pharmacother 79:62–70. https://doi.org/10.1016/j.biopha.2016.01.044

Valiuliene G, Stirblyte I, Cicenaite D, Kaupinis A, Valius M, Navakauskiene R (2015) Belinostat, a potent HDACi, exerts antileukaemic effect in human acute promyelocytic leukaemia cells via chromatin remodelling. J Cell Mol Med 19(7):1742–1755. https://doi.org/10.1111/jcmm.12550

Valiuliene G, Stirblyte I, Jasnauskaite M, Borutinskaite V, Navakauskiene R (2017) Anti-leukemic effects of HDACi belinostat and HMTi 3-Deazaneplanocin A on human acute promyelocytic leukemia cells. Eur J Pharmacol 799:143–153. https://doi.org/10.1016/j.ejphar.2017.02.014

Velicescu M, Weisenberger DJ, Gonzales FA, Tsai YC, Nguyen CT, Jones PA (2002) Cell division is required for de novo methylation of CpG islands in bladder cancer cells. Cancer Res 62(8):2378–2384

Vernier M, Bourdeau V, Gaumont-Leclerc MF, Moiseeva O, Bégin V, Saad F, Mes-Masson AM, Ferbeyre G (2011) Regulation of E2Fs and senescence by PML nuclear bodies. Genes Develop 25(1):41–50

Villar-Garea A, Fraga MF, Espada J, Esteller M (2003) Procaine is a DNA-demethylating agent with growth-inhibitory effects in human cancer cells. Cancer Res 63(16):4984–4989

Vitkeviciene A, Baksiene S, Borutinskaite V, Navakauskiene R (2018) Epigallocatechin-3-gallate and BIX-01294 have different impact on epigenetics and senescence modulation in acute and chronic myeloid leukemia cells. Eur J Pharmacol 838:32–40. https://doi.org/10.1016/j.ejphar.2018.09.005

Vitkeviciene A, Skiauteryte G, Zucenka A, Stoskus M, Gineikiene E, Borutinskaite V, Griskevicius L, Navakauskiene R (2019) HDAC and HMT inhibitors in combination with conventional therapy: a novel treatment option for acute promyelocytic leukemia. J Oncol 2019:11. https://doi.org/10.1155/2019/6179573

Vogelstein B, Lane D, Levine AJ (2000) Surfing the p53 network. Nature 408(6810):307–310

Vrana JA, Decker RH, Johnson CR, Wang Z, Jarvis WD, Richon VM, Ehinger M, Fisher PB, Grant S (1999) Induction of apoptosis in U937 human leukemia cells by suberoylanilide hydroxamic acid (SAHA) proceeds through pathways that are regulated by Bcl-2/Bcl-X L, c-Jun, and p21 CIP1, but independent of p53. Oncogene 18(50):7016–7025

Wang X, Huang S, Chen JL (2017) Understanding of leukemic stem cells and their clinical implications. Mol Cancer 16. https://doi.org/10.1186/s12943-016-0574-7

Wang P, Wang Z, Liu J (2020) Role of HDACs in normal and malignant hematopoiesis. Mol Cancer 19(1):5

Watts GS, Oshiro MM, Junk DJ, Wozniak RJ, Watterson SJ, Domann FE, Futscher BW (2004) The acetyltransferase p300/CBP-associated factor is a p53 target gene in breast tumor cells. Neoplasia (New York, NY) 6(3):187

Wong JC, Hong R, Schreiber SL (2003) Structural biasing elements for in-cell histone deacetylase paralog selectivity. J Am Chem Soc 125(19):5586–5587

Wu D, Shao K, Zhou Q, Sun J, Wang Z, Yan F, Liu T, Wu X, Ye B, Huang H, et al (2018) Autophagy and ubiquitin-mediated proteolytic degradation of PML/Rarα fusion protein in matrine-induced differentiation sensitivity recovery of ATRA-Resistant APL (NB4-LR1) cells: in vitro and in vivo studies. Cell Physiol Biochem 48(6):2286–2301

Xie YL, Zhang B, Jing L (2018) Mir-125b blocks bax/cytochrome c/caspase-3 apoptotic signaling pathway in rat models of cerebral ischemia-reperfusion injury by targeting p53. Neurol Res 40(10):828–837

Yam CH, Fung TK, Poon RYC (2002) Cyclin A in cell cycle control and cancer. Cell Mol Life Sci 59(8):1317–1326

Yoo CB, Chuang JC, Byun HM, Egger G, Yang AS, Dubeau L, Long T, Laird PW, Marquez VE, Jones PA (2008) Long-term epigenetic therapy with oral zebularine has minimal side effects and prevents intestinal tumors in mice. Cancer Prev Res 1(4):233–240

Yoon JH, Kim HJ, Kwak DH, Park SS, Jeon YW, Lee SE, Cho BS, Eom KS, Kim YJ, Lee S, et al (2017) High WT1 expression is an early predictor for relapse in patients with acute promyelocytic leukemia in first remission with negative PML-RARα after anthracycline-based chemotherapy: a single-center cohort study. J Hemat Oncol 10(1):30

Zhang J, He P, Xi Y, Geng M, Chen Y, Ding J (2015) Down-regulation of G9a triggers DNA damage response and inhibits colorectal cancer cells proliferation. Oncotarget 6(5):2917

Zhang L, Yan L, Zhang Y, Wu Nh, Shen Yf (2010) Role of acetylated p53 in regulating the expression of map2 in retinoic acid-induced P19 cells. Chin Med Sci J 25(2):71–75

Zhang X, Min KW, Wimalasena J, Baek SJ (2012) Cyclin D1 degradation and p21 induction contribute to growth inhibition of colorectal cancer cells induced by epigallocatechin-3-gallate. J Cancer Res Clin Oncol 138(12):2051–2060

Zhou L, Cheng X, Connolly BA, Dickman MJ, Hurd PJ, Hornby DP (2002) Zebularine: a novel DNA methylation inhibitor that forms a covalent complex with DNA methyltransferases. J Mol Biol 321(4):591–599

Part II
Proteomics in Leukemia

Chapter 4
Proteome in Leukemic vs. Differentiated Leukemia Cells

During cells response to external stimulus, signals are transmitted to the nucleus from the surface of the the cell during the processes that regulate the signalling events, during which the cascade of protekinases and proteinphosphatases are initiated and other regulatory proteins are activated (Hunter 1995; Karin et al. 1997). Long-term effects may alter the activity of receptors in the proliferating cells and effect on the differentiation or apoptosis pathways. Leukemic cell lines are applied as an *in vitro* model in leukemia, enabling the molecular mechanisms of cellular proliferation or differentiation processes to be investigated. Differentiation inducers (DMSO, ATRA, dbcAMP, etc.) can induce NB4 and HL-60 to granulocytic differentiation. The forbolmyristat acetate (PMA), vitamin D3 and sodium butyrate can induce differentiation to monocytes/marophages lineage of NB4, HL-60, and THP-1 cells (Breitman et al. 1980; Collins et al. 1977; Collins 1987). During differentiation, cell growth is stopped and specific proteins that are involved in phenotype formation are activated.

The cell differentiation process into granulocytes can be devided into several stages: the first—initiation of differentiation, the second—differentiation into granulocytes or monocytes depending on the used inducer, and the third- maturation, when the cell that has lost its ability to multiply, begins to produce the specific proteins. The inducer of granulocytic differentiation all-trans retinoic acid (ATRA) activates genes by binding to the ATRA receptors in the nuclei of cells through the all-trans retinoic acid response elements (RARE). Some of these genes' products may be involved, directly or indirectly, in the process of differentiation induction (Sporn et al. 1994; Pawson 1995), others can be expressed in the cytoplasm or the nuclei of differentiated. Matured granulocytes induced to differentiation by

R. Navakauskienė et al., *Epigenetics and Proteomics of Leukemia*,
https://doi.org/10.1007/978-3-030-68708-3_4

ATRA, similar to neutrophils, underwent the programmed cell death (apoptosis). The signalling pathways and molecular mechanisms that are involved in the granulocytic differentiation of leukemic cells that leads to apoptosis remain unclear. The mechanisms of cellular proliferation, differentiation, and apoptosis involve various protein-modifying enzymes, including protein phosphatases and protein kinases that are responsible for phosphorylation or dephosphorylation of proteins (on threonine, serine, or tyrosine residues). The phosphorylation of tyrosine residues is particularly specific at the stage of transmission of the external signal to the DNA, so it is important not only to identify newly synthesized proteins, but also to determine where and when they are modified (e.g., thyrosine phosphorylated).

Global functional proteomic analysis helps to understand the molecular mechanisms of diseases, providing new possibilities for detecting specific regulatory proteins and new targets for the rational treatment of leukemias and cancer.

Proteomic changes in leukemia cells, promoted to granulocytic differentiation by only ATRA or together with epigenetic modifiers were investigated our study (Navakauskiene et al. 2002, 2003a,b, 2004a,b, 2012, 2014; Kulyte et al. 2001, 2002; Borutinskaite et al. 2011, 2005; Borutinskaite and Navakauskiene 2015; Treigyte et al. 2000b,a, 2004; Valiuliene et al. 2015).

Methods for computerized analysis of proteome maps have been developed (Matuzevicius et al. 2008). Such proteomic analysis serves to elucidate the network of protein interactions and proteins responsible for myeloid cell development and leukemia, as well as to discover new targets for rational cancer therapy.

4.1 Proteomic Analysis of Cytoplasmic and Nuclear Proteins in Human Hematopoietic CD34+, AML Cell Line KG1 and Mature Neutrophils

Cytoplasmic and nuclear proteins we isolated from cells of different state of hematopoietic differentiation: primary hematopoietic CD34+, cancerous cells arrested at the stage of incomplete differentiation KG1 (AML cell line) and healthy mature neutrophils. The isolated cytoplasmic and nuclear proteins were fractionated in the 2DE system (Figs. 4.1 and 4.2).

After protein visualization a comparative analysis of the protein maps was performed and proteins that could serve as potential markers of leukemia were selected for mass spectrometry analysis. Proteins were selected based on changes in their expression in different states of cell differentiation. The summarized analysis data are presented in Table 4.1. The computational methods for protein expression analysis in different hematopoietic cells were applied and fold change between protein expression calculated. The network of identified cytoplasmic proteins distinctive for human hematopoietic CD34+, incompletely differentiated KG1 cells, and mature human neutrophils is presented in Fig. 4.3.

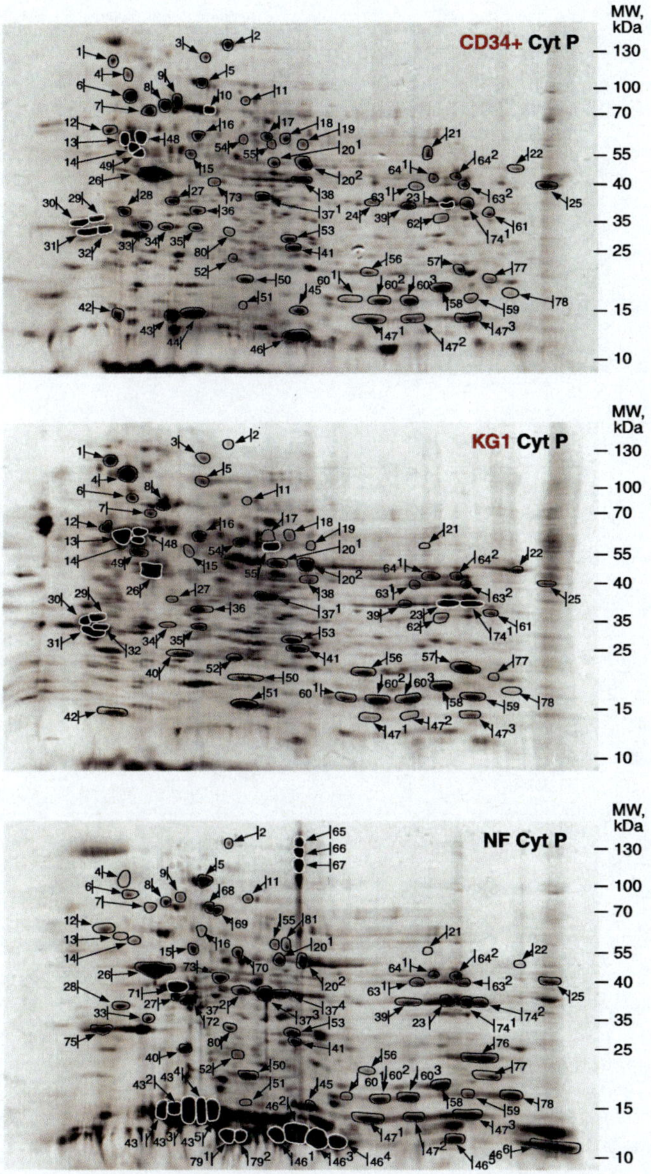

Fig. 4.1 Proteomic analysis of cytoplasmic proteins isolated from human hematopoietic CD34+, incompletely differentiated KG1 cells, and mature human neutrophils (NF). The cytoplasmic proteins (Cyt P) are isolated and fractionated in the 2DE system. After visualization of the proteins by staining with Coomassie blue, protein comparative analysis was performed to select proteins with different expression in different cell differentiation states. These proteins were prepared for mass spectrometry analysis, identified and quantified by computational analysis (Table 4.1)

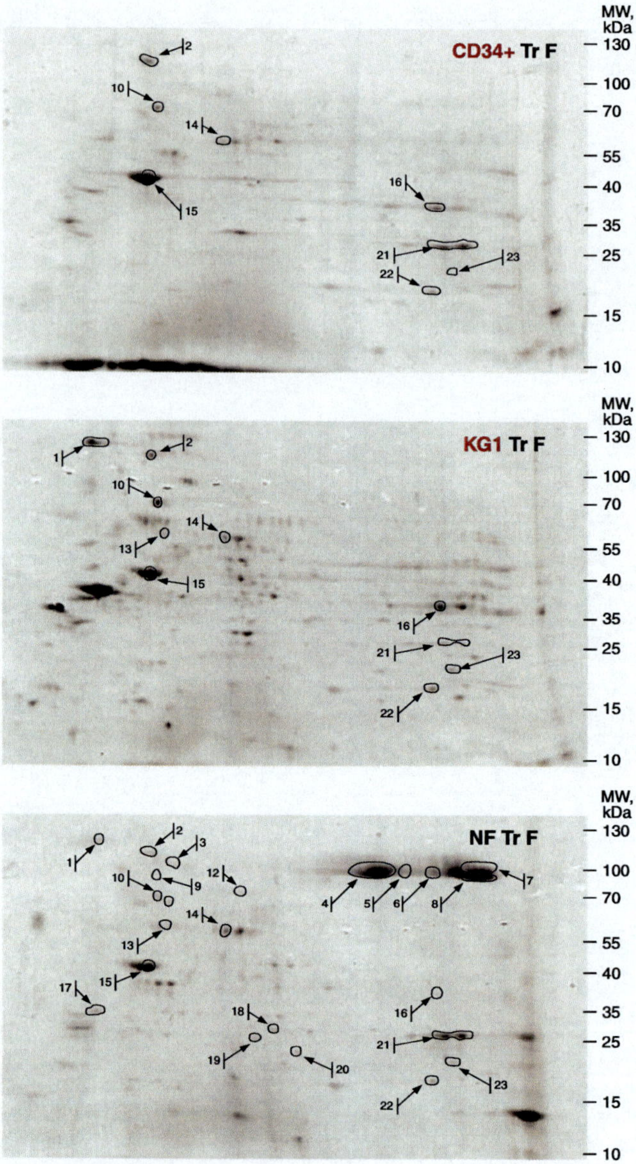

Fig. 4.2 Proteome analysis of nuclear proteins isolated from human hematopoietic CD34+, incompletely differentiated AML cell line KG1, and mature human neutrophils (NF). Nuclear proteins (Tr F) were isolated and fractionated in the 2DE system. After visualization of the proteins by staining with Coomassie blue, protein comparative analysis was performed to select proteins with different expression in different hematopoietic cell differentiation states. These proteins were analyzed by mass spectrometry, identified and quantified by computational analysis (Table 4.2)

Table 4.1 The summarized search results (by `PepIdent`, EMBL, `MS-Tag` software) of comparative expression analysis of cytoplasmic proteins from KG1, Neutrophils (NF) and CD34+

				Experimental		Modeling		Avg. vol. ratio		
Spot No.	Access number	Protein name (abbreviation)	Match, %	Mw, kDa	pI	Mw, kDa	pI	KG1/ CD34+	NF/ CD34+	NF/ KG1
1	Q9Y3P9	Rab GTPase-activating protein 1	13	120	4.85	121.7	5.1	3.1	-	-
	P18206	Vinkulin	23	134	5.75	123.3	5.5	−3.7	−3.0	1.2
2	Q14289	Protein-tyrosine kinase 2-beta	13	134	5.75	115.8	5.9	−3.7	−3.0	1.2
3	Q9UPQ0	LIM and calponin homology domains-containing protein 1	27	120	5.55	121.1	6.1	1.7	-	-
	O94779	Contactin-5	17	120	5.55	120.6	6.0	1.7	-	-
	Q9HBR0	Putative sodium-coupled neutral amino acid transporter 10	14	120	5.55	119.7	5.51	1.7	-	-
4	Q562E7	WD repeat-containing protein 81	6	110	5.0	95.6	5.0	7.4	1.4	−5.3
5	Q9Y6U3	Adseverin	8	108	5.6	80.4	5.5	−1.1	2.0	2.2

(continued)

Table 4.1 (continued)

Spot No.	Access number	Protein name (abbreviation)	Match, %	Experimental		Modeling		Avg. vol. ratio		
				Mw, kDa	pI	Mw, kDa	pI	KG1/ CD34+	NF/ CD34+	NF/ KG1
6	P11021	78 kDa glucose-regulated protein (GRP-78)	51	90	5.0	72.4	5.07	−2.9	−3.4	−1.2
7	P13796	Plastin-2	24	70	5.3	70.2	5.2	−2.7	−3.2	−1.2
8	P11142	HSP7C	18	80	5.35	70.8	5.37	−1.4	−2.4	−1.7
9	P08133	Annexin A6 (ANX A6)	35	85	5.4	76.1	5.42	-	−2.5	-
	P38646	Stress-70 protein	16	85	5.4	73	5.9	-	−2.5	-
10	P02768	Serum albumin (ALBU)	27	80	5.55	69.3	5.9	–	-	-
11	Q9UKJ8	Disintegrin and metalloproteinase domain-containing protein 21	14	85	5.8	80.8	6.2	−1.0	1.2	1.2
12	Q96I76	G patch domain-containing protein 3	23	63	4.85	59.3	4.9	1.6	1.3	−1.2
13	Q9H7C4	Syncoilin	14	60	4.9	55.5	4.6	2.0	−4.6	−9.1
14	P68366	Tubulin alpha-4A chain	18	55	4.95	50	4.95	−1.1	−3.2	−2.8
15	A6NHL2	Tubulin alpha chain-like 3	14	55	5.47	49.9	5.7	−1.2	1.4	1.7
16	P30101	Protein disulfide-isomerase A3	14	63	5.5	56.7	6.0	1.7	−2.3	−4.0

17	P14868	Aspartyl-tRNA synthetase (SYDC)	45	62	6.0	57.4	6.11	2.6	-	-
18	P14868	Aspartyl-tRNA synthetase (SYDC)	30	60	6.2	57.4	6.11	−1.1	-	-
19	P14868	Aspartyl-tRNA synthetase (SYDC)	31	57	6.99	57.4	6.11	−1.4	-	-
20[1]	P06733	Alpha-enolase	25	47	6.99	47.1	7.0	2.5	1.7	−1.5
20[2]	P06733	Alpha-enolase	25	47	6.99	47.1	7.0	1.3	1.5	1.1
21	Q15646	59 kDa 2'-5'-oligoadenylate synthase-like protein	17	55	8.3	59.0	8.0	−3.9	−5.9	−1.5
22	Q9H2G9	Golgin-45	38	45	9.0	45.2	8.74	1.3	−2.7	−3.4
23[1]	P04406	Glyceraldehyde-3-phosphate dehydrogenase (GAPDH)	16	37	8.58	36.0	8.6	1.5	1.3	− 1.2
23[2]	P04406	Glyceraldehyde-3-phosphate dehydrogenase (GAPDH)	12	39	8.58	36.0	8.6			
24	P07355	Annexin A2	48	38	7.6	38.8	7.37	-	-	-

(continued)

Table 4.1 (continued)

Spot No.	Access number	Protein name (abbreviation)	Match, %	Experimental Mw, kDa	Experimental pI	Modeling Mw, kDa	Modeling pI	Avg. vol. ratio KG1/CD34+	Avg. vol. ratio NF/CD34+	Avg. vol. ratio NF/KG1
25	Q9UBV7	Beta 1-4-galactosyltransferase 7	20	40	10.0	37.4	9.2	−1.8	−1.4	1.3
26	Q96HG5	Actin, cytoplasmic 1, Actin, beta (ACTB)	29	43	5.3	42.0	5.29	−1.8	−1.1	1.7
27	P27216	Annexin A13 (ANX 13)	65	38.5	5.4	35.6	5.47	−2.0	1.1	2.1
28	P08758	Annexin A5 (ANX 5)	62	36.5	4.9	35.9	4.94	-	-1.1	-
29	P06753	Tropomyosin alpha-3 chain (TPM3)	55	35	4.6	32.8	4.68	1.6	-	-
30	P62258	14-3-3 protein epsilon	16	35.5	4.75	29.1	4.6	−1.0	-	-
31	P63103	14-3-3 protein zeta/delta	40	30	4.65	27.7	4.7	1.2	-	-
32	P31946	14-3-3 protein beta/alpha	36	32	4.8	28.0	4.8	−1.1	-	-
33	Q15024	Exosome complex exonuclease (RRP42)	24	35	5.1	31.8	5.1	-	−1.9	-
34	Q9UL46	Proteasome activator complex subunit 2	28	35	5.4	27.3	5.4	−2.3	-	-

35	Q06323	Proteasome activator complex subunit 1	30	35	5.47	28.7	5.8	1.8	-	-
36	P30519	Heme oxygenase	28	36.5	5.47	36.0	5.3	1.6	-	-
37[1]	P04083	Annexin A1	59	39	6.0	38.0	6.6	1.7	2.3	1.4
37[2]	P09525	Annexin A4 (ANXA4)	83	38	5.65	36.0	5.8	-	-	-
	P04083	Annexin A1	18	38	5.65	36.0	5.8	-	-	-
37[3]	P09525	Annexin A4 (ANXA4)	60	38	6.0	36.0	5.8			
	P04083	Annexin A1	34	38	6.0	36.0	5.8			
37[4]	P09525	Annexin A4 (ANXA4)	60	38	6.15	36.0	5.8	-	-	-
	P04083	Annexin A1	43	28	6.15	36.0	5.8	-	-	
38	P63261	γ-actin, Actin, cytoplasmic 2 (ACTG)	44	41	6.99	41.7	5.29	-1.4	-	-
39	P04406	Glyceraldehyde-3-phosphate dehydrogenase (GAPDH)	11	36	8.0	36.0	8.57	-1.1	1.6	1.7
40	P09211	Glutathione S-transferase P	45	25	5.45	23.3	5.4	-	-	-2.3
41	P60174	Triosephosphate isomerase	36	26	6.3	26.6	6.4	1.5	-1.1	-1.6
42	Q5TAB7	Protein ripply2	41	15	4.8	14.0	4.4	1.7	-	-
43[1]	P06702	Protein S100-A9	86	11	5.4	13.2	5.7	-	-	-
43[2]	P06702	Protein S100-A9	76	11	5.4	13.2	5.7	-	3.2	

(continued)

Table 4.1 (continued)

Spot No.	Access number	Protein name (abbreviation)	Match, %	Experimental Mw, kDa	Experimental pI	Modeling Mw, kDa	Modeling pI	Avg. vol. ratio KG1/CD34+	Avg. vol. ratio NF/CD34+	Avg. vol. ratio NF/KG1
43^3	P06702	Protein S100-A9	92	11	5.45	13.2	5.7	-	-	-
43^4	P06702	Protein S100-A9	72	11	5.55	13.2	5.7	-	-	-
43^5	P06702	Protein S100-A9	86	11	5.6	13.2	5.7	-	-	-
44	Q86SG5	Protein S100-A7A (S1A7A)	62	15	5.5	11.3	6.9	-	-	-
45	Q05315	Eosinophil lysophospholi-pase	36	15	6.3	16.4	6.8	-	-1.5	-
46^1	PP05109	Protein S100-A8	52	12	6.85	10.8	6.51	-	-	-
46^2	PP05109	Protein S100-A8	60	12	6.95	10.8	6.51	-	1.8	-
46^3	PP05109	Protein S100-A8	67	12	7.2	10.8	6.51	-	-	-
46^4	PP05109	Protein S100-A8	52	12	7.3	10.8	6.51	-	-	-
46^5	PP05109	Protein S100-A8	52	12	8.58	10.8	6.51	-	-	-
46^6	PP05109	Protein S100-A8	52	12	10.0	10.8	6.51	-	-	-
47^1	P07737	Profilin-1	37	14	7.6	15.0	8.4	-2.5	1.4	3.5
47^2	P07737	Profilin-1	41	14	8.0	15.0	8.4	-2.2	1.2	2.6
47^3	P07737	Profilin-1	30	14	8.6	15.0	8.4	-2.3	1.3	2.9
48	P68366	Tubulin alpha-4A chain	25	63	5.1	50.6	4.95	-4.4	-	-
49	Q15084	Protein disulfide-isomerase A6 (PDIAG)	26	50	5.1	48.4	4.95	-1.0	-	-

(continued)

50	P24666	Low molecular weight phosphotyrosine protein phosphatase, LMW-PTPase (PPAC)	68	19	5.8	18.4	6.30	2.1	2.1	1.0
51	Q01469	Fatty acid-binding protein, epidermal (FABP5)	73	14	5.85	15.4	6.6	10.1	1.0	-9.7
52	P61019	Ras-related protein Rab-2A (RAB2A)	45	22.5	5.75	23.7	6.08	2.1	1.6	-1.3
53	P18669	Phosphoglycerate mutase 1 (PGAMI)	49	26	6.2	28.9	6.67	1.5	1.6	1.0
54	P14868	Aspartyl-tRNA synthetase, cytoplasmic (SYDC)	28	58	5.8	57.4	6.11	3.2	-	-
55	P54577	Tyrosyl-tRNR synthetase, cytoplasmic (SYYC)	19	57	6.1	59.4	6.61	3.0	-1.2	-3.6
56	P30086	Phosphatedyethanolamine – b.p. 1 (PEBP-1)	33	21	7.4	21.0	7.00	2.2	-1.3	-2.9
57	Q06830	Peroxiredoxin-1	62	21	8.7	22.3	8.77	4.1	-	-
58	P23528	Coffilin 1	32	17	8.5	18.7	8.22	-1.1	-1.1	-1.0

(continued)

Table 4.1 (continued)

Spot No.	Access number	Protein name (abbreviation)	Match, %	Experimental		Modeling		Avg. vol. ratio		
				Mw, kDa	pI	Mw, kDa	pI	KG1/ CD34+	NF/ CD34+	NF/ KG1
59	O60361	Putative nucleoside diphosphate kinase (NDK8)	27	16	8.8	15.6	8.76	3.2	1.3	-2.5
60[1]	P62937	Peptidyl-prolyl cis-trans isomerase A (PPIA)	55	15	7.4	18.2	7.68	1.5	1.3	-1.2
60[2]	P62937	Peptidyl-prolyl cis-trans isomerase A (PPIA)	50	15	7.6	18.2	7.68	1.7	1.1	-1.6
60[3]	P62937	Peptidyl-prolyl cis-trans isomerase A (PPIA)	52	15	8.0	18.2	7.68	1.5	-1.5	-2.2
61	P40926	Malate dehydrogenase	25	35.5	8.9	35.0	8.9	-1.1	-	-
62	Q13155	Aminoacyl tRNA synthase complex-interacting multifunctional protein 2	16	35	8.4	35.3	8.4	1.2	-	-

63[1]	P04075	Fructose-bisphosphate aldolase A (ALDOA)	33	40	8.0	39.4	8.3	1.6	1.2	-1.3
63[2]	P04075	Fructose-bisphosphate aldolase A (ALDOA)	31	40	8.5	39.8	8.3	1.4	1.8	1.4
64[1]	Q15120	Pyruvate dehydrogenase kinase (PDH)	32	42	8.1	46.8	8.5	2.1	2.1	1.0
64[2]	Q15120	Pyruvate dehydrogenase kinase (PDH)	15	42	8.45	47.0	8.46	3.9	1.2	-3.2
65	Q86UK5	Limbin	71	140	6.95	148.8	6.5	-	-	-
66	Q9UHW9	Solute carrier family 12 member 6	12	130	6.95	128.6	6.64	-	-	-
67	P52333	Tyrosine-protein kinase JAK3	9	115	6.95	126.8	6.77	-	-	-
68	Q96T33	Merlin (MERL)	12	75	5.6	72.5	6.11	-	-	-
69	Q96T33	Merlin (MERL)	14	70	5.65	69.8	6.11	-	-	-
70	P50395	Rab GDP dissociation inhibitor beta	28	55	5.6	51.0	6.11	-	-	-

(continued)

Table 4.1 (continued)

Spot No.	Access number	Protein name (abbreviation)	Match, %	Experimental		Modeling		Avg. vol. ratio		
				Mw, kDa	pI	Mw, kDa	pI	KG1/ CD34+	NF/ CD34+	NF/ KG1
71	Q5T288	Annexin A8-like pro- tein 1 (AXA81)	38	39	5.4	37.0	5.45	-	-	-
72	P12429	Annexin A3 (ANXA3)	57	37	5.45	36	5.6	-	-	-
73	P40121	Macrophage-capping protein (CAPG)	34	40	5.6	38.7	5.8	-	2.5	-
74[1]	P04406	Glyceraldehyde-3-phosphate dehydroge-nase (GAPDH)	29	36	8.58	36.2	8.5	1.8	1.3	-1.3
74[2]	P04406	Glyceraldehyde-3-phosphate dehydroge-nase (GAPDH)	29	36	8.7	36.2	8.5	-	-	-
75	Q04917	14-3-3 protein eta (1433F)	32	35.5	4.75	28	4.76	-	-	-
76	P80188	Neutrophil gelatinase-associated lipocalin (NGAL)	44	23	9.0	22.7	9.01	-	-	-

77	O60225	Protein SSX5 (SSX5)	44	20	9.3	21.6	9.35	1.8	3.0	5.4
78	P49913	Cathelicidin antimicrobial peptide (CAMP)	33	18	9.5	19.5	9.48	-1.0	4.5	4.5
79[1]	P80511	Protein S100-A12 (Neutrophil S100 protein)	52	13	5.6	10.5	5.83	-	-	-
79[2]	P80511	Protein S100-A12 (Neutrophil S100 protein)	48	11.5	5.65	10.5	5.83	-	-	-
80	Q96BR5	Hcp beta-lactamase-like protein C1orf163	40	26	6.1	26.4	5.66	-	1.7	-
81	P11413	Glucose-6-phosphate 1-dehydrogenase (G6PD)	17	60	5.65	59.0	6.39	-	-	-

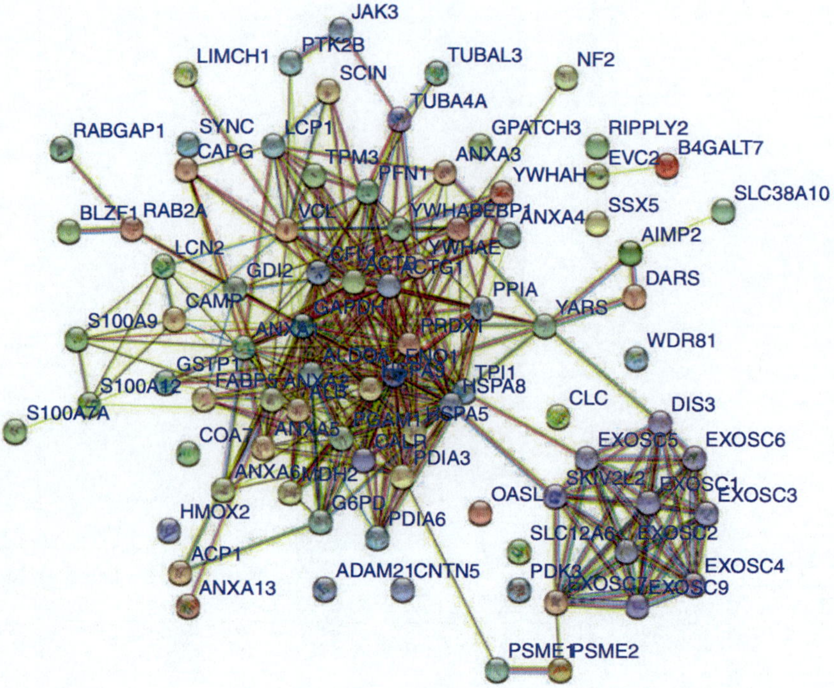

Fig. 4.3 Proteomic analysis of cytoplasmic proteins isolated from human hematopoietic CD34+, incompletely differentiated KG1 cells, and mature human neutrophils. The network of identified cytoplasmic proteins distinctive for human hematopoietic CD34+, incompletely differentiated KG1 cells, and mature human neutrophils

It is noticeable that in the cytoplasm (Figs. 4.1 and 4.3) cancerous, non-differentiated KG1 cells contain increased levels of the following proteins: Rab GTPase-activating protein 1, WD repeat-containing protein 81, Syncoilin, Aspartyl-tRNA synthetase cytoplasmic (SYDC), proteasome activator complex subunit 1, fatty acid-binding protein epidermal (FABP5), and others. Proteins with markedly reduced expression in cancer cells have also been identified: CAMP (Cathelicidin antimicrobial peptide), annexin A13, 14-3-3 protein, tubulin alfa-4A chain and several others. These proteins are related to the cellular signaling system (Rab-GTPase-activating protein, 14-3-3, etc.), to the maintenance of cell structure (sinkoiline, annexin, tubulin, etc.), are also involved in cell signaling and other cellular functions. Nuclear proteins isolated from different stages of hematopoietic differentiation were fractionated in the 2DE system (Fig. 4.2), followed by protein 2DE map comparison and mass spectrometry analysis of selected proteins. Proteins for analysis were selected on changes in their expression in different states of cell differentiation. The identified nuclear proteins are presented in Table 4.2 and the network of identified nuclear proteins typical for human hematopoietic CD34+, incompletely differentiated AML cell line KG1, and mature human neutrophils was established (Fig. 4.4).

Table 4.2 The summarized search results (by `PepIdent`, `EMBL`, `MS-Tag` software) of comparative expression analysis of identified nuclear proteins isolated from CD34+, KG1 cells, and human neutrophils (NF)

Spot No.	Access number	Protein name (abbreviation)	Match, %	Experimental Mw, kDa	Experimental pI	Modeling Mw, kDa	Modeling pI	Avg. vol. ratio KG1/CD34+	Avg. vol. ratio NF/CD34+	Avg. vol. ratio NF/KG1
1	Q92539	Phosphatidate phos-phatase (LPIN2)	20	125	4.7–4.85	100193	5.18	-	-	−6.7
	P57740	Nup107	4	125	4.7–4.85	106300	5.3	-	-	−6.7
2	Q92598	Heat shock protein 105 kDa	5	110	5.35	96866	5.3	−1.8	−2.3	−1.3
3	Q9BWU0	Kanadaptin	7	105	5.4	88815	5.1	-	-	-
4	P02788	Lactotransferrin	18	100	7.7	78183	8.5	-	-	-
	Q9P0L2	Serine/threonine-protein kinase (MARK1)	6	100	7.7	89003	9.4	-	-	-
5	P02788	Lactotransferrin	3	100	7.9	78183	8.5	-	-	-
6	P02788	Lactotransferrin	18	95	8.2	78183	8.5	-	-	-
	Q9P0L2	Serine/threonine-protein kinase (MARK1)	7	95	8.2	89003	9.4	-	-	-
7	P02788	Lactotransferrin	7	100	9.3	78183	8.5	-	-	-
	O94929	Actin-binding LIM protein 3	7	100	9.3	77802	8.9	-	-	-
8	P02788	Lactotransferrin	16	95	9.3	78183	8.5	-	-	-
	Q9P0L2	Serine/threonine-protein kinase (MARK1)	6	95	9.3	89003	9.4	-	-	-
9	Q04864	Proto-oncogene c-Rel	13	75	5.4	68520	5.6	-	-	-
10	P22460	Potassium voltage-gated channel subfamily A member 5	22	68	5.4	67228	5.7	1.3	−2.4	−3.2
11	Q8WX92	Negative elongation factor B	7	65	5.6	65698	5.8	-	-	-
12	Q86VW2	Guanine nucleotide exchange factor (GEFT)	14	70	6.0	63843	6.2	-	-	-

(continued)

Table 4.2 (continued)

Spot No.	Access number	Protein name (abbreviation)	Match, %	Experimental		Modeling		Avg. vol. ratio		
				Mw, kDa	pI	Mw, kDa	pI	KG1/ CD34+	NF/ CD34+	NF/ KG1
13	O95376	Protein ariadne-2 homolog	20	60	5.4	57819	5.4	-	-	−1.1
14	P17014	Zinc finger protein 12	10	58	5.75	58223	7.1	-	1.8	1.8
15	P60709	ACTB	18	43	5.3	42	5.29	−1.4	−2.2	−1.5
16	Q13394	Protein mab-21-like 1	15	35	8.2	40957	8.9	1.0	−6.1	−6.2
	P25105	Platelet-activating factor receptor	28	35	8.2	39204	9.2	1.0	−6.1	−6.2
17	Q9UMX6	Guanylyl cyclase-activating protein 2	22	33	4.8	23420	4.7	-	-	-
18	B5MD39	Putative γ-glutamyltrans-ferase light chain 3	23	29	5.9	24102	5.8	-	-	-
19	Q86V88	Magnesium-dependent phosphatase 1	22	25	5.75	20109	6.0	-	–	-
20	Q5W111	Chronic lymphocytic leukemia deletion region gene 6 protein	10.	23	6.5	21666	6.2	-	-	-
	Q9Y6G5	COMM domain-containing protein 10	12	23	6.5	22967	6.1	-	-	-
	Q9BUE0	Mediator of RNA polymerase II transcription subunit 18	11	23	6.5	23663	6.1	-	-	-
21	P18847	Cyclic AMP-dependent transcription factor ATF-3	18	25	8.7	20576	8.8	−5.8	−1.0	5.7
22	Q9Y421	Protein FAM32A	32	15	8.2	13178	10	1.1	−1.5	−1.6
	P03973	Antileukopro-teinase	16	15	8.2	14326	9.1	1.1	−1.5	−1.6
23	Q52LT0	Putative GTP-binding protein FLJ12595	6	21	8.5	16442	8.7	3.2	2.5	−1.3

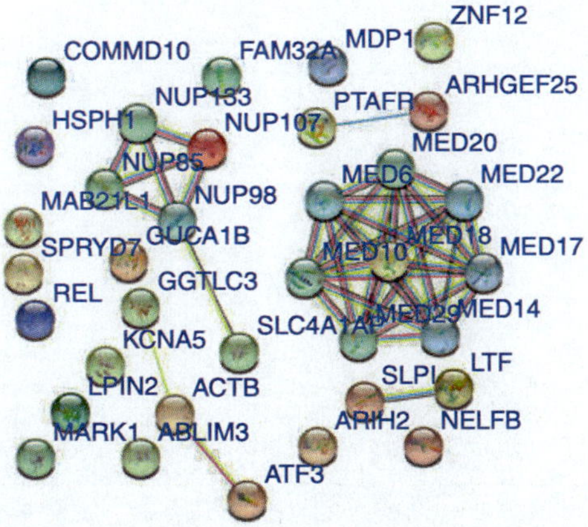

Fig. 4.4 Proteome analysis of nuclear proteins isolated from human hematopoietic CD34+, incompletely differentiated AML cell line KG1, and mature human neutrophils. The network of identified nuclear proteins typical for human hematopoietic CD34+, incompletely differentiated AML cell line KG1, and mature human neutrophils

Several specific proteins have been identified that are specific only to healthy mature neutrophils. Multiple protein expression is enhanced in KG1 cancerous non-differentiated cells: putative GTP-binding protein, Protein mab-21-like 1, etc. For example, the expression of the periodic AMP-dependent transcription factor ATF-3 is reduced in KG1 leukemic cells. Changes in the expression of these proteins could be related to blood cell differentiation.

4.2 Proteomic Analysis of APL Cells Induced to Differentiation

Our initial study (Navakauskiene et al. 2003a,b; Treigyte et al. 2000b,a, 2004; Borutinskaite et al. 2005) was concentrated on identification of proteins that can be involved in cell differentiation and proliferation in acute promyelocytic HL-60 and NB4 cells. The differentiation of leukemic cells was stimulated by using ATRA. The protein profile increase was observed afterwards.

The proteomics in NB4 cells was studied when induction of granulocytic differentiation with following growth inhibition was induced with all-trans retinoic acid and HDAC inhibitor BML-210. The alterations in protein expression at various times of treatment (2, 4, 8, 24, 48, 72, 96 h) were noticed (Fig. 4.5) (Borutinskaite et al. 2005).

Our studies demonstrated potential of combination of HDACi BML-210 with all-trans retinoic acid for differentiation induction in leukemic NB4 cells: the largest amount of differentiated cells after ATRA treatment was observed at 72–96 h (70–80%), and the combined treatment considerably improves granulocytic differentiation of NB4 cells (up to 90%).

Total soluble and less soluble proteins were isolated, fractionated, and identified by mass spectrometry. The less soluble (or insoluble) protein fraction of

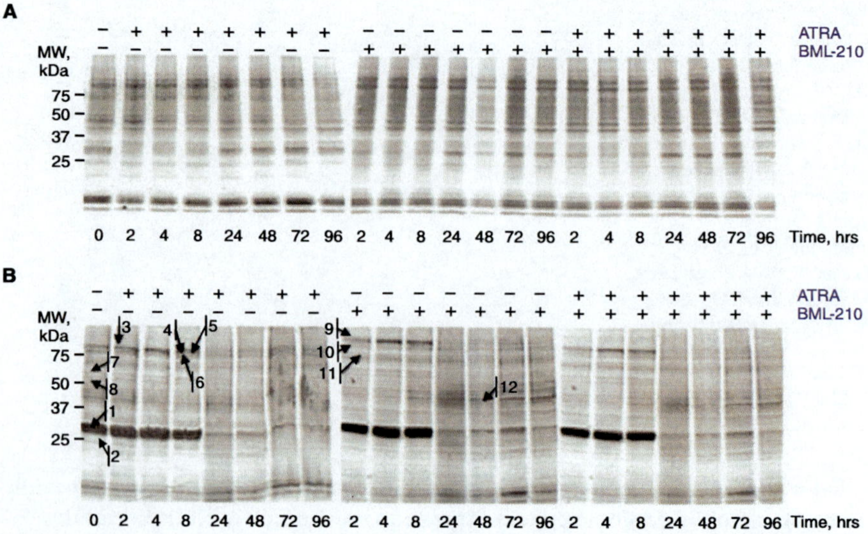

Fig. 4.5 Expression of proteins in NB4 cells treated with all-trans retinoic acid (ATRA) and histone deacetylase inhibitor BML-210. Total soluble (A) and insoluble (B) protein fractions were dissociated from control and treated with 1 μM ATRA and 10 μM BML-210 separately or in combination (1 μM ATRA and 5 μM BML-210) NB4 cells during the indicated period of time. Isolated proteins were fractionated by SDS-PAGE on an 8–16% acrylamide gradient gel and then stained with brilliant Blue G-Colloidal. Migration of the molecular size marker proteins is shown to the left (kDa values). The position of bands that were cut to MALDI-MS analysis is designated by arrows and numbers in the images. According Borutinskaite et al. (2005)

Fig. 4.6 Expression of proteins in NB4 cells treated with all-trans retinoic acid (ATRA) and histone deacetylase inhibitor BML-210. Represents the identified protein network. According Borutinskaite et al. (2005)

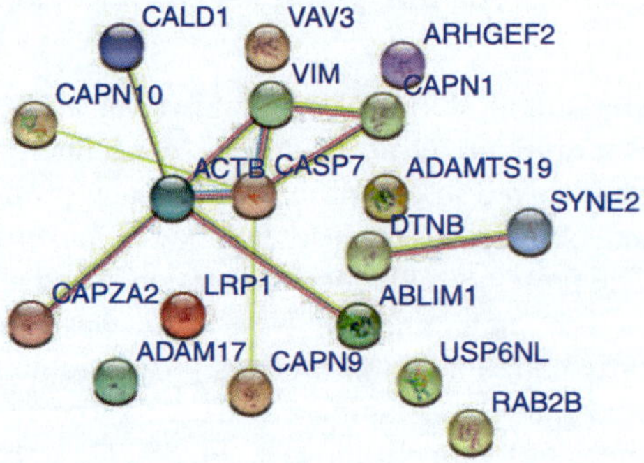

NB4 cells showed biggest differences in expression of proteins during treatments with ATRA and HDAC inhibitor BML-210. Using mass spectrometry, there were inspected proteins of cells with cancer and membrane or membrane-associated proteins. We identified (Figs. 4.6 and 4.7) such proteins like caspase-7, nesprin-2, RAB, USP6NL protein, Vav-3, ADAMTS-19, lipoprotein receptor-related protein, ADAM-17, actin-binding LIM protein 1, caldesmon (CDM) splice isoform 3, caldesmon, calpain 10, dystrobrevin-β, vimentin, ADAMTS-17, GEF-H1, calpain

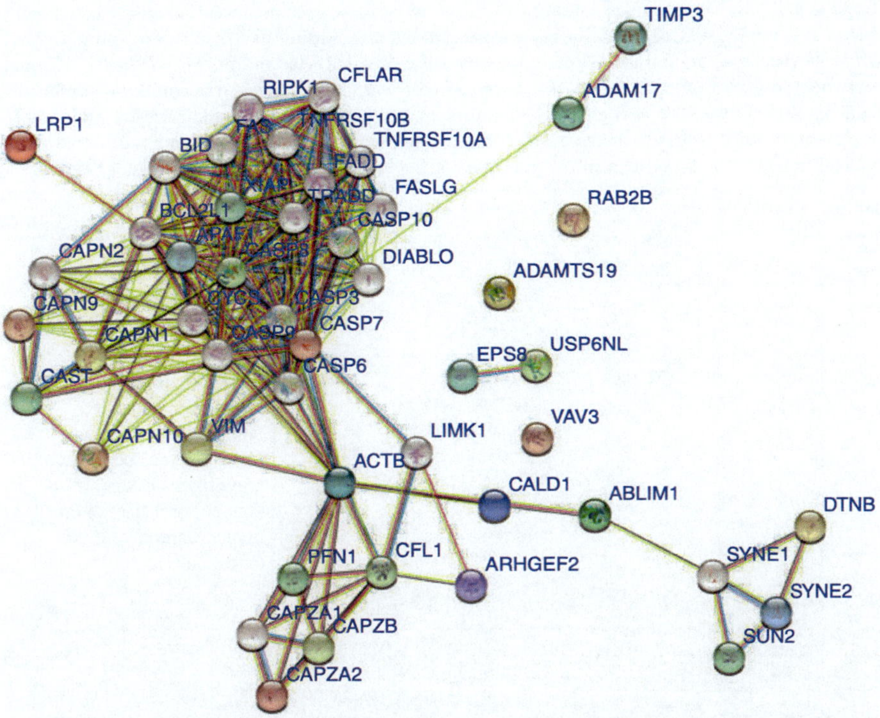

Fig. 4.7 Expression of proteins in NB4 cells treated with all-trans retinoic acid (ATRA) and histone deacetylase inhibitor BML-210. Represents the involvement of identified proteins into large scale of proteins network inclusive different cellular processes like apoptosis and cytoskeleton reorganization. According Borutinskaite et al. (2005)

9, calpain 1, actin, F-actin capping protein alfa-subunit (Borutinskaite et al. 2005). All functions of identified proteins are listed in Table 4.3.

The high expression of small GTPase, Rab2B was detected in insoluble fraction of untreated and treated cells for 8 h, and their decreased level was observed after 24 h of treatment. This indicate that Rab2B expression may be related with granulocytic differentiation or apoptosis of NB4 cells. It was shown that the main function of Rab2B protein is vesicle transport and membrane fusion regulation. Actin (ACTB) and actin-associated proteins (nesprin-2, actin-binding LIM protein, vimentin and caldesmon) were found in insoluble fraction of NB4 cells. These proteins are important for cytoskeletal reorganization during cell growth and differentiation/apoptosis. The higher expression of aforementioned proteins was in soluble fraction compared with insoluble fraction of NB4 cells.

Also we found two proteins (ADAM17 and ARHG2) that were upregulated both in control cells and induced to differentiation with following apoptosis by ATRA and HDAC inhibitor BML-210. It was showed that ADAMs have adhesive and proteolytic characteristics which result in regulation of such processes as

Table 4.3 The characteristics and functions of identified proteins in NB4 cells treated with ATRA and inhibitor BML-210. Proteins were estimated according to `PeptIdent` program and `EXPASY` database, experimental (gel position) and calculated Mw and pI, MOWSE SCORE and sequence coverage (`PeptIdent`) for each protein were represented: 1st column—number of band in gels (Fig. 4.5B); 2nd column—the protein Mw and pI, calculated from the amino acid sequence in the `ExPASy` database; 3rd column—the name of identified proteins; 4th column—general function(s) of a protein in `Uniprot` database. According Borutinskaite et al. (2005)

Band No.	Protein pI	Protein Mw, kDa	Protein name, entry name, entry	Function
1	8.65	28.029	Caspase-7, CASP7_HUMAN, P55210	Involved in the activation cascade of caspases responsible for apoptosis execution.
	6.5	30.269	Nesprin-2, SYNE2_HUMAN, Q8WXH0	As a component of the LINC (LInker of Nucleoskeleton and Cytoskeleton) complex participates in the connection between the nuclear lamina and the cytoskeleton.
2	4.7	26.101	Ras-related protein Rab-2B, RAB2B_HUMAN, Q8WUD1	Needed for protein transport from the endoplasmic reticulum to the Golgi complex.
	12.41	22.305	USP6 N-terminal-like protein, US6NL_HUMAN, Q92738	Acts as a GTPase-activating protein for RAB5A and RAB43. Involved in receptor trafficking.
3	6.65	97.775	Guanine nucleotide exchange factor, VAV3_HUMAN, Q9UKW4	Exchange factor for GTP-binding proteins, may be significant for integrin-mediated signaling.
4	8.13	100.198	A disintegrin and metalloproteinase with thrombospondin motifs 19, ATS19_HUMAN, Q8TE59	Uncharacterized. ADAMTS19 hypermethylation is related with transcriptional downregulation and reduces the *in vitro* migration capabilities of CRC cells[a].

(continued)

Table 4.3 (continued)

Band No.	Protein		Protein name, entry name, entry	Function
	pI	Mw, kDa		
	4.86	99.174	Prolow-density lipoprotein receptor-related protein 1, LRP1_HUMAN, Q07954	Involved in endocytosis and in phagocytosis of apoptotic cells.
5	5.6	78.542	Disintegrin and metalloproteinase domain-containing protein 17, ADA17_HUMAN, P78536	Acts as an activator of Notch, TNF-alpha and other signaling pathways.
	8.88	87.644	Actin-binding LIM protein 1, ABLM1_HUMAN, O14639	May act as scaffold protein. May be significant during the development of the retina. Has been suggested to play a role in axon guidance.
6	6.66	64.256		
	5.6	93.56	Caldesmon, CALD1_HUMAN, Q05682	Interacts with actin, myosin, two molecules of tropomyosin and with calmodulin. Also plays a significant role during cellular mitosis and receptor capping.
7	6.41	57.984	Calpain-10, CAN10_HUMAN, Q9HC96	Calcium-regulated non-lysosomal thiol-protease which catalyzes limited proteolysis of substrates involved in cytoskeletal remodeling and signal transduction.
	8.82	64.424	dystrobrevin-β, DTNB_HUMAN, O60941	Dystrobrevin-β is a component of the dystrophin-associated protein complex (DPC), which is responsible for the lateral transmission of the sarcomere-generated contractile force to the sarcolemma and the ECM.
8	5.06	53.52	Vimentin, VIME_HUMAN, P08670	Vimentins are class-III intermediate filaments, comprises the cytoskeleton.

(continued)

Table 4.3 (continued)

Band No.	Protein		Protein name, entry name, entry	Function
	pI	Mw, kDa		
9	7.77	96.496	Disintegrin and metalloproteinase domain-containing protein 17, ADA17_HUMAN, P78536	Acts as an activator of Notch, TNF-alpha and other signaling pathways.
	8.73	101.173	Rho guanine nucleotide exchange factor 2, ARHG2_HUMAN, Q92974	Activates Rho-GTPases by promoting the exchange of GDP for GTP. May be involved in epithelial barrier permeability, cell motility and polarization, dendritic spine morphology, antigen presentation, leukemic cell differentiation, cell cycle regulation, innate immune response, and cancer.
10	5.37	79.096	Calpain-9, CAN9_HUMAN, O14815	Calcium-regulated non-lysosomal thiol-protease which catalyzes limited proteolysis of substrates involved in cytoskeletal remodeling and signal transduction.
11	5.49	81.89	Calpain-1 catalytic subunit, CAN1_HUMAN, P07384	Calcium-regulated non-lysosomal thiol-protease which catalyzes limited proteolysis of substrates involved in cytoskeletal remodeling and signal transduction.
12	5.29	41.605	Actin. cytoplasmic 1, ACTB_HUMAN, P60709	Actin plays key functions, such as cell motility and contraction, regulate gene transcription and motility and repair of damaged DNA.
	7.61	35.024	F-actin capping protein alfa-subunit, CAZA2_HUMAN, P47755	Involved in actin cytoskeleton organization.

[a]https://clinicalepigeneticsjournal.biomedcentral.com/articles/10.1186/s13148-015-0158-1

cellular fate, proliferation, and growth. ADAM-17 is known as the one of the best investigated ADAM enzymes and is extensively expressed in different tissues including the brain, kidney, heart, and skeletal muscle. ADAM-17 is an important player in inflammation through TNF-alpha pathway, in the development of the nervous system, as it activates the neural cell adhesion, in carcinogenesis due to the enzyme which sheds growth factors required for tumor progression and growth. Another protein in the same band after fractionation in SDS/PAGE is ARHG2 (GEF-H1) that is involved in cellular processes, e.g., cell motility, cell-cycle regulation, polarization, epithelial barrier permeability, and cancer (Kashyap et al. 2019).

The other exciting protein that we identified—DTNB (dystrobrevin-β) protein in insoluble fraction. It was shown that dystrobrevin-β interacts with dystrophin short form DP71 and syntrophins SNTG1 and SNTG2. It was demonstrated that DTNB-knockout mice were healthy and had no abnormality, however the level of DTNB binding proteins were reduced. This leads to the conclussion that DTNB can be an anchor or scaffold for dystrophin and syntrophins/other associating proteins at the basal membranes of kidney and liver.

Also we indentified the two proteins (calpain 1 and calpain 9) that belong to the calcium ion-dependent papain-like protease (Calpain) family. It is known that these proteins are critical mediators of the action of calcium and are tightly regulated by an endogenous inhibitor, calpastatin. It was showed that calpain can regulate the dystrophin, heat shock protein (HSP90), actin, caspases, phospholipase A/B/C, and other protein functions (Patterson et al. 2011).

In conclusion, in this study we have found differently expressed proteins in the soluble and less soluble (insoluble) fraction of the NB4 cells. These changes in protein expression can be associated with early changes in chromatin structure during the process of differentiation or apoptosis after the treatment with HDAC inhibitor BML-210 and ATRA. Identification of new proteins involved in the differentiation process can be helpful in finding new therapies for APL treatment (Borutinskaite et al. 2005).

4.3 Proteome Profile in APL Cells Induced with Histone Deacetylase Inhibitor BML-210

The study was dedicated to identify proteins whose expression changed after NB4 cells treatment with HDAC inhibitor BML-210 (Borutinskaite and Navakauskiene 2015). BML-210 at concentration up to 20 μM was the reason of anti-proliferative and slight cytotoxic effects on NB4 cells in a dose- and time-dependent manner with accumulation of cells in G0/G1 cycle phase. We found out that after 20 μM BML-210 treatment apoptotic population in cell culture was about 90% after 48 h of treatment. Proliferating (control) NB4 cells and cells treated with 20 μM BML-210 for 24 h were lysed and soluble cell proteins were fractionated in 2DE gels and analysed by mass spectrometry (Fig. 4.8). Description of proteins identified by mass spectrometry in NB4 cells treated with HDACi BML-210 and fractionated by 2DE is presented in Appendix D (Borutinskaite and Navakauskiene 2015).

We observed variations in protein expression profile while treating with BML-210. Computational analysis of protein expression changes are presented as fold change in comparisson with control cells (Table 4.4). The expression of proteins whose ratio G2/G1>0 is increased in treated (G2) cells more than in untreated (G1).

In total, 35 proteins were identified: 16 were down-regulated such as endoplasmin, ENPL; heat shock 84 kDa, HSP90B; 14-3-3 protein, 1433F; proliferating cell nuclear antigen, PCNA; calreticulin, CALR; and 19 proteins up-regulated (such as chloride intracellular channel protein 1, CLIC1; thioredoxin domain-containing protein 12, TXD12; lactoylglutathione lyase, LGUL) after treatment with BML-210. It is known that α/β-tubulin, β-actin, cofilin-1, myosin regulatory light chain 12A, tropomyosin, and gelsolin are involved in cell growth and/or homeostasis regulation. Other identified proteins participate in metabolism processes (disulphide isomerase, α-enolase), in protein folding (the heat shock proteins like endoplasmin, HSP90B, GRP75, GRP78 and CH60). Proteins of one more group, such as PCNA, nucleophosmin, 14-3-3 protein, prohibitin, guanine nucleotide-binding protein subunit α-11, chloride intracellular channel protein 1, and nucleoside diphosphate kinase A are in charge of signal transduction, apoptosis, cell differentiation and communication processes (Fig. 4.9).

One of the proteins with detected expression changes is HSP90. It was observed that HDAC inhibitors can provoke hyperacetylation of HSP90 and its inactivation, resulting to the degradation of proteins that need the chaperone function of HSP90 (including some oncoproteins) (Bali et al. 2005). Furthermore it was demonstrated that HSP90 inhibition correlated with growth arrest followed by differentiation and apoptosis (Nawarak et al. 2009; Heller et al. 2009). Calreticulin function was investigated by the other group (Sheng et al. 2014). They have found that high expression of calreticulin was positively related with tumor and metastasis and that calreticulin regulated cell proliferation, migration and invasion of pancreatic cancer cells in a MEK/ERK pathway dependent manner. 14-3-3 proteins have a significant role in a broad spectrum of vital regulatory processes, especially regarding signal transduction, apoptosis, cell cycle progression, and DNA replication. It was shown that treatment of leukemic cells with HDAC inhibitor suberoylanilide hydroxamic acid (SAHA) provoked cofilin phosphorylation, rised of the vimentin and paxillin expression, also lowered the expression of stathmin (Grebenova et al. 2012).

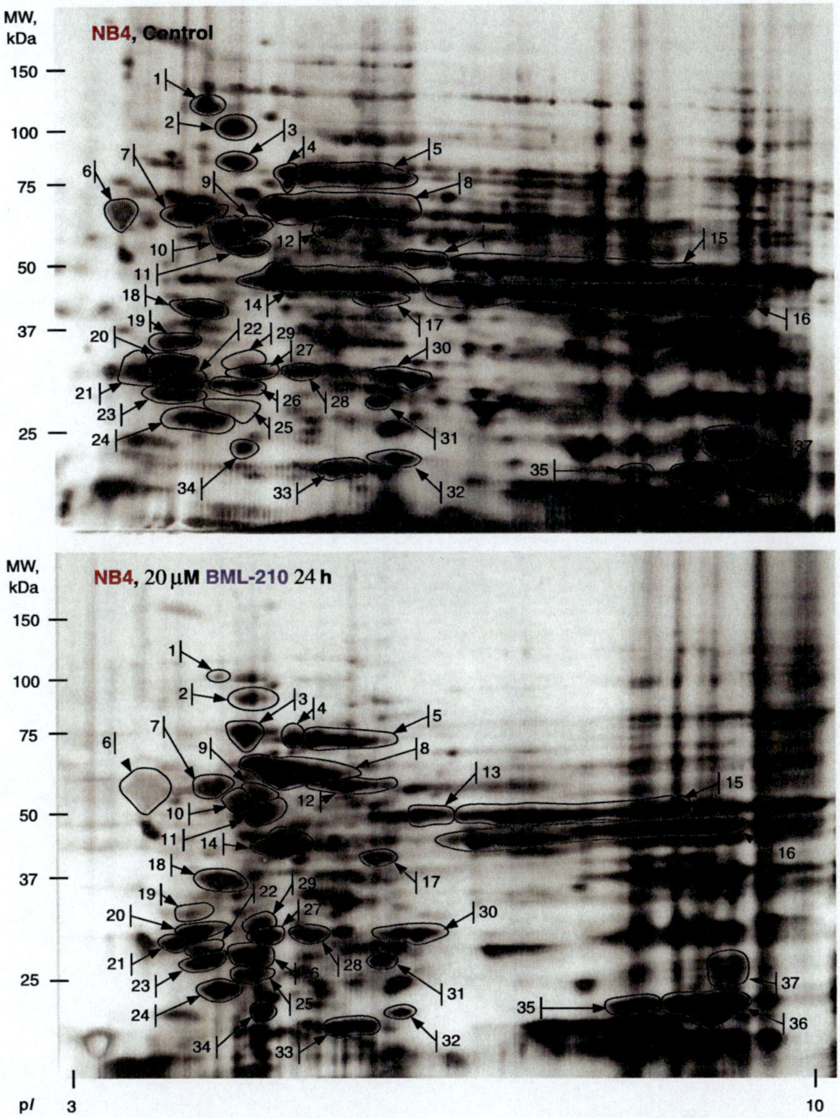

Fig. 4.8 Proteomic analysis of proteins after NB4 cell treatment with HDAC inhibitor BML-210. Proteins from untreated NB4 cells and cells treated with 20 μM BML-210 for 24 h fractionated in 2-DE system and visualized by Coomasie staining. Identified proteins are indicated in the Table 4.4. Migration of the molecular size marker proteins is shown at the center (kDa)

Table 4.4 Characterization of proteins identified in control and treated with HDAC inhibitor BML-210 NB4 cells. Computational analysis of protein expression changes are presented as fold change in comparisson with control cells (Ratio G2/G1). The expression of proteins whose Ratio G2/G1>0 is increased in treated (G2) cells than in untreated (G1)

No	Protein name	Ratio G2/G1	pI	MW	No.	Protein name	Ratio G2/G1	pI	MW
1	ENPL	−7.82	4.7	110.7	20	TPM4	−1.24	4.5	31.3
2	HSP90B	−1.30	4.9	96.9	21	1433E	−1.24	4.3	30.2
3	GRP78	1.55	4.9	78.9	22	1433F	−2.02	4.6	29.3
4	HSP72	1.17	5.2	74.9	23	1433B	1.16	4.5	28.1
5	GRP75	−1.56	5.4	74.9	24	ML12A	−1.25	4.7	25.6
6	CALR	−1.07	4.2	59.3	25	LGUL	2.40	4.9	26.8
7	PDIA1	−1.61	4.7	60.0	26	GDIR1	1.90	4.9	28.9
8	CH60	−1.15	5.2	62.0	27	EFHD2	1.40	5.0	30.4
9	TBA1B	1.25	5.0	56.7	28	PHB	1.46	5.2	30.6
10	TBB5	−1.44	4.9	54.1	29	CLIC1	2.70	5.0	31.9
11	PDIA6	2.20	5.0	51.3	30	TSG6	1.02	5.7	30.2
12	PDIA3	1.51	5.5	58.2	31	HSPB1	1.90	5.6	27.7
13	ENOA	1.27	5.7	49.4	32	COF1	−1.72	5.7	23.4
14	ACTB	−1.95	5.1	43.9	33	NDKA	1.34	5.5	22.6
15	ENOA	1.14	6.8	47.8	34	TXD12	2.26	4.9	23.7
16	NDUAA	−1.22	7.2	44.3	35	COF1	1.83	7.1	23.1
17	GNA11	1.30	5.6	41.3	36	COF1	1.76	8.4	22.7
18	NPM	1.07	4.7	38.6	37	TAGL2	−1.21	8.9	25.6
19	PCNA	−2.49	4.6	33.9					

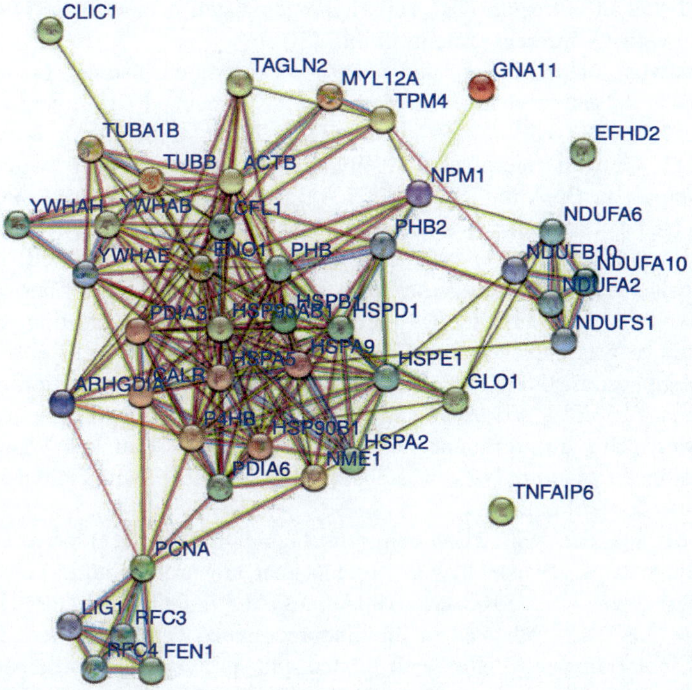

Fig. 4.9 Involvement of identified proteins in control and treated with HDAC inhibitor BML-210 NB4 cells into the cellular processes network

4.4 HDAC Inhibitor Belinostat Modulates Protein Profile in NB4 Cells

One of the most promising agents acting as epigenetic drug such as HDAC inhibitors—belinostat. It is an innovative and potent hydroxamate-type HDAC inhibitor that demonstrates potency to block enzymatic activity of 1-st and 2-nd class of HDACs (Khan et al. 2008). Belinostat exerts its anti-deacetylase action via its hydroxamic acid moiety binding to zinc ion in enzymes catalytic domain and blocking substrate access (Witter et al. 2007). In the others (Gravina et al. 2012) studies and we observed (Savickiene et al. 2014; Valiuliene et al. 2015; Valiulienė et al. 2016; Valiuliene et al. 2017; Vitkeviciene et al. 2019) its activity resulting in cell cycle arrest, apoptosis initiation, and cell proliferation inhibition. In addition, our group indicated that belinostat promotes APL granulocytic differentiation AML (Savickiene et al. 2014). Regarding belinostat's activity on APL cells, we performed proteomic analysis after chromatin immunoprecipitation with hyper-acetylated histone H4 (H4hyperAc) and established proteins being in the active complex with H4hyperAc (Valiuliene et al. 2015). Totaly we identified 68 proteins.

We found that in untreated NB4 cells hyperacetylated histone H4 (H4hyperAc) associated with 45 different proteins (Table 4.5).

The network of proteins related with hyperacetylated histone H4 in control NB4 cells is presented in Fig. 4.10 and proteins associated with hyperacetylated histone H4 in NB4 cells treated with HDAC inhibitor belinostat is presented in Fig. 4.11. Only in control cells, H4hyperAc was found related with proteins that participate in DNA replication (POLA2), transcription (GCOM1, POLR2M, NELFE, NCL), translation (RPL7) and RNA splising (SCNM1). Also it was estimated that H4hyperAc is associated with proto-oncogene SPECC1, regulator of apoptosis ADAMTSL4, together with proteins involved in various signaling cascades: NFκB, JAK2/STAT4, Ras and Hedgehog signal transduction pathways. Noteworthy, it was observed that H4hyperAc in control NB4 cells is associated with nucleophosmin (NPM), protein who is responsible for regulation of tumor suppressors TP53/p53 and ARF and is shown to be overexpressed in actively proliferating cells, like different cancer and stem cells (Lim and Wang 2006). Somewhat lesser extent of NPM was observed in complexes with H4hyperAc after treatment with 2 μM belinostat.

After 6 h treatment with 2 μM belinostat (Table 4.5, Fig. 4.11) it was estimated that H4hyperAc is associated with proteins that are pro-apoptotic and required for apoptotic response (S100A9, S100A8, LGALS7, GOLGA3, PPT1). Tumor suppressor APC was indicated in immunoprecipitated complexes, too. It should be noted that H4hyperAc is as well related with proteins that participate in the defense against oxidative stress (TXNRD2) and access of all-trans retinoic acid to the nuclear retinoic acid receptors regulation (CRABP1). We also found that after 6 h treatment of NB4 cells with 2 μM belinostat hyperacetylated histone H4 was no longer associated with proteins involved in gene transcription and/or translation. Nevereless it was found to associate with proteins, that are usually detected in cytosolic fraction as components of neutrophil extracellular traps (NETs). It is known that NETs are in association with DNA specific proteins, such as histones and antimicrobial proteins, that form an extracellular mesh able to trap and kill pathogens and they are released during a cell death that depends on ROS produced by the NADPH-oxidase complex (Valiuliene et al. 2015). After NB4 cells treatment with belinostat we identified calprotectin (S100A8 and S100A9) associated with hyperacetylted histone H4. Calprotectin is essential for the neutrophilic NADPH oxidase activation. Calprotectin is a protein complex composed of two calcium-binding proteins (S100A8 and S100A9) that are abundantly found in neutrophils cytosolic fraction and have shown to have apoptosis inducing activity. We also found a probable serine protease TMPRSS11A associated with hyperacetylated histone H4, which is in agreement with data, showing that NETs contain serine proteases, as they may execute antimicrobial functions in those structures. Taking all together, we assume that belinostat has cell death inducing activity in some manner may relate to NETs formation. Although, it is already known that belinostat triggers apoptosis in myeloid cells (Savickiene et al. 2014), not NETosis (cell death when NETs are released), the possibility that belinostat intervenes in NETs formation may not be rejected completely.

Table 4.5 Summary of identified NB4 cells proteins identified in complexes with hyperacetylated histone H4 in control (C) and belinostat (Bel) treated cells. According Valiuliene et al. (2015)

No.	Accession	Gene name	Score	C/Bel ratio	Function
1	Q5QNW6	HIST1H2AH	9505.53	0.77105	Core component of nucleosome
2	Q99878	HIST1H2AJ	8305.59	0.59452	Core component of nucleosome
3	P33778	HIST1H2BB	42815.45	1	Core component of nucleosome
4	P58876	HIST1H2BD	10401.13	0.51171	Core component of nucleosome
5	P57053	HIST2H2BF	45639.36	Bel	Core component of nucleosome
6	P84243	H3F3A	10974.41	1.10517	Core component of nucleosome
7	Q6NXT2	H3F3C	828.6	0.69768	Core component of nucleosome
8	P62805	HIST1H4A	17505.7	0.84366	Core component of nucleosome
9	P16401	HIST1H1B	615.33	1.1853	Nucleosomal condensation
10	P16403	HIST1H1C	2448.72	1.05127	Nucleosomal condensation
11	Q71UI9	H2AFV	5543.68	1.23368	Replaces conventional H2A in a subset of nucleosomes
12	P57053	H2BFS	10401.13	0.5886	Replaces conventional H2A in a subset of nucleosomes
13	P0C0S5	H2AFZ	7646.1	Bel	Replaces conventional H2A in a subset of nucleosomes
14	Q14181	POLA2	152.92	C	DNA replication
15	Q9BZD3	GCOM1	224.06	C	Component of Pol II(G) complex
16	P0CAP2	POLR2M	284.39	C	Component of Pol II(G) complex
17	P18615	NELFE	266.18	C	Represses RNA polymerase II transcript elongation
18	P51504	ZNF80	444.63	Bel	Transcriptional regulation
19	P18124	RPL7	235.63	C	Translation aparatus regulation
20	P47914	RPL29	622.4	0.92312	Translation apparatus regulation
21	Q9BWG6	SCNM1	620.47	C	RNA splicing
22	Q8WXA9	SREK1	148.46	Bel	Regulation of alternative splicing
23	P19338	NCL	132.66	C	Pre-rRNA transcription and ribosome assembly
24	P02788	LTF	275.73	C	Antimicrobial and anti-inflammatory activity
25	P61626	LYZ	751.9	3.56085	Bacteriolysis
26	P06702	S100A9	2362.67	Bel	Antimicrobial activity. Phagocyte migration promotion. Apoptosis
27	P05109	S100A8	1869.7	Bel	Antimicrobial activity. Phagocyte migration promotion. Apoptosis
28	P60709	ACTB	3806.68	1.82212	Cell motility
29	P63261	ACTG1	1713.96	0.34301	Cell motility
30	Q562R1	ACTBL2	664.65	Bel	Cell motility
31	A6NHL2	TUBAL3	110.36	C	Microtubule element
32	Q71U36	TUBA1A	452.48	C	Microtubule element

(continued)

Table 4.5 (continued)

No.	Accession	Gene name	Score	C/Bel ratio	Function
33	P07437	TUBB	620.75	2.2034	Microtubule element
34	Q9BQS8	FYCO1	61.54	C	May mediate microtubule plus end-directed vesicle transport
35	Q13326	SGCG	315.83	C	Component of sarcoglycan complex
36	Q9NY65	TUBA8	123.57	Bel	Microtubule element
37	Q9BQE3	TUBA1C	54.26	Bel	Microtubule element
38	O15144	ARPC2	666.79	Bel	Regulation of actin polimerization
39	Q96A32	MYLPF	737.84	Bel	Myosin light chain
40	Q6UY14	ADAMTSL4	247.27	C	Positive regulation of apoptosis
41	P47929	LGALS7	392.7	Bel	Apoptosis regulation. Pro-apoptotic
42	Q08378	GOLGA3	7.44	Bel	Golgi str. maintenance. Cleavage product necessary for apoptotic response
43	P50897	PPT1	124.76	Bel	Lysosomal degradation. DNA fragmentation during apoptosis
44	Q5M775	SPECC1	380.22	C	Proto-oncogene
45	P25054	APC	30.36	Bel	Tumor supressor
46	Q04760	GLO1	253.31	C	Involved in the regulation of TNF-induced transcriptional activity of NFκB
47	Q5T200	ZC3H13	22.03	C	Downregulation of NFκB pathway
48	O95989	NUDT3	215.22	C	Signal transduction. Negatively regulates ERK1/2 pathway
49	Q99665	IL12RB2	269.83	C	Signaling component coupling to the JAK2/STAT4 pathway. Promotes the proliferation of T-cells as well as NK cells
50	Q8IV04	TBC1D10C	834.75	0.8781	Ras signaling pathway inhibition
51	Q96NH3	C6orf170	1777.55	C	Controls ciliary morphology. Involved in Hedgehog signal transduction
52	P06748	NPM1	612.64	1.85893	Regulates tumor supressors TP53/p53 and ARF. Chaperone
53	Q9NNW7	TXNRD2	158.36	Bel	Implication in the defenses against oxidative stress
54	P29762	CRABP1	133.35	Bel	Regulates access of retinoic acid to the nuclear retinoic acid receptors
55	P17066	HSPA6	121.58	Bel	Chaperone
56	P48741	HSPA7	78.01	Bel	Chaperone
57	P11142	HSPA8	144.69	Bel	Chaperone. Repressor of transcriptional activation
58	P55735	SEC13	122	C	May be involved in protein transport
59	P62987	UBA52	755.27	0.77105	Proteosomal degradation, chromatin structure maintenance, gene expression regulation and stress response

(continued)

Table 4.5 (continued)

No.	Accession	Gene name	Score	C/Bel ratio	Function
60	P0CG47	UBB	241.72	C	Proteosomal degradation, chromatin structure maintenance, gene expression regulation and stress response
61	Q6ZMR5	TMPRSS11A	860.74	Bel	Pobable serine protease
62	P00738	HP	1190.67	Bel	Makes hemoglobin accessible to degradative anzymes
63	Q6S8J3	POTEE	346.61	C	Protein and ATP binding
64	A5A3E0	POTEF	369.2	2.13828	Protein and ATP binding
65	P0CG39	POTEJ	107.66	1.46228	Protein and ATP binding
66	Q9BTF0	THUMPD2	238.14	C	RNA binding. Methyltransferase activity
67	Q68CQ7	GLT8D1	206.26	C	Glycosyltransferase
68	A6NIV6	LRRIQ4	145.7	Bel	Leucine-rich repeats and IQ motif containing

Regarding effect of belinostat on cell growth, differentiation, gene and protein expression, and on epigenetic modifications, which was identified by us, belinostat could have a potential value in APL therapy.

4.5 Proteomic Maps of Leukemia Cells Induced to Granulocytic Differentiation and Apoptosis

In our study published in 2004 year (Navakauskiene et al. 2004b) we tried to evaluate protein level changes in proliferating HL-60 cells compared to cells induced for apoptosis using etoposide/Z-VAD(OH)-FMK. Programmed cell death has a significant role in the development and maintenance of homeostasis within all cells. It is widely acknowledged that the physiological form of cell death in neutrophils is apoptosis. For quite a long time it was considered that aged neutrophils die within a short period of time by spontaneous apoptosis under healthy conditions, for the sake of maintaining of homeostatic cell numbers (Geering and Simon 2011). The aim of our study was to evaluate apoptosis—associated protein patterns in HL-60 cells that were provoked to apoptosis with etoposide and also with or without the presence of the broad caspase and apoptosis inhibitor Z-VAD(OH)-FMK. Cellular cytosolic and nuclear proteins were fractionated by 2DE and changes in protein expression were detected (Fig. 4.12). In our studies that are presented in following chapters (Treigyte et al. 2000a,b; Navakauskiene et al. 2012, 2004a), we have shown that the synthesis of new proteins and protein modification takes place when HL-60 cells are induced into granulocytic differentiation. We found some quantitative and qualitative differences in the cytosolic and nuclear protein patterns of control cells, cells treated with etoposide alone or together with Z-VAD(OH)-

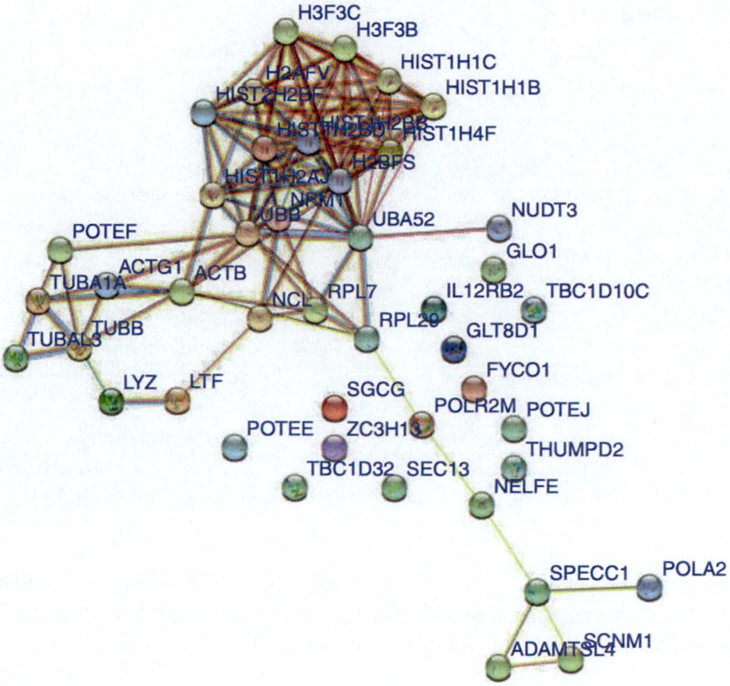

Fig. 4.10 Proteins identified in association with hyperacetylated histone H4 in control NB4 cells. Untreated NB4 cells were subjected to ChIP—MS analysis. Association network of identified proteins was studied and represented using STRING database (http://string.embl.de). According Valiuliene et al. (2015)

FMK for 6 h (Fig. 4.12). Analyses of cytosolic protein reference maps of control and drug-induced HL-60 cells revealed some new protein spots appeared, whereas the relative amount of others proteins markedly diminished or vanished after induction of apoptosis. The number of both cytosolic and nuclear polypeptides was different in HL-60 cells treated for 6 h with etoposide (60% apoptotic cells), by comparison with the control or 6 h etoposide/Z-VAD(OH)-FMK-treated cells (5–7% apoptotic cells) (Fig. 4.12). Proteins newly synthesized in HL-60 cells treated for apoptosis and absent in pan-caspase inhibitor-treated cells should have an apoptotic origin. Such proteins (marked by arrows in Fig. 4.12, 68 μM etoposide, CytP) had mostly an acidic pI 4.5–5.1, with molecular masses between 25 kDa and 150 kDa.

Some proteins with acidic and neutral pI 4.5–7.0, which were detected in the cytoplasm both etoposide and etoposide/Z-VAD(OH)-FMK treated cells were translocated exclusively into the nucleus of etoposide treated cells (Fig. 4.12, NuP, marked by arrows). These proteins could be involved in regulation of genes required in apoptosis process. We implied that variations of protein synthesis and protein modifications which were observed in differentiating HL-60 cells reflect the

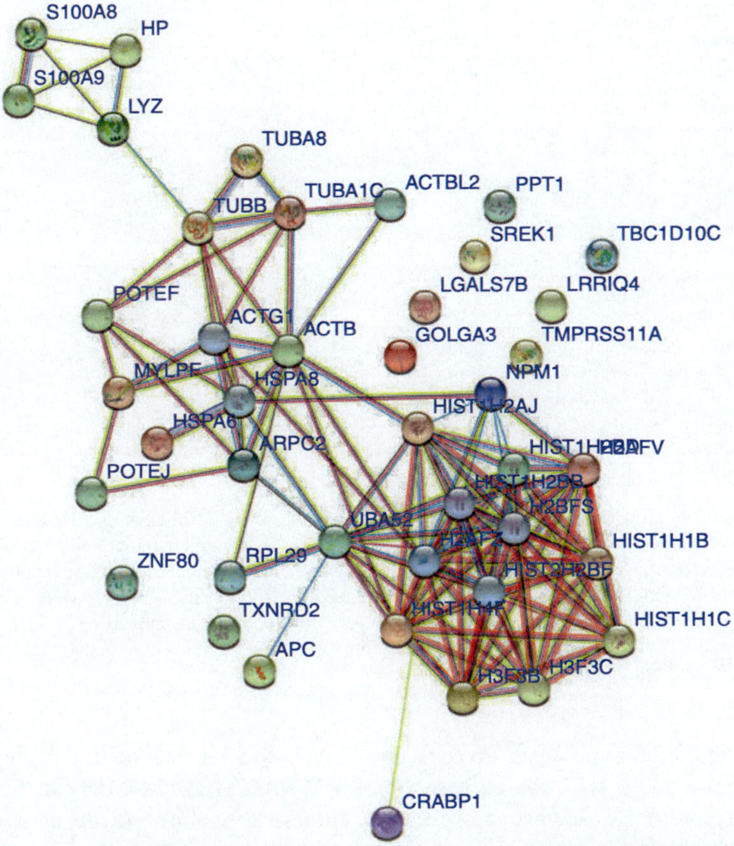

Fig. 4.11 Proteins identified in association with hyperacetylated histone H4 in belinostat treated NB4 cells. 2 μM belinostat treated NB4 cells were subjected to ChIP—MS analysis. Association network of identified proteins was studied and represented using STRING database (http://string.embl.de). According Valiuliene et al. (2015)

formation of the differentiated granulocyte phenotype and the apoptotic process. Our current research demonstrates that some proteins are highly upregulated in the cytoplasm and afterwards accumulated in the nuclei after treatment of HL-60 cells with etoposide. These proteins could participate in regulation of genes needed for apoptosis. Proteins estimated both after treatment with etoposide alone and together with pan-caspase inhibitor Z-VAD(OH)-FMK show that they do not represent a downstream event of caspase activation. We suggest that these proteins may be required for the regulation of apoptosis.

Taken together, our present study shows that some proteins are strongly upregulated in the cytoplasm and subsequently accumulated in the nuclei after treatment of HL-60 cells with etoposide. These proteins could be involved in regulation of

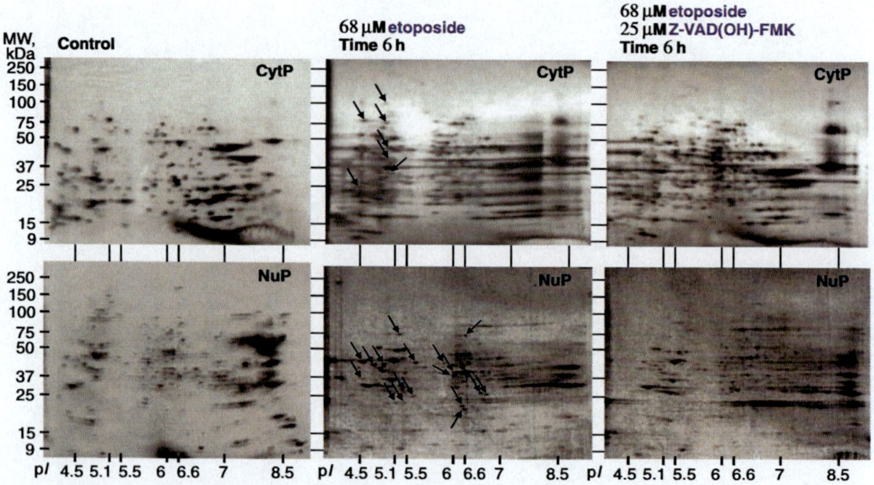

Fig. 4.12 Two-Dimensional electrophoretic maps of cytosolic and nuclear proteins of proliferating and induced to apoptosis HL-60 cells. Cytosolic (CytP) and nuclear (NuP) proteins of HL-60 cells treated for 6 h with 68 μM of etoposide alone or in the presence of 25 μM pan-caspase inhibitor Z-VAD(OH)-FMK were fractionated by 2D electrophoresis. The gels were stained by silver. Arrows show the position of proteins of apoptotic cell origin, appearing in cells treated with etoposide, but absent in broad caspase inhibitor-treated cells. According Navakauskiene et al. (2004b), License No 4796010620810

genes required for apoptosis. Proteins identified both after treatment with etoposide alone and together with pan-caspase inhibitor Z-VAD(OH)-FMK indicate that they do not represent a downstream event of caspase activation. Taken as a whole, obtained during our research results demonstrate that important alterations in both cytosolic and nuclear proteins take place in a highly regulated manner that leads to programmed cell death. Selective activation of some caspases is required to provoke apoptosis in HL-60 cells.

References

Bali P, Pranpat M, Bradner J, Balasis M, Fiskus W, Guo F, Rocha K, Kumaraswamy S, Boyapalle S, Atadja P, Seto E, Bhalla K (2005) Inhibition of histone deacetylase 6 acetylates and disrupts the chaperone function of heat shock protein 90—A novel basis for antileukemia activity of histone deacetylase inhibitors. J Biol Chem 280(29):26729–26734. https://doi.org/10.1074/jbc.C500186200

Borutinskaite V, Navakauskiene R (2015) The histone deacetylase inhibitor BML-210 influences gene and protein expression in human promyelocytic leukemia NB4 cells via epigenetic reprogramming. Int J Mol Sci 16(8):18252–18269. https://doi.org/10.3390/ijms160818252

Borutinskaite V, Magnusson KE, Navakauskiene R (2005) Effects of retinoic acid and histone deacetilase inhibitor Bml-210 on protein expression in NB4 cells. Biologija 4:88–93

Borutinskaite VV, Magnusson KE, Navakauskiene R (2011) α-Dystrobrevin distribution and association with other proteins in human promyelocytic NB4 cells treated for granulocytic differentiation. Mol. Biol. Rep. 38(5):3001–3011. https://doi.org/10.1007/s11033-010-9965-9

Breitman TR, Selonick SE, Collins SJ (1980) Induction of differentiation of the human promyelocytic leukemia cell line (HL-60) by retinoic acid. Proc Natl Acad Sci USA 77(5):2936–2940. https://doi.org/10.1073/pnas.77.5.2936

Collins SJ (1987) The HL-60 promyelocytic leukemia cell line: proliferation, differentiation, and cellular oncogene expression. Blood 70(5):1233–1244

Collins SJ, Gallo RC, Gallagher RE (1977) Continuous growth and differentiation of human myeloid leukaemic cells in suspension culture. Nature 270(5635):347–349

Geering B, Simon HU (2011) Peculiarities of cell death mechanisms in neutrophils. Cell Death Differ 18(9, SI):1457–1469. https://doi.org/10.1038/cdd.2011.75

Gravina GL, Marampon F, Giusti I, Carosa E, Di Sante S, Ricevuto E, Dolo V, Tombolini V, Jannini EA, Festuccia C (2012) Differential effects of PXD101 (belinostat) on androgen-dependent and androgen-independent prostate cancer models. Int J Oncol 40(3):711–720. https://doi.org/10.3892/ijo.2011.1270

Grebenova D, Roeselova P, Pluskalova M, Halada P, Roesel D, Suttnar J, Brodska B, Otevrelova P, Kuzelova K (2012) Proteins implicated in the increase of adhesivity induced by suberoylanilide hydroxamic acid in leukemic cells. J Proteomics 77:406–422. https://doi.org/10.1016/j.jprot.2012.09.014

Heller T, Asif AR, Petrova DT, Doncheva Y, Wieland E, Oellerich M, Shipkova M, Armstrong VW (2009) Differential proteomic analysis of lymphocytes treated with mycophenolic acid reveals caspase 3-induced cleavage of rho GDP dissociation inhibitor 2. Ther Drug Monit 31(2):211–217. https://doi.org/10.1097/FTD.0b013e318196fb73

Hunter T (1995) Protein-kinases and phosphatases—the yin and yang of protein-phosphorylation and signaling. CELL 80(2):225–236. https://doi.org/10.1016/0092-8674(95)90405-0

Karin M, Liu ZG, Zandi E (1997) AP-1 function and regulation. Curr Opin Cell Biol 9(2):240–246. https://doi.org/10.1016/S0955-0674(97)80068-3

Kashyap AS, Fernandez-Rodriguez L, Zhao Y, Monaco G, Trefny MP, Yoshida N, Martin K, Sharma A, Olieric N, Shah P, Stanczak M, Kirchhammer N, Park SM, Wieckowski S, Laubli H, Zagani R, Kasenda B, Steinmetz MO, Reinecker HC, Zippelius A (2019) GEF-H1 signaling upon microtubule destabilization is required for dendritic cell activation and specific anti-tumor responses. Cell Rep 28(13):3367+. https://doi.org/10.1016/j.celrep.2019.08.057

Khan N, Jeffers M, Kumar S, Hackett C, Boldog F, Khramtsov N, Qian X, Mills E, Berghs SC, Carey N, etal (2008) Determination of the class and isoform selectivity of small-molecule histone deacetylase inhibitors. Biochem J 409(2):581–589

Kulyte A, Navakauskiene R, Treigyte G, Gineitis A, Magnusson KE (2001) Parallel assessment of tyrosine phosphorylation and nuclear targeting of proteins. Biotechniques 31(3):510, 512–4, 517

Kulyte A, Navakauskiene R, Treigyte G, Gineitis A, Bergman T, Magnusson KE (2002) Characterization of human alpha-dystrobrevin isoforms in HL-60 human promyelocytic leukemia cells undergoing granulocytic differentiation. Mol Biol Cell 13(12):4195–4205. https://doi.org/10.1091/mbc.E02-03-0128

Lim MJ, Wang XW (2006) Nucleophosmin and human cancer. CAncer Detect Preven 30(6):481–490. https://doi.org/10.1016/j.cdp.2006.10.008

Matuzevicius D, Zurauskas E, Navakauskiene R, Navakauskas D (2008) Improved proteomic characterization of human myocardium and heart conduction system by using computational methods. Biologija 4:283–289

Navakauskiene R, Treigyte G, Pivoriunas A, Savickiene J (2002) Cell cycle inhibitors in retinoic acid- and etoposide-mediated biological responses. Biologija 2:64–67

Navakauskiene R, Kulyte A, Treigyte G, Gineitis A, Magnusson KE (2003a) Translocation of transcription regulators into the nucleus during granulocyte commitment of HL-60 cells. Biochem Cell Biol.-Biochim et Biol Cell 81(4):285–295. https://doi.org/10.1139/O03-055

Navakauskiene R, Treigyte G, Kulyte A, Magnusson KE (2003b) Proteomic analysis by MALDI-TOf mass spectrometry and its application to HL-60 cells. Biologija 3:63–65

Navakauskiene R, Treigyte G, Gineitis A, Magnusson KE (2004a) Identification of apoptotic tyrosine-phosphorylated proteins after etoposide or retinoic acid treatment of HL-60 cells. Proteomics 4(4):1029–1041. https://doi.org/10.1002/pmic.200300671

Navakauskiene R, Treigyte G, Savickiene J, Gineitis A, Magnusson KE (2004b) Alterations in protein expression in HL-60 cells during etoposide-induced apoptosis modulated by the caspase inhibitor ZVAD.fmk. In: Diederich M (ed) Signal Transduction Pathways, Chromatin Structure, and Gene Expression Mechanisms as Therapeutic Targets, Fdn Rech Canc & Sang; Novartis Luxembourg; Q8 Petr, Annals of the New York Academy of Sciences, vol 1030, pp 393–402. https://doi.org/10.1196/annals.1329.0049

Navakauskiene R, Treigyte G, Borutinskaite VV, Matuzevicius D, Navakauskas D, Magnusson KE (2012) Alpha-dystrobrevin and its associated proteins in human promyelocytic leukemia cells induced to apoptosis. J Proteomics 75(11):3291–3303. https://doi.org/10.1016/j.jprot.2012.03.041

Navakauskiene R, Borutinskaite VV, Treigyte G, Savickiene J, Matuzevicius D, Navakauskas D, Magnusson KE (2014) Epigenetic changes during hematopoietic cell granulocytic differentiation—comparative analysis of primary CD34+cells, KG1 myeloid cells and mature neutrophils. BMC Cell Biol 15:4. https://doi.org/10.1186/1471-2121-15-4

Nawarak J, Huang-Liu R, Kao SH, Liao HH, Sinchaikul S, Chen ST, Cheng SL (2009) Proteomics analysis of A375 human malignant melanoma cells in response to arbutin treatment. Biochim Biophys Acta Proteins Proteomics 1794(2):159–167. https://doi.org/10.1016/j.bbapap.2008.09.023

Patterson C, Portbury AL, Schisler JC, Willis MS (2011) Tear me down role of calpain in the development of cardiac ventricular hypertrophy. Circ Res 109(4):453–462. https://doi.org/10.1161/CIRCRESAHA.110.239749

Pawson T (1995) Protein modules and signaling networks. Nature 373(6515):573–580. https://doi.org/10.1038/373573a0

Savickiene J, Treigyte G, Valiuliene G, Stirblyte I, Navakauskiene R (2014) Epigenetic and molecular mechanisms underlying the antileukemic activity of the histone deacetylase inhibitor belinostat in human acute promyelocytic leukemia cells. Anti-Cancer Drugs 25(8):938–949. https://doi.org/10.1097/CAD.0000000000000122

Sheng W, Chen C, Dong M, Zhou J, Liu Q, Dong Q, Li F (2014) Overexpression of calreticulin contributes to the development and progression of pancreatic cancer. J Cell Physiol 229(7):887–897. https://doi.org/10.1002/jcp.24519

Sporn MB, Roberts AB, Goodman DS (1994) The retinoids: biology, chemistry, and medicine. Raven Press, New York

Treigyte G, Navakauskiene R, Kulyte A, Gineitis A, Magnusson KE (2000a) Characteristics of cytosolic proteins and changes in their tyrosine phosphorylation during HL-60 cell differentiation. Biologija 2:32–35

Treigyte G, Navakauskiene R, Kulyte A, Gineitis A, Magnusson KE (2000b) Tyrosine phosphorylation of cytoplasmic proteins in proliferating, differentiating, apoptotic HL-60 cells and blood neutrophils. Cell Mol Life Sci 57(13–14):1997–2008. https://doi.org/10.1007/PL00000681

Treigyte G, Savickiene J, Navakauskiene R (2004) Identification of O- and N-glycosylated nuclear proteins of HL-60 cells induced to granulocytic differentiation. Biologija 2:49–51

Valiuliene G, Stirblyte I, Cicenaite D, Kaupinis A, Valius M, Navakauskiene R (2015) Belinostat, a potent HDACi, exerts antileukaemic effect in human acute promyelocytic leukaemia cells via chromatin remodelling. J Cell Mol Med 19(7):1742–1755. https://doi.org/10.1111/jcmm.12550

Valiulienė G, Treigytė G, Savickienė J, Matuzevičius D, Alksnė M, Jarašienė-Burinskaja R, Bukelskienė V, Navakauskas D, Navakauskienė R (2016) Histone modifications patterns in tissues and tumours from acute promyelocytic leukemia xenograft model in response to combined epigenetic therapy. Biomed. Pharmacother. 79:62–70. https://doi.org/10.1016/j.biopha.2016.01.044

Valiuliene G, Stirblyte I, Jasnauskaite M, Borutinskaite V, Navakauskiene R (2017) Anti-leukemic effects of HDACi belinostat and HMTi 3-Deazaneplanocin A on human acute promyelocytic leukemia cells. Eur J Pharmacol 799:143–153. https://doi.org/10.1016/j.ejphar.2017.02.014

Vitkeviciene A, Skiauteryte G, Zucenka A, Stoskus M, Gineikiene E, Borutinskaite V, Griskevicius L, Navakauskiene R (2019) HDAC and HMT inhibitors in combination with conventional therapy: a novel treatment option for acute promyelocytic leukemia. J. Oncol. 2019:11. https://doi.org/10.1155/2019/6179573

Witter DJ, Belvedere S, Chen L, Secrist JP, Mosley RT, Miller TA (2007) Benzo[b]thiophene-based histone deacetylase inhibitors. Bioorg Med Chem Lett 17(16):4562–4567. https://doi.org/10.1016/j.bmcl.2007.05.091

Chapter 5
Protein Phosphorylation in Leukemia

Phosphorylation and dephosphorylation of cellular proteins are involved in many biologically significant cellular processes and cell functioning. Protein phosphorylation by different protein kinases is a key process that regulates proliferation, differentiation, and apoptosis in the cell's response to external stimuli. Tyrosine phosphorylation of the signaling protein is a main modification for transduction of the outer signal to the nuclei. Our research focused on the identification of newly synthesized or modified by tyrosine phosphorylation nuclear and cytoplasmic proteins during granulocytic differentiation of leukemic cells. These proteins may be involved in cell signaling at the downstream stage of differentiation and may be responsible for nuclear reorganization at the stage of leukemic cell maturation into neutrophils. In our studies, we identified a number of transcription factors (c-Myb, C/EBPβ, STAT3, STAT 5a, STAT 5b, MAPK, NFκB p50, and NFκB p65) whose expression levels change during induced granulocytic differentiation. Proteomic analysis of proteins revealed that the transcription factors STAT3, STAT 5b, NFκB p65 and C/EBPβ are tyrosine phosphorylated during leukemic cell granulocytic differentiation (Navakauskiene et al. 2004a,b; Savickiene et al. 2010; Navakauskiene et al. 2003b). Tyrosine phosphorylation is one of the most important modifications implicated in the regulation of cellular functioning. Therefore the tyrosine phosphorylated proteins could have a potential as drug targets. In our studies we attempt to investigate more precisely the biological importance of tyrosine phosphorylated proteins in AML cell differentiation and apoptosis.

In vitro, differentiated leukemic cells, like normal blood neutrophils, die by apoptosis, but the regulatory mechanisms involved in granulocytic differentiation during apoptosis are not yet understood. To elucidate which proteins are specific for mature neutrophils in differentiated leukemic cells and which in apoptotic cells, we investigated protein modifications, particularly tyrosine phosphorylation of nuclear and cytoplasmic proteins in proliferating, differentiating and apoptotic AML cells, blood neutrophils and primary hematopoietic CD34+ cells (Treigyte et al. 2000b; Kulyte et al. 2002; Navakauskiene et al. 2002, 2000, 2003a,b, 2004a,b, 2012).

R. Navakauskienė et al., *Epigenetics and Proteomics of Leukemia*,
https://doi.org/10.1007/978-3-030-68708-3_5

For the analysis of protein modifications the proteins have been exposed to the apoptosis-inducing chemotherapeutic agent—topoisomerase II inhibitor etoposide. To identify proteins involved in apoptosis, the cytosolic and nuclear proteins and their tyrosine phosphorylation were analyzed, as well as O- and N-glycosylation of cytoplasmic and nuclear proteins were identified (Treigyte et al. 2003, 2004) during granulocytic differentiation. The purpose of our project was to indentify the proteins that are involved in signaling events in the course of differentiation and apoptosis of leukemia cells. The main tasks of the study are: (1) Determination of nuclear and cytosol proteins in proliferating, differentiated and apoptosis-dependent cells of promyelocytic leukemia HL-60 cells; (2) Evaluation of the tyrosine phosphorylation of nuclear and cytosol proteins in HL-60 cells during granulocytic differentiation and apoptosis; (3) Comparative analysis of the protein binding of the cells and their tyrosine phosphorylation in the combination of the chemotherapy with etoposide; (4) Use of a global proteomic analysis to identify proteins that are specific to leukemic cells; (5) Changes in protein expression in apoptotic leukemic cells.

5.1 Tyrosine Phosphorylation of Cytosolic Proteins during Differentiation of Leukemic Cells

As a model *in vitro*, we used all-trans retinoic acid mediated- promyelocyticleukemia HL-60 cell granulocytic differentiation. Our preceding research has shown that after induction of granulocytic differentiation of HL-60 cells the number of differentiated cells in the population increases up to 60% after 96 h of treatment. The cells remain viable up to 72 h from the beginning of differentiation induction, and subsequently their viability is significantly reduced. The number of apoptotic cells in such a cell population begins to increase from 48 h (the commitment phase of differentiation), and after 120 h of treatment reaches 50–55% of the population (Treigyte et al. 2000c,a; Navakauskiene et al. 2004a). We determined that phosphorylation increased after 48 h of treatment with ATRA when differentiation process started leading to apoptosis of the cells (Fig. 5.1).

After HL-60 cell induction with ATRA cytoplasmic proteins isolated from proliferating, differentiated HL-60 cells and human mature neutrophils were fractionated by two-dimensional electrophoresis, visualized with silver staining, and immunoanalysis for tyrosine phosphorylation was performed (Navakauskiene et al. 2004a,b). After total cytoplasmic protein visualization we identified more than 300 proteins (Fig. 5.2). In this study, we showed that only 70, 60, 40, and 30 kDa protein groups were tyrosine phosphorylated (Fig. 5.3). We detected only slight protein phosphorylation level changes after 24 h of treatment with ATRA treatment, the major degree of phosphorylation was detected after 96 h of treatment with ATRA. The phosphorylation level that occurred immediately after ATRA treatment could be associated with differentiation initiation processes. At the commitment stage (48 h after differentiation induction) and especially in the later stages of

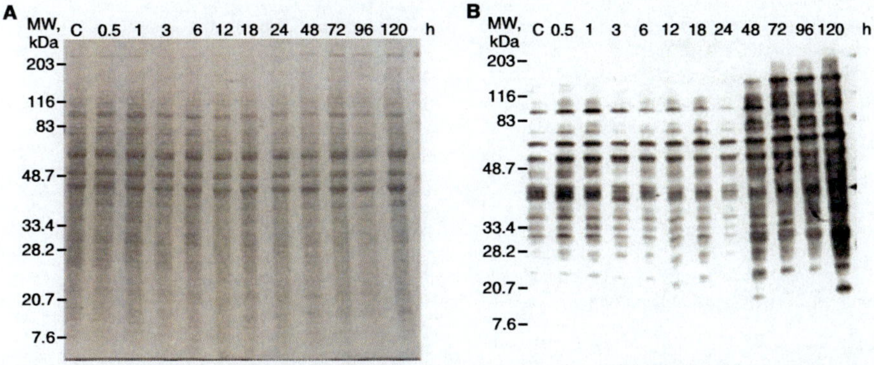

Fig. 5.1 Tyrosine phosphorylation of cytoplasmic proteins in HL-60 cells induced to granulocytic differentiation. (**A**) Cytoplasmic proteins isolated from untreated and various time treated with 1 μM ATRA, fractionated in 7–18% gradient SDS/PAGE and visualized with colloidal CBB-250. (**B**) Cytoplasmic proteins isolated from untreated and various time treated with 1 μM ATRA were analyzed with antibodies against tyrosine phosphorylation. According Treigyte et al. (2000c), License No 4796570093384

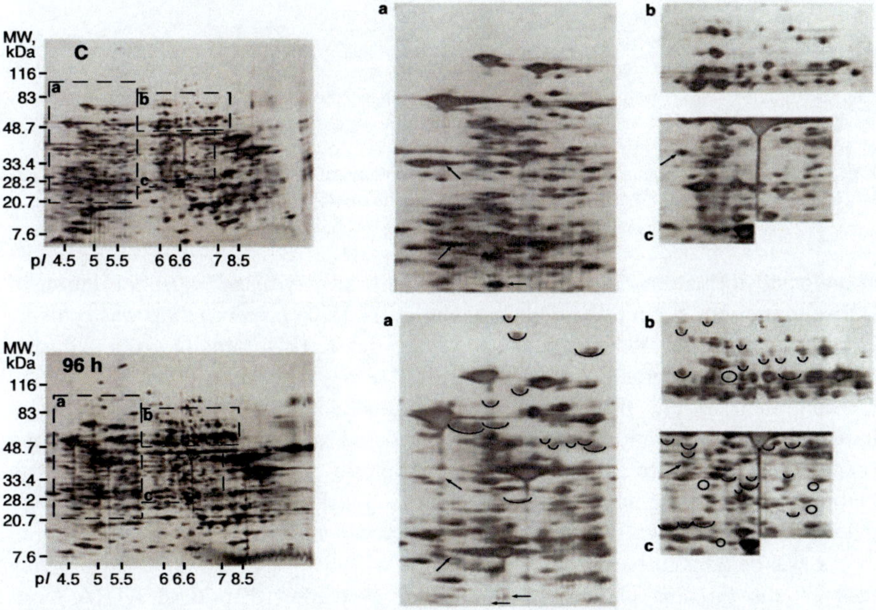

Fig. 5.2 The total cytoplasmic proteins in control and differentiated with ATRA HL-60 cells and mature human neutrophils. The cytosolic proteins of proliferating (**A**) and differentiated HL-60 cells (96 h with ATRA) (**B**) were fractionated by 2DE and visualized by staining with silver. Semicircles indicate newly synthesized proteins in the cytoplasm of differentiated cells. Circularly labeled proteins represent the significant decrease (or disappearance) during cellular differentiation compared to proliferating cells. Arrow-labeled proteins represent decrease in amount during differentiation. According Treigyte et al. (2000c), License No 4796570093384

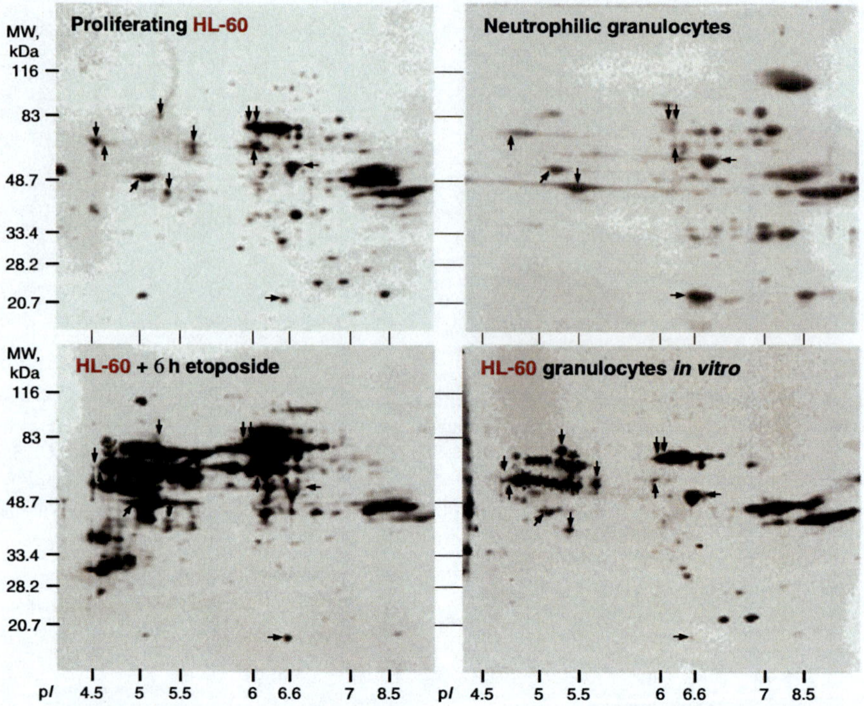

Fig. 5.3 The tyrosine phosphorylated cytoplasmic proteins in control, differentiated with ATRA HL-60 cells, mature human neutrophils, and HL-60 cells treated with etoposide

granulocyte formation (72, 96, and 120 h), intensive tyrosine phosphorylation of cytosolic proteins occurs. We detected that the 140 kDa protein group was tyrosine phosphorylated after 48 h treatment and later, the 200 kDa protein group—during the formation of the granulocyte phenotype.

More than 30 new proteins were found and about 10 proteins disappear in induced to differentiation HL-60 cells compared with control cells. We found caspase-dependent proteins specific for apoptotic process after treatment with etoposide and caspase inhibitor. Our findings suggest that many of proteins investigated are involved in cell signal transduction cascades.

As we demonstrated in the studies that will be presented in the following chapters the tyrosine phosphorylation occur even after 30 min of ATRA treatment (Navakauskiene et al. 2003a; Kulyte et al. 2001, 2002). The tyrosine phosphorylated proteins that overlap in 96 h ATRA treated cells and in human neutrophils could be responsible for terminal differentiation of granulocytes. The other tyrosine phosphorylated proteins specific only for 96 h differentiated HL-60 cells could be related to apoptosis processes following the terminal granulocytic differentiation in cell population.

5.2 Tyrosine Phosphorylation of Nuclear and Cytosolic Proteins in Differentiated and Apoptotic AML Cells

It is known that when due to all-trans retinoic acid treatment HL-60 cells differentiate into granulocytes *in vitro*, similar to normal blood neutrophils, they die after initiation of apoptosis process. To date, there are no clear regulatory mechanisms for the administration of granulocytic differentiation during apoptosis. In order to elucidate, which processes are included in the differentiation of HL-60 cells and which-apoptosis, we used a chemotherapeutic agent-topoisomerase II inhibitor, etoposide, which causes cellular apoptosis without differentiation (Navakauskiene et al. 2004a,b; Kulyte et al. 2002; Navakauskiene et al. 2000, 2002, 2003b,a, 2004a,b, 2012).

During cellular differentiation or apoptosis processes the disturbance of cell cycle occurs. We showed that after 68 μM etoposide treatment of HL-60 cells, cells accumulated in the G1 phase after 3 h, and after 18 h, the subG1 fraction consists of the entire cell population (Navakauskiene et al. 2004a,b). After orange/ethidium bromide staining, we found that after 2 h of 68 μM etoposide treatment number of apoptotic HL-60 cells increased up to 35% and after 6 h—up to 85%. Another characteristic of apoptosis is the fragmentation of the DNA. We determined DNA fragmentation in time dependent manner both after 68 μM etoposide (after 3 h) and ATRA (after 96 and 120 h) treatment (Navakauskiene et al. 2004a,b). The objective of this research was to determine tyrosine-phosphorylated nuclear and cytosolic proteins of apoptotic origin. A common hematopoietic cell model, the human promyelocytic cell line HL-60 was applied to investigate the protein tyrosine phosphorylation associated with chemically induced apoptosis, and granulocytic differentiation leading apoptosis. We evaluated the heterogeneity of nuclear and cytosolproteins in proliferating, differentiated into neutrophils and etoposide-induced apoptosis-induced leukemic cells using a two-dimensional electrophoresis protein fractionation system.

We found that total level of cytosolic and nuclear proteins (Fig. 5.4) quantity during HL-60 granulocytic differentiation increased, also the increased protein levels were detected after treatment with etoposide. After imunoanalysis we observed that tyrosine phosphorylated proteins increased with etoposide apoptosis-induced HL-60 cells (Fig. 5.5). In these cells, 1 h after etoposide exposure to cytosol, new tyrosine phosphorylated proteins were detected as 17 kDa, pI 8.3 (no 18), 20 kDa, pI 7.0–7.5 (no 16, 17), 52 kDa, pI 5.5–5.6 (No 3, 4), 55 kDa, pI 4.4 (No 15), 59 kDa, pI 5.6 (No 7) 65 kDa, pI 6.2–6.7 (No 11–13), and 73 kDa, pI 4.7 (No. 14). Some of these tyrosine phosphorylated proteins, No. 3, 4, 7, 11–13, are also found in the nucleus; only in the nucleus as the newly tyrosine phosphorylated proteins were observed 20 kDa, pI 4.8–5.0 (No. 8–10), and 62 kDa, pI 4.9 (No. 2). After HL-60 cell treatment with etoposide we detected newly tyrosine phosphorylated proteins the modification of which significantly enhanced at a later stage of the apoptosis process (6 h). These proteins are No. 5, 6 (27–29kDa, pI 8) and No. 1 (52 kDa, pI 4.8) (Fig. 5.5). In addition, we found 17 newly tyrosine phosphorylated proteins

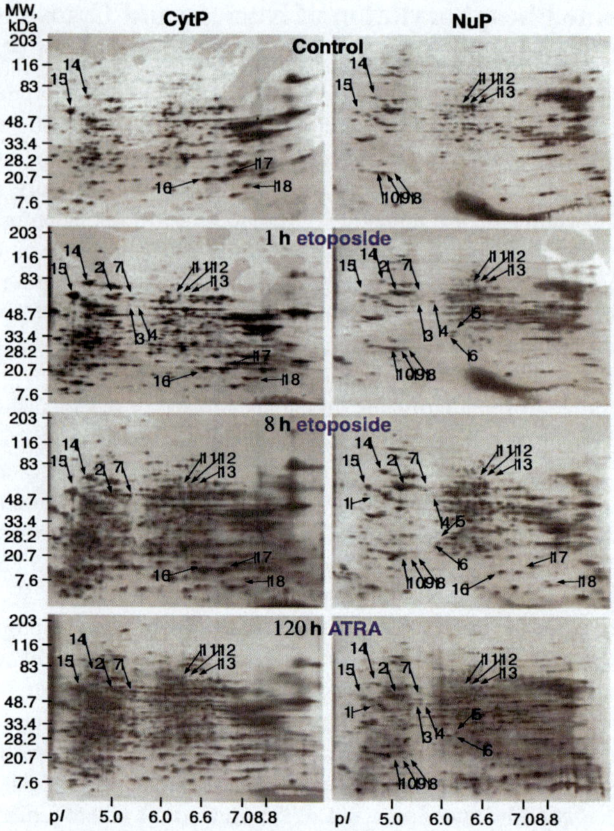

Fig. 5.4 The changes in cytosolic and nuclear proteins levels of proliferating, induced to granulocytic differentiation with ATRA and to apoptosis with etoposide HL-60 cells. Cytosolic and nuclear proteins were isolated from proliferating HL-60 cells (Control), from 68 μM etoposide-exposed cells for 1 and 6 h, and from differentiated (120 h) with ATRA cells. Proteins were fractionated by two-dimensional electrophoresis (2DE) and stained with silver. Arrows indicate those proteins that are tyrosine phosphorylated in apoptotic cells—both in etoposide-induced and in ATRA-induced to granulocytic differentiation cells. According Navakauskiene et al. (2004a), License No 4796570649985

in the nucleus and 11 in the cytosol. The summarized search results of tyrosine-phosphorylated proteins of apoptotic HL-60 cell origin are presented Table 5.1.

In the cytosol of differentiated cells, the major part of tyrosine phosphorylated proteins are acidic (pI 4.5–6.0), their molecular mass is about 48–89 (ATRA, 120 h). Our observed newly tyrosine phosphorylated nuclear and cytosolic proteins undergo modification throughout the granulocytic differentiation process. These data indicate that tyrosine phosphorylation of nuclear and cytosolic proteins throughout HL-60 cell differentiation are not only related to the formation of differentiated granulocyte phenotype, but also to a large extent to the induction or execution of

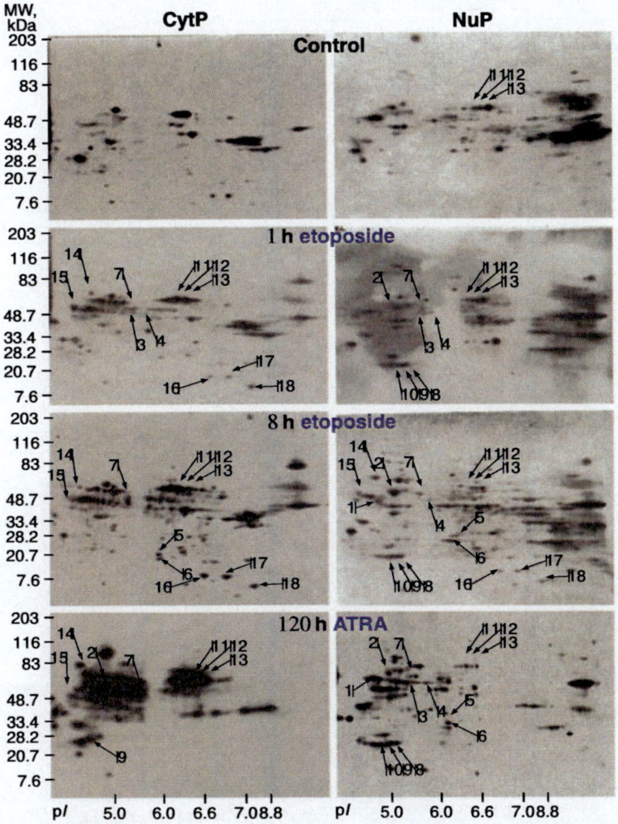

Fig. 5.5 Tyrosine phosphorylation of cytosolic and nuclear proteins in proliferating, etoposide-induced apoptotic, and differentiated HL-60 cells. Cytoplasmic and nuclear proteins were isolated from proliferating HL-60 (control), etoposide-exposed, and differentiated cells (120 h after ATRA induction). Proteins were fractionated with 2DE, transferred to a PVDF membrane, and tyrosine phosphorylated proteins were detected with anti-phosphotyrosine antibodies. Arrows indicate newly tyrosine phosphorylated proteins in HL-60 cells induced for apoptosis with etoposide and I terminal stages of granulocytic differentiation. According Navakauskiene et al. (2004a), License No 4796570649985

the apoptotic pathway. Some tyrosine-phosphorylated proteins detected by us after etoposide-induced and granulocytic differentiation leading apoptosis are involved in signal transduction pathways, contribute to cell transcriptional and translational control and participate in cell cytoskeleton organization.

One of these identified proteins, the nuclear factor kappa B (NFκB) is a transcription factor, and its transcriptional activity is modulated by phosphorylation. NFκB regulates cell proliferation, migration, differentiation, and apoptosis, and even constitutive activation of NFκB was connected with cancer (Fuchs 2010). Nonetheless, specific aspects of NFκB activation pathways imply that it can also

Table 5.1 The summarized search results of tyrosine-phosphorylated proteins of apoptotic HL-60 cell origin. According Navakauskiene et al. (2004a), License No 4796570649985

No.	Mw	pI	PY	Description of matching, %
1.	52	4.8	+++	Chain 1: ATP syntase beta chain, 28%
				Nucleolar autoantigen No55, 9.2%
				Transcription initiation factor IIE, alpha subunit (TFIIE), 7.7%
2.	62	4.9	+++	Chain 1: 60 kDa heat shock protein, 28%
				Src substrate cortactin, 23%
				Retinoic acid-induced protein, 7.9%
3.	52	5.5		Heterogeneous nuclear ribonucleoprotein H (hnRNP H), 16.3%
				Eucaryotic translation initiation factor 5 (eIF), 7.2%
				Heterogeneous nuclear ribonucleoprotein H (hnRNP F), 6.5%
4.	52	5.6		Heterogeneous nuclear ribonucleoprotein H hnRNP H), 13.4%
				Transcription factor RELB (I-Rel), 5.4%
				Splice isoform of P22102 (PUR2_HUMAN), 10.2%
5.	29	5.8	+++	Nuclear protein Hcc-1 (HSPC316), 30%
				Protein 1–4 (ATP binding protein associated with cell differentiation), 15.9%
6.	27	5.8	+++	Proteosome subunit β type, 12%
				Zinc finger protein 138, 20%
7.	59	5.6	+++	Transcription factor p65 (Nuclear factor NFκB p65 subunit), 7.6%
8.	20	5.0	+++	Chromobox protein homolog 3 (HP1) 15.6%
9.	19.9	4.9	+++	Chain 1: Calveolin-1 alpha, 15.7%
				Guanylyl cyclase activating protein 2 (GCAP), 12.1%
10.	20	4.8	+++	25K GTP-binding protein (GP), 22.5%
				Chromobox protein homolog 3 (HP1), 22.5%
11.	65	6.2	+++	Stress induced-phosphoprotein 1 (ST11), 13%
				Protein kinase C, beta-II type, 12.9%
				WD-repeat protein 1 (actin interacting protein 1), 15.7%
				Heat shock factor protein 4 (HSF 4), 4.9%
				M-phase induced phosphatase 1 (EC 3.1.3.48), 4.4%
12.	65	6.6	+++	Stress induced-phosphoprotein 1 (ST11), 13%
				Kinesin light chain 2 (KLC1), 10.8%
				Tyrosin protein kinase LYK, 8.1%
				Tyrosyl-TRNA synthetase, 15%
13.	65	6.7	+++	3-phosphoinositide dependent protein kinase-1, %?
				Heterogeneous ribonucleoprotein L (hnRNP L), 35.8%
				Kinesin light chain 1 (KLC1), 10.5%
				Kinesin light chain 2 (KLC1), 10.8%
				Lamin A/C (70 kDa), 3.5%
14.	73	4.75	+++	Chain 1: 78 kDa Glucose-regulated protein, 19%
				Chain 1: Furin, 11.1%
15.	55	4.4	+++	Chain 1: Protein disulfide isomerase, 34.8%
				Putative transcription factor Ovo-like 1 (hOvo 1), 14%
16.	20	7	+	Peptidyl-prolyl cis-trans isomerase A (Cyclophilin A), 44.5%
17.	20	7.5	+	Peptidyl-prolyl cis-trans isomerase A (Cyclophilin A), 44.5%
				Phosphatidylethanolamine-binding protein (PEBP), 10.2%
18.	17	8.3	+	Profilin I, 38.1%
				Chain 1: Group IID secretory phospholipase A2, 10.4%

have a tumor-suppressor function. In the present research we have determined that the NFκB p65 subunit is modified by tyrosine phosphorylation in HL-60 cells induced to apoptosis. The post-translational modification could influence NFκB subunit-specific target genes throughout apoptosis process.

As a tyrosine phosphorylated protein in our study we determined GTP-binding protein that appears to be essential for the interaction with respective effector proteins; peptidyl-prolyl isomerases (PPIases) that are a cluster of cytosolic enzymes, cyclophilin A (CypA), etc. We have also detected some proteins which participate in cell transcriptional and translational control, such as hnRNP proteins, several stress proteins, like glucose-related and heat-shock proteins. As tyrosine-phosphorylated protein we identified HP1 protein, the modification of this protein occurs instantly after induction of apoptosis and at the terminal stage of granulocytic differentiation. The other group study (Shimada and Murakami 2010) demonstrates that phosphorylation of HP1/Swi6 provides a dynamic pathway for the differential gene regulation by epigenetics and chromatin remodeling. Based on these we can argue that modification on tyrosine residues makes and impact on heterochromatin-induced repression, gene silencing for differentiation, and apoptosis.

Additional category of tyrosine-phosphorylated and apoptosis-associated proteins are those proteins which contribute to cell cytoskeleton restructuring. The profilin is a cytoskeleton protein that emerges as a critical regulator of actin dynamics, is regulated by phosphatidylinositol bisphosphate (PIP2) and composes complex with G-actin (Pinto-Costa and Sousa 2019). We discovered that profilin undergoes a tyrosine phosphorylation in the nucleus after 8 h of etoposide treatment when 85% of apoptotic cells occur in the population (Navakauskiene et al. 2004a,b). Profilin may be a significant link between signal transduction pathways and cytoskeletal dynamics throughout processes of differentiation and apoptosis. Dynamic rearrangements of the actin cytoskeleton are essential to the morphological alterations noticed both in apoptosis and granulocytic differentiation, for example, enabling cell motility and phagocytic capacity. Currently it was demonstrated that differential profilin2a (brain specific isoform) phosphorylation is a sensitive mechanism which regulates its neuronal functions and the dysregulation of profilin2a phosphorylation may be associated with neurodegeneration (Walter et al. 2020).

Other important protein identified by us was caveolin-1 that undergoes tyrosine phosphorylation throughout etoposide-induced apoptosis and ATRA-mediated granulocytic differentiation in the HL-60 cells. The data presented by us indicate the role of profilin and caveolin in apoptosis is potentially related with actin cytoskeleton regulation. The other studies (Tsuji et al. 2005) demonstrated the importance of tyrosine phosphorylation of different caveolin isoforms in human T cell leukemia cell lines.

In conclusion, our data indicate that tyrosine-phosphorylated nuclear and cytosolic proteins are included in execution mechanisms of apoptosis, while early tyrosine-phosphorylated proteins may participate in the induction of apoptosis.

5.3 Role of Dystrobrevin-α in Leukemia Cells During Proliferation and Differentiation

Investigating the cytoplasmic and nuclear tyrosine phosphorylated proteins during granulocytic differentiation, we demonstrated for the first time the tyrosine phosphorylation of the cytoplasmic protein dystrobrevin-α (DB-α, for the nomenclature see Table 5.2) and its isoforms in human hematopoietic cells and its rapid (within 30 min) tyrosine phosphorylation after HL-60 cell treatment with ATRA (Fig. 5.6) (Kulyte et al. 2002; Borutinskaite et al. 2011; Navakauskiene et al. 2012). This protein and its isoforms have so far only been detected in muscle and nerve cells, and their function in the hematopoietic cell system was unknown.

In our previous studies (Borutinskaite et al. 2005, 2011; Kulyte et al. 2002) one of identified proteins which expression level was changed after differentiation induction with ATRA and/or apoptosis induction in NB4 and HL-60 cells was dystrobrevin. In Table 5.2 the dystrobrevin gene and protein nomenclature is presented. Dystrobrevin, the mammalian orthologue of the Torpedo 87 kDa postsynaptic protein, is a member of the dystrophin-associated protein complex (DAPC). DAPC has also been suggested to compose a putative cellular signaling complex by conferring the scaffold for multiple signaling proteins (Grady et al. 1999; Constantin 2014).

In our studies we found a few isoforms of dystrobrevin-α in leukemia cell NB4, HL-60 lines. We observed DTN-1, dystrobrevin-α, dystrobrevin-β, dystrobrevin-γ, dystrobrevin-ϵ and dystrobrevin-ζ as differently distributed in different cell

Table 5.2 The dystrobrevin gene and proteins' isoforms nomenclature

Isoforms of the protein dystrobrevin-α corresponding to gene DTNA	Isoforms of the protein dystrobrevin-β corresponding to gene DTNB
Isoform 1 (id: Q9Y4J8-1), DTN-1, 83.901 kDa	Isoform 1 (id: O60941-1), DTN-B1, 71.356 kDa
Isoform 2 (id: Q9Y4J8-2), dystrobrevin-α, 77.653 kDa	Isoform 2 (id: O60941-2), dystrobrevin-associated protein A0, 64.603 kDa
Isoform 3 (id: Q9Y4J8-3), DTN-2, 65.123 kDa	Isoform 3 (id: O60941-3), DTN-B2, 64.424 kDa
Isoform 4 (id: Q9Y4J8-4), dystrobrevin-β, 64.720 kDa	
Isoform 5 (id: Q9Y4J8-5), dystrobrevin-γ, 58.874 kDa	
Isoform 6 (id: Q9Y4J8-6), dystrobrevin-ϵ, 43.621 kDa	
Isoform 7 (id: Q9Y4J8-7), DTN-3, Alpha-dystrobrevin-3, dystrobrevin-δ, 42.394 kDa	
Isoform 8 (id: Q9Y4J8-8), dystrobrevin-ζ, 22.108 kDa	

compartments: we determined 75 kDa isoform present in membrane fraction of the cell or nuclear, the other most common isoforms were detected in cytosolic protein fraction (Borutinskaite et al. 2011).

After treatment with all-trans retinoic acid and BML-210 alone or while combining those two we detected different distribution depending on the treatment and treatment time—we detected isoforms with molecular weight between 50 and 75 kDa after treatment with HDACi BML-210. These isoforms were not detected in control proliferating or after ATRA/ATRA + BML-210 treatments (Borutinskaite et al. 2005).

Nuclear and cytoplasmic proteins were separated from HL-60 cells that were either proliferating (control) or differentiating (ATRA from 30 min to 96 h). The analysis of dystrobrevin-α expression in cytoplasmic fraction of control HL-60 cell and cells treated with all-trans retinoic acid displayed 2 losely migrating bands with molecular masses of ~75–85 kDa. Furthermore, a band of ~45 kDa was also determined. By comparison, in the nuclear fraction of control and differentiating HL-60 cells we detected three to five distinct dystrobrevin isoform proteins that differed in size from 55 to 75 kDa and also two bands of ~45 and 95 kDa (Fig. 5.6). The degrees of dystrobrevin-α in the nuclear and cytosolic fractions were basically steady during the processes of cell proliferation and granulocytic differentiation (Kulyte et al. 2002).

The exact role of dystrobrevin in cell proliferation or apoptosis is unknown, but dystrobrevin is suggested to be involved in signal transduction. The dystrophin complex is essential for muscle function, suggesting that dystrobrevin may be a substrate for downstream tyrosine phosphorylation. The schematic presentation of the Dystrophin-associated protein complex is presented in Fig. 5.7. Previously, our investigations showed that dystrobrevin is tyrosine phosphorylated after all-trans retinoic acid treatment for leukemia cells (Kulyte et al. 2002). The src protein family is involved in signal transduction at the neuromuscular junction, so there is a possibility that the kinases of src family are able to phosphorylate dystrobrevin. Also it was demonstrated that dystrophin is a substrate for various protein kinases, for example, casein kinase II, CaM kinase, and c protein kinase, both *in vitro* and *in vivo* (Senter et al. 1995). It is known that two separate classes of dystrobrevin, α and β, attach directly to dystrophin and are significant components of DAPC that connect the cytoskeleton to the extracellular matrix. In addition, DAPC was acknowledged to be molecularly heterogeneous. It exists in many tissues and plays a significant role in brain development, synapse formation and plasticity as well as water and ion homeostasis. The impact of dystrobrevin in other types of cell is still investigated insufficiently.

During dystrobrevin-α protein (Q9Y4J8) analysis, 8 named isoforms were produced by alternative splicing: DTN-1 represented by isoform with the molecular weights of 84 kDa (Q9Y4J8-1), dystrobrevin-α by 77.6 kDa (Q9Y4J8-2), DTN-2 by 65.15U (Q9Y4J8-3), dystrobrevin-β by 64.7U (Q9Y4J8-4), dystrobrevin-γ by 59 kDa (Q9Y4J8-5), dystrobrevin-ϵ by 43.6 kDa (Q9Y4J8-6), DTN-3 by 42.4 kDa (Q9Y4J8-7), and dystrobrevin-ζ by 22 kDa (Q9Y4J8-8) proteins, respectively (Nawrotzki et al. 1998; Sadoulet-Puccio et al. 1996; Bohm and Roberts 2009).

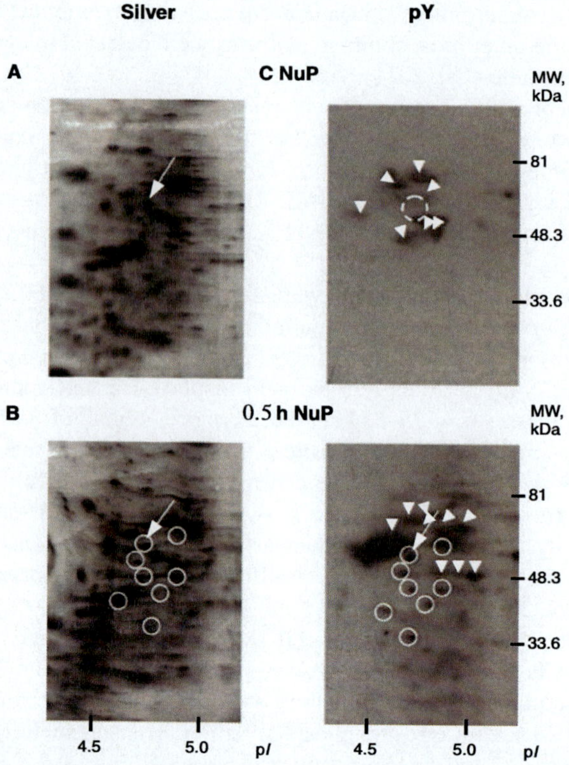

Fig. 5.6 Two-dimensional electrophoretic patterns of total and tyrosine-phosphorylated proteins with acidic pI isolated from the differentiating HL-60 cells nuclear. Total nuclear proteins were isolated from proliferating HL-60 cells (**A**) and from cells induced to granulocytic differentiation with ATRA for 30 min (0.5 h) (**B**). The isolated proteins were fractionated by 2DE, visualized by silver staining (left panel) or checked for tyrosine phosphorylation (pY). In both images, the arrow indicates a protein that was tyrosine-phosphorylated in the nuclei after induction of differentiation and by mass spectrometry analysis was identified as human dystrobrevin-γ. The arrowheads in pY indicate proteins that were consistently found to be tyrosine-phosphorges show nuclear proteins that were not tyrosine phosphorylated in proliferating cells, but underwent tyrosine phosphorylation after 30 min treatment with ATRA. Migration of the molecular size marker proteins is indicated to the right (kDa values). According Kulyte et al. (2002), License No 1025167-1

To describe the distribution of dystrobrevin-α isoforms in various cell fractions, hydrophilic and hydrophobic proteins, also nuclear and cytosolic proteins were separated from proliferating NB4 cells (Borutinskaite et al. 2011). Proteins were additionally fractionated with SDS-PAGE, transferred to PVDF membrane, and studied with a polyclonal antibody against dystrobrevin-α.

In the hydrophobic protein fraction of whole NB4 cell lysates, which consists of all membrane fractions, the antibody shown 2 closely migrating bands with the molecular weight of nearly 75 and 80 kDa, representing the dystrobrevin-

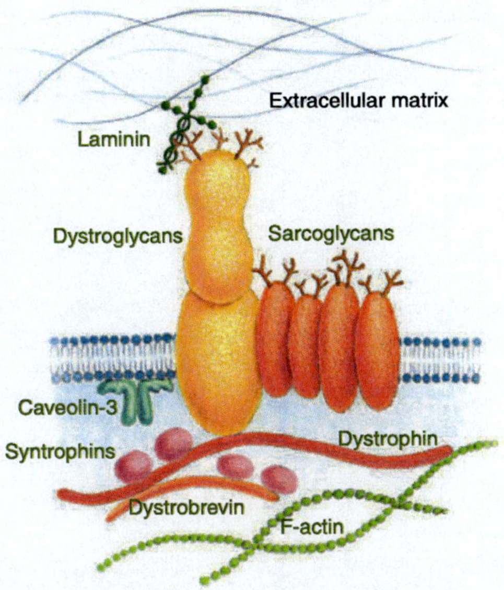

Fig. 5.7 The schematic presentation of the Dystrophin-associated protein complex

α splice isoform (Fig. 5.8B). dystrobrevin-α, dystrobrevin-β, dystrobrevin-γ and dystrobrevin-ζ isoforms existed both in the cytosolic (Fig. 5.8D) and the hydrophilic fractions (Fig. 5.8A), however in different proportions. Merely the dystrobrevin-α isoform existed in the hydrophobic nuclear fraction (Fig. 5.8C), while the hydrophilic nuclear fraction consisted of dystrobrevin-α and dystrobrevin-β isoforms (Fig. 5.8E). The band with 70 kDa molecular weight in hydrophilic fractions can be dystrobrevin-β gene product (DTN-B1). In conclusion, we showed that dystrobrevin-α isoforms are expressed in NB4 cells, and that the degree of dystrobrevin-α isoforms expression differs in different cell fractions, such as nucleus, cytosol, and the hydrophobic cell compartments. Thus three to five dystrobrevin isoforms existed in the hydrophilic fraction of the whole cell lysates, in the nucleus and the cytoplasm. Nonetheless, the hydrophobic fraction contains only the biggest protein isoforms (dystrobrevin-α) that proves their relation with membrane structures.

To describe the human dystrobrevin-γ expression throughout granulocytic differentiation of HL-60 cells we isolated nuclear and cytoplasmic proteins from cells that were proliferating (control) or differentiating (ATRA for 30 min to 96 h). Total expression of cytoplasmic and nuclear dystrobrevin-γ was estimated by applying the antibodies against dystrobrevin (Fig. 5.9). In the cytosol the antibody has shown 2 closely migrating bands with molecular weights of nearly 75–88 kDa. In comparison, in the nuclear fraction we identified three different isoforms proteins which size was from 51 to 70 kDa (Fig. 5.9A). The level of dystrobrevin-γ in the nuclear fraction was basically steady throughout both proliferation and granulocytic differentiation.

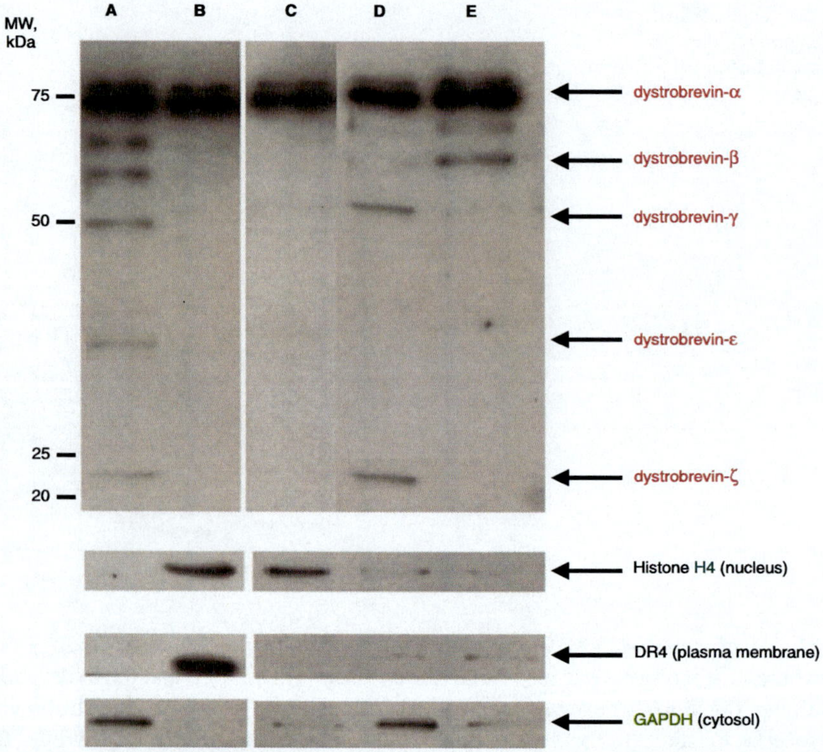

Fig. 5.8 Expression of dystrobrevin-α isoforms in NB4 cells. Hydrophobic and hydrophilic proteins from the NB4 cell line were solubilised and extracted using the Mem-PER Kit. The total hydrophilic (soluble proteins) NB4 cell fraction (**A**), the total hydrophobic (membrane proteins) fraction (**B**), the nuclear hydrophobic (**C**) fraction, the nuclear hydrophilic (**E**) and the cytosolic (**D**) fractions were fractionated in 8–18% SDS-PAGE gel and visualized by Western blot with polyclonal antibodies against dystrobrevin-α. Proteins regarded as markers for different cell compartments were detected using protein-specific antibodies and HRP-conjugated secondary antibodies with chemiluminescent detection. Migration of the molecular size marker proteins is indicated to the left (kDa values). According Borutinskaite et al. (2011), License No 4796590532046

We also examined dystrobrevin-α isoforms in HL-60 cells throughout proliferation process and during the three main phases of granulocytic differentiation (i.e., at 30 min and 24 and 96 h of treatment with ATRA, accordingly representing the beginning, commitment, and maturation phases). Four distinct proteins were identified that varied in size (51–70 kDa) and had different pI values that represent dystrobrevin-α isoforms (Fig. 5.9B). Dystrobrevin-γ isoform I (Fig. 5.9B) existed in the nuclei of both proliferating and differentiated cells, and the largest relative amount of dystrobrevin-γ isoform I was noticed in terminally differentiated granulocytes (Fig. 5.9B, maturation). By comparison, the amount of dystrobrevin-γ isoform III was negligible in the course of proliferation, at the beginning of

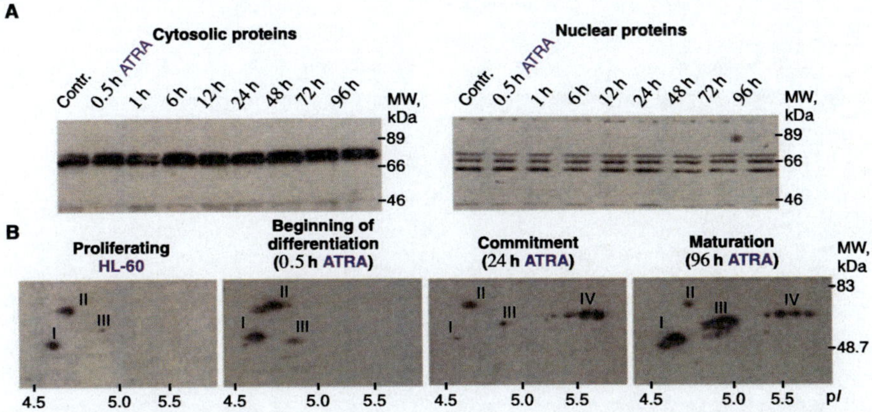

Fig. 5.9 Expression of human dystrobrevin-α isoforms in HL-60 cells undergoing granulocytic differentiation. Cytosol and nuclear proteins were isolated from proliferating and differentiated (ATRA treatment for 30 min, 24 h, and 96 h) HL-60 cells. The isolated proteins were fractionated by SDS-PAGE on a 7–15% acrylamide gel gradient (**A**) or 2-DE (**B**), blotted on PVDF membranes, and analyzed with polyclonal antibodies against human dystrobrevin-γ. The numerals I–IV denotes different isoforms of human dystrobrevin-α detected in the nucleus of the HL-60 cells. Migration of the molecular size marker proteins is indicated to the right (kDa values). According Kulyte et al. (2002), License No 1025167-1

differentiation, and during the commitment phase, whereas the degree growed dramatically during granulocytic maturation. Eventually, isoform IV was identified only at the beginning of differentiation, and its molecular weight and pI value were very similar to those of isoform II.

During the experiments we noticed that in proliferating and differentiating leukemia cells only one dystrobrevin the isoform (dystrobrevin-γ) underwent tyrosine phosphorylation (Fig. 5.10) (Kulyte et al. 2002). Our results support an observation, that the phosphorylation state of dystrobrevin-γ in the nucleus is changed throughout differentiation process after leukemia cell line treatment with ATRA (Kulyte et al. 2002). The tyrosine phosphorylation of the dystrobrevin-γ observed in the differentiating cells (circles in Fig. 5.10, pY) was not noticed in the proliferating control cells (dashed circle in Fig. 5.10, pY).

Presented results demonstrate that dystrobrevin-γ is expressed in human promyelocytic HL-60 cells, and, in the nucleus is changed by tyrosine phosphorylation followed by differentiation which is induced by treatment with ATRA (Kulyte et al. 2002). By using antibodies against human dystrobrevin-γ, we estimated distribution of this protein distinct dystrobrevin-γ, i.e., in two cell compartments—the nucleus and the cytosol. We also demonstrated co-localization with actin and myosin light chain in association with human dystrobrevin-γ in proliferating HL-60 cells, differentiated to mature granulocytes, and in human neutrophils.

It is supposed that dystrophin-related and dystrophin-associated protein complexes play a significant role in the plasma membrane stability and maintenance

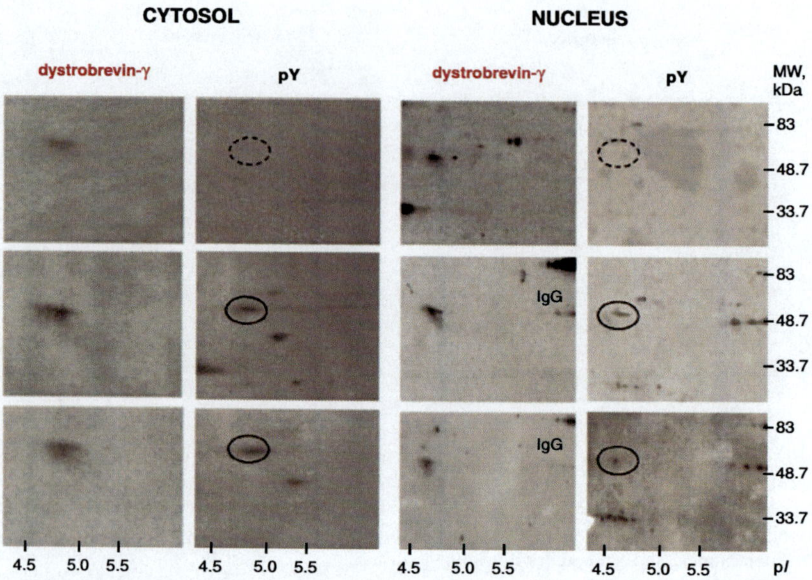

Fig. 5.10 Human dystrobrevin-γ tyrosine phosphorylation in proliferating and differentiating HL-60 cells. Proteins isolated from proliferating and differentiating HL-60 cells were used for immunoprecipitation with human dystrobrevin-γ. Co-immunoprecipitated proteins were subjected to 2-DE fractionation, silver stained (Silver) or blotted on PVDF membranes. The proteins on the membrane were detected with antibodies against dystrobrevin-γ. The membranes were subsequently reprobed with anti-phosphotyrosine antibodies to ascertain tyrosine phosphorylation (pY). The dashed circle and the circles indicate the absence or presence of tyrosine-phosphorylated dystrobrevin respectively. Migration of the molecular size marker proteins is indicated to the right (kDa values). According Kulyte et al. (2002), License No 1025167-1

during muscle contraction and relaxation, and they have also been involved in cell signaling events. These proteins are perfectly placed to transduce signals from the extracellular matrix to the cytoskeleton and there exists proof of direct phoshorylation of dystrophin and the DAPC proteins. The presence of multiple isoforms with various tissue specificities allow to argue that dystrobrevin has different functions (Sadoulet-Puccio et al. 1996). The experiments conducted by us show that there are various dystrobrevin-γ isoforms in the cytoplasm and nucleus of leukemic cells, which in turn signify that these isoforms possess functional rather than structural properties. The signaling role is hinted by the fact that occurrence of tyrosine phosphorylation of dystrobrevin is specific to the phase of differentiation (Kulyte et al. 2002).

5.4 Dystrobrevin-α in Proliferating, Differentiated and Apoptotic Leukemia Cells

In this research we treated NB4 cells with 1 μM ATRA alone or in combination with HDAC inhibitor 5 μM BML-210 to provoke cell differentiation into granulocyte-like cells. As we demonstrated earlier (Savickiene et al. 2006), 1 μM ATRA alone and in combination with 5 μM BML-210 can stimulate differentiation up to 80% after 96 h of treatment. The degrees of dystrobrevin-α isoforms were basically steady in all fractions during both proliferation and granulocytic differentiation of NB4 cells, i.e., after treatment with ATRA alone or plus the HDACi BML-210 for 96 h. When leukemic cells were treated with higher concentrations of HDACi they underwent apoptosis. Therefore after NB4 treatment with 10 μM of BML-210 cells became apoptotic. Figure 5.11 demonstrates the alterations in the expression of dystrobrevin-α in proliferating and induced to differentiation human leukemic cells. Its expression increases during apoptosis processes.

To determine proteins interacting with dystrobrevin-α in leukemia cells we immunoprecipitated protein dystrobrevin with total leukemia cell proteins isolated from proliferating and stimulated to granulocytic differentiation with ATRA and HDAC inhibitor BML-210 alone or in combination, fractionated in 2DE system (Fig. 5.12). The chosen proteins were analyzed by mass spectrometry (Borutinskaite et al. 2011). Nearly 100–120 protein spots were observed on every gel (Fig. 5.12). According to quantitative and statistical analyses, about 45% of these spots were variously expressed in the untreated (control) cells and the treated cells. Some spots expressed in different ways were detected by

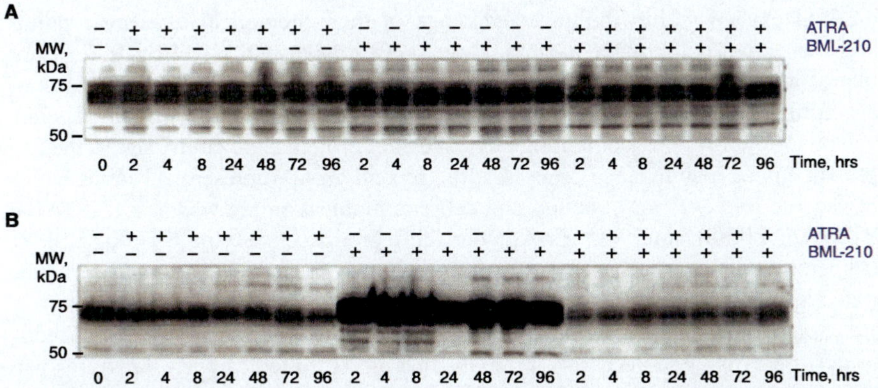

Fig. 5.11 Expression of human dystrobrevin-α in NB4 cells treated with ATRA and HDAC inhibitor BML-210. Total soluble (**A**) and insoluble (**B**) proteins were isolated from proliferating and differentiating (0.5–96 h) NB4 cells. The proteins were fractionated by SDS-PAGE on a 7–15% acrylamide gel gradient and transferred on to immobilon PVDF membrane. The membranes were analyzed with polyclonal antibodies against human dystrobrevin-α. According Borutinskaite et al. (2005)

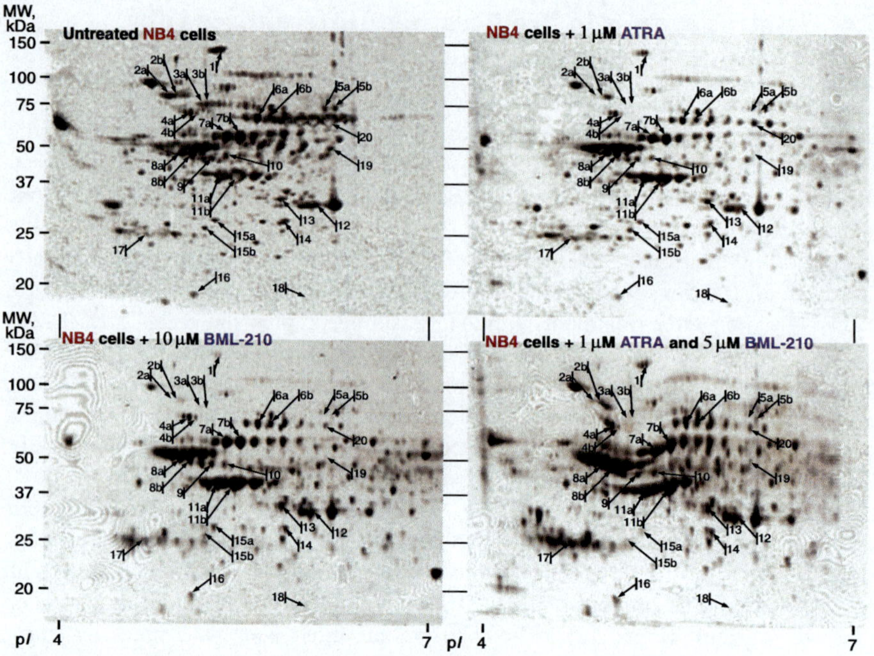

Fig. 5.12 Proteins associated with dystrobrevin-α in proliferating NB4 cells and cells induced to granulocytic differentiation with ATRA and HDAC inhibitor BML-210 alone or in combination. According Borutinskaite et al. (2011), License No 4796590532046

applying MALDI-TOF-MS and ESI-MS-MS analysis. Proteins detected (Table 5.3) on 2DE existed within the expected range of their theoretical isoelectric points and molecular sizes. After these analyses both known and novel dystrobrevin-α interacting proteins were identified. These proteins could be divided into groups according to their function: the first group containing b-Actin, α/β-tubulin, desmin, gelsolin, tropomyosin, stathmin are cytoskeleton proteins and contribute to the cell growth and/or maintenance processes; the second group represents proteins which participate in signal transduction and cell communication processes, e.g., STAG1, RIBA, prohibitin, and AnnexinA4; the third group consists of HSP90, HSP70, chaperonin-60, and BiP, which can be participate in protein metabolism, signal transduction, and gene transcription regulation.

These dystrobrevin associated proteins are able to attach directly or indirectly to dystrobrevin-α via accessory proteins, that are recruited to dystrobrevin as part of multiprotein complexes (Fig. 5.13). All the thirteen detected dystrobrevin-α associated proteins (STAG1, Gp96, HSP90β, BiP, HSP70-9, HSP70-8, chaperonin 60, desmin, α/β-tubulin, β-actin, tropomyosin, and prohibitin) contain the coiled-coil (CC) domain. We argue that these proteins directly interact with dystrobrevin-α

Table 5.3 Dystrobrevin-α associated proteins identified by mass spectrometry. G1—untreated NB4 cells; G2—NB4 cells + 1 μM ATRA; G3—NB4 cells + 10 μM BML-210; G4—NB4 cells + 1 μM ATRA + 5 μM BML-210. According Borutinskaite et al. (2011), License No 4796590532046

Spot No.	Theor. Mw, kDa	Theor. pI	Ratios G2/G1	Ratios G3/G1	Ratios G4/G1	Accession number (GI), protein name	Domains	Biological activity
1	140	5.3	−1.08	−3.6	1.2	40352781, Cohesin subunit SA-1 (STAG1)	Coiled-coil	Component of cohesin complex, a complex required for the cohesion of sister chromatids after DNA replication; may also play a role in spindle pole assembly during mitosis.
2a; 2b	90	4.7	−3.87	−23.7	−4.9	15010550, Heat shock protein Gp96 precursor	HATPase Coiled-coil	Molecular chaperone that functions in the processing and transport of secreted proteins.
3a; 3b	83	5.0	−1.57	−7.5	1.8	17865718, Hsp90-1, beta	Coiled-coil	Molecular chaperone. It has ATPase activity, interacts with TP53/p53.
4a; 4b	72	5.1	−10.2	−8.7	−5.7	14916999, 78 kDa glucose-regulated protein precursor (BiP, GRP78)	Coiled-coil	It plays a role in facilitating the assembly of multimeric protein complexes inside the ER.
5a; 5b	73	5.9	−9.76	−9.6	−6.3	24234688, Heat shock 70 kDa protein 9B (GRP75)	Coiled-coil	Implicated in the control of cell proliferation and cellular aging. May also act as a chaperone.
6a; 6b	70	5.4	1.54	1.6	1.7	123648, Heat shock cognate 71 kDa protein	Coiled-coil	Chaperone.
7a; 7b	61	5.7	−1.16	1.3	1.5	306890, Chaperonin (HSP60)	Coiled-coil	Implicated in mitochondrial protein import and macromolecular assembly. May facilitate the correct folding of imported proteins. May also prevent misfolding and promote the refolding and proper assembly of unfolded polypeptides generated under stress conditions in the mitochondrial matrix.

(continued)

Table 5.3 (continued)

Spot No.	Theor. Mw, kDa	pI	Ratios G2/G1	G3/G1	G4/G1	Accession number (GI), protein name	Domains	Biological activity
8a	50	4.7	−22.7	−9.8	−7.6	18088719, Beta-tubulin	TUBULIN Coiled-coil	Tubulin is the major constituent of microtubules. It binds two moles of GTP, one at an exchangeable site on the beta chain and one at a non-exchangeable site on the alpha-chain.
8b	50	4.9	−3.93	−3.3	−1.8	2843123, Alpha-tubulin	TUBULIN Coiled-coil	Tubulin is the major constituent of microtubules. It binds two moles of GTP, one at an exchangeable site on the beta chain and one at a non-exchangeable site on the alpha-chain.
9	47	5.0	1.31	1.3	1.4	61104905, Heat shock pro-tein 90Ad	HATPase	ATP binding.
10	53	5.2	1.40	1.4	1.5	21358854, Desmin	Coiled-coil	Desmin are class-III intermediate filaments found in muscle cells. In adult striated muscle they form a fibrous network connecting myofibrils to each other and to the plasma membrane from the periphery of the Z-line structures.
11a; 11b	41	5.5	1.66	1.8	2.5	14250401, Beta-actin	ACTIN Coiled-coil	Actins are highly conserved proteins that are involved in various types of cell motility and are ubiquitously expressed in all eukaryotic cells.
12	33	5.6	1.60	1.6	1.7	39645467, Annexin A4	ANX	Calcium/phospholipid-binding protein, which promotes membrane fusion and is involved in exocytosis.
13	34	5.5	1.64	2.4	2.4	55958543, heterogeneous nuclear ribonucleo-protein K	Nuclear localization signal, RNA binding	One of the major pre-mRNA-binding proteins. Binds tenaciously to poly(C) sequences. Likely to play a role in the nuclear metabolism of hnRNAs, particularly for pre-mRNAs that contain cytidine-rich sequences. Can also bind poly(C) single-stranded DNA.

(continued)

Table 5.3 (continued)

Spot No.	Theor. Mw, kDa	pI	Ratios G2/G1	G3/G1	G4/G1	Accession number (GI), protein name	Domains	Biological activity
14a; 14b	29	5.5	1.42	1.9	1.9	66267315, prohibitin	PHB Coiled-coil	Prohibitin inhibits DNA synthesis. It has a role in regulating proliferation. As yet it is unclear if the protein or the mRNA exhibits this effect. May play a role in regulating mitochondrial respiration activity and in aging.
15a;	29	4.7	1.43	1.7	1.6	19072649, TPMsk3	Coiled-coil	Binds to actin filaments in muscle and non-muscle cells.
15b	26	4.8	1.06	1.3	1.9	55665778, tropomyosin 3	Coiled-coil	Binds to actin filaments in muscle and non-muscle cells. Plays a central role, in association with the troponin complex, in the calcium dependent regulation of vertebrate striated muscle contraction. Smooth muscle contraction is regulated by interaction with caldesmon. In non-muscle cells is implicated in stabilizing cytoskeleton actin filaments.
16	17	4.8	2.36	2.6	2.8	2286103, RIBA	RAS	Binds and exchanges GTP and GDP.
17	22	4.6	1.97	2.3	2.4	55960303, gelsolin	Signal peptide GEL	Calcium-regulated, actin-modulating protein that binds to the plus (or barbed) ends of actin monomers or filaments, preventing monomer exchange (end-blocking or capping). It can promote the assembly of monomers into filaments (nucleation) as well as sever filaments already formed.
18	17	5.7	1.05	1.1	1.6	5031851, Stathmin		Involved in the regulation of the microtubule (MT) filament system by destabilizing microtubules. Prevents assembly and promotes disassembly of microtubules. Phosphorylation at Ser-16 may be required for axon formation during neurogenesis.
19	57	5.8	2.08	2.6	2.6	2245365, ER-60 (disulfide isomerase)	Signal peptide	Catalyzes the rearrangement of -S-S- bonds in proteins.
20	68	5.6	1.35	1.3	1.5	10800417, BRCA1 associated protein		Deubiquitinating enzyme which may be involved in BRCA1 signal transduction pathway.

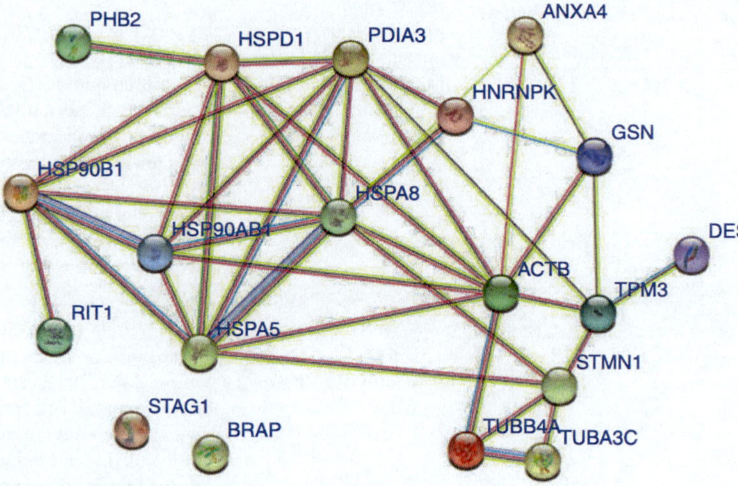

Fig. 5.13 The network of dystrobrevin-α-associated proteins. These are the group of biologically related proteins, generated on the basis of an experimentally derived dataset. The graphic layout and network display were performed using the Ingenuity pathway analysis finder (www.Ingenuity.com) or Human Protein Reference Database by providing the Unigene/Swiss-Prot accession number of the proteins those were identified from gels (Fig. 5.12) and are listed in Table 5.3. According Borutinskaite et al. (2011), License No 4796590532046

via the coiled-coil domain. Consequently, the proteins with CC domain are able to act as a bridge associating dystrobrevin-α with the other dystrobrevin-α interacting proteins.

In this study (Borutinskaite et al. 2011), we have identified proteins that interact with dystrobrevin-α in promyelocytic leukemia cells after treatment with differentiating chemical agents (ATRA, BML-210). Identified proteins are essential in cytoskeleton reorganization, signal transduction, and other cellular processes. Therefore, the dystrobrevin-α may act as a significant signaling and structural protein throughout proliferation and differentiation processes of human cancer cells.

5.5 Dystrobrevin-α in Induced to Apoptosis Leukemia Cells

Earlier we have studied spatial localization and the biochemical properties of dystrobrevin-α and other isoforms in human promyelocytic HL-60 cells (Kulyte et al. 2002). As a result we demonstrated newly that dystrobrevin-α and its splice isoforms exist in human hematopoietic cell system, and that it is observed both in the nucleus and the cytoplasm. We also estimated a distinct distribution pattern of dystrobrevin-γ, including co-localization with actin, in promyelocytes, differentiated mature granulocytes, and human neutrophils. Additionally, dystrobrevin-γ was tyrosine phosphorylated short time later the onset of all-trans retinoic acid stimulated differentiation.

Further we investigated the proteins that associate with dystrobrevin-α during leukemia cells induced apoptosis by using etoposide (Navakauskiene et al. 2012). For the stimulation of the HL-60 cells apoptosis the etoposide was applied, which is an inhibitor of the topoisomerase II. Inhibition of the stimulated apoptosis by etoposide was conducted by using Z-VAD(OH)-FMK, which is a peptide blocking activity of etoposide-induced caspase. The alterations of dystrobrevin-α in proliferating and apoptotic cells were estimated by aplying the computational methods.

To investigate alterations in the expression of dystrobrevin-α and its splice isoforms in the nucleus and the cytoplasm of proliferating and apoptosis-induced human promyelocytic leukemia HL-60 cells, nuclear and cytosolic proteins were fractionated by SDS/PAGE, transferred onto PVDF Immobilon-P membranes and then investigated with antibodies against to dystrobrevin-α to evaluate the relative presentation of dystrobrevin-α isoforms expression in the nucleus and the cytoplasm (Fig. 5.2). In fact, there were a few quantitative variations in the expression of the cytoplasmic dystrobrevin-α isoform with a molecular weight of 75 kDa, dystrobrevin-β of 62 kDa and dystrobrevin-γ of 58 kDa after treatment with 68 μM Et, appearing even with presence of caspase inhibitor Z-VAD(OH)-FMK (Fig. 5.14 CytP). We observed that dystrobrevin-α and its splice isoforms expression changed in the nucleus too (Fig. 5.14 NuP). The dystrobrevin-α isoform (75 kDa) was hardly detectable in proliferating cells but rised dramatically in the nucleus of apoptotic cells (9, 18 h of Et treatment). Both β- and dystrobrevin-γ (62 and 58 kDa, respectively) existed at high levels in the nucleus at the start of apoptosis (0.5–3 h Et), but lowered when apoptosis in cell population reached approximately 70% (Fig. 5.14, 18 h Et). When Z-VAD(OH)-FMK was involved the levels of the dystrobrevin-α isoforms (75 and 62 kDa) were nearly the same as f the level of proliferating control cells; that of dystrobrevin-γ (58 kDa) was basically kept throughout the stimulation of apoptosis. Overall, the first conclusion made by us may be that the expression level of dystrobrevin-α isoform was opposite to the dystrobrevin-β and dystrobrevin-γ, and second that dystrobrevin-α and dystrobrevin-γ were influenced by Et-induced apoptosis.

A proteomic approach was applied to determine the proteins co-precipitating with dystrobrevin-α. Proteins of total leukemia cell were separated from proliferating HL-60 cells and cells provoked to conduct apoptosis with etoposide for 3 h and 6 h. After immunoanalysis dystrobrevin spot volume was evaluated (Fig. 5.15). We identified that the dystrobrevin-α spot volume raised 1.3-fold after 3 h treatment with etoposide 2.1-fold after 6 h treatment with etoposide in comparisson to the level in proliferating control leukemia cells. The raised level of dystrobrevin-α in apoptotic cells correlates with apoptotic processes.

Proteins associated with dystrobrevin-α in leukemia cells during apoptosis were investigated by treating the HL-60 cells with 68 μM etoposide or 25 μM Z-VAD(OH)-FMK separately or in combination (Fig. 5.16). The precipitated com-

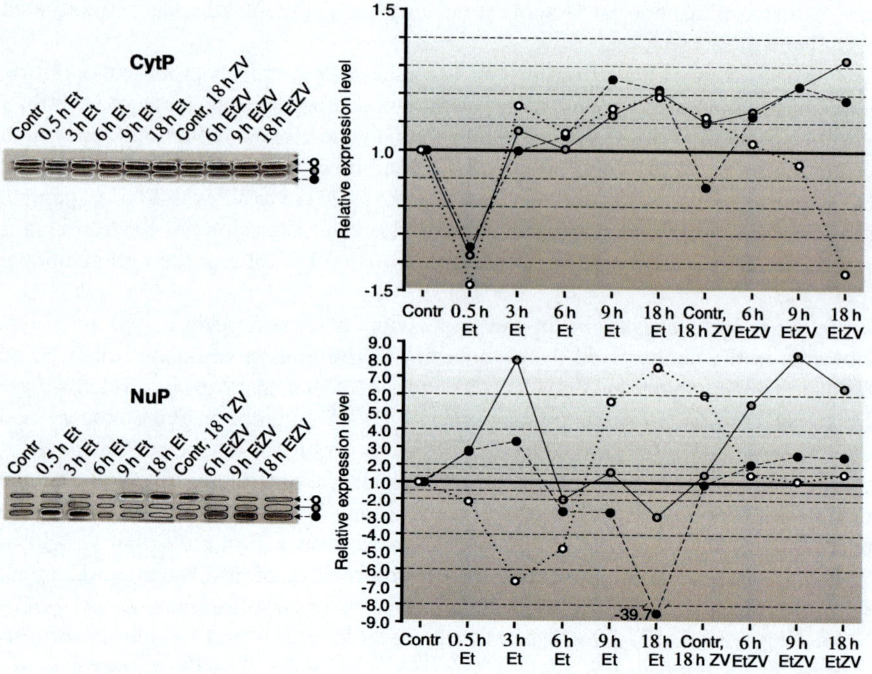

Fig. 5.14 Dystrobrevin expression and the quantification of the levels in the nucleus and cytosol of HL-60 cells during apoptosis. Cytosolic (CytP) and nuclear proteins (NuP) were isolated from HL-60 cells, that were proliferating (Contr), induced to undergo apoptosis with 68 μM of etoposide (Et) for different times (0.5, 3, 6, 9 and 18 h Et), treated with 25 μM Z-VAD(OH)-FMK alone for 18 h (Contr, 18 h ZV) or together with etoposide (6, 9, 18 h EtZV), subsequently fractionated by SDS-PAGE electrophoresis; finally proteins were transferred onto PVDF membranes and immunoblotted with anti-dystrobrevin (DB) antibodies. Protein concentrations were estimated before SDS electrophoresis and equal amounts of sample protein were loaded onto the gel. The relative protein expression changed with the different treatments in the cytosol and nucleus; a positive ratio denotes increase in the level of expression and a negative ratio—a decrease; the dotted line with white markers represents dystrobrevin-α, the solid line with grey markers—dystrobrevin-β and the dashed line with black markers—dystrobrevin-γ. The ratios of protein expression levels were computed by the new image analysis algorithms implemented in `Matlab`. According Navakauskiene et al. (2012), License No 4797021401553

plexes were later resolved by 2-DE by applying pH range 4–7 and visualized with silver staining (Fig. 5.16).

Proteins co-precipitating with dystrobrevin-α were detected by applying the mass spectrometry. The expression degrees of the detected proteins were estimated by applying the computer-assisted image analysis of the gels (Table 5.4). After this we determined relations with structural and transport proteins (tropomyosin, vimentin, myosin), membrane (ADAM21, syntrophin, potassium voltage-gated channel), ER-Golgi (TGN51, eIF38), and nuclear (Lamins, ribonucleoprotein C1/C2) proteins. These proteins are involved in distinct cellular structures and cellular processes,

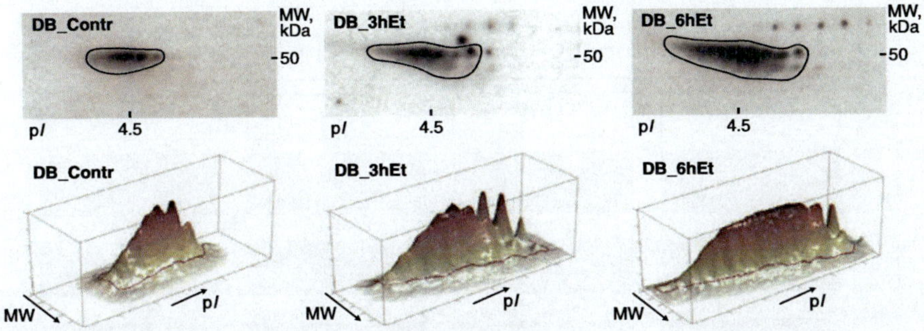

Fig. 5.15 Dystrobrevin level evaluation in proliferating and apoptotic leukemia cells. According Navakauskiene et al. (2012), License No 4797021401553

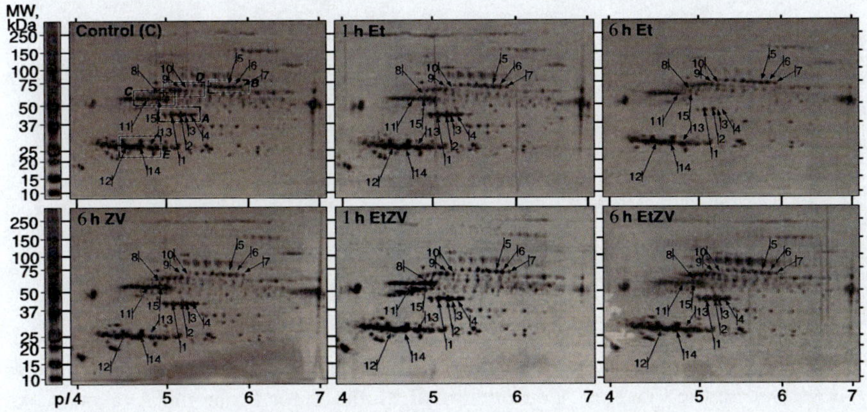

Fig. 5.16 Proteins co-precipitating with dystrobrevin-α in proliferating, induced to apoptosis with etoposide alone or with etoposide inhibitory treatment with Z-VAD(OH)-FMK cells. According Navakauskiene et al. (2012), License No 4797021401553

suggesting the dystrobrevin-α interaction—networks in a key controller during apoptosis of human promyelocytic leukemia cells.

Five protein groups (Fig. 5.16 in C marked as A, B, C, D, and E) were used during the computer-assisted analysis to estimate the protein-level alterations in cells treated by various methods. In one experiment there were compared the C, 1hEt and 6hEt gel images (Fig. 5.17, Comparison 1) and in the other, 6hEt wih 6hEtZV (Fig. 5.17, Comparison 2). Results of detailed computer analysis of 15 identified areas of interest are also summarized in Table 5.4. Fold change of average normalized spot volumes together with experimental values of molecular weight (MW) and isoelectric point (pI) were evaluated. Detected areas are marked with corresponding labels in Fig. 5.16. Figure 5.17 shows 5 enlarged areas of analyzed images which include all 15 studied protein spots. Positions of enlarged areas in full gel images are depicted with rectangles in C gel of Fig. 5.16. Alterations

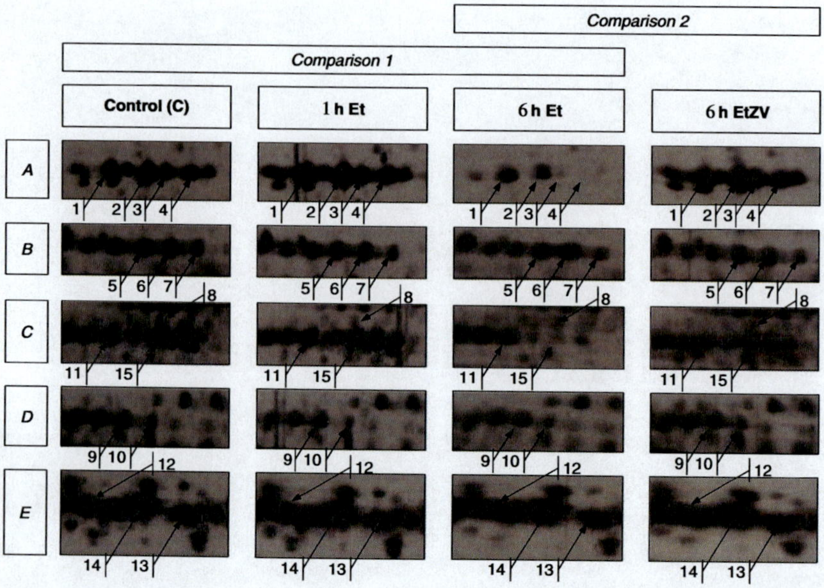

Fig. 5.17 Dystrobrevin associated proteins with supreme level changes during apoptosis. Spot labels are the same as in Fig. 5.16 and Table 5.4. Comparison 1 and 2 focus on the effects of etoposide (Et) alone or in combination with caspase inhibitor Z-VAD(OH)-FMK (ZV). According Navakauskiene et al. (2012), License No 4797021401553

in spot volume were represented as ratios of averages of normalized spot volumes as described in the Chapter "Computational Methods for Proteome Analysis". An increase in spot abundance is represented with positive fold change and a decrease, with a negative fold change. Alterations between protein spots from gels C and 1hEt, 1hEt and 6hEt, C and 6hEt of the first experimental group and between 6hEt and 6hEtZV of the second experimental group are summarized in Table 5.4.

This research was conducted to study the role of dystrobrevin-α and its splice isoforms and relation with other proteins throughout the progression of human promyelocytic leukemia cell line HL-60 apoptosis process. We first determined that there were noticeable qualitative and quantitative differences in the expression of cytoplasmic and nuclear dystrobrevin-α, not only throughout cell differentiation process (Kulyte et al. 2002), but through apoptosis as well (Borutinskaite et al. 2011; Navakauskiene et al. 2012).

Earlier, we have determined dystrobrevin-α relation with actin and myosin light chain both in human promyelocytic leukemia cells and human neutrophils (Kulyte et al. 2002). There is no sufficient data on proteins related with human dystrobrevin-α and its splice isoforms. Since there is an evidence that actin cytoskeleton is involved in regulation of many signaling processes, such as receptor affinity, cell cycle control, apoptosis, and others, our achieved results could provide new information on how dystrobrevin-α potentially contributes to signal transduction in

Table 5.4 The summarized search results (by `PepIdent`, `EMBL`, `MS-Tag` software) of proteins co-precipitated with dystrobrevin-α in HL-60 cells induced to apoptosis. According Navakauskiene et al. (2012), License No 4797021401553

Spot No.	Access. No.	Protein name	Match, %	Theor.		Exp.		Av. vol. ratio (fold change)			
				Mw, kDa	pI	Mw, kDa	pI	1 ⇝ 2	2 ⇝ 3	1 ⇝ 3	3 ⇝ 6
1	P09493	Tropomyosin 1 alpha chain (Alpha tropomyosin)	30	40	5.2	44	5.1	1.21	−1.60	−1.32	1.71
		Tropomyosin beta chain (Tropomyosin 2, Beta-ropomyosin)	25.7								
		Tropomyosin alpha 3 chain (Tropomyosin 3, Tropomyosin gamma)	26.1								
		Tropomyosin alpha 4 chain (Tropomyosin 4, TM30p1)	23.6								
2	P07910	Heterogenous nuclear ribonucleoproteins C1/C2	18.1	40	5.1	44	5.2	1.21	−1.59	−1.31	2.05
3	O75821	Eukaryotic translation initiation factor 3 subunit 4 (eIF-3delta)	23.1	40	5.2	44	5.2	1.18	−2.94	−2.49	3.16
4	O43493	Trans-Golgi network protein TGN51	27.6	40	5.4	43	5.3	1.08	−3.96	−3.68	4.12
5	Q03252	Lamin B2	22	70	5.4	73	5.8	−1.12	−2.53	−2.83	2.61
6	P02545	Lamin A/C	23	70	5.5	73	5.8	−1.18	1.39	1.18	−1.32
7	P02545-2	Splice isoform Lamin C of Lamin A/C	19	70	5.6	74	5.9	−1.18	1.19	1.01	−1.16
8	P22459	Potassium voltage-gated channel subfamily A member 4 (HK1, HPCN2)	21.4	62	4.9	63	4.9	−1.48	1.64	1.10	−1.41
9	Q9UKJ8	ADAM 21	23.6	65	5.1	70	5.2	−1.02	1.04	1.02	−1.24
10	Q9UKJ8	ADAM 21	19.4	65	5.2	69	5.3	−1.59	1.31	−1.21	−1.01
11	Q9NSN8-2	Splice isoform 2 of gamma-1-syntrophin	17	55	4.9	59	4.8	−1.17	−1.09	−1.28	−1.17
12	P12829	Myosin	30.6	22	4.7	25	4.6	−1.14	1.03	−1.11	1.04
13	Q93038	Tumor necrosis factor receptor	20.2	25	4.8	24	4.8	−1.10	1.24	1.12	1.08
	P25942-5	Splice isoform II of Tumor necrosis factor receptor superfamily member 5 precursor (CD40L receptor, B-cell surface antigen CD40)	31.5								
14	Q9Y696	Chloride intracellular channel protein 4	22.3	25	4.7	25	4.7	1.04	1.03	1.07	1.03
15	P08670	Vimentin	37	57	4.9	58	4.9	1.15	−2.29	−1.99	2.19

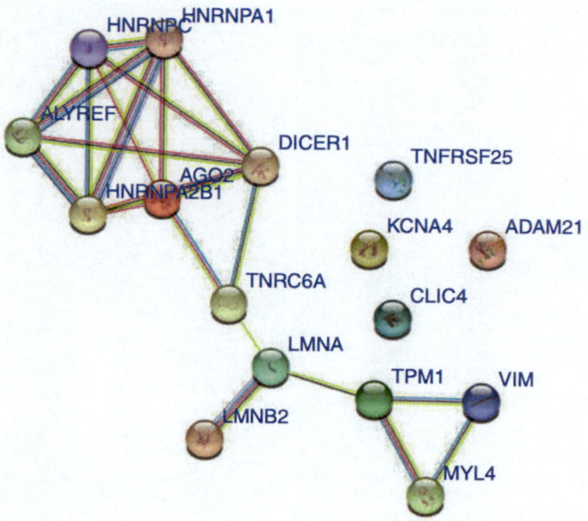

Fig. 5.18 The dystrobrevin-α and its associated proteins during leukemia cell apoptosis processes networking scheme. According Navakauskiene et al. (2012), License No 4797021401553

myeloid or phagocyte cells during cell differentiation and apoptosis progression. We here detected some proteins that co-precipitate with dystrobrevin-α both in control and apoptosis-induced HL-60 cells. Still the fifteen detected proteins are only the most abundant and distinguished of those co-precipitating with dystrobrevin-α. We identified myosin, tropomyosin, ADAM 21, chloride intracellular channel protein 4, heterogeneous ribonucleoprotein C/C1, potassium voltage-gated channel subfamily A member 4 (HK1), Trans-Golgi network protein TGN51, Eukaryotic translation initiation factor 3 subunit 4, lamin A/C and B2, tumor necrosis factor receptor, and vimentin. This group of proteins may be seen arbitrary, hence we applied the computer-assisted analysis to estimate both the alterations in the degrees of expression and the pattern formation of proteins related with dystrobrevin-α.

The network scheme between dystrobrevin and associated proteins during leukemia cell apoptosis processes is shown in Fig. 5.18.

Based on our findings (Navakauskiene et al. 2012) we designed the dystrobrevin-α (DTNA) interaction network that represents a wide spectrum of proteins that in direct or indirect relation with dystrobrevin could participate in apoptosis, and in many other cellular processes as well. Overall, our outcomes supply novel information regarding potential involvement of dystrobrevin-α and its splice isoforms in a complex with other proteins in apoptosis of myeloid cells and suggest further support to signaling role for the dystrophin-associated protein complex in various cellular compartments and cellular processes.

References

Bohm SV, Roberts RG (2009) Expression of members of the dystrophin, dystrobrevin, and dystrotelin superfamily. Crit Rev Eukaryot Gene Expr 19(2):89–108

Borutinskaite V, Magnusson KE, Navakauskiene R (2005) Effects of retinoic acid and histone deacetilase inhibitor Bml-210 on protein expression in NB4 cells. Biologija 4:88–93

Borutinskaite VV, Magnusson KE, Navakauskiene R (2011) α-Dystrobrevin distribution and association with other proteins in human promyelocytic NB4 cells treated for granulocytic differentiation. Mol Biol Rep 38(5):3001–3011. https://doi.org/10.1007/s11033-010-9965-9

Constantin B (2014) Dystrophin complex functions as a scaffold for signalling proteins. Biochim Biophys Acta Biomembr 1838(2, SI):635–642. https://doi.org/10.1016/j.bbamem.2013.08.023

Fuchs O (2010) Transcription factor nf-κb inhibitors as single therapeutic agents or in combination with classical chemotherapeutic agents for the treatment of hematologic malignancies. Curr Mol Pharmacol 3(3):98–122

Grady R, Grange R, Lau K, Maimone M, Nichol M, Stull J, Sanes J (1999) Role for alpha-dystrobrevin in the pathogenesis of dystrophin-dependent muscular dystrophies. Nat Cell Biol 1(4):215–220. https://doi.org/10.1038/12034

Kulyte A, Navakauskiene R, Treigyte G, Gineitis A, Magnusson KE (2001) Parallel assessment of tyrosine phosphorylation and nuclear targeting of proteins. Biotechniques 31(3):510+

Kulyte A, Navakauskiene R, Treigyte G, Gineitis A, Bergman T, Magnusson KE (2002) Characterization of human alpha-dystrobrevin isoforms in HL-60 human promyelocytic leukemia cells undergoing granulocytic differentiation. Mol Biol Cell 13(12):4195–4205. https://doi.org/10.1091/mbc.E02-03-0128

Navakauskiene R, Savickiene J, Treigyte G (2000) Protein tyrosine phosphorylation modulates apoptosis and differentiation of human promyelocytic leukemia HL-60 cells. Biomedicine 2(1):23–29

Navakauskiene R, Treigyte G, Pivoriunas A, Savickiene J (2002) Cell cycle inhibitors in retinoic acid- and etoposide-mediated biological responses. Biologija 2:64–67

Navakauskiene R, Kulyte A, Treigyte G, Gineitis A, Magnusson KE (2003a) Translocation of transcription regulators into the nucleus during granulocyte commitment of HL-60 cells. Biochem Cell Biol Rev. Biochim. Biol. Cell. 81(4):285–295. https://doi.org/10.1139/O03-055

Navakauskiene R, Treigyte G, Kulyte A, Magnusson KE (2003b) Proteomic analysis by MALDI-TOf mass spectrometry and its application to HL-60 cells. Biologija 3:63–65

Navakauskiene R, Treigyte G, Gineitis A, Magnusson KE (2004a) Identification of apoptotic tyrosine-phosphorylated proteins after etoposide or retinoic acid treatment of HL-60 cells. Proteomics 4(4):1029–1041. https://doi.org/10.1002/pmic.200300671

Navakauskiene R, Treigyte G, Savickiene J, Gineitis A, Magnusson KE (2004b) Alterations in protein expression in HL-60 cells during etoposide-induced apoptosis modulated by the caspase inhibitor ZVAD.fmk. In: Diederich M (ed) Signal Transduction Pathways, Chromatin Structure, and Gene Expression Mechanisms as Therapeutic Targets, Fdn Rech Canc and Sang; Novartis Luxembourg; Q8 Petr, Annals of the New York Academy of Sciences, vol 1030, pp 393–402. https://doi.org/10.1196/annals.1329.0049

Navakauskiene R, Treigyte G, Borutinskaite VV, Matuzevicius D, Navakauskas D, Magnusson KE (2012) Alpha-dystrobrevin and its associated proteins in human promyelocytic leukemia cells induced to apoptosis. J Proteomics 75(11):3291–3303. https://doi.org/10.1016/j.jprot.2012.03.041

Nawrotzki R, Loh N, Ruegg M, Davies K, Blake D (1998) Characterisation of alpha-dystrobrevin in muscle. J Cell Sci 111(17):2595–2605

Pinto-Costa R, Sousa MM (2019) Profilin as a dual regulator of actin and microtubule dynamics. Cytoskeleton 77:76–83. https://doi.org/10.1002/cm.21586

Sadoulet-Puccio HM, Khurana TS, Cohen JB, Kunkel LM (1996) Cloning and characterization of the human homologue of a dystrophin related phosphoprotein found at the torpedo electric organ post-synaptic membrane. Hum Mol Genet 5(4):489–496

Savickiene J, Borutinskaite VV, Treigyte G, Magnusson KE, Navakauskiene R (2006) The novel histone deacetylase inhibitor BML-210 exerts growth inhibitory, proapoptotic and differentiation stimulating effects on the human leukemia cell lines. Eur J Pharmacol 549(1–3):9–18. https://doi.org/10.1016/j.ejphar.2006.08.010

Savickiene J, Treigyte G, Gineitis A, Navakauskiene R (2010) A critical role of redox state in determining HL-60 cell granulocytic differentiation and apoptosis via involvement of PKC and NF-kappa B. In Vitro Cellular and Developmental Biology-Animal 46(6):547–559. https://doi.org/10.1007/s11626-010-9296-0

Senter L, Ceoldo S, Petrusa MM, Salviati G (1995) Phosphorylation of dystrophin—effects on actin-binding. Biochem Biophys Res Commun 206(1):57–63. https://doi.org/10.1006/bbrc.1995.1009

Shimada A, Murakami Y (2010) Dynamic regulation of heterochromatin function via phosphorylation of HP1-family proteins. Epigenetics 5(1):30–33. https://doi.org/10.4161/epi.5.1.10605

Treigyte G, Navakauskiene R, Kulyte A, Gineitis A, Magnusson KE (2000a) Characteristics of cytosolic proteins and changes in their tyrosine phosphorylation during HL-60 cell differentiation. Biologija 2:32–35

Treigyte G, Navakauskiene R, Kulyte A, Gineitis A, Magnusson KE (2000b) Tyrosine phosphorylation of cytoplasmic proteins in proliferating, differentiating, apoptotic HL-60 cells and blood neutrophils. Cell Mol Life Sci 57(13–14):1997–2008. https://doi.org/10.1007/PL00000681

Treigyte G, Navakauskiene R, Kulyte A, Gineitis A, Magnusson KE (2000c) Tyrosine phosphorylation of cytoplasmic proteins in proliferating, differentiating, apoptotic HL-60 cells and blood neutrophils. Cell Mol Life Sci 57(13–14):1997–2008. https://doi.org/10.1007/PL00000681

Treigyte G, Savickiene J, Navakauskiene R (2003) Changes in O- and N- glycosylation of cytoplasmic proteins in proliferating HL-60 cells and in those induced to granulocytic differentiation. Biologija 3:36–38

Treigyte G, Savickiene J, Navakauskiene R (2004) Identification of O- and N-glycosylated nuclear proteins of HL-60 cells induced to granulocytic differentiation. Biologija 2:49–51

Tsuji Y, Hatanaka M, Maeda T, Seya T, Takenaka H, Shimizu A (2005) Differential-expression and tyrosine-phosphorylation profiles of caveolin isoforms in human T cell leukemia cell lines. Int J Mol Med 16(5):889–893

Walter LM, Franz P, Lindner R, Tsiavaliaris G, Hensel N, Claus P (2020) Profilin2a-phosphorylation as a regulatory mechanism for actin dynamics. FASEB J 34(2):2147–2160. https://doi.org/10.1096/fj.201901883R

Chapter 6
Computational Methods for Proteome Analysis

6.1 Automatic 2DEG Image Analysis

In the following section, we formulate the efficiency and reliability of 2DE gel analysis evaluation criteria, and we propose automatic 2DEG image analysis strategy based on these criteria. The section begins with the discussion of a two-dimensional gel electrophoresis study importance in protein research and 2DE automation of gel image analysis options. The solutions for 2DE gel analysis automation problems are analyzed in the second subsection. In the third subsection, requirements for automatic 2DEG image analysis system are formulated. In the fourth subsection, a detailed discussion of the communication system model and the original 2DEG image matching model based on it is presented. The requirements for the new model are named in the fifth subsection. The section concludes with the introduction of new automatic 2DEG image analysis strategy.

Section generalizes results presented in Matuzevičius (2010b,a); Pivoriūnas et al. (2010); Matuzevicius et al. (2008); Navakauskiene et al. (2012); Treigytė et al. (2014); Navakauskiene et al. (2014).

6.1.1 2DEG Image Analysis Automation Problem

The importance of two-dimensional electrophoresis in the life sciences, more precisely protein studies, is large (Pomastowski and Buszewski 2014; Vlahou 2008; Dowsey et al. 2010). Although the 2DE technology has shortcomings (Zech et al. 2011), however, due to significant 2DE possibilities, this technology still is irreplaceable, and more importantly, technology alone complements the data provided by other techniques (Oliveira et al. 2014; Kim and Cho 2019).

This subsection focuses on the analysis of 2DE gels and its automation capabilities—it is shown that 2DE can be improved (i.e., some of its shortcomings

R. Navakauskienė et al., *Epigenetics and Proteomics of Leukemia*,
https://doi.org/10.1007/978-3-030-68708-3_6

eliminated) by automating 2DE gel analysis after application of electronic and informatics means (de Jesus et al. 2019).

6.1.1.1 Aims and Significance

The aim of expression proteomics is to find qualitative (what has changed) and quantitative (how much has changed) changes between samples under comparisons (Dowsey et al. 2010; Rabilloud et al. 2010). Since 2DE is is a technology used in proteomics, *2DE gel analysis goal* can be formulated as a recovery of proteomics relevant information from the experimental group of 2DE gels. Then *2DEG Image Analysis Objective* is to create the automated conditions for obtaining of information relevant in proteomics using electronics, signal processing, and bioinformatics means.

Relevant Information is a structured, processed, correct data that are necessary and can be used to achieve proteomics purposes. Because 2DE is an imperfect process (the technology has shortcomings), leading to variations in it, the data collected becomes unreliable or even erroneous, and therefore extracted information significance is declining. Using specialized tools for the analysis of gel images and by applying the theory of experimental design (when planning biological and technological copies), the negative impact of 2DE deficiencies can be reduced (Valledor and Jorrín 2011; Zech et al. 2011; Zhang et al. 2013). In addition, the application of advanced bioinformatics tools for interpreting biological information (Laptik and Navakauskas 2005; Nickelson et al. 2006; Treigys et al. 2008; Jegelevicius et al. 2002), validation and generation from experimental data, enables a greater variety of information to be extracted in comparison to a simple gel analysis (Navakauskas 2005; Valledor and Jorrín 2011; Moche et al. 2013).

Therefore, the goals of *2DE gel analysis automation are to increase extracted amount of information from 2DE significant in proteomics* (main purpose) and *extract this information efficiently*—at the lowest possible cost (secondary objective, significant only after the first). To implement automation gels are digitized and analyzed by numerical methods (Kostopoulou et al. 2014; Dowsey et al. 2010; Morris et al. 2009). There are only few digitization methods; therefore, there is not much freedom here to increase efficiency. In image analysis the situation is very different. Although a number of theories have been presented, wide range of means to process images were developed, but fully automatic analysis of 2DEG images is not implemented yet (Wu and Zhang 2011; Millioni et al. 2012). Due to various distortions in gels, the analysis of gels performed with the aid of the systems is only partly automated—researcher intervention is required at certain stages. So to create the automatically functioning system and operate efficiently is an important task.

The main goal—to get as much information as possible—is achievable in two main directions of automation: (a) reducing the negative impact of 2DE technological shortcomings and (b) adapting advanced electronic and IT tools for 2DE data processing.

The negative effects of the shortcomings of the 2DE technology can be reduced by the following means: (a) combining data from biological and technological copies; (b) using narrow and different pI range gels and then combining them; (c) more precisely defining pI and MM parameters; and (d) by increasing the variety of dyes used to label proteins.

Replicates of biological and technological processes are used for cancellation of variations, arising during sample preparation (e.g., sample rate inaccuracies) and 2DE procedure (e.g., structural irregularities of the acrylamide gel, temperatures, and voltage fluctuations). Repetition of the same experiment and then use of statistical methods for data analysis improve identifiability of important protein expression changes. Narrow pI gels can distribute proteins into separate groups better if in a regular gel protein spots are too merged. When multiple distributions of different pI areas are used, a loss of overall proteome image occurs; thus by numerical methods, individual gels can be combined into one. During protein spot parameterization, their pI and MM values are determined. They are determined by comparing the searched protein spots in gel to the identified protein spots that have known parameters. Due to geometric distortions in gels, sought proteins and the positions of the standards are compared inaccurately; thus straightening of the gel images by numerical methods would increase precision of determination of these parameters. With the possibility of digitizing gels, the variety of dyes to be used increases, because fluorescent dyes that are invisible to the naked eye can be used. Broadening of paint choice is important because due to wide paint range, different options for protein visualization can be chosen; in addition, there is no single best dye.

The development and application of advanced electronic and IT tools for 2DE gel analysis will enable to (a) perform fully automatic analysis of 2DE gels and thus eliminate the influence of human factors, which arises from subjective and individual visual information estimates and errors; (b) visualize gels without geometric distortions, without changing background, pseudo-coloring multiple gels in one image, improve contrast, and depict intensity on a three-dimensional surface; (c) perform not only qualitative analysis (what has changed) but also quantitative (how much it has changed); (d) create protein expression maps by combining the different sample gels into one; (e) combine and process large quantity gels and data from large-scale experiments; (f) align very poor quality gels obtained by combining 2DE processes actions; (g) prepare the proteins to be identified by MS and coordinate list for the robot to automatically remove proteins from gel; (h) prepare research reports; (i) accumulate experiment results in databases; (j) combine the results of 2DE with the data from other fields obtained while investigating gene functions, metabolic pathways, cells, organs, and diseases, by creating computer cells, organs, and organism models for the purposes of systems biology.

The economic benefits of automating gel analysis will come from (a) replacement of the expensive work of experienced specialists—biochemists with cheaper computer resources, increasing work comfort because the comparison of gels is a tedious process; (b) reduction of analysis time, therefore, faster results can be obtained—

faster disease diagnostics and faster drug development—and opportunities to carry out larger-scale research arise at the same time; (c) reduction of errors detected too late, caused by the human factor; and (d) higher values driven by the greater amount of information and capacity available adapt it to a larger number of objectives.

6.1.1.2 Strategies of Analysis

Main steps in all 2DE gel image analysis strategies (Natale et al. 2011; Dowsey et al. 2010) contain (a) data input; (b) isolation and quantitative analysis of protein spots; (c) 2DEG image matching or protein spot pairing; (d) analysis of changes; and (e) collection or output of information. Specific strategies differ only in the order of mentioned steps.

First step in computerized 2DEG image analysis is gel image input. The proteins distributed during 2DE may be visualized (marked, painted) with various materials; therefore for digitization of gels, a case-by-case dedicated device must be used. The following scanners are available on the market for 2DEG image input: document scanners (adapted for wet gel application), laser scanners, special equipment suitable for fluorescent inks, and phosphor scanners. The most versatile and good enough are the usual document scanners. They are suitable for gels highlighted with visible dye scan, but their lower dynamic range creates spot saturation problem, which is then sometimes attempted to be corrected by processing images. Laser scanners have a wider dynamic range but, however, are expensive and can scan gels that use dye excitation spectrum coinciding with the spectrum of the scanner source. In a case of different fluorescent dye-labeled proteins (differential electrophoresis), it is necessary to use a special fluorescent laser scanner. Phosphor scanners are used for X-ray slide radiolabelled protein samples. More detailed 2DE overviews of gel scanners are given by Miller et al. (2001) and Miura (2001, 2003).

In the second step, the initial pre-processing of 2DEG images is performed (Goez et al. 2018; Laptik and Navakauskas 2009). The edges of gel images that do not contain protein spots are removed; areas with gel defects such as tears or informative marking are marked, all that performing by hand. Moreover, this step aims to remove scans, and dust affected noise by image filtering methods. Often at this stage, background that results from the coloring of side effects is removed.

Third and fourth steps are very important for analysis of gel images. These are the steps which influence the final result steps the most, and they require the most attention of the researcher (because of unresolved complete automation problem, researcher must intervene in the semi-automated processes by entering additional data or checking results). The consistency of these steps depends on the specific strategies. In either case, one of the steps is to extract the protein spots and the other—to pair protein spots. Such procedure is needed wiling to find protein spot pairs when protein relative amounts are different.

The first systems initially, due to limited computational resources, identified protein spots as follows. Image segmentation methods (Serackis and Navakauskas 2008; Laptik and Navakauskas 2007; Savelonas et al. 2012; Kostopoulou et al.

2015; Nhek et al. 2015; Sengar et al. 2016; Rashwan et al. 2015; Fernandez-Lozano et al. 2016; Marczyk 2017) were used to estimate contours, centers, and quantitative parameters (optical density, total optical density, normalized optical densities), and then these parameter vectors were used to identify the respective protein spots in different gel images. In subsequent algorithms, according to matched protein spot pairs, 2DEG images were additionally transformed using geometric transformations (Glasbey and Mardia 1998; Gustafsson et al. 2002) in order to depict gels under comparison fused into one place. Only after application of geometric transformations corresponding to the same protein the spots are physically present in exactly the same places. Due to the large amounts of gels and geometric and intensity distortions, fully automatic analysis is imperfect, so the investigator has to intervene—to correct the segmentation errors or specify starting points for pairing spots.

When more powerful computers were developed and computing resources significantly became cheaper, another strategy was applied—initially images were aligned so that the respective spots coincide, and then protein spots were identified, and their main parameters determined. Image alignment was based not according to spots parameters but according to the similarities of the image pixels (Rodriguez et al. 2014). Thus a considerable number of repeated calculations of similarity estimates (Penney et al. 1998) and geometric transformations had to be performed (Mateika and Martavicius 2008). In gel images the different types of geometric distortions exist; therefore it is not enough to apply only rigid transformations. Non-rigid transformations (Zagorchev and Goshtasby 2006; Kohlrausch et al. 2005) have more degrees of freedom; therefore a lot more repetitions of transformation applications and image alignment quality estimations need to be done (Keller and Averbuch 2006). This type of gel image alignment in literature is referred as direct gel alignment. If large geometric distortions are present in the images, the researcher has to point out the initial matching point locations from which the alignment process begins.

If someone chooses to perform image segmentation first and only then alignment, a comparison of those areas that are considered insignificant for segmentation is avoided. This can speed up the process of alignment; however improper segmentation will result in misalignment errors. In case of performing image alignment before segmentation, the effects of segmentation errors on alignment result are avoided, but alignment process becomes more complex and takes longer on its own (Smilansky 2001). Until the images are aligned, segmentation is possible only for each gel individually, while when images are geometrically aligned, points are segmented according to methods utilizing information from all gels at once. After 2DE gel image alignment, most often a common gel image is created (by combining all the gels), then protein spots are identified, and amounts of protein are determined for individual gels. If a common gel is created by averaging n gels, noise is reduced $\sqrt{n}$ times; therefore the reliability of small spot identification increases. Gel alignment according to segmentation results has "inherent" disadvantage—segmentation errors affect image alignment and the rest processing stages (if the investigator does not correct the errors). As the number of gels analyzed everyday increases, the likelihood of segmentation errors increases as well; thus in large-scale experiments, this strategy is not appropriate.

In the fifth step, analysis of matched and segmented images is performed: searching for differences between images (appearing and disappearing, magnified and reduced spots) (Zhang et al. 2006), evaluating differences quantitatively (Moritz and Meyer 2003; Morris et al. 2008; Brauner et al. 2014), and performing reliability analysis of changes (Fodor et al. 2005; Ahmad et al. 2006; Biron et al. 2006; Corzett et al. 2006). Hunsucker and Duncan (2006)'s and Krogh et al. (2007)'s efforts are directed to eliminate errors; the aim is to reject the null hypothesis that the mean relative protein quantity does not change and accept an alternative. By checking many hypotheses, the false discovery rate (FDR) is controlled. The analyses are performed in order to determine the required minimum number of experiment repetitions necessary for an accurate estimate of changes (Thornbury 1994). Based on the received data, the required proteins (spots) are selected for following mass spectrometry, and possibly a list of spot cutout coordinates for the protein extractor robot is composed (Mahnke et al. 2006; Quadroni and James 1999; Alterovitz et al. 2006).

In the sixth step, data is collected, processed, and visualized applying bioinformatics tools (Vijayendran et al. 2007). After identification of proteins by mass-spectrometric methods, data can be loaded into gel analysis system that associates them with gel images (Lo et al. 2006; Pleisner et al. 1999). It is essential for the researchers to link the protein changes with factors (Becher et al. 2006; Eravci et al. 2007) that may have led to changes—this is one of the tasks of expression proteomics. Aim is in extraction of as much relevant biological information as possible, linking it to genomics, transcriptomics, and metabolomics data and utilization in modeling of biological systems (Simutis 1997). That is what systems biology—an evolving field of science—seeks to achieve.

6.1.1.3 Challenges

For biological and technological reasons, 2DE gels that need to be compared differ in the arrangement of protein spots, their intensity, noise level, and even the resulting integrity infringements. Comparison of such gels does not become the simplest process. Although 2DE automation of gel analysis is currently significantly advanced, however there are no fully automatic systems, capable of analyzing gels affected by large distortions. They require the investigator intervention to select analysis parameters, input data entrance, check-up, and correction of analysis errors. This subsection discusses the reasons for the differences between the 2DE gels and how these differences complicate the automation of gel analysis.

The biological causes of the differences between gels are external and intrinsic factors targeting the cell or the whole organism and altering the proteome. But these are the differences that are being sought to determine by 2DE method. Biological causes lead to the appearance or disappearance of protein spots, their increase or decrease, and change of their position in 2DE gels due to post-translational modifications.

Technological reasons are inaccuracies up to 2DE (taking samples during sample preparation) and during 2DE time (2DE technology shortcomings). Due to the above factors, it is not possible to obtain completely uniform gels repeating the experiment, which adversely affects the resulting biological information. Inaccuracies in sample preparation lead to changes in gels, changes similar to those of a biological nature. Outsider (not protein) substances may give rise to derivatives distorting protein spots. The 2DE technology disadvantages are process parameter inaccuracies, instabilities in protein distribution and staining as well as polyacrylamide gel properties, loss of protein molecules, conditions for human factors, indeterminate small amounts of proteins during staining, a nonlinear relationship between protein content and protein spot brightness, gel digitization inaccuracies, and mechanical gel breaks.

For these reasons, geometric and intensity distortions occur. If they are large, it is very difficult to match 2DE gels. Merged, oversaturated (stained with silver) and vanishingly small protein spots complicate segmentation, and quantities of poorly segmented spots are wrongly estimated. Noise and protein-like derivatives impair segmentation, too. Segmentation errors affect not only quantitative analysis. They harm the alignment of gel images when the images are aligned according to segmentation results. Due to poorly matched images and changing (or/and poorly removed) background, incorrect results of change analysis are occurring.

For reliable 2DEG image analysis, it is very important to create and apply noise- and distortion-resistant gel image matching, segmentation, and change estimation techniques.

6.1.2 Solution Ways of Automation Problem

It is worth starting to develop an automatic 2DEG image analysis strategy from discussion of problem-solving methods. In general problem-solving process is not formalized; there are no algorithms to guarantee optimal or near-optimal solution to the problem. Solutions to complex problems are a consequence of the harmonious functioning of human mental processes. Good complex solution to the problem is called the creative solution.

The reason for the need of a creative solution to 2DEG image automatic analysis is that the automatic 2DEG image analysis systems are not yet fully developed. Even in recent scientific publications, researchers state that users of 2DEG image analysis systems must manually adjust the system settings and correct errors. It has been over 30 years since the development of the electrophoresis technology, and approximately at that time, gel analysis algorithms have been started to develop. Unfortunately, so far fully automated analysis is not available, so the reasons why it is not available need to be clarified. Maybe it will not be possible to do, or there is a need to look for a completely new solution, starting at the most abstract level of the problem.

Although there is no formal method of solving any problem, there are necessary (but insufficient) conditions that must be met in order to increase the probability of finding a good solution. There are various technical creation techniques that indicate the course of reasoning (Kasperavičius 1997). One technique that most accurately identifies the stages of problem-solving is proposed and described by Baku engineer G. Altshuler—invention problem-solving algorithm ARIZ. In these and similar methods, the greatest importance is given to the analysis of the problem and compilation of the greatest possible list of decisions. Problem analysis requires trying to formulate an objective in terms of perspectives in different ways, consider possible solutions, break down the problem into various smaller problems, and break it down into already solved sub-problems. Motivating to find solutions to problems in other fields of science, Kurt Gödel proved the theorem that it is not possible to completely formalize complex systems. At its core is the assertion that in any system or theory, one can find the statement whose correctness is impossible to prove on the basis of those declared in that system or theory axioms, statements, or rules. This can only be done by supplementing it by new statements, rules, or axioms. Cybernetics pioneer Norbert Wiener in his work "Cybernetics: either the management and communication of animals and in machines" (Wiener 1965) states that "if there are physiological problems the reason is basically mathematical, then 10 physiologists ignoring math, will be able to solve the problem exactly as much as one a physiologist who ignores mathematics." He also noted that in one area the solution to the problem is stuck, while in another the same problem-solving has already become a classic. Knowledge is not exploited due to different terms used in different fields of science.

Why does the reachability of the goal depend on the goal formulation? Creating goal criteria can make problem-solving more difficult due to selection of the wrong set of criteria. Human seeks to formulate everything as simple and concise as possible; therefore fewer but stricter criteria may be chosen instead of more and simpler criteria. Choosing a smaller set of criteria, but requiring them to be more thoroughly describing the purpose, the criteria usually have to be stricter, i.e., the smaller set of criteria must include the entire larger set of simpler criteria group. As a result, stricter criteria may be formulated which will be impossible to accomplish, or more challenges will need to be solved than there are sufficient to achieve the objective. Someone can choose to formulate less criteria also due to required better understanding of the problem, in order to formulate precise criteria. If the problem is not well analyzed and not all necessary tasks are known exactly, then a choice is made to solve more tasks, and thus it guarantees the achievement of the goal. Clarity of the problem allows you to set minimum criteria and find effective solution to problems.

If the problem is being solved not starting from the essential decision-making, the situation may occur in which a solution will be sought in a defined area of possible solutions where there isn't simply any right decisions. Many search steps will then be performed, and best solution found in that area will not work. Peter Drucker, Business Management professor, described a similar situation as doing effectively what should not be done at all. Development of 2DEG image analysis strategy is a definition of the range of possible solutions to the problem. Knowledge that a solution exists and approximately perception where to search for that solution increase the probability that the solution will be found and reduces the solution search costs. In order to properly define the scope of solutions, problem analysis needs to be done at the most abstract level possible.

Having a solution to a problem, someone can aim to improve it. Increase of the effectiveness of problem-solving is necessary because the resources to achieve the goal—energy, time, and others—are limited. For example, selection of function in integrated circuits, depending on the number of elements, required to perform the function depends on the cost of the scheme and reliability. Therefore logic circuits are degraded—the same function will be performed with a smaller number of elements (Kirvaitis 1999)—resulting in the lower cost of the chain, greater reliability of the chain, and lower energy consumption.

6.1.3 Requirements for Automatic Analysis System

Knowing that the solvability of a problem depends on the description of the objective, in this subsection the purpose and the tasks of automatic 2DEG image analysis are formulated according to the previous material performed specific 2DE use analysis—what information needs to be extracted from 2DE gels.

When solving a big problem, usually the whole task is broken down into the smaller challenges that need to be addressed in order to solve the whole problem. Usually a smaller task can be solved only after solving the task before it. Then criteria estimated for achieving summarized goal in each smaller task help to solve the task and get such results that do not diminish the probability of solving the other tasks. In biochemical research process, 2DE is one of the intermediate tasks—2DE results are mainly used for mass spectrometry (MS) planning and conducting research. This means that the data obtained during analysis of 2DEG images must not impair the results of MS. If between the 2DE and MS processes researcher makes some additional decisions, 2DEG image analysis results must have negative influence on the researcher's decisions.

The most important data to be obtained from the analysis of 2DE gels are changes—which protein spots have changed. Additionally researches are interested in change size, which also determines the reliability of the change. Needed, but much smaller amount of data is obtained on the correspondence of the remaining protein spots between gels. Therefore, in order to achieve the goal of 2DE gel analysis—extract as much useful information as possible from the gels—it is necessary to focus processes on reliable critical information extraction. The amount of information will decrease if provided data possesses errors, so it is more important to provide less but more reliable data than try to obtain more possible data. Unreliable data that may contain errors should be inferred or, in individual cases, provided to the investigator for additional revision.

If someone wants to perform a fully automated 2DE gel analysis process, all means are needed to detect (worst case) and correct (at best) human factor errors. If manual handling was done prior to the automatic gel analysis step, then there may be ingrained human errors—improperly prepared and digitized gels, gel lesions, and lack of initial data on gel input. Because, after automatic 2DE gel image analysis, human will not check the results, the initial errors may lead to complete erroneous analysis results. The automated system must be able to cope with mentioned errors.

When comparing 2DEG image analysis systems in terms of segmentation, often calculations on how many protein spots were excreted by one or another system and the researcher are done. Because some protein spots are more significant than others (interest is in change), it is necessary to assess whether all spots that have changed also have been found. Hence, the comparison in terms of absolute number of found protein spots is not objective.

Later, in the development of an automatic 2DEG image analysis strategy, we will formulate actions that should be taken in order to comply with main requirements of 2DE gel analysis and then with the less important requirements.

6.1.4 2DEG Image Matching Modeling

In communication theory employed generalized information transmission system model was chosen to describe the 2DEG image matching process because the model defines the conditions for the efficiency of information transmission. So if 2DEG image matching can be called information transfer process, then it will be possible to benefit from those already developed and well-researched conditions for the transmission of information in order to create effective 2DEG image matching scheme.

Firstly it must be confirmed that for image matching analysis, the summarized communication system model (see Fig. 6.1) described by Claude E. Shannon in "A Mathematical Theory of Communication" can be applied.

The communication system model consists of five parts—information source (message sender), transmitter, communication channel, receiver, and the recipient of the message. Based on this model, the sender of the message forms message or message sequence; the transmitter converts the message into a signal which can be transmitted over a communication channel; the receiver receives the signal and reconstructs it into a message; the recipient of the message receives the message, interprets, and accepts part of the information. Now let's discuss how image matching involves the transmission of information. The task of 2DEG image matching is to find the appropriate gel places—where protein spots in one gel correspond to another gel protein spots. After finding spots of the same protein in different gels (in this case, it is assumed that one protein in the gel forms one spot),

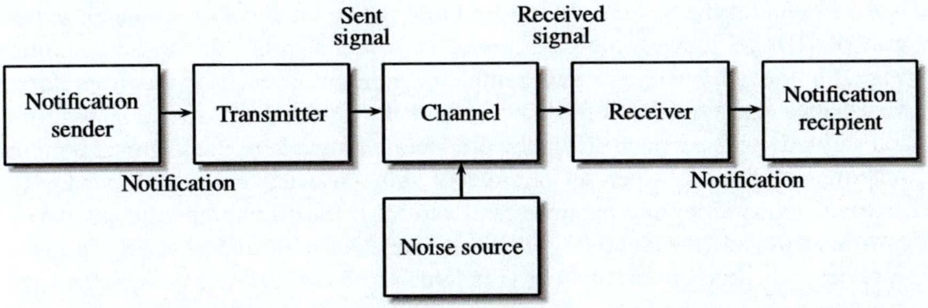

Fig. 6.1 Generalized communication system model

the relative amount of that protein in each gel together with overall amount of it can be estimated. The search of matching spots is the transmission of message about the first gel area to a second gel processing algorithm that, according to the received message, has to find the corresponding area of the second gel. Let's in detail discuss each information transfer step performed during the 2DEG image matching.

1. *Source of information (sender)* forms a message or the sequence of messages to be transmitted to the recipient of the information. Notice is formed by selecting it from a set of possible messages. In the case of 2DEG images, the message is information about a specific gel location (dot or vanishingly small area), while the set of messages consist of all gel sites in the active part of the gel. The width of the active part of the gel is defined by the length of the immobilized pH gradient(IPG) strip, while the height is the distance from the IPG strip at the top of the gel to the local gel at the bottom, beyond which the protein can no longer enter. During the same experiment, obtained gels that have to be compared are done using the same materials (except for a few specific cases); therefore the active parts of the gels are ideally identical. This ideal case means that each site of the active part of one gel is denoted by vanishingly small gel area S_{G1n} with center coordinates described by a unique parameter pair (pI, M_M), which will be also found in the active part of the another gel. Hence, the message (a pair of parameters specifying the gel location is selected) from the set $\mathcal{S}_{G1}$ $((pI, M_M))$ is equal to the set $\mathcal{S}_{G2}$ in which the message recipient can find a matching location based on the received parameter pair S_{G2m}, i.e., $\mathcal{S}_{G1} = \mathcal{S}_{G2}$. This condition is important because it means that it is possible to convey a message about each gel location—find the appropriate locations on the gels and thus match gels.

   In the following text, when referring to a gel, the active gel part will be meant. Selection of a message will be marked $(pI, M_M) = \mathcal{S}_{G1}(S_{G1n})$, and the message acceptance $S_{G2m} = \mathcal{S}_{G2}(pI, M_M)$. In practice, for each gel site S_{Gi}, parameter pairs (pI, M_M) cannot be measured directly. Parameters can be determined only when, at the site, after electrophoretic protein breakdown, there is a protein whose (pI, M_M) parameters are known. Because of these reasons, the data about the distribution of protein spots will be transmitted instead of the pair (pI, M_M).

2. *Sender* prepares the message to transmit through communication channel. In the case of 2DE gel images, the gel is coded by a visual code—the protein mixture spread in the gel is like a protein sample mapping in 2DE gel space. Therefore, visual data that encodes information about the selected gel site S_{G1n} will be transmitted. Let's suppose that in the production of two gels, the identical protein mixtures containing a certain number of proteins were used, with randomly distributed quantities and parameters pI and MM. Electrophoretic breakdown of proteins will yield the identical gels (ideal case). Due to the distribution of protein parameters in proteins arranged in a random order in the gel, it is possible to form such an area in the gel that will be unique throughout the gel, and the larger the area, the larger the description of protein spot arrangement structure uniqueness. Because the area is unique in the first gel, it will be unique in the second gel. The main goal of the 2DEG image matching task is to associate the gel area S_{G1n} with the corresponding area of another gel S_{G2m}; therefore it is enough to form a gel area of sufficient size in the selected area of first gel, to transfer it, and to recognize the same area in the other gel, thus finding a corresponding place on the second gel.
3. *Channel* is the medium on which the signal is transmitted from the transmitter to the receiver. The signal emitted on the channel due to the effect of *noise* is distorted. 2DE gels have geometric and intensity noise distortions leading to similarities between the areas concerned and differences in location. Intensity distortions are altered spot forms, appearing and/or disappearing spots, increased and/or decreased spots, and extraneous derivatives appearing in the images. Intensity distortions are caused by technical reasons (with the development of 2DE technology, the distortions caused by it decrease) and biological causes (in the samples existing differences that are targeted for identification by 2DE gel analysis). Geometric distortions are protein spot shifts. Typically, geometric distortions occurring in many small areas of the image are not large and are monotonous, but global distortions can be big. Geometric distortions are caused by technical reasons.

   When receiving and decoding a signal, it will be important to know about the channel noise parameters and how they influence distortions of the 2DE gel images. Intensity distortions due to biological causes are randomly distributed in the gel, i.e., they change in random spot intensity. Intensity distortions caused by technical reasons usually change the image evenly—change background and increase or decrease total gel brightness. However intensity distortions—saturated or over-saturated large protein spots stained with silver, gel tears, and other extraneous visual derivatives—may appear at random places, too. The main feature of geometric distortions is that, regardless of their size, the position of the protein spots will remain relatively the same. This is due to the fact that during the first step of protein breakdown by the pI, a uniform increasing pH gradient is formed, and there are equilibrium states that reached their migration by the protein stops. In the second stage of protein separation by MM, their migration trajectories do not intersect.

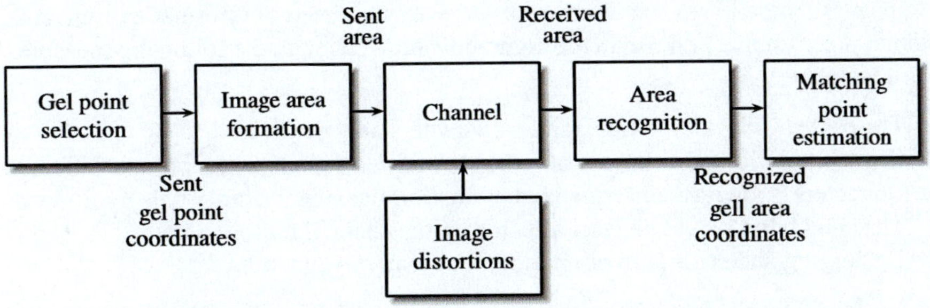

Fig. 6.2 2DEG image matching model

Precisely describing the 2DEG image matching according to the communication system model, it must be noted that noise of biological origin occurs in the transmitter during encoding message in the visual data, and the technical reasons become apparent in the channel. But for the solution of the 2DEG image matching problem, distortions' origin does not need to be identified, so all noise can be attributed to the signal-distorting noise appearing in the channel.

4. *Receiver* receives the signal and decodes it. In the case of 2DEG images, the received signal is the area of the gel image that is distorted by noise. The region obtained from the first gel must be found by decoding the signal in the second gel. Since $P_1^G = P_2^G$, so really the second gel has an area corresponding to the accepted area. In this situation the main search principle is by applying the selected similarity function to compare the accepted area with all areas of the second gel and select the location of the gel where the similarity is greatest.
5. *Message recipient* by the receiver-recognized (decoded) area determines the location of S_{G2m} on the second gel that corresponds to place S_{G1n} from the first gel.

2DEG image matching model based on the communication system model is shown in Fig. 6.2. Knowing that 2DEG image matching can be modeled by communication system elements, we can apply the principles of efficient information transfer in order to investigate 2DEG image match and requirements that ensure effective implementation of matching.

6.1.5 Requirements for 2DEG Image Matching Model

In order to achieve the required 2DE image matching goal, it is necessary to find overlapping areas in images—successfully transfer each area formed in the first image, and then recognize it in the second image. Image matching efficiency will be higher, and the faster all areas will be transmitted and identified.

In communication systems aiming to increase the speed of information transmission in the presence of noise in the communication channel, the following measures shall be taken:

1. The probabilities of the transmitted characters are equalized.
2. Cross-similarity (correlation) of to be transmitted signals that represent different characters is lowered and equalized—most similar signal similarity is minimized.
3. The signal that would be least affected by the channel noise is selected.
4. Debugging and error correction tools are being implemented.

During 2DEG image alignment, the transmitted information has the following features:

- Data on *N* gel image areas are transmitted.
- The data for each one are transmitted only once.
- When transmitting image areas after geometric distortions, information about the inter-spot position remains, so the position between the areas is known.
- Due to unique nature of the spot arrangement, some areas are more characteristic than others.
- Image areas are exposed to noise unequally.

So, to effectively match the 2DEG images, it is necessary to:

- Once pass each image area one by one (1 tool).
- Extract features from the transmitted image area that describe the area the best and distinguish it from other areas. It is important that the extracted features would be least exposed to noise. Selecting specialized similarity function or by extracting appropriate features, it is possible to optimize area recognition (2 and 3 tools).
- Pass image areas of the size that maximize signal to noise ratio. More characteristic areas and their relative position remain (4 tool).

Requirement (2DEG Image Matching)

1^{R}. Area selection way that would select from the first image the appropriate (given size and gel image place) area for the following transfer
2^{R}. Similarity function that evaluates the similarity of gel image areas according to the features that were least affected by the noise
3^{R}. Matching way that takes into account the fact that the area is transmitted once
4^{R}. An error correction way that takes into account the condition of mutual situation of the spots □

6.1.6 Automatization of 2DEG Image Analysis

The automatic 2DEG image analysis development should follow the purpose of the analysis—what data need to be obtained in order to complete the analysis. In order to find proteins spots whose expression has changed, it is necessary to (a) find spots of the same protein in all the gels being compared and (b) estimate the relative amounts of each spot and their changes between gels, in order to determine the reliability of the changes.

Two procedural ways (strategies) can be carried out wiling to find correspondent spots in the gels: (a) initially protein spots are isolated in individual gels, and then the spots are aligned; (b) initially 2DEG images are matched, and then overlapping spots are found knowing that their positions in the images are similar. The advantages and disadvantages of these strategies will be discussed below, aiming to select the most appropriate for 2DEG image analysis automation.

If 2DEG images are segmented firstly and then searched for spot pair match, then false segmentation (not all spots are found; spots include extraneous derivatives; spots groups are not properly separated into individual spots or individual spots over segmented) will increase the probability of false spot pairing, and that will consequently yield incorrect change assessment. If the images are initially aligned and then segmented, then in the misalignment locations, spots will be incorrectly segmented yielding incorrect spot pair match. Hence, the reliability of the first step influences the reliability of results obtained from the whole analysis; therefore the choice of the first step must be based on the highest probability of correct processing.

The probability of image matching was defined by describing the match by communication system model. If the gel parameters are the same, i.e., mixtures of different proteins were distributed (to cover the gel with spots), gels will be able to match (transmit and recognize the signal for each gel site). The reliability of gel matching is enhanced by the fact that the gel areas maintain the mutual position, resulting in ability to check overlap of more noise-affected areas using reliably matched areas. Spot separation reliability by segmenting each gel image separately is much lower. There will always be vanishingly small spots in the gels, so in these places spots will not be identified. Decision if there is a spot in any area of the gel can be made by comparing several spots of the same areas from different gels—if other gels contain a spot, then all those areas can be used for protein spot identification. If there is a derivative in one area that resembles a small spot but there are no spots in other areas, then a decision—no spot at that location—will not change MS test progress, because there will be not enough material in place to identify the protein.

A possible variation of the strategy "matching–segmentation" can be "segmentation–matching-segmentation." Then segmentation results as additional information are used during matching, while afterward matching results refine the segmentation. In this case segmentation data will complement the sets of attributes by which similar areas are recognized.

Based on the listed assumptions, automatic 2DEG image analysis strategy is proposed (Fig. 6.3). A first essential step in the analysis is the alignment of the images and the second the analysis of protein expression variations.

Image matching consists of three main steps: individual gel preparation, initial image alignment, and final image matching.

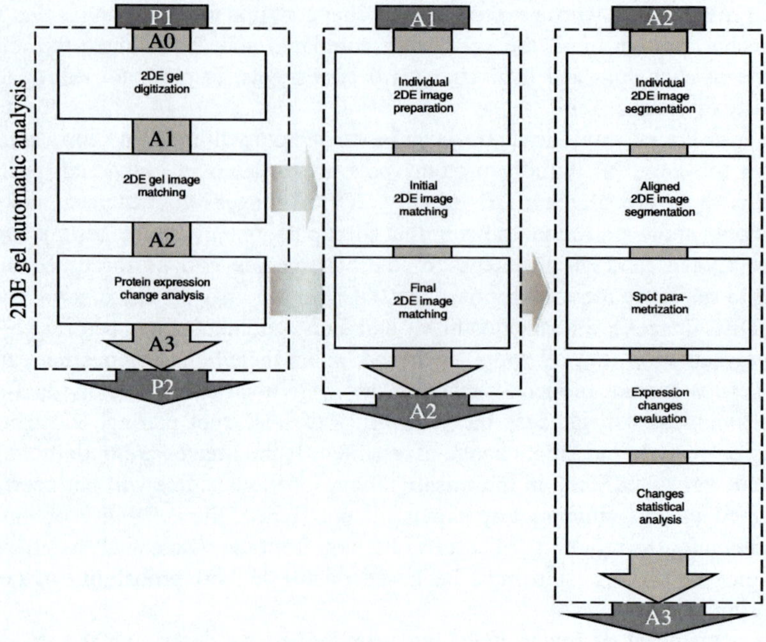

Fig. 6.3 Automatic 2DEG image analysis strategy

2DEG image analysis, performed in order to find expression variations, starts with the segmentation of individual 2DEG images. Then segmentation results are adjusted according to the all gel group segmentation results. Each segmented spot is parameterized to estimate the relative amount of protein and then to use the relative amounts to assess changes in expression. Because gels contain intensity distortions (due to 2DE technology flaws), found expression changes must be processed by statistical means.

Section Generalization

1. After applying the communication system model for 2DE gel image matching process description, conditions for effective information transfer can be used to create an efficient 2DEG image matching technique.
2. For effective 2DEG image matching, the most important are:
   - Area selection way that selects from the first image the appropriate area for the following transfer
   - Similarity function that evaluates the similarity of gel image areas according to the features that were least affected by the noise
   - Matching way that takes into account the fact that the area is transmitted once
   - Error correction way that takes into account the condition of mutual situation of the spots

3. The probability of successful automatic analysis of 2DEG images will be higher if images will be aligned before segmentation.
4. 2DEG image segmentation reliability will be higher if at the same time separate and aligned gel images are segmented.
5. The proposed 2DEG image analysis strategy is appropriate to automate the process.

6.2 2DEG Image Matching

The purpose of two-dimensional electrophoresis gel (2DEG) image matching is to eliminate geometric distortions in gel images. During creation of an automatic 2DEG image analysis strategy (see Sect. 6.1), it has been shown that it is necessary and possible to eliminate relative and reduce absolute geometric distortions.

In this section we specify the previously created abstract 2DEG image matching strategy. The first subsection details the 2DE gel image matching strategy. Later in separate subsections, while offering and studying solutions, we focus on the three new matching technique steps: preparation of individual 2DEG images and initial and final 2DEG image matching.

In the section generalization of results published in Matuzevičius (2010b); Matuzevičius et al. (2010); Matuzevičius and Navakauskas (2010); Pivoriūnas et al. (2010); Matuzevičius (2010a); Navakauskiene et al. (2012); Treigytė et al. (2014); Navakauskiene et al. (2014) is given.

6.2.1 Specification of 2DEG Image Matching

During the development of a generalized matching strategy, the importance to start process from image locations with a highest probability to match was demonstrated. Then it becomes necessary to make step-by-step decisions about the match of remaining places in the images by evaluating both the similarity in those areas and earlier found positions of matches. Such strategy allows achievement of the maximum reliability of the matching compared to other strategies. In order a split match search process and introduce additional debugging procedures that increase the reliability of the overlap, two important stages of image matching were identified. The first stage is the initial image matching, and the second is the final image matching. Because there is a requirement to reduce the absolute geometric distortions affecting pI and MM, in addition, gels with the smallest geometric distortions are selected.

A detailed 2DEG image matching procedure is provided in Fig. 6.4. Next to each stage of the matching strategy, the more detailed processing steps are depicted. The first major stage is *individual 2DE gel image preparation*. Individual images are prepared at the beginning in order to reduce absolute geometric distortions. Only

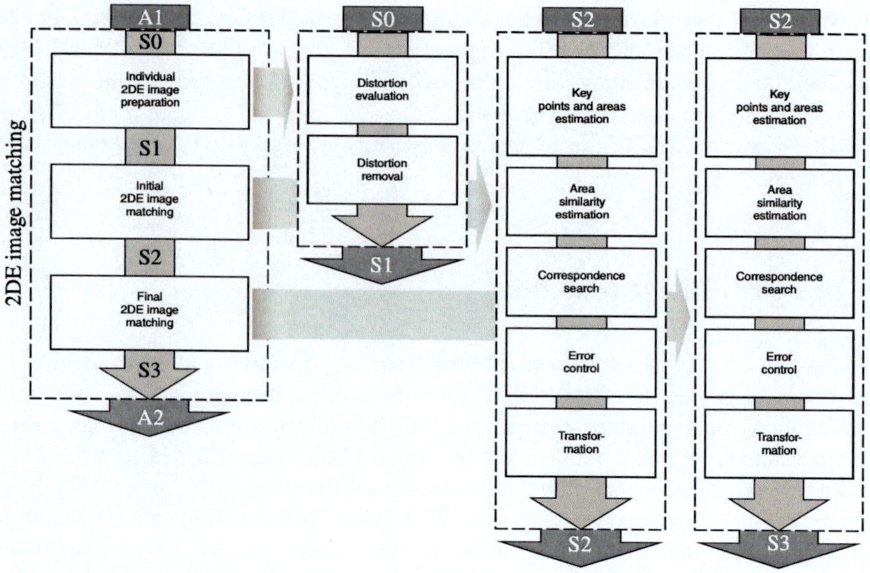

Fig. 6.4 Elaborated stage of 2DEG image matching

relative geometric distortions that determine the respective locations of physical discrepancies in the images will be deleted at a later stage of image matching, resulting in the same absolute distortions in all 2DE gel images—the same as in the image, according to which all the remaining 2DEG images align.

Unremoved absolute geometric distortions degrade the estimates of protein pI and MM parameters Specifically, the horizontal geometric distortions determine the values of pI, and vertical ones determine the values of MM. Because of horizontal and vertical distortions, the parameters of proteins of the same pI or MM are arranged in the gel not in the one (same) vertical or horizontal direction. The parameters of unknown proteins are determined based on known proteins pI and MM parameters during the comparison of their x and y coordinates; therefore it is essential to eliminate that part of the absolute distortion which transforms vertical and horizontal lines into curves.

Absolute geometric distortions cannot be fully determined because reference point against which the magnitude of distortions can be assessed does not exist and there are no gels that are completely free of distortions and with which it would be possible to compare each distorted gel. The reduction of the most negative effects making absolute geometric distortions is possible in two ways: (a) when image alignment is performed, all images are aligned to a gel with the least geometric distortions, and (b) in the absence of such a gel, to detect and reduce horizontal and vertical geometric distortions using indirect ways. According to the listed alternatives, during the preparation of 2DE gel images, screening of individual gels by the smallest geometric distortions and removal of geometric distortions from individual gels (Fig. 6.4S0–S1 stages) are performed.

The next step in 2DEG image matching is *initial matching* (Fig. 6.4S1–S2 steps). The aim of the matching is to find matches with highest reliability in the 2DE gels. The initial image matching process consists of identification of significant gel areas, evaluation of similarity of areas from different gels, matching, and error search (debugging).

The benefits of finding matches in relevant places in the image already in detail are stated. This is a way to increase the reliability and to reduce the required amount of calculations.

Similarity or distance (as opposed to similarity) function is required to assess the similarity of the areas being compared. It would directly evaluate areas of the image similarity on a pixel basis or indirectly—in each area extracting feature vectors and determining the distance between them. In Sect. 6.2.3 three algorithms to estimate similarity of areas are constructed and investigated.

Debugging (error search) is a mean to use additional information to determine matches. Some information about 2DEG images can be used more efficiently without assessment of the area similarity but checking match compliance; therefore debugging step is added. Depending on the stage in which it is decided to use more information and increase the calculation amount, at the same time seeking a minimum number of overall calculations, it is necessary to balance between the following strategies: (a) perform similarity evaluation of complex areas, but use simple debugging; (b) perform simple similarity assessment for areas, but use complex error search. The quantity of total calculations must be adapted according to the probability of reaching the target—to allocate only necessary computational resources in order to find matches.

The initial alignment of gel images ends with image geometric transformation according to established equivalents. Rigid transformation is used at this stage because it is mildly affected by small displacement errors when transformation parameters are set from large set of equivalents. The algorithm uses an affine geometric transformation. Moreover, elastic transformation is not necessary at this stage because during the final match, all matches are searched again, just the search area is significantly narrowed.

The last stage of image alignment is *the final matching of 2DE images* (Fig. 6.4 S2–S3 steps). It is based on the initial matching stage; thus both have a lot in common. The difference between the initial and the last stages of image alignment is in the use of information—a narrowed search area of matches is used, and the similarity measure of the regions is based on the geometric arrangement of protein spots. When all matches are found, elastic image warping using thin plate spline transformation is performed.

6.2.2 Preparation of Individual 2DEG Images

In protein separation by 2DE, not all distorting factors of the experiment are manageable, or they can be insufficiently precisely controlled. One of the consequences is the unevenness of protein movement—during the separation of proteins by mass, at the edges and in the middle of the gel molecules, movement rates are not identical, resulting in higher or lower protein vertical displacements of the spot location from a specific horizontal indicator which denotes the molecular weight (Aittokallio et al. 2005). Such distortions prevent visibly the gel protein spots from being precisely associated with the mass marker on the side and set the protein MM. These distortions are attributed to absolute distortions, which are usually not eliminated by image matching.

Absolute geometric distortions can be reduced directly or indirectly. An indirect way is when during correct follow-up, sequence of processing distortions is reduced automatically without any additional actions. A direct way is when specialized actions specifically for vertical or horizontal geometric distortion removal are performed.

Distortions can be indirectly reduced during 2DEG image matching by properly selecting the gel to which all other experimental group gels are aligned. If the gel least affected by geometric distortions is selected, then after matching all gels will have the same absolute geometric distortions, i.e., each gel distortions will be reduced to the minimum attributed to that group gels. During regular experiment a big group of gels is created (Luhn et al. 2003); therefore, as the number of gels increases, the probability to find a gel that will have small distortions in the group increases. Hence, it is very important to select the 2DE gel that is least distorted. Also it is important to be able to perform this selection automatically in order to speed up processes in laboratories and to reduce probability and number of human error (Dowsey and Yang 2008).

In order to *estimate vertical geometric distortions*, it is necessary to highlight those locations and structures that could be affected by geometric distortions in the image and proportionally indicate vertical effect size. Steep distortions are most visible in horizontal linear vertical shifts of structures, so it is necessary to distinguish these horizontal structures and determine their inclination to the horizontal respect. Horizontal structures can be distinguished in elongated protein spots. Moreover the characteristic 2DE gel background elements, background derivatives that are not formed from proteins but caused by impurities in the samples, can be noticed. These background structures are commonly linear in nature but fuzzy, so they need to be found and strengthened. Gels also contain protein aggregates formed due to horizontally close protein spots. Often protein spots are interconnected. All the derivatives listed would be arranged in a horizontal line in the absence of geometric distortions; therefore, after estimation of the deviations of the latter structures from horizontal, gel distortion can be assessed.

Means are required that can extract only those 2DE gel structures that provide adequate indications of distortions. But a lot of listed structures are indistinct,

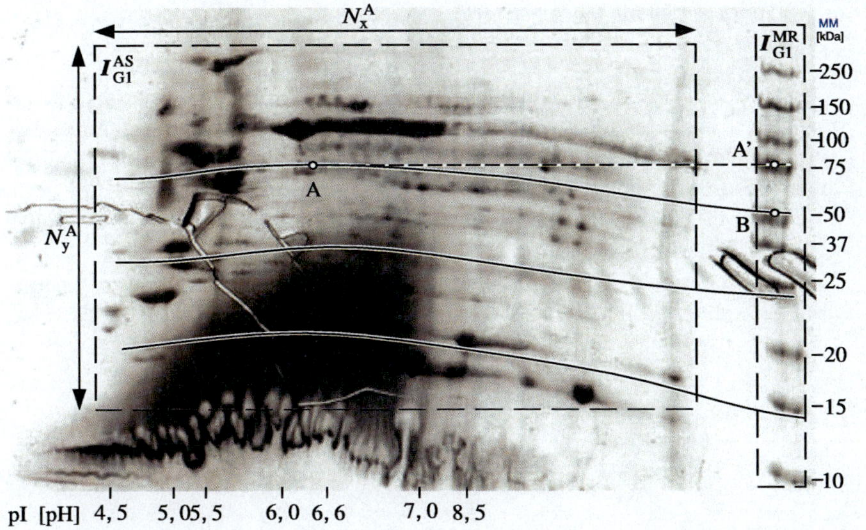

Fig. 6.5 2DE gel I_{G1} with vertical geometrical distortions: I_{G1}^{AS} , active gel area with separated proteins; I_{G1}^{MR} , molecular mass marker. Three curves indicate the same molecular mass areas

affected by noise (noise depends on vertical structures); therefore, in the light of the comments, the generalized separation process is as follows: horizontal structures are highlighted, structures are approximated by curves, and curve tilt is estimated. Algorithm 6.1 provides the details for all steps.

Algorithm 6.1

Gel vertical geometrical distortions estimation

1. Asymmetric median filtering
2. Horizontal image smoothing
3. Contour estimation using modified Canny detector
4. Short contour segments removal
5. Contour polynomial approximation
6. Local distortion estimation
7. Tilt coefficient filtering
8. Image geometrical distortion evaluation

1. *Filtering with an asymmetric median filter.* In the first step, gel image $I_G(x, y)$ is filtered by the median filter (Matuzevičius and Navakauskas 2005; Gonzalez et al. 2003). The median filter can selectively filter the impulsive noise. At this stage, the median filter is used to suppress vertical contrasting lines:

$$I_{\mathrm{G}}^{(1)}(x, y) = \underset{(i,j)\in S_{xy}}{\mathrm{median}} \left\{I_{\mathrm{G}}(i, j)\right\} \quad \forall x \in [1, N_{\mathrm{x}}],\ y \in [1, N_{\mathrm{y}}], \tag{6.1}$$

Therefore, the selected median filter window S_{xy} is 9×1 pixels (columns×rows).

2. *Horizontal image smoothing*. In order to highlight horizontal gel structures, they are "stretched" horizontally. It is achieved by strongly smoothing the image in the direction of x axis. Using asymmetric filter which is a low-pass filter in the horizontal direction, and all-pass in the vertical direction, the desired result is achieved. Filtering is implemented in the frequency domain.

   The foundation of linear filtering in time and frequency domain is a convolution theorem that can be written (Gonzalez et al. 2003; Sonka et al. 2007):

$$\begin{aligned} I_{\mathrm{G}}^{(2)}(x, y) &= I_{\mathrm{G}}^{(1)}(x, y) * K(x, y), \\ &\mathcal{F} \downarrow \uparrow \mathcal{F}^{-1} \\ I_{\mathrm{G}}^{(2)\mathcal{F}}(u_{\mathrm{x}}, u_{\mathrm{y}}) &= I_{\mathrm{G}}^{(1)\mathcal{F}}(u_{\mathrm{x}}, u_{\mathrm{y}}) \cdot K^{\mathcal{F}}(u_{\mathrm{x}}, u_{\mathrm{y}}), \end{aligned} \tag{6.2}$$

   here $*$ is convolution operation. According to this expression, the convolution operation between the original image $I_{\mathrm{G}}^{(1)}(x, y)$ and filter kernel $K(x, y)$ may be replaced by the multiplication of their Fourier transforms $I_{\mathrm{G}}^{(1)\mathcal{F}}(u_{\mathrm{x}}, u_{\mathrm{y}})$ and $K^{\mathcal{F}}(u_{\mathrm{x}}, u_{\mathrm{y}})$ respectively.

   It is convenient to create a filter of the required characteristics in the frequency domain. An ideal low-pass filter has such frequency transfer function:

$$K^{\mathcal{F}}(u_{\mathrm{x}}, u_{\mathrm{y}}) = \begin{cases} 1, & \text{when } \|(u_{\mathrm{x}} - u_{\mathrm{x0}}, u_{\mathrm{y}} - u_{\mathrm{y0}})\| \le u_{\mathrm{c}}; \\ 0, & \text{otherwise}, \end{cases} \tag{6.3}$$

   here u_{c} is cut-off frequency.

   So filtration can be done by the following processing sequence:

$$I_{\mathrm{G}}^{(2)}(x, y) = \mathrm{Re}\left[\mathcal{F}^{-1}\left(\mathcal{F}\left(I_{\mathrm{G}}^{(1)}(x, y)\right) \cdot K^{\mathcal{F}}(u_{\mathrm{x}} - N_{\mathrm{x}}/2, u_{\mathrm{y}} - N_{\mathrm{y}}/2)\right)\right], \tag{6.4}$$

   here $K^{\mathcal{F}}(u_{\mathrm{x}} - N_{\mathrm{x}}/2, u_{\mathrm{y}} - N_{\mathrm{y}}/2)$ is such frequency transfer functions $K^{\mathcal{F}}(u_{\mathrm{x}}, u_{\mathrm{y}})$ that have swapped places of I with III and II with IV quarters (see Fig. 6.6) and Re(.) is a real part of complex number.

   The gel image should be smoothed horizontally but left unchanged vertically; thus the following asymmetric frequency transfer function is created:

$$K^{\mathcal{F}}(u_{\mathrm{x}}, u_{\mathrm{y}}) = \begin{cases} 1, & \text{when } |\, u_{\mathrm{x}} - u_{\mathrm{x0}} \,| \le u_{\mathrm{xc}}; \\ 0, & \text{otherwise}, \end{cases} \tag{6.5}$$

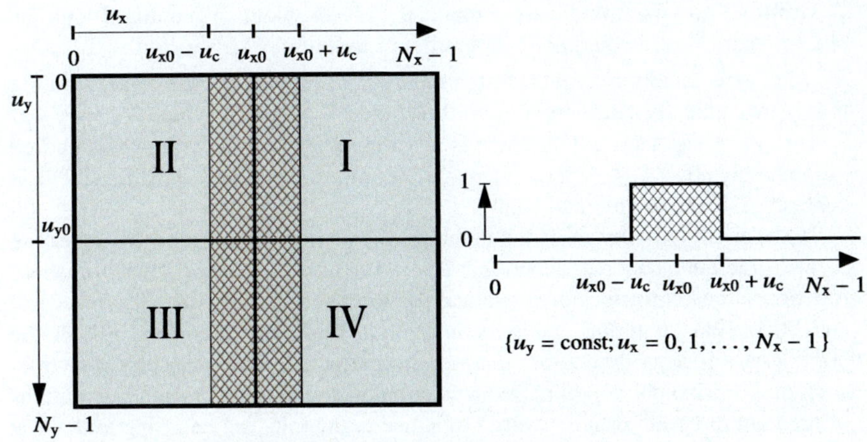

Fig. 6.6 Frequency response function for horizontal smoothing

here $u_x \in [0, 2^{\lceil \log_2 N_x \rceil} - 1]$, $u_y \in [0, 2^{\lceil \log_2 N_y \rceil} - 1]$; $2^{\lceil \log_2 N_x \rceil}$, $2^{\lceil \log_2 N_y \rceil}$ is number of columns and rows in array, calculated taking filtered image column and row numbers and increasing their value to the nearest 2^n values; $(u_{x0}, u_{y0}) = (2^{\lceil \log_2 N_x \rceil}/2, 2^{\lceil \log_2 N_x \rceil}/2)$ is the center of filter; and $u_{xc} = 0,035 \cdot u_{x0}$ is chosen cut-off frequency.

3. *Contour extraction with a modified Canny detector.* The aim of this step is to extract the characteristic contours of linear structures. Canny edge detector is chosen because it is optimal for isolation of stepped edges affected by white noise (Canny 1986; Gonzalez et al. 2003; Sonka et al. 2007). Detector's optimality is defined by the following criteria:
   - Good detection—the algorithm must mark real edges in the image as much as possible.
   - Good localization—marked edges should be as close as possible to the locations of the actual edges.
   - Clear answer—the actual edge must be marked only once, and noise in the image should not lead to erroneous edge marking.

Having gradients $\nabla_x I(x, y)$ and $\nabla_y I(x, y)$ in x and y directions, the total gradient at each pixel is computed as follows:

$$\nabla^{\parallel} I(x, y) = \sqrt{\nabla_x I(x, y)^2 + \nabla_y I(x, y)^2}, \tag{6.6}$$

while its direction:

$$\nabla^{\angle} I(x, y) = \tan^{-1} \frac{\nabla_y I(x, y)}{\nabla_x I(x, y)}. \tag{6.7}$$

Omitting non-maximum values marks the places where the contours could be placed. A location is marked if its gradient modulus $\nabla^{\|} I(x, y)$ in the direction $\nabla^{\angle} I(x, y)$ is locally maximal. In the next step, the contours are marked in the places available for contours by two thresholds ∇_1 and ∇_2, where $\nabla_1 > \nabla_2$. If $\nabla^{\|} I(x, y) \geq \nabla_1$, then in that place is a strong edge, and the contour is marked immediately. If $\nabla_2 \leq \nabla^{\|} I(x, y) < \nabla_1$, contour is marked only then if it is connected to already marked contours.

In the modified version of the detector, the contour is not marked according to the gradient size, but according to its direction—the two-level threshold function is customizable for contour direction. Because the image has vertical and horizontal contours, and only horizontal ones are required, only in the horizontal direction dominant contours are extracted. Dual threshold function, excluding horizontal contours, reduces contour cracking due to the fluctuations of gradient direction usually caused by noise in the image. According to the first threshold θ_1, contour is outlined if it is close to the horizontal axis or the direction of the gradient (edge normal) is deviated not more than 90° from the threshold. When the range of gradient directions is in the interval $\nabla^{\angle} I(x, y) \in [0°, 180°)$, then the first threshold rule is met if:

$$\left|\nabla^{\angle} I(x, y) - 90°\right| \leq \theta_1 . \tag{6.8}$$

The other edges are marked with contours if they meet the condition of the second threshold and touches the outlines already marked:

$$\theta_1 < \left|\alpha - 90°\right| \leq \theta_2 . \tag{6.9}$$

The first and second threshold values are selected as follows: $\theta_1 = 15°$ and $\theta_2 = 30°$.

4. *Short contour segments removal.* Short contours are removed, because they are misaligned or have very little effect on distortions compared to the long ones (they are not representative of distortions or do not reflect true distortions). There are short contours that are not sloping due to geometric deformations in the image, but due to the large inflated shape of a protein spot. Contours are removed if their length is less than $N_x^{AS}/15$, N_x^{AS}—gel active area width in pixels.
5. *Polynomial approximation of contours.* In this step, the contours remaining after the removal of the short segments in the fourth step are approximated by a third degree polynomials:

$$\widehat{C(x)} \triangleq y(x) = b_0 + b_1 x + b_2 x^2 + b_3 x^3 . \tag{6.10}$$

The best approximation is found by the least squares method—searching for such polynomial coefficients for which the square error between the modeled (approximated) points and the simulated points is the smallest. The problem can be solved by finding an excess system of equations, formed for the contour segment of n points, the approximate solution:

$$\begin{bmatrix} 1 & x_1 & x_1^2 & x_1^3 \\ 1 & x_2 & x_2^2 & x_2^3 \\ 1 & x_3 & x_3^2 & x_3^3 \\ \vdots & \vdots & \vdots & \vdots \\ 1 & x_n & x_n^2 & x_n^3 \end{bmatrix} \begin{bmatrix} b_0 \\ b_1 \\ b_2 \\ b_3 \end{bmatrix} = \begin{bmatrix} y_1 \\ y_2 \\ y_3 \\ \vdots \\ y_n \end{bmatrix} . \tag{6.11}$$

System of equations is written in matrix form:

$$\boldsymbol{Xb} = \boldsymbol{y} . \tag{6.12}$$

If the matrix $\boldsymbol{X}^T\boldsymbol{X}$ is non-degenerative (invertable), the approximate solution of the system of excess equations is unique and optimal with respect to least squares error:

$$\min_{\boldsymbol{b}} \|\boldsymbol{Xb} - \boldsymbol{y}\| . \tag{6.13}$$

An approximate solution is found by:

$$\boldsymbol{b} = \left(\boldsymbol{X}^T\boldsymbol{X}\right)^{-1}\boldsymbol{X}^T\boldsymbol{y} . \tag{6.14}$$

6. *Local distortions estimation.* In order to evaluate local horizontal 2DEG image geometric distortions, it is necessary to determine horizontal image structure distortions—they will be the same. Therefore by determining the inclination of the horizontal structures, we will have estimates of geometric distortions at the gel place of interest. By extraction in the image horizontal contours, we obtained elementary horizontal structure models (polynomials). Thus the inclination of horizontal structures due to the geometric deformations at a particular point x will be the same as the coefficient of inclination of the polynomial tangent at that point, i.e., the first sequence derivative:

$$\widehat{C(x)}' = b_1 + 2b_2x + 3b_3x^2 . \tag{6.15}$$

Derivative of the function for approximating horizontal contours $\widehat{C(x)}'$ defines vertical geometric deformations at a specific point, so they can be related to the total gel geometric distortions.

7. *Filtering the tilt factors.* Since the contours which are more inclined than the actual distortions at the gel placed under investigation occur (e.g., the contour is locally curved due to being separated from a large upper or lower edge of the protein spot rather than along its central edge line), the filtering of the tilt coefficients was implemented. Filtering is performed calculating median of tilt coefficients of the number of adjacent polynomials entering the filter area defined by the window. Use median filter window $\boldsymbol{S}_{xy}$ size $1 \times N_{\mathrm{y}}^{\mathrm{AS}}/10$—one pixel width and one tenth of the active gel area height. In this way, filtering

eliminates randomly occurring large inclinations, and as the filtering is performed at each pixel, the inclinations are determined even in areas of the image where no contours were extracted.

8. *Estimation of geometric image distortions.* After filtering, having distortion estimate at each pixel, it becomes possible to estimate global 2DEG image distortion. For this purpose, tilt coefficients are integrated in the horizontal direction—starting from the center of the gel active area part moving toward the left and right sides of the gel and drawing virtual curve. Displacement of the point on the curve in the vertical direction depends on the tilt coefficients at that point. This way ten horizontal curves (that were started in the center of the gel, and their initial points in the center of the gel were arranged in a vertical direction at equal distances) are drawn.

   If there is a need to evaluate global vertical distortions in the gel and accordingly select the highest-quality gel with the least distortion, in each curve the distance between its most up and down remote points is determined, then the largest distance is selected, and the highest-quality gel is selected that one which has the smallest aforementioned distance.

   If there is no suitable gel in the whole experimental group of gels, upright distortion can be reduced by transforming the image in a way that its horizontal curves are transformed to equally distanced horizontals. To do this, it is necessary to move vertically each curve points in order to coincide its y coordinate with the point, from which curve was started to be drawn, in the coordinate y.

   After the image rotation by an angle of 90° and repeating the same procedure again, the whole 2DEG image slanted geometrical distortions will be estimated. Horizontal distortions of the gel will be reduced in exactly the same way as vertical ones.

   Image quality was judged by vertical geometric distortions because they dominate the 2DEG images. Horizontal distortions, degrading the pI determination, are much smaller. It is intrinsic of 2DE technology.

   Results of an experimental study to determine vertical distortions in gels are presented in Fig. 6.7. In this figure, gels from each experimental group are arranged in ascending order of vertical geometric distortions. The gel in the first place is chosen as having the highest quality. In the figure additionally are marked gels that were selected by the expert (considering them being the best ones). The other gels are then aligned according to the selected ones.

Subsection Final Notes

1. Due to vertical geometric distortions in 2DEG images, it is not possible to determine the exact molecular weight of the protein by directly associating the location of the protein spot with the mass indicator. Experimental groups of 2DE gels usually contain gels which are undistorted or slightly distorted, so the selection of a gel with the smallest vertical geometric distortions and aligning it with the others will result in least distorted set of images.
2. The highest-quality 2DE gels selected by the expert and the algorithm coincided in the group of gels with greater distortions. In the selection of gels from the

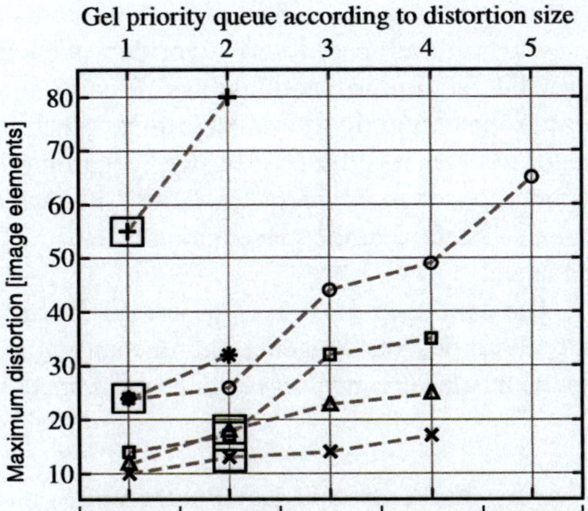

Symbol	1	2	3	4	5
+	55	80			
✱	24	32			
○	24	26	44	49	65
□	14	17	32	35	
△	12	18	23	25	
×	10	13	14	17	

Fig. 6.7 Results of selecting 2DE gel with minimal geometric distortions: symbols—separate 2DE experimental groups; squared symbol—expert decision

group of having less distortions, including those having small difference of geometric distortions, the gels selected by the expert were in the second place of the priority queues generated by the algorithm.

6.2.3 *Initial 2DEG Image Matching*

Purpose of the initial image matching is to find matches between gels with the highest reliability of overlap and eliminate major geometric distortions.

The benefits of initial matching have been theoretically demonstrated and detailed in Sect. 6.1. The correctly performed initial matching limits the spread of local errors throughout the gel, as in the next stage after it, the matching is allowed only in a small distance and therefore does not affect the search for match in other locations. By the way, due to the greatly reduced search area, at the time of final matching, computationally intensive matching methods can be used already as fewer possible matches will be required to check, resulting in nonsignificant increase of a total search time. During initial matching global and major local geometric distortions are eliminated, so later it is possible to apply in such areas similarity assessment methods that are sensitive to geometry distortions; therefore otherwise their use could not be successful.

This subsection provides three initial 2DEG image matching algorithms that have been developed according to the general principles specified in the image matching strategy (see Sect. 6.1). Algorithms are presented in the order in which they were developed. Every algorithm is created aiming to remove the set of the previous algorithm shortcomings. Finally, it is presented the comparison of the initial matching algorithm experiments that shows how the characteristics of the algorithms changed after decisions regarding improvement. The algorithms differ from each other in parts P1–P2–P3–P4, while parts P0–P1 and P4–P5 are the same; therefore in the presentation of each algorithm, only the differences will be analyzed.

The first step P0–P1 is described in Sect. 6.2.3.1. Then, before moving to algorithm design, the general error management principles (parts P3–P4) that will guide the development of reliable search methods for matching areas are presented.

6.2.3.1 Estimation of Key Points and Regions

Significant pixels are the areas of the image that provide mostly information about the image or part of it around the point. At significant points, it is possible to distinguish such feature vectors that will be in a furthest distance from others in the feature space. In the rest of the image, extracted feature vectors are more concentrated in clusters, and the distances between them are smaller. Important property of the significant point extracted feature vectors is that they are more resistant to noise—only stronger noise can impact the vector to approach to the nearest feature vector. Another important property is when a place is significant, then it can be described by smaller number of features, i.e., unique places are described and identified much more efficiently than the rest of the image space. Matching of images intends to find relevant image area locations and geometrically transform images so that the concerned places will physically coincide. During recognition of relevant locations in the images, the most likely correctly matched will be those image locations that are significant (because of their significance, they are the most resistant to noise). Image matching by significant places is more effective not only due to reduced probability of missed matches but also due to the reduced need of calculation resources.

Usefulness of extracted significant area groups directly depends on meeting the following conditions:

$$\mathrm{card}(\mathcal{S}^\star_{\mathrm{G1}}) \ll \mathrm{card}(\boldsymbol{I}_{\mathrm{G1}}); \tag{6.16a}$$

$$\mathrm{card}(\mathcal{S}^\star_{\mathrm{G2}}) \ll \mathrm{card}(\boldsymbol{I}_{\mathrm{G2}}); \tag{6.16b}$$

$$\mathcal{S}^\star_{\mathrm{G1}} \cap \mathcal{S}^\star_{\mathrm{G2}} \simeq \mathcal{S}^\star_{\mathrm{G1}} \cup \mathcal{S}^\star_{\mathrm{G2}}, \tag{6.16c}$$

here $\mathcal{S}^\star_{\mathrm{G1}}$, $\mathcal{S}^\star_{\mathrm{G2}}$ are important regions in $\boldsymbol{I}_{\mathrm{G1}}$ and $\boldsymbol{I}_{\mathrm{G2}}$ 2DEG images; $\mathrm{card}(\mathcal{S}^\star_{\mathrm{G1}})$ is the number (cardinal) of pixels in the set of gel important areas; and $\mathrm{card}(\boldsymbol{I}_{\mathrm{G2}})$ is the number of pixels in the gel image $\boldsymbol{I}_{\mathrm{G2}}$.

Significant areas in 2DEG images are protein spot centers. When aligning gel images, the corresponding protein spots have to overlap; therefore, in order to find the appropriate gel places with the maximum reliability, it is necessary to search for similar places by features extracted at protein spot centers. Based on the developed 2DEG image matching strategy (see Sect. 6.1), initial matching actually seeks to find similar gel places with the highest probability of matching. Protein spot centers are significant places because they exhibit strong central symmetry (CS). Image intensity distortions (noise) have the least effect on the CS point position, and the point can only be lost if it disappears completely. The position of the CS point in the gels shifts due to geometric distortions, but local geometric distortions in 2DE gel images are small compared to global ones. For the before-mentioned reasons, protein spot centers are searched, and only their feature vectors for similarity search are extracted.

Their central symmetry property was used to find the centers of the protein spots. 2DEG image segmentation studies have shown that compared to other methods of calculating central symmetry—phase symmetry and radial symmetry—Johansson rotational symmetry detector (Johansson 2004) is the best in description and localization of the protein spot center. According to the Johansson method, the centers of protein spots are locations of second-order rotational symmetry peaks.

The central symmetry places are not only the centers of protein spots (black spots) but also places between protein spots (symmetrical white spots). The second step of the algorithm evaluates the curveness at each point of central symmetry of the image as two-dimensional functions. In the protein places, there are deflections, while inflections are present at the white spots. The curvature of a function is indicated by its second-order derivative—at deflection places it is negative; thus white spots (not at protein spots centers) can be sorted out according to this condition. Image second-order derivative is usually acquired using Laplacian of Gaussian filter that removes image noise and performs differentiation.

In order to find the most significant image locations, in the third step, the points of central symmetry that are in least probable spot places are removed. There are various foreign derivatives in 2DEG images that are similar to protein spots. Frequently their intensity varies very little. Such places are excluded under minimum image intensity change criterion—if in the central point of symmetry and around it image intensity change is less than the set threshold, the point is removed from the list of significant points.

Features are extracted for each significant point. They are calculated in an area of a certain size around a significant point. The features will not necessarily be extracted during the search for similar areas in images—areas can be compared directly using the selected similarity function. In any case, it is necessary to know the size areas for feature extraction and comparison. An area of appropriate size formed around a significant point is significant area.

The size of the area should be chosen considering it to be as large as possible, thus the more unique it is in that gel, but at the same time, it has more noise. Noise makes the areas less similar. Geometric distortions in small areas are very small, but intensity distortions can vary widely—high if a very small area was formed at high intensity distortions or may have no intensity distortions at all. That is why when selecting the size of the area, someone needs to seek smaller geometric distortions and greater area specificity. Therefore, an area of approximately five medium protein spot size was selected as the most appropriate.

6.2.3.2 Control of Errors During the Search of Matching Areas

Error control is an important process for increasing the reliability of results. Control of errors involves processes from detecting erroneous data to their removal until all errors are fixed. Error control is a certain feedback indicating the state of the goal achievement—whether tasks required to achieve the goal were done successfully.

Because the aim is to create such matches in the image search method that has a minimal probability of error firstly, it is necessary to define the conditions for avoiding errors. Most likely correct identification of matching areas will be in cases when those areas are maximally similar to each other and maximally stand out from the other areas. A condition necessary to find the common points from these two sets is that those sets have common points. This condition has been established in the process of extracting significant points by introduction of an expression (6.16c).

Let's suppose that the areas $S^{\star}_{G1n}$ and $S^{\star}_{G2m}$ are from the sets of corresponding gel areas $\mathbb{S}^{\star}_{G1}$, $i \in [1, N_{G1}]$, and $\mathbb{S}^{\star}_{G2}$, $j \in [1, N_{G2}]$. Here N_{G1} and N_{G2} are the number of areas formed in the respective gels when areas were formed at each significant point, and dist(., .) is distance between selected areas with respect to selected metrics (metrics must properly express 2DEG image areas of similarity or dissimilarity; hence the problem of metric selection must be decided, too). The areas $S^{\star}_{G1n}$ and $S^{\star}_{G2m}$ will coincide most likely if:

$$\mathrm{dist}\left(S^{\star}_{G1n}, S^{\star}_{G2m}\right) \to \min; \tag{6.17a}$$

$$\mathrm{dist}\left(S^{\star}_{G1n}, S^{\star}_{G1i}\right) \to \max; \tag{6.17b}$$

$$\min_{\substack{j\in[1,N_{G2}]\\ j\neq m}} \mathrm{dist}\left(S^{\star}_{G2m}, S^{\star}_{G2j}\right) \to \max\,. \tag{6.17c}$$

If $\mathrm{dist}\left(S^{\star}_{G1n}, S^{\star}_{G2m}\right) = 0$, then conditions can be rewritten by:

$$\mathrm{dist}\left(S^{\star}_{G1n}, S^{\star}_{G2m}\right) \to \min(=0); \tag{6.18a}$$

$$\min_{\substack{i\in[1,N_{G1}]\\ i\neq n}} \mathrm{dist}\left(S^{\star}_{G2m}, S^{\star}_{G1i}\right) \to \max; \tag{6.18b}$$

$$\min_{\substack{j\in[1,N_{G2}]\\ j\neq m}} \mathrm{dist}\left(S^{\star}_{G1n}, S^{\star}_{G2j}\right) \to \max\,. \tag{6.18c}$$

Upon request (simplifying the maximum reliability conditions) that the distance between points from different sets would be shorter than the distance to the other points from the same set (matching areas would be more mutually similar than they are similar to the most similar areas of the same sets), then (6.18) can be approximated by the following conditions:

$$\mathrm{dist}\left(S^{\star}_{\mathrm{G1}n}, S^{\star}_{\mathrm{G2}m}\right) < \min_{\substack{i\in[1,N_{\mathrm{G1}}]\\ i\neq n}} \mathrm{dist}\left(S^{\star}_{\mathrm{G2}m}, S^{\star}_{\mathrm{G1}i}\right); \tag{6.19a}$$

$$\mathrm{dist}\left(S^{\star}_{\mathrm{G1}n}, S^{\star}_{\mathrm{G2}m}\right) < \min_{\substack{j\in[1,N_{\mathrm{G2}}]\\ j\neq m}} \mathrm{dist}\left(S^{\star}_{\mathrm{G1}n}, S^{\star}_{\mathrm{G2}j}\right). \tag{6.19b}$$

During the development of matching search algorithm and willing to speed up the search while maintaining the basic properties of fault resistance, the tolerance conditions (6.19) are implemented by the search of similarity between image areas in both directions. The area from the first gel has to have the most similar area in the second gel, and then the area found in the second gel is searched in order to find the most similar area in the first gel. At the end the area found in the first gel must be the same as the one from which the search has begun.

Introduction of simplified terms in a search of matches explains the splitting of image matching into initial and final matching stages. If the conditions of (6.17) were followed, then the search would be performed consistently, starting with the most reliable match found and gradually selecting less and less reliable matches, before each such processing discarding the sets that were already have been matched. The application of (6.19) prevents reliability rating of matches, so that all matches meeting the new conditions had the same increased reliability in the group. The finding of this group is the result of an initial matching.

6.2.3.3 First Algorithm for the Search of Matching Areas

The kernel of a first matching area search algorithm (Algorithm 6.2) consists of a combination of two measures of similarity. The similarity measures used are standardized cross-correlation and mutual information. These similarity measures are among the most commonly used in image matching tasks. The combination of the two measures of similarity aims at reducing search time for overlapping areas and increasing results reliability. In the first step, using a normalized cross-correlation, candidate areas or, in other words, excluded areas, which are unlikely to overlap, are selected. In the second stage, using mutual information, the most similar area is selected. Normalized cross-correlation is calculated quickly, but it is not very suitable for gels to assess the similarity of the images due to their intensity and geometry distortions. Mutual information is more resistant to noise, but it is receptive to calculations. The combination of both methods allows reducing of both methods' disadvantages and exploitation of their individual advantage.

Algorithm 6.2

Matching area search

1. One significant area $S^{\star}_{G1n}$ of the first gel image is taken.
2. Applying a normalized cross-correlation to the second gel image, a group of areas most similar to the area $S^{\star}_{G1n}$ is found.
3. Using mutual information from the areas of the second gel group, one of the most similar areas is selected.
4. The same but in reverse direction search is done when applying normalized cross-correlation; a group of areas in the first gel image is found that are most similar to the area in the second gel image.
5. Using mutual information from the areas of the first gel group, one of the most similar areas is selected.
6. The correspondence between the areas of the two gels is recorded, if the area found in the last step coincides with the area from which search started, i.e., with area $S^{\star}_{G1n}$.

An initial gel image processing was performed in order to improve the assessment of area similarity. It increases the contrast of gel images according to the initial processing principles of the proposed matching algorithm.

Normalized cross-correlation for initial similar group search was used because of its rapid performance in the frequency domain:

$$\boldsymbol{I}^{\mathrm{R}} = \mathcal{F}^{-1}\left(\frac{\mathcal{F}(\boldsymbol{I}_{\mathrm{G2}}) \cdot \mathcal{F}^{*}(S^{\star}_{\mathrm{G1}n})}{\left|\mathcal{F}(\boldsymbol{I}_{\mathrm{G2}}) \cdot \mathcal{F}^{*}(S^{\star}_{\mathrm{G1}n})\right|}\right), \tag{6.20}$$

here $\mathcal{F} and \mathcal{F}^{-1}$ are direct and inverse Fourier transforms and $\mathcal{F}^{*}(S^{\star}_{\mathrm{G1}n})$ is complex conjugate of $\mathcal{F}(S^{\star}_{\mathrm{G1}n})$, i.e., area $S^{\star}_{\mathrm{G1}n}$ rotation by 180°.

The peaks in the correlation image $\boldsymbol{I}^{\mathrm{R}}$ indicate the range $S^{\star}_{\mathrm{G1}n}$ offset, i.e., places where the correlation is locally maximum. For a further search of a similar area when mutual information similarity function will be used, preselection of areas that meets the following condition is done:

$$S^{\star}_{\mathrm{G2}j} = \underset{x,y}{\arg\max\text{ local}}\left(I^{\mathrm{R}}(x, y) > 0,5 \cdot I^{\mathrm{RT}}\right), \tag{6.21}$$

here arg max local is search of local image maximum coordinates.

Using mutual information function from area set $S^{\star}_{\mathrm{G2}j}$, one most similar area is selected by:

$$S^{\star}_{\mathrm{G2}m} = \underset{j\in[1,N_{\mathrm{G2}}]}{\arg\min}\left(H\left(S^{\star}_{\mathrm{G1}n}\right) + H\left(S^{\star}_{\mathrm{G2}j}\right) - H\left(S^{\star}_{\mathrm{G1}n}, S^{\star}_{\mathrm{G2}j}\right)\right), \tag{6.22}$$

here $H(.,.) = -\sum_i \sum_j p_{ij}(.,.) \ln p_{ij}(.,.)$ is joint entropy and $H(.) = -\sum_i p_i(.) \ln p_i(.)$ is marginal entropy.

Found area $S^{\star}_{G2m}$ is most similar to the area $S^{\star}_{G1n}$ with regard to the two similarity criteria. To find the area most similar to $S^{\star}_{G2m}$ in the first gel, the same steps are followed. If this is the area $S^{\star}_{G1n}$, then a coincidence between images of both 2DE gels is registered.

Important preliminary results of this study—the initial matching advantages of the strategy over conventional iterative image matching methods (based on geometric image transformation and estimation of matching by similarity throughout the whole picture)—were confirmed. The combination of the two similarity measures significantly increased search speed, and the two-way search for similarities immediately rejected a considerable amount of coinciding but false spot pairs. However, when the noise in 2DEG images increased, fewer and fewer correct matches were found. The resistance to noise could be increased by following examination of the main image similarity assessment methods and exploration of the possibility to develop an alternative similarity assessment method that utilizes multilayer perceptron.

6.2.3.4 Second Algorithm for the Search of Matching Areas

This section focuses on the study of distance functions, willing to determine suitable assessment of the similarity of 2DEG image areas. Because the algorithm presented in the previous section obtains promising results, it was decided to improve this initial alignment strategy. The main focus is on the study of distance functions and creation of an alternative method suitable for the evaluation of gel area similarity. An alternative algorithm is constructed using multilayer perceptron. Research is organized based on the idea that the assessment of image similarity must be specialized for specific images. The image of each specific area of the gel is distinguished by its characteristic information and noise. Examining wisely the used distance measurements, we aim to develop an alternative algorithm which is able to find the similarities inherent in the 2DEG images and is as resistant as possible to noise—intensity and geometric distortions.

A Set of Distance Metrics

The distance between the two image areas $\boldsymbol{x}$ and $\boldsymbol{y}$ is denoted $\text{dist}(\boldsymbol{x}, \boldsymbol{y})$. Usually the distance (dissimilarity) is a following function $\text{dist} : N \times N \to \mathbb{R}$, where N is the set that for all $x,\ y \in N$ following properties are valid: $\text{dist}(x,\ y) \geq 0$ (non-negativity); $\text{dist}(x,\ y) = \text{dist}(y,\ x)$ (symmetry); $\text{dist}(x,\ x) = 0$ (identity); and $\text{dist}(x,\ z) \leq \text{dist}(x,\ y) + \text{dist}(y,\ z)$ (triangle inequality).

We will examine how each distance function is able to evaluate the 2DEG image area dissimilarity and would therefore be suitable for searching for similar areas. A search for a similar areas is required to match images.

A summary of the distance functions selected for the comparison tests is provided in Table 6.1. Distances (1)–(5) and (6b)–(9b) (they are alternative realizations of

Table 6.1 Distance measures selected for comparison

Distance measure	dist(x, y)	Alternatives	
Squared Euclidean	$\sum_i (x_i - y_i)^2$	–	(F1)
Cosine	$1 - \frac{\sum_i x_i y_i}{\sqrt{\sum_i x_i x_i} \cdot \sqrt{\sum_i y_i y_i}}$	–	(F2)
Pearson correlation	$1 - \frac{\sum_i (x_i - \bar{x})(y_i - \bar{y})}{\sqrt{\sum_i (x_i - \bar{x})^2 \sum_i (y_i - \bar{y})^2}}$	–	(F3)
Spearman's rank correlation	$\frac{6 \sum_{i=1}^{n} \left(\text{rank}(x_i) - \text{rank}(y_i)\right)^2}{n\left(n^2 - 1\right)}$	–	(F4)
Chebyshev	$\max_i \lvert x_i - y_i \rvert$	–	(F5)
Histogram intersection	$1 - \sum_i \min\left(p_i(x), p_i(y)\right)$	$1 - \sum_i \min\left(\hat{x}_i, \hat{y}_i\right)$	(F6a,b)
Jeffrey divergence	$\sum_i p_i(x) \ln \frac{2p_i(x)}{p_i(x) + p_i(y)} + \sum_i p_i(y) \ln \frac{2p_i(y)}{p_i(x) + p_i(y)}$	$\sum_i \hat{x}_i \ln \frac{2\hat{x}_i}{\hat{x}_i + \hat{y}_i} + \sum_i \hat{y}_i \ln \frac{2\hat{y}_i}{\hat{x}_i + \hat{y}_i}$	(F7a,b)
Bhattacharyya (Hellinger)	$\sqrt{1 - \sum_i \sqrt{p_i(x) p_i(y)}}$	$\sqrt{1 - \sum_i \sqrt{\hat{x}_i \hat{y}_i}}$	(F8a,b)
χ^2	$\sum_i \frac{\left(p_i(x) - p_i(y)\right)^2}{2\left(p_i(x) + p_i(y)\right)}$	$\sum_i \frac{\left(\hat{x}_i - \hat{y}_i\right)^2}{2\left(\hat{x}_i + \hat{y}_i\right)}$	(F9a,b)
Normalized mutual information	$\frac{1}{2}\left(1 - \frac{H(x) + H(y)}{H(x, y)}\right)$	–	(F10)

Here: rank(.)—position of the data after sorting in ascending order;
$p(x)$, $p(y)$—probability density functions of discrete random variables x and y;
$\sum_i p_i(x) = 1$; $\sum_i p_i(y) = 1$; $\sum_i \hat{x}_i = 1$; $\sum_i \hat{y}_i = 1$; $\bar{x} = \frac{1}{n} \sum_i x_i$; $\bar{y} = \frac{1}{n} \sum_i y_i$;
$H(x)$, $H(y)$—marginal entropies; $H(x, y)$—joint entropy;
$H(.) = -\sum_i p_i(.) \ln p_i(.)$; $H(x, y) = -\sum_i \sum_j p_{ij}(x, y) \ln p_{ij}(x, y)$.

distances (6a)–(9a)) are calculated directly from the points in the 2DEG image area, thus enabling the evaluation of the spot arrangement geometric features. Distances (6a)–(9a) and (10) are from information theory and are calculated from each image probabilistic distributions. The Euclid and Chebyshev distances are separate Minkowski distance cases: $\sqrt[p]{\sum_i \lvert x_i - y_i \rvert^p}$, when $p = 2$ and $p = \infty$, correspondingly.

Dataset Generation for Comparison of Distance Metrics

The content of the image determines the effectiveness of the distance measurements. The measure must be sensitive for those visual features that are characteristic to the images under comparison. Only then will it be possible to find similar areas in the images more reliably. If a set of images is composed so that it contains examples

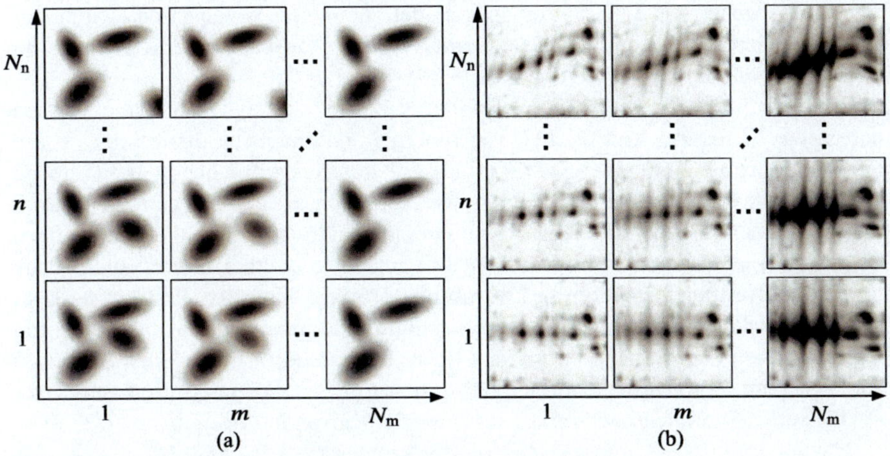

Fig. 6.8 Examples from created 2DEG image dataset: **(a)** example from synthetic image dataset $\mathcal{Z}_1$; **(b)** example from semi-synthetic image dataset $\mathcal{Z}_2$

that are ranked according to a uniformly varying similarity, then measured distance between the first and each subsequent image would successively increase. However, the group of natural 2DEG image areas cannot be ordered according to similarity because a suitable tool for that does not exist. In order to solve this problem, we created datasets of synthetic and semi-synthetic image possessing basic features of gel images.

Set of synthetic 2DEG images $\mathcal{Z}_1$. The 2D Gaussian function as a protein spot model was used in the development of synthetic 2DEG image examples. This model can represent forms of most protein spots. Two-dimensional Gaussian function, modeling a single protein spot, is written as follows:

$$\Psi_{2G}(x, y; \boldsymbol{\theta}_{2G}) = I^{A} \cdot \exp\left(-\frac{x'^2}{2\sigma_x^2}\right) \cdot \exp\left(-\frac{y'^2}{2\sigma_y^2}\right), \tag{6.23}$$

here $\boldsymbol{\theta}_{2G} = \{I^{A}, \theta_R, x_0, y_0, \sigma_x, \sigma_y\}$; $x' = \cos\theta_R \cdot (x - x_c) + \sin\theta_R \cdot (y - y_c)$; $y' = -\sin\theta_R \cdot (x - x_c) + \cos\theta_R \cdot (y - y_c)$; θ_R is the rotation with respect to its center angle of the spot model; (x_c, y_c) is the coordinates of the center of the spot model; I^{A} is the maximum intensity of the spot without a background; σ_x, σ_y is spot width in horizontal and vertical directions.

The set of synthetic 2DEG images $\mathcal{Z}_1$ consists of elementary image sets $\boldsymbol{Z}_{n,m}$. Elemental image sets $\boldsymbol{Z}_{n,m}$ have synthetic images that differ by only two parameters—the coordinates of the center of the spot model $(x_c(n), y_c(n))$ and intensity maximum $I^{A}(m)$. Images from different elementary sets differ in all model parameters. Elementary image sets $\boldsymbol{Z}_{n,m}$ as $N_n \times N_m = 25 \times 25$ size arrays were created monotonically changing parameter pair $\{(x_c(n), y_c(n)), I^{A}(m)\}$, depending on indices (n, m). The indices (n, m) indicate a specific synthetic image in the elementary image set $\boldsymbol{Z}_{n,m}$. Other spot model parameters in the same elementary set are constant. So the synthetic 2DEG images differ from each other in the position of the spots and intensity (see Fig. 6.8a).

A set of semi-synthetic 2DEG images $\mathcal{Z}_2$. These image sets were created according to natural 2DEG image geometric and intensity distortions. These distortions are a major cause in 2DEG images that makes difference in 2DEG image areas that should be identical. Having corresponding areas $\boldsymbol{S}_{\mathrm{G}1n}$ and $\boldsymbol{S}_{\mathrm{G}2m}$ from different gels, we generate intermediate ones gradually changing the images from $\boldsymbol{S}_{\mathrm{G}1n}$ to $\boldsymbol{S}_{\mathrm{G}2m}$. Gradual transition from one image to another is implemented by the geometric transformation and transition of image intensity. Performing both processes independently, i.e., gradually changing the parameters of the geometric transformation and drawing up various linear derivatives of the original images, a set $\mathcal{Z}_2$ according to essential principles similar to $\mathcal{Z}_1$ is generated. Both sets sizes are the same. Semi-synthetic image set example is given in Fig. 6.8b.

Having two 2DEG image areas of the same size $\boldsymbol{S}_{\mathrm{G}1n}$ and $\boldsymbol{S}'_{\mathrm{G}2m}$ ($\boldsymbol{S}'_{\mathrm{G}2m}$ is area $\boldsymbol{S}_{\mathrm{G}2m}$ after matching with area $\boldsymbol{S}_{\mathrm{G}1n}$), taken from the original 2DEG images, intensity shift is obtained by calculating linear image derivatives:

$$\boldsymbol{Z}_{1,m} = \boldsymbol{S}_{\mathrm{G}1n}\frac{N_{\mathrm{m}} - m}{N_{\mathrm{m}} - 1} + \boldsymbol{S}'_{\mathrm{G}2m}\frac{m - 1}{N_{\mathrm{m}} - 1}\,. \tag{6.24}$$

Here N_{m} is the number of images in the sequence to be obtained ($N_{\mathrm{m}} = 25$); $m \in [1, N_{\mathrm{m}}]$ is intermediate image index; $\boldsymbol{Z}_{1,1} = \boldsymbol{S}_{\mathrm{G}1n}$; and $\boldsymbol{Z}_{1,N_{\mathrm{m}}} = \boldsymbol{S}'_{\mathrm{G}2m}$.

Geometric distortions are modeled by extracting overlap vectors, according to which performed $f : \boldsymbol{S}_{\mathrm{G}2m} \rightarrow \boldsymbol{S}'_{\mathrm{G}2m}$, images $\boldsymbol{S}_{\mathrm{G}1n}$ and $\boldsymbol{S}_{\mathrm{G}2m}$ are matched. Everyone previously generated intermediate image $\boldsymbol{Z}_{1,N_{\mathrm{m}}}$ is geometrically transformed into an image$\boldsymbol{Z}_{N_{\mathrm{n}},N_{\mathrm{m}}} = \boldsymbol{S}_{\mathrm{G}2m}$ in equal selected ($N_{\mathrm{n}} = 25$) steps. Parameters $\boldsymbol{\theta}_{\mathrm{PLS}}(n)$ of thin spline models (see Eq. (6.32)) are selected according to the geometric transformation index $n \in [2, N_{\mathrm{n}}]$. From a set $\mathcal{Z}_2$ using available geometric transformation parameters, a set of semi-synthetic images $\boldsymbol{Z}_{n,m}$ is generated:

$$\boldsymbol{Z}_{n,m} = \Psi_{\mathrm{PLS}}\big(\boldsymbol{Z}_{1,m}; \boldsymbol{\theta}_{\mathrm{PLS}}(n)\big)\,, \tag{6.25}$$

here Ψ_{PLS} is a thin spline transform and $\boldsymbol{\theta}_{\mathrm{PLS}}(n)$ is transform parameters.

Comparison of Distance Metrics

The distance functions are compared in order to select ones that best stipulate the identification of matched in 2DE gel areas. Seven experiments for distance comparison were prepared and performed (results of the distance ratings obtained during them are presented in Table 6.2). The best result in each of the group (table row) is denoted by 1, and ascending indices mean deteriorating properties.

Three main types of experiments were used to compare distances: (a) investigation of distance calculation duration (first row in Table 6.2); (b) investigation

Table 6.2 Ranking of 2DE gel image sets pairwise comparison results

Comparison way	Distance measure, dist(x, y)													
	F1	F2	**F3**	F4	F5	F6a	**F6b**	F7a	F7b	F8a	F8b	F9a	**F9b**	F10
Processing time	4	6	7	14	1	11	3	10	8	12	5	9	2	13
SOM/1DM/$\mathcal{Z}_1$	8	6	1	5	3	13	2	11	9	10	13	14	4	7
SOM/1DM/$\mathcal{Z}_2$	7	6	1	4	2	10	3	12	9	11	14	13	5	8
SOM/1DM/$\mathcal{Z}_*$	8	7	1	5	2	13	3	10	6	11	12	14	4	9
SOM/2DM/$\mathcal{Z}_*$	8	7	1	5	2	13	3	10	6	11	12	14	4	9
SOM/3DM/$\mathcal{Z}_*$	6	2	1	4	3	9	5	12	8	13	14	11	7	10
Expert/1DM/$\mathcal{Z}_N$	8	2	1	4	9	10	3	13	6	14	11	12	5	7

of Kohonen self-organizing feature map SOM (Dzemyda et al. 2008) classifier (rows 2–6); (c) comparison with expert-provided decisions (row 7). SOM application is motivated by the ability of this classifier to sort examples according to their similarity (or dissimilarity), which in this case corresponds to distance measurements. Because images in compiled synthetic and semi-synthetic sets are sorted according to the true similarity change (by measuring true similarity, we can find matching areas in gels under comparison), thus best distance measures or their groups will be those that SOM classifier arranged in the same order as they were lined up during generation.

To better research the informativeness of individual distance measures or groups of them during comparison of samples by similarity, five experiments with SOM were carried: using 1, 2, or 3 distance measures (features) (in Table 6.2 denoted by 1AM, 2AM, or 3AM), which were calculated with synthetic, semi-synthetic, and all datasets (labeled $\mathcal{Z}_1$, $\mathcal{Z}_2$, and $\mathcal{Z}_*$, correspondingly).

Distance values dist($\mathbf{Z}_{1,1}$, $\mathbf{Z}_{n,m}$), when $n \in [1, N_\mathrm{n}]$ and $m \in [1, N_\mathrm{m}]$ calculated between $\mathbf{Z}_{1,1}$ sample and remaining samples in that set, were input to SOM classifier. It was found that distance measurements can be clearly compared by SOM classification of samples into three classes: 1 class samples are closest to $\mathbf{Z}_{1,1}$ because their distance to the sample value must be the smallest (here proximity is understood according to the location in the dataset); 2 class includes moderate distance samples; while 3 class represents the farthest from $\mathbf{Z}_{1,1}$ samples—$\mathbf{Z}_{1,N_\mathrm{m}}$, $\mathbf{Z}_{N_\mathrm{n},1}$, $\mathbf{Z}_{N_\mathrm{n},N_\mathrm{m}}$, and close to them samples. During the experiments, each set of samples was classified into three classes and then assessed whether samples have been assigned to the appropriate classes as given in the description.

Ratings were assigned to distance measures as follows: (a) distances dist($\mathbf{Z}_{1,1}$, $\mathbf{Z}_{n,m}$), here $n \in [1, N_\mathrm{n}]$ and $m \in [1, N_\mathrm{m}]$ for all sample sets, were calculated; (b) 1, 2, or 3 distance values (depending on the test; in the study of 2 or 3 distance groups, all possible distance combinations up to 16 distances were constructed) were passed to SOM; (c) SOM classification and evaluation, whether the set of samples is classified according to the description, were done; (d) for distance measures that have contributed to the correct classification of the set, a score was awarded; (e) the distance with the highest score was ranked the best ("1"). For each experiment type, distances were ranked separately. Distance ratings for five experiments are presented in Table 6.2 rows 2–6.

Experts' ranking was performed as follows: (a) expert by hand noted protein spots corresponding to the same proteins in the compared gels (when matching gel images, those places must necessarily overlap); (b) the 2DEG image areas were taken from the marked locations and that the respective areas are similar was recorded—in such a way, sample set $\mathcal{Z}_N$ was created; (c) during testing the distances between all collected areas were calculated; (d) if the use of distance measure yielded the minimum distance between recorded similar areas, then a score was given for the distance; and (e) the distance with the highest score was ranked the best ("1"). Results of this experiment are provided in Table 6.2 row 7.

Based on the results of the experiments, the following group of three distance measures was selected (grayed in Table 6.2): Pearson correlation distance (3), the intersection of the histograms, calculated from the normalized pixel intensity values (6b), and the distance χ^2, calculated from the pixel intensities (9b). Selected distances will be used in further experiments. Spearman's rank correlation (4) and Chebyshev (5) distances yield good results, but they were not selected because their calculation requires big computational resources or due to very different test results when tested with generated samples and with samples provided by the expert.

6.2.3.5 Development of MLP Detector for Area Similarity Estimation

The choice of possible multilayer perceptron (MLP) detector structures was limited by the available training dataset specifics. For MLP the training examples and responses are required. It was possible to create a dataset that does not contain an absolute but relative value of the area similarity, i.e., data on the actual $\text{dist}(\boldsymbol{S}_{\text{G1}n}, \boldsymbol{S}_{\text{G2}m})$ and $\text{dist}(\boldsymbol{S}_{\text{G1}n}, \boldsymbol{S}_{\text{G2}r})$ distances was not available. Therefore the distance $\text{dist}(\boldsymbol{S}_{\text{G1}n}, \boldsymbol{S}_{\text{G2}m})$ and $\text{dist}(\boldsymbol{S}_{\text{G1}n}, \boldsymbol{S}_{\text{G2}r})$ values were fed into the MLP input. If the actual distance between the first pair areas is larger ($\boldsymbol{S}_{\text{G1}n}$ area is more similar to $\boldsymbol{S}_{\text{G2}m}$ than to $\boldsymbol{S}_{\text{G2}r}$ area), then the MLP response had to be "1" and "−1" otherwise.

The training sample set consisted of three sample sources—synthetic, semi-synthetic, and natural 2DEG images ($\mathcal{Z}_1$, $\mathcal{Z}_2$, $\mathcal{Z}_N$, respectively). Such set of samples covers a large part of distortion in the 2DEG images. The sample set $\mathcal{Z}_N$ was created based on the expert-provided information on overlapping areas of the gel.

Let's mark by $\boldsymbol{S}_{\text{G1}i}$, $i \in [1, N_{\text{G1}}]$, and $\boldsymbol{S}_{\text{G2}j}$, $j \in [1, N_{\text{G2}}]$, areas that were separated in I_{G1} and I_{G2} 2DEG images. A part (N_{mtch}) of these areas $\boldsymbol{S}_{\text{G1}i}$ and $\boldsymbol{S}_{\text{G2}j}$, when $i = j$ ($i, j \in [1, N_{\text{mtch}}]$, $N_{\text{mtch}} \leq N_{\text{G1}}$, and $N_{\text{mtch}} \leq N_{\text{G2}}$), is marked by expert as matching in I_{G1} and I_{G2} 2DEG images. Then $\text{dist}(\boldsymbol{S}_{\text{G1}i}, \boldsymbol{S}_{\text{G2}j})$ ($i = j$; $i, j \in [1, N_{\text{mtch}}]$) will be the distance between the areas that must overlap when the gels match, which means that this distance must be kept to a minimum in order to correctly identify this pair of areas as matching. Distances between all areas that do not match $\text{dist}(\boldsymbol{S}_{\text{G1}i}, \boldsymbol{S}_{\text{G2}j})$, $i \neq j$ were also calculated.

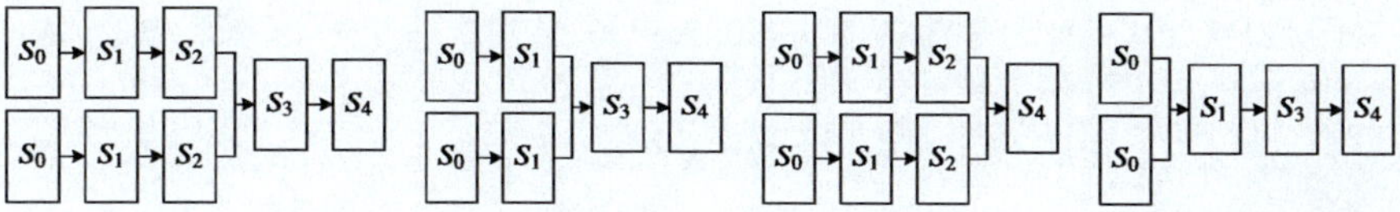

Fig. 6.9 Generalized representations of investigated MLP structures: **(a)** MLP-A; **(b)** MLP-B; **(c)** MLP-C; **(d)** MLP-D

The MLP training dataset is composed as follows: (a) for all examples dist($\boldsymbol{S}_{G1i}$, $\boldsymbol{S}_{G2j}$) ($i = j;\ i, j \in [1, N_{\text{mtch}}]$), the response value “1” was assigned; (b) for examples dist($\boldsymbol{S}_{G1i}$, $\boldsymbol{S}_{G2j}$) ($i \neq j;\ i \in [1, N_{G1}];\ j \in [1, N_{G2}]$), a response value “$-1$” was assigned.

There were three original and one standard MLP structure created (shown in Fig. 6.9). Blocks in flowcharts denote the artificial neuron layer. The input layer consists of three inputs to provide the value of each distance. There can be up to three hidden layers with possible sizes—$N^{(1-3)} = \{5, 10, 15, 20, 30, 40\}$ neurons in each layer. Output layer consists of a single neuron. MLP activation function in the first hidden layer is the logarithm sigmoid (LS, $\Phi_{\text{LS}}^{(1)}$) because the distance values fed to the network are positive and normalized. In the third hidden layer, a tangent sigmoid (TS) transfer function ($\Phi_{\text{TS}}^{(3)}$) is used, because the MLP network response sign will indicate which image pair similarity is higher. Selection of the second hidden layer transfer function is unrestricted by input and output data; thus the use of two functions—$\Phi_{\text{LS}}^{(2)}$ ir $\Phi_{\text{TS}}^{(2)}$ was tested.

The first three MLP structures (named by MLP-A, MLP-B, and MLP-C) are such that their inputs are not initially fed to a common layer, but into two parallel branches with completely identical parameters, that are connected only in the third hidden layer. The fourth investigated structure of MLP-D reassembles the regular multilayer structure.

All MLP structures were tested with 57 380 image pairs randomly divided into the training (60%), validation (20%), and testing (20%) sets. Each MLP was initialized according to the Nguyen and Widrow method and taught by the Levenberg and Marquard algorithm. Employed training parameters are as follows: maximum number of epochs, 100; training error, 0; maximum number of consecutive validation failures, 5; minimal gradient value, 10^{-10}; $\mu_{\text{init}} = 0,001$; $\mu_{\text{dec}} = 0,1$; $\mu_{\text{inc}} = 10$; and $\mu_{\text{max}} = 10^{10}$. The best result of each MLP structure use was selected:

- In MLP-A case: $N^{(0)} = 3$, $N^{(1)} = 5$, $N^{(2)} = 5$, $N^{(3)} = 30$, $N^{(4)} = 1$, with $\Phi_{\text{LS}}^{(1)}$, $\Phi_{\text{TS}}^{(2)}$, $\Phi_{\text{TS}}^{(3)}$, and $\Phi_{\text{S}}^{(4)}$ functions.
- In MLP-B case: $N^{(0)} = 3$, $N^{(1)} = 40$, $N^{(2)} = 15$, $N^{(3)} = 1$, with $\Phi_{\text{LS}}^{(1)}$, $\Phi_{\text{TS}}^{(2)}$, and $\Phi_{\text{S}}^{(3)}$ functions.

- In MLP-C case: $N^{(0)} = 3$, $N^{(1)} = 30$, $N^{(2)} = 5$, $N^{(3)} = 1$, with $\Phi^{(1)}_{\mathrm{LS}}$, $\Phi^{(2)}_{\mathrm{TS}}$, and $\Phi^{(3)}_{\mathrm{S}}$ functions.
- In MLP-D case: $N^{(0)} = 3$, $N^{(1)} = 30$, $N^{(2)} = 10$, $N^{(3)} = 1$, with $\Phi^{(1)}_{\mathrm{LS}}$, $\Phi^{(2)}_{\mathrm{TS}}$, and $\Phi^{(3)}_{\mathrm{S}}$ functions.

These networks were further tested by selecting matching areas only from natural 2DEG images.

Estimation of Area Similarities using MLP

The matching area pair MLP selects in a following way:

1. Image area $S^{\star}_{\mathrm{G1}n}$ from the first 2DE gel is prepared for comparison with all possible areas $S^{\star}_{\mathrm{G2}j}$, $j \in [1, N_{\mathrm{G2}}]$ in another gel image.
2. Three chosen distance measures (see Sect. 6.2.3.4) are used for calculation of distances between all areas under comparison.
3. According to the calculated distance values, MLP detector selects a pair of image areas between which distance is smallest (areas most similar).
4. The area $S^{\star}_{\mathrm{G2}m}$ found in the second gel in accordance to the same principles is comparable to all areas $S^{\star}_{\mathrm{G1}i}$, $i \in [1, N_{\mathrm{G1}}]$.
5. The location of the match found on the alignment gels is marked and will be used to geometrically transform images.

Result of Automatic 2DEG Image Matching

The test dataset formation is based on the same principle as natural image training dataset—according to specified by the expert gel image matches. The test dataset images are from three different biochemical experiments in which proteins of specific cells (HL-60 cells, the conductive part of the human heart cells, and mesenchymal stem cells) were studied. The first image group consists of 186,992 2DEG, the second has 76,722, and the third experimental group has 101,007 image pairs.

The results of the experiments are presented in Table 6.3. During the experiments each best neural network from each principal structure was compared as well as the

Table 6.3 Percentages of successful natural 2DE gel image similarity comparisons

Origin of 2DE gel sets	Single feature classifier			Multilayer Perceptron			
	(F3)	(F6b)	(F9b)	A	B	C	D
HL-60 cells	99.965	99.985	99.987	99.983	99.990	99.961	99.973
Human heart cells	98.915	99.014	98.811	99.363	99.589	99.159	99.253
Mesenchymal stem cells	99.919	99.864	99.868	99.959	99.974	99.930	99.954

result of each distance measure selected for use with MLP in order to find matching areas. Results as a ratio of the number of correctly found matching pairs in each test group having known number of matching pairs are expressed in percentages. The best results are marked by gray background. Comparative experiment results show that the network structures that are formed by splitting the input layer into separate branches and merging them into one layer only at the end of network are superior in deciding on gel image area similarity. A network of such structure initially concentrates in distance values between all pairs while postponing the decision which pair is more similar, to the last neuron layer.

Subsection Generalization

1. An automatic search for similar areas procedure has been created.
2. From the analyzed 14 distance metrics as the most informative were selected Pearson correlation, the intersections of the histograms, and the χ^2 distance.
3. To detect a 2DEG image areas, matching a multilayer perceptron was proposed.
4. The results of experiments show that after applying the selected MLP network structure and distance measures, area matching identifibility increases.

6.2.3.6 Third Algorithm for the Search of Matching Areas

In order to speed up the search procedure, new algorithm uses a completely new principle during assessment of area similarity. Moreover, to reduce the number of errors when the level of distortion increases, additional debugging step is added.

The core of the algorithm is the search for similar areas of the 2DEG image according to the features that are extracted by Lowe technique (Lowe 2004). Accordingly, at each significant point, 128 feature vectors are formed, while matches are found according to previously presented algorithms—the distances between all vectors from different images are calculated, and matching is registered between those vectors that have the smallest distances between images when searched in both directions.

At the beginning of the Lowe technique, the significant pixels are found. Then at each point dominant image direction and scale are determined. They are used to normalize the features in order to describe each point independently from those parameters. Lowe's descriptors are calculated at each significant point found in the presented method in Sect. 6.2.3.1, while the omission of the feature normalization step results in far more correct matches found before the error detection step stage. Attribute normalization can be omitted if both 2DEG images are scanned at the same resolution and possess the same orientation on the scanner.

Lowe's descriptor is a group of histograms formed by local gradients. Each descriptor consists of $4 \times 4 \times 8 = 128$ attributes—it is an array of 4×4 histograms, in which each histogram consists of 8 columns describing the size of the gradient in eight directions. The array of histograms is produced from 16×16 pixel size image

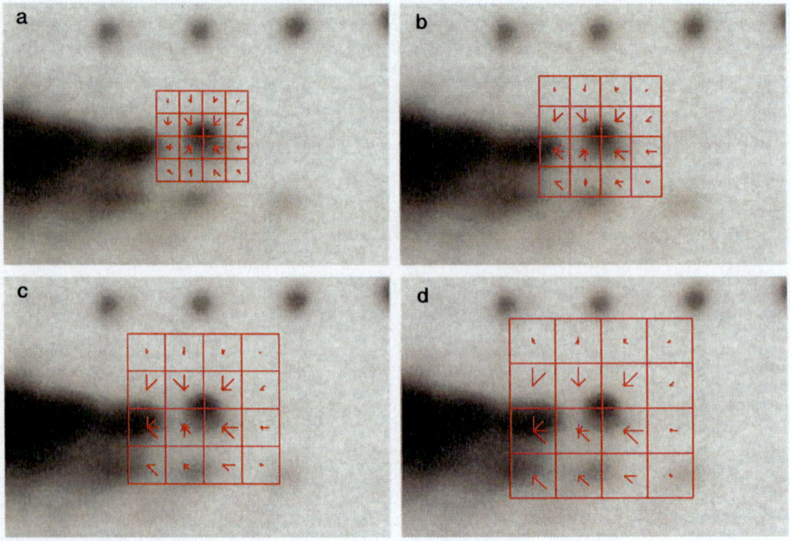

Fig. 6.10 Descriptors of different scale but having the same position (same orientation and centered at the same spot). Descriptors visualized using `VLFeat` library (Vedaldi and Fulkerson 2008)

area by combining the size of the gradient and its direction data weighting them by a Gaussian function. Four descriptors of different scale but having the same orientation and calculated at the same location of gel image (centered at the same spot) are shown in Fig. 6.10.

The search for matches is performed similarly to the principle of the previous algorithm—those matches that were maximal in both directions between regions $S^{\star}_{G1i}$ and $S^{\star}_{G2j}$ are selected. But in this algorithm an additional debugging procedure based on the RANSAC algorithm is performed at the end.

Random Sample Consen-sus (RANSAC) algorithm often is used for the selection of data that correspond to a particular model in case of large data groups. Data selection (sampling) could be needed due to a large number of erroneous data or due to mixture of data belonging to different models. RANSAC algorithm in 2DEG image matching task was used in order to reject (detected) erroneously found matches. During RANSAC algorithm data that belongs to the same model of geometric transformation are selected.

When matches (matching areas) are found in the 2DEG images, some of them may be incorrect. It was found that with increase of noise (intensity distortions) level in 2DEG images during bidirectional search, number of mismatches are on the rise and sometimes quite strongly. Therefore before the geometric transformation of images, erroneous matches need to be removed.

RANSAC can eliminate erroneous data in case it constitutes up to 50% from all the data. Because during 2DEG images matching errors occur due to noise in the images, erroneously matched area places are irregular, and RANSAC algorithm will not assign them to the model. There is a vanishingly low probability that noise exposure will result in a large data group belonging to the wrong model. Under these circumstances, the RANSAC algorithm is very suitable for the correct match processing.

Most of the geometric distortions in the 2DEG images can be removed by use of projection transformation. RANSAC algorithm is efficient, with a large amount of data and a small number of model parameters; therefore instead of complex and plastic transformation model, we have chosen to use a projection transformation model that requires four point pairs to determine the match. In a homogeneous coordinate system, the projection transformation is defined by a matrix of six coefficients.

Subsection Final Notes

Initial 2DEG image matching is important because:

1. It removes large geometric distortions from 2DEG images
2. Area for the final matching search is reliably reduced
3. Final matching error risk is broken down in parts—a matching error made in the final matching will not affect matches found in other fields
4. Number of comparisons required to determine a pair of similar image areas is reduced

6.2.4 Final 2DEG Image Matching

The final 2DEG image matching process is similar to the initial matching as it is announced in the circumstantiation of the whole matching method (Fig. 6.4).

Main differences between final and initial matching are as follows:

1. During the extraction of significant points, more points are left because no final rejection of points according to the minimum height is performed.
2. Additional means to assess the matching areas are used—the focus is on the geometric information of the images rather than the intensity.
3. The search for matching areas is conducted in a small area around the significant area. Such search restriction transfers the information from the initial to the final matching stage.
4. During the error control, checking for overlaps between mapping vectors is done.
5. Instead of non-rigid geometric transformation, the plastic transformation is performed.

6.2.4.1 Geometrical Transformation of 2DEG Image

At the end of the final 2DEG image matching, different from initial stage, geometric transformation is performed. As in the initial matching stage found matches are reliable, a non-rigid transform here can be used. Thus a thin plate spline transformation is used to complete 2DEG image matching (Bookstein 1989).

The thin plate spline represents superposition of eigenvectors of the surface curvature energy matrix. The splines can be used to model surface deformations. Therefore, from a pair of splines corresponding to deformations along x and y axis, it is possible to form mapping $\Psi_{PLS} : \mathbb{R}^2 \rightarrow \mathbb{R}^2$ that links landmark sets in two images. The spline mapping can be broken down into linear parts (affine transformations) and principal distortions that are geometrically independent and dissimilar deformations of decreasing areas. In this case, the mapping function is broken down into elementary parts called the main solution of the biharmonic equation:

$$\Upsilon_U(d) = d^2 \log d^2 , \tag{6.26}$$

here $d = \sqrt{x_i^2 - y_i^2}$ is the distance from start of Descartes coordinate system.

Let us discuss the use of thin plate splines for geometric transformation. Suppose sets of points $S_{G1n} = \{x_{G11}, y_{G11}; \ldots; x_{G1n} y_{G1n}\}$ are available in the first image and $S_{G2n} = \{x_{G21}, y_{G21}; \ldots; x_{G2n}, y_{G2n}\}$ in the second one. Such function Ψ_{PLS} is required that would represent the points S_{G1n} into S_{G2n}. At the beginning the distances $d_{ij} = \|S_{G1i} - S_{G1j}\|$ between all possible permutations of points S_{G1n} are calculated, and then (6.26) is applied. The results are recorded as $n \times n$ size matrix:

$$\boldsymbol{X_0} = \begin{bmatrix} 0 & \Upsilon_U(d_{1,2}) & \cdots & \Upsilon_U(d_{1,n}) \\ \Upsilon_U(d_{2,1}) & 0 & \cdots & \Upsilon_U(d_{2,n}) \\ \vdots & \vdots & \ddots & \vdots \\ \Upsilon_U(d_{n,1}) & \Upsilon_U(d_{n,2}) & \cdots & 0 \end{bmatrix} . \tag{6.27}$$

The coordinates of the points are written in another matrix:

$$\boldsymbol{A}_{G1} = \begin{bmatrix} 1 & x_{G11} & y_{G11} \\ 1 & x_{G12} & y_{G12} \\ \vdots & \vdots & \vdots \\ 1 & x_{G1n} & y_{G1n} \end{bmatrix} \tag{6.28}$$

and from them we compose:

$$\boldsymbol{X} = \left[\begin{array}{c|c} \boldsymbol{X_0} & \boldsymbol{A}_{\mathrm{G1}} \\ \hline & 0\,0\,0 \\ \boldsymbol{A}_{\mathrm{G1}}^{T} & 0\,0\,0 \\ & 0\,0\,0 \end{array}\right], \tag{6.29}$$

here "$\cdot^T$" is matrix transposition operator.

The coordinates of the reference image points $S_{\mathrm{G2}n}$ are written in the following matrix:

$$\boldsymbol{A}_{\mathrm{G2}} = \left[\begin{array}{cccc|c} x_{\mathrm{G21}} & x_{\mathrm{G22}} & \cdots & x_{\mathrm{G2}n} & 0\,0\,0 \\ y_{\mathrm{G21}} & y_{\mathrm{G22}} & \cdots & y_{\mathrm{G2}n} & 0\,0\,0 \end{array}\right]^{T}. \tag{6.30}$$

Parameter vector $\boldsymbol{\theta}_{\mathrm{PLS}}$ elements $(\theta_1, \theta_2, \ldots, \theta_n)$ and $(\theta_{\mathrm{a}}, \theta_{\mathrm{ax}}, \theta_{\mathrm{ay}})$ become:

$$\boldsymbol{\theta}_{\mathrm{PLS}} = \left[\theta_1\ \theta_2\ \ldots\ \theta_n \mid \theta_{\mathrm{a}}\ \theta_{\mathrm{ax}}\ \theta_{\mathrm{ay}}\right]^T = \boldsymbol{X}^{-1}\boldsymbol{A}_{\mathrm{G2}}\,. \tag{6.31}$$

$\boldsymbol{\theta}_{\mathrm{PLS}}$ elements will be used as mapping function $\Psi_{\mathrm{PLS}}(\cdot)$ coefficients:

$$\Psi_{\mathrm{PLS}}(x, y) = \theta_{\mathrm{a}} + \theta_{\mathrm{ax}}x + \theta_{\mathrm{ay}}y + \sum_{i=1}^{n} \theta_i \Upsilon_{\mathrm{U}}\left(\|S_i^{\mathrm{a}} - (x, y)\|\right). \tag{6.32}$$

The function $\Psi_{\mathrm{PLS}}(x, y)$ is searched for thin plate spline mapping that maps the set of points $S_{\mathrm{G1}n}$ with $S_{\mathrm{G2}n}$.

6.2.4.2 Experimental Evaluation of the Final 2DEG Image Matching

After the matching of final images, the protein spots in the 2DE gels must coincide. In order to fully automate the image analysis process, image matching needs to be reliable—even in growing noise level case, the probability of mismatching of the images must be minimal. At the same time, another aspect is also important for successful automation—it is desirable that human errors (introduced in the start of matching, e.g., when the gels were scanned using different resolution or the scan did not cover the entire gel) be detected during the matching.

During the experimental studies, the following 2DEG image matching methods and programs were compared:

1. New, 2DEG image matching technique introduced in this section
2. A multiresolution image registration (MIR) stepwise 2DEG image matching technique (Veeser et al. 2001);
3. A robust automated image normalization (RAIN) 2DEG image matching technique (Dowsey et al. 2008).

Table 6.4 2DEG image matching technique investigation results

Matching experiment types	2DE image 16 pairs matching		
	MIR	RAIN	Proposed technique
Natural gel image	4 (25%)	12 (75%)	15 (94%)
Gel with gel part	0 (0%)	0 (0%)	10 (63%)
Different resolution gels	4 (25%)	10 (63%)	14 (88%)
Matching mean duration	1.5 [s]	14 [min]	21 [s]

Objectives of the matching experiments are the following:

1. Create examples of gel image pairs.
2. Study natural 2DEG image matching.
3. Investigate the gel image alignment with part of another gel image.
4. Study different resolution gel image matching.
5. Estimate the duration of calculations.

The results on the experiments using 16 gel image pairs are presented in Table 6.4.

Section Generalization

1. Developed and implemented 2DE gels with the smallest geometric distortion selection method allow automation base gel selection—in groups of 2DEG images with larger than 20 pixels (3% gel height) minimal vertical geometric distortions, in a new way, chosen gel always coincided with the one chosen by the expert.
2. Developed and investigated by multilayer perceptron and three distance functions (Pearson correlations, histogram intersections, and χ^2) based method for assessing the similarity of image areas allows to achieve 99,990% true matching.
3. A proposed locally adaptive two-way matching method sets a minimum similarity threshold for confirmation of equivalence.
4. Adapted to detect errors, RANSAC algorithm takes into account conditions for maintaining the relative position of protein spots.
5. Created and implemented automatic 2DEG image matching technique works with natural and semi-artificial gel images more efficiently than the analogous MIR and RAIN methods:

   a. The new method automatically matches 94% natural 2DEG images—that is, 20% more than RAIN and 70% more than MIR methods.
   b. When aligning the 2DEG image with part of the other gel image, new technique matches 63% images—other methods do not perform such a task.
   c. While matching 2DEG images with 20% resolution differences, the new technique succeeds to match 88% images—that is, 25% more than by the use of RAIN and 63% more than by MIR method.

d. The new technique matches 2DEG images by an average of 21 s—that is, 40 times faster than RAIN (14 min) and 14 times slower than the MIR (1.5 s) methods.

6.3 Proteomic Data Analysis

In Sect. 6.1 presented automatic 2DEG image analysis strategy (Fig. 6.3) enlightened two essential steps: 2DEG image matching (P1–P2:A1–A2) and protein expression analysis (P1–P2:A2–A3). The purpose of two-dimensional electrophoresis gel image segmentation is to parameterize protein spots in order to determine qualitative and quantitative protein expression characteristics. Information about the changes is important for subsequent experiments, performed after 2DE gel analysis, planning, and formulation of hypotheses regarding protein functions.

This section is dedicated to the solutions of most problematic protein expression analysis part—2DEG image segmentation. In the first subsection, three new 2DEG image segmentation into meaningful area algorithms and their research results are presented. In the second and third subsections, protein spot modeling studies involving protein spot reconstruction and parameterization are presented.

In the section generalization of results published in Matuzevičius (2010b); Serackis (2008); Matuzevičius and Navakauskas (2008); Matuzevičius et al. (2007); Serackis et al. (2006); Matuzevičius and Navakauskas (2005); Matuzevicius et al. (2008); Serackis et al. (2010); Pivoriūnas et al. (2010); Navakauskiene et al. (2012); Treigytė et al. (2014); Navakauskiene et al. (2014) is given.

6.3.1 Segmentation of 2DEG Images

6.3.1.1 First Algorithm for 2DEG Image Segmentation

The algorithm is based on the watershed transformation. In order to improve the results of the segmentation, i.e., aiming at really one spot in each area, even where protein spots are strongly merged, two-level watershed transformation was applied. Also noise reduction and contrast enhancement operations were performed. 2DEG image segmentation algorithm is shown in Fig. 6.11.

The first step suppresses the noise, when the gel image $I_G(x, y)$ is filtered by the median filter (Gonzalez et al. 2003). The nonlinear median filter has the ability to selectively filter out impulsive noise that generates false local minima and misleads watershed transformation in those places. The filter window $\boldsymbol{S_{xy}}$ of 3×3 point size:

$$I_G^{(1)}(x, y) = \underset{(i,j)\in S_{xy}}{\text{median}}\, I_G(i, j) \quad \forall x \in [1, N_x],\ y \in [1, N_y], \tag{6.33}$$

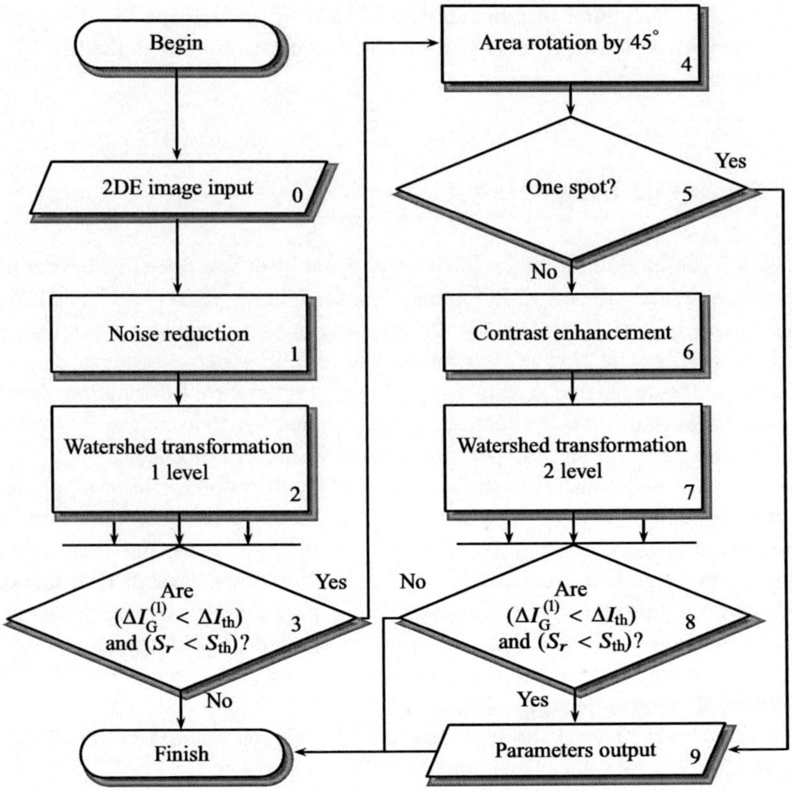

Fig. 6.11 Image splitting into segments using two-level watershed transformation algorithm

here N_x, N_y is image size (number of columns and rows).

In the second step, the first level of watershed transformation is applied—outlines are drawn through the most prominent pixels so that the local minima are separated from each other, resulting in protein spots being separated.

There is usually one spot in each highlighted area; however, there are also those in which there are no spots or more than one. In areas which are very small, i.e., the area of S_r is less than the boundary area S_{th}, and where the difference of point intensity maximum and minimum values $\Delta I_G^{(1)}(x, y)$, $\forall(x, y) \in S_{Gr}$ is less than the selected ΔI_{th}, there are no spots, so they are removed *in the third step*.

In *steps 4 and 5*, the areas where there could be several merged spots are selected. Area projections are analyzed if they have pronounced curves; then such area is taken for further processing.

In selected areas, during *step 6* the contrast is increased—this allows more accurate distinguishing of the spots. Contrast was not increased during the previous processing stages, due to emerging of noisy image components alltogether. Let us discuss in detail the morphological operations applied for contrast increase.

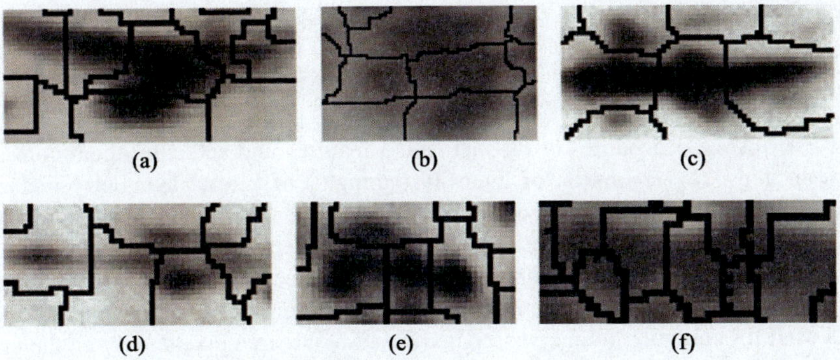

Fig. 6.12 Results of 2DEG image splitting into segments

Using the disk-shaped structural element $\boldsymbol{D}$, the image area $\boldsymbol{S}_{\mathrm{G}r}$ contrast enhancement is based on the extension $\boldsymbol{S}_{\mathrm{G}r}^{\oplus} = \boldsymbol{S}_{\mathrm{G}r} \oplus \boldsymbol{D}$ and erosion $\boldsymbol{S}_{\mathrm{G}r}^{\ominus} = \boldsymbol{S}_{\mathrm{G}r} \ominus \boldsymbol{D}$ morphological operations. These operations are used to perform opening (denoted by $\circ$) and closing (denoted by $\bullet$) morphological operations:

$$\boldsymbol{S}_{\mathrm{G}r} \circ \boldsymbol{D} = (\boldsymbol{S}_{\mathrm{G}r} \ominus \boldsymbol{D}) \oplus \boldsymbol{D}\,; \tag{6.34}$$

$$\boldsymbol{S}_{\mathrm{G}r} \bullet \boldsymbol{D} = (\boldsymbol{S}_{\mathrm{G}r} \oplus \boldsymbol{D}) \ominus \boldsymbol{D}\,. \tag{6.35}$$

Contrast is enhanced by the following expression:

$$\boldsymbol{S}_{\mathrm{G}r}^{\mathrm{kontr}} = \boldsymbol{S}_{\mathrm{G}r} + \left[\boldsymbol{S}_{\mathrm{G}r} - (\boldsymbol{S}_{\mathrm{G}r} \circ \boldsymbol{D})\right] - \left[\boldsymbol{S}_{\mathrm{G}r} - (\boldsymbol{S}_{\mathrm{G}r} \bullet \boldsymbol{D})\right]. \tag{6.36}$$

In the seventh step, the second-level watershed transformation on selected and in contrast areas is performed.

In the eight step, using the same criteria as in third step, i.e., S_{th} and ΔI_{th}, areas having spots are selected again.

During the last step—*ninth step* of segmentation algorithm—parameters of the areas that were found in the fifth and eighth steps are provided.

Results of 2DEG image segmentation experiments using proposed algorithm are presented in Fig. 6.12. Typical examples that are selected confirm the proposed segmentation algorithm is able to segment quite complex combinations of spots: merged (Fig. 6.12d,e), blurred (Fig. 6.12b,d), and prolonged (Fig. 6.12a). However in Fig. 6.12c part some small protein spots were not distinguished.

After studying the algorithm, it was found that (a) for better segmentation results (one spot in one area), an adaptive selection of parameter thresholds is required, but it is difficult to compose reliable feedback for that purpose; (b) the information on edges of the spots are not fully exploited when segmenting strongly merged spots; and (c) small spots that are strongly adhering to a large spot are inseparable. In order to address mentioned shortcomings, a second algorithm has been developed.

6.3.1.2 Second Algorithm for 2DEG Image Segmentation

Extraction of Image Features

The established procedure for distinguishing features and selecting them consists of four steps: (a) generation of intensity (strength) of central symmetry features map $I_{\mathrm{G}}^{\mathrm{rot}}(x, y)$; (b) watershed contouring in the calculated feature map (watershed transformation); (c) feature extraction in each watershed formed area; and (d) application of cluster validation techniques in order to select significant spots. Features extracted from selected area can be broadly divided in three groups depending on data used for computation, i.e., computed based on all area pixels, only on the area contour pixels, or on the central point of an area containing maximum of central symmetry. Let's discuss each processing step in detail.

Initially, 2DEG image $I_{\mathrm{G}}(x, y)$ is processed by a small ($\boldsymbol{S}_{xy}$—3×3 pixels) size median filter, and then from the filtered image, central symmetry intensity feature map is calculated. This is done by counting the second-order symmetries using the Johansson method (Johansson 2004). Received central symmetry map is of the complex type, but the data needed for the following calculations are concentrated in the real part of $I_{\mathrm{G}}^{\mathrm{rot}}(x, y)$. Applying watershed transformation $\mathcal{W}$ for an inverted map

$$\left\{\mathcal{S}_{\mathrm{G}}^{\mathcal{W}}, \mathcal{C}_{\mathrm{G}}^{\mathcal{W}}\right\} \triangleq \mathcal{W}\left(-\boldsymbol{I}_{\mathrm{G}}^{\mathrm{rot}}\right) \tag{6.37a}$$

set of areas is received:

$$\mathcal{S}_{\mathrm{G}}^{\mathcal{W}} \equiv \boldsymbol{S}_r^{\mathcal{W}} = \left(\boldsymbol{x}^{\bullet(r)}, \boldsymbol{y}^{\bullet(r)}\right) \quad \forall r \in [1, N^{\mathcal{W}}] \tag{6.37b}$$

together with set of contours:

$$\mathcal{C}_{\mathrm{G}}^{\mathcal{W}} \equiv \boldsymbol{C}_r^{\mathcal{W}} = \left(\boldsymbol{x}^{\circ(r)}, \boldsymbol{y}^{\circ(r)}\right) \quad \forall r \in [1, N^{\mathcal{W}}], \tag{6.37c}$$

here $\left(\boldsymbol{x}^{\bullet(r)}, \boldsymbol{y}^{\bullet(r)}\right)$ is image element indices for each rth region $\boldsymbol{S}_r^{\mathcal{W}}$ that was formed by watershed transformation; $(\boldsymbol{x}^{\circ(r)}, \boldsymbol{y}^{\circ(r)})$ is pixel indices of each rth contour $\boldsymbol{C}_r^{\mathcal{W}}$; and $N^{\mathcal{W}}$ is the number of found areas (contours).

Inverted map $\boldsymbol{I}_{\mathrm{G}}^{\mathrm{rot}}$ watershed transformation splits into $N^{\mathcal{W}}$ areas, each r of which has one local minimum specifying the area center of symmetry. Because most protein spots are symmetrical on their own with respect to the center, the peak of the symmetry map and the center of the spot coincides. The weakest central symmetry is between the spots, so through those places the transformation of the watershed draws a distinctive contour. Further calculations will require local symmetry locales maximum coordinates:

$$\left(x_{\mathrm{T}}^{\bullet(r)}, y_{\mathrm{T}}^{\bullet(r)}\right) = \underset{S_r^{\mathcal{W}}}{\arg\max}\, I_{\mathrm{G}}^{\mathrm{rot}}\left(S_r^{\mathcal{W}}\right), \tag{6.38}$$

here $\left(x_{\mathrm{T}}^{\bullet(r)}, y_{\mathrm{T}}^{\bullet(r)}\right)$ is the coordinates of the maximum found in each area of the central symmetry map, defined by coordinates $S_r^{\mathcal{W}} = \left(x^{\bullet(r)}, y^{\bullet(r)}\right)$. Because these places are the most probable locations of spot centers, some features will be extracted in these places.

The used central symmetry detector provides the same meaning for light structures on a dark background (convex image places) and dark structures on a light background (sunken image places) which are equally symmetrical. To exclude areas in which there are no spots, it is necessary to distinguish between inculvations and inflections. Function curvature can be derived from its second derivative—in curved (convex) places, the second derivative is negative, and bent positive. Spots in the gel image $I_{\mathrm{G}}(x, y)$ can be seen as concavities when the 2DEG image is visualized as a surface in three-dimensional space, so the second derivative in the spot regions will be positive; therefore regions having the negative second derivative in their centers may be rejected. This criterion is often used for image segmentation algorithms. Because, when differentiating the image, the noise is enhanced, the image needs to be smoothed beforehand. The Gaussian kernel is used as the smoothing filter. Based on convolution associativity properties, firstly the Laplace operator is applied to the Gaussian kernel, and then the convolution of the new kernel and image is calculated. After applying Laplace operator to the Gaussian kernel, the resulting kernel is called Laplacian of Gaussian (LoG).

Some 2DEG image areas r are removed from further stages of analysis. The area r is removed if the image $I_{\mathrm{G}}(x, y)$ second-order derivative at points $\left(x_{\mathrm{T}}^{\bullet(r)}, y_{\mathrm{T}}^{\bullet(r)}\right)$ using the Laplacian of Gaussian filter is negative, i.e., $I_{\mathrm{G}}^{\mathrm{LoG}}\left(x_{\mathrm{T}}^{\bullet(r)}, y_{\mathrm{T}}^{\bullet(r)}\right) < 0$. At this step, without using much computational resources, we discard insignificant areas that would otherwise require much more computing resources in detail analysis according to the steps described below.

The initial set of features consists of:

- The maximum and minimum gel image intensity values (in each of the remaining areas) with indices $r \in [1, \widehat{N}^{\mathcal{W}}]$:

$$I_{\mathrm{G}}^{\bullet(r)\top} = \max I_{\mathrm{G}}\left(S_r^{\mathcal{W}}\right); \tag{6.39a}$$

$$I_{\mathrm{G}}^{\bullet(r)\perp} = \min I_{\mathrm{G}}\left(S_r^{\mathcal{W}}\right), \tag{6.39b}$$

- The maximum and minimum gel image intensity values from contour of each area:

$$I_{\mathrm{G}}^{\circ(r)\top} = \max I_{\mathrm{G}}\left(C_r^{\mathcal{W}}\right); \tag{6.40a}$$

$$I_{\mathrm{G}}^{\circ(r)\perp} = \min I_{\mathrm{G}}\left(C_r^{W}\right), \tag{6.40b}$$

- Symmetry-based features of protein spots—the central symmetry feature intensity map $I_{\mathrm{G}}^{\mathrm{rot}}(x, y)$, Laplacian of Gaussian $I_{\mathrm{G}}^{\mathrm{LoG}}(x, y)$, Loy and Zelinsky radial symmetry $I_{\mathrm{G}}^{\mathrm{LZ}}(x, y)$ (Loy and Zelinsky 2003), and Kovesi phase symmetry $I_{\mathrm{G}}^{\mathrm{K}}(x, y)$ (Kovesi 2020).

Based on the initial set of spots, 16 characteristics of protein spot were formed:

P1. The difference between maximum and minimum values of the intensities for each 2DEG image $I_{\mathrm{G}}(x, y)$ area in the range r:

$$X_1 = I_{\mathrm{G}}^{\bullet(r)\top} - I_{\mathrm{G}}^{\bullet(r)\perp}. \tag{6.41}$$

P2. The ratio between minimum and maximum values of the intensities for each 2DEG image $I_{\mathrm{G}}(x, y)$ area in the range r:

$$X_2 = I_{\mathrm{G}}^{\bullet(r)\perp} \Big/ I_{\mathrm{G}}^{\bullet(r)\top}. \tag{6.42}$$

P3. The normalized difference between maximum and minimum values of the contour intensities for each 2DEG image $I_{\mathrm{G}}(x, y)$ area in the range r:

$$X_3 = \left(I_{\mathrm{G}}^{\circ(r)\top} - I_{\mathrm{G}}^{\circ(r)\perp}\right) \Big/ I_{\mathrm{G}}^{\circ(r)\top}. \tag{6.43}$$

P4. The normalized according to feature **P1** difference between maximum and minimum values of the contour intensities for each 2DEG image $I_{\mathrm{G}}(x, y)$ area in the range r:

$$X_4 = \left(I_{\mathrm{G}}^{\circ(r)\top} - I_{\mathrm{G}}^{\circ(r)\perp}\right) \Big/ X_1. \tag{6.44}$$

P5. The normalized according to feature **P1** difference between area and contour intensities for each 2DEG image $I_{\mathrm{G}}(x, y)$ area in the range r:

$$X_5 = \left(I_{\mathrm{G}}^{\bullet(r)\top} - I_{\mathrm{G}}^{\circ(r)\top}\right) \Big/ X_1. \tag{6.45}$$

P6. The standard deviation of area intensities for each 2DEG image $I_{\mathrm{G}}(x, y)$ area in the range r:

$$X_6 = \operatorname{std} I_{\mathrm{G}}\left(S_r^{W}\right). \tag{6.46}$$

P7. The entropy of area intensities for each 2DEG image $I_{\mathrm{G}}(x, y)$ area in the range r:

$$X_7 = \text{entropy}\, I_G\left(S_r^W\right). \tag{6.47}$$

P8. The mode of area intensities for each 2DEG image $I_G(x, y)$ area in the range r:

$$X_8 = \text{mode}\, I_G\left(S_r^W\right). \tag{6.48}$$

P9. The maximum value of symmetry feature intensity map $I_G^{rot}(x, y)$ for each area in the range r:

$$X_9 = \max I_G^{rot}\left(S_r^W\right). \tag{6.49}$$

P10. The difference between maximum and minimum values of the symmetry feature intensity map $I_G^{rot}(x, y)$ for each area in the range r:

$$X_{10} = \max I_G^{rot}\left(S_r^W\right) - \min I_G^{rot}\left(S_r^W\right). \tag{6.50}$$

P11. The ratio between minimum and maximum values of the symmetry feature intensity map $I_G^{rot}(x, y)$ for each area in the range r:

$$X_{11} = \min I_G^{rot}\left(S_r^W\right) \Big/ \max I_G^{rot}\left(S_r^W\right). \tag{6.51}$$

P12. The normalized according to feature **P10** difference between maximum and minimum values of the symmetry feature intensity map $I_G^{rot}(x, y)$ for each area in the range r:

$$X_{12} = \left(\max I_G^{rot}\left(C_r^W\right) - \min I_G^{rot}\left(C_r^W\right)\right) \Big/ X_{10}. \tag{6.52}$$

P13. $I_G^{LoG}(x, y)$ image value in places where $I_G^{rot}(x, y)$ has local maximum:

$$X_{13} = I_G^{LoG}\left(x_T^{\bullet(r)}, y_T^{\bullet(r)}\right). \tag{6.53}$$

P14. The maximum value of image $I_G^{LoG}(x, y)$ for each area in the range r:

$$X_{14} = \max I_G^{LoG}\left(S_r^W\right). \tag{6.54}$$

P15. Maximum value of Loy and Zelinsky radial symmetry detector (Loy and Zelinsky 2003) for each area in the range r:

$$X_{15} = \max I_G^{LZ}\left(S_r^W\right). \tag{6.55}$$

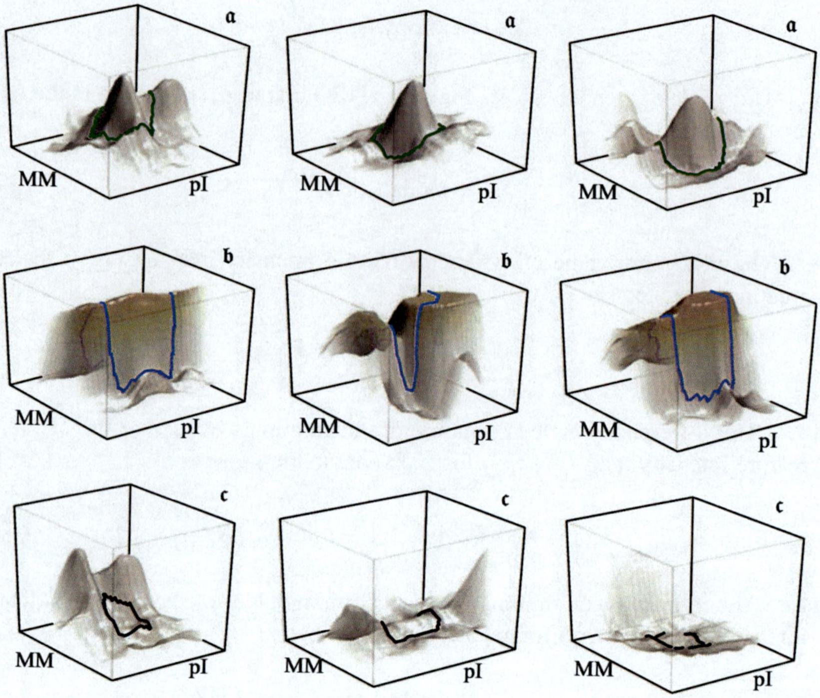

Fig. 6.13 Examples from three spot classes

P16. Maximum value of Kovesi phase symmetry detector (Kovesi 2020) for each area in the range r:

$$\boldsymbol{X}_{16} = \max I_{\mathrm{G}}^{\mathrm{K}}\left(\boldsymbol{S}_r^{\mathcal{W}}\right) . \tag{6.56}$$

The extracted features are further ranked according to their significance (informativeness) to describe areas for the presence of spot and then are selected for classifier use.

Feature Selection

From all the areas r, 230 areas were selected by hand and pre-classified into three classes: $\mathfrak{a}$, area contains one space; $\mathfrak{b}$, contains only part of the spot; and $\mathfrak{c}$, there are no spots in the area (see Fig. 6.13). Further it is necessary to select the set of attributes (find those attributes) on the basis of which it will be possible to classify the selected areas into the same classes as they were made by hand. We will use clustering means to select such a group.

Table 6.5 Summary of feature ranking

Features		Ranks						
ID	Symb.	S	CH	D	A	C	STV	Σ
P1.	X_1	11	9	13	9	9	10	10
P2.	X_2	15	9	7	13	13	12	15
P3.	X_3	2	9	4	2	2	2	2
P4.	X_4	7	9	9	14	4	15	8
P5.	X_5	12	9	11	16	13	16	16
P6.	X_6	10	9	14	12	11	13	12
P7.	X_7	16	16	16	5	16	4	13
P8.	X_8	1	1	1	1	1	1	1
P9.	X_9	9	13	12	11	14	11	11
P10.	X_{10}	13	11	15	7	7	7	9
P11.	X_{11}	5	6	8	8	6	8	6
P12.	X_{12}	3	5	2	3	5	3	3
P13.	X_{13}	6	7	5	6	12	6	7
P14.	X_{14}	8	8	6	4	10	5	5
P15.	X_{15}	14	14	10	15	8	14	14
P16.	X_{16}	4	3	3	10	9	9	4

The initial normalization of the data (features) was performed transforming it to zero mean and unit dispersion set. The distances between the feature vectors were calculated according to Euclidean metrics. Combinations were made of the 2–16 features, and according to them clusters by hand were formed. The quality of the formed clusters was further assessed—the vectors in the cluster must be as close as possible while the clusters must be as distant as possible. The quality of clusters was assessed by standard cluster validity indices.

Six cluster validity indices were used: $\mathscr{E}_S$, Silhouette; $\mathscr{E}_{CH}$, Calinsky and Harabasz; $\mathscr{E}_D$, Dunn; $\mathscr{E}_A$, separation; $\mathscr{E}_C$, Hubert and Levin or C index; and $\mathscr{E}_{STV}$, cluster. For all indices except for $\mathscr{E}_C$, a higher value indicates a better cluster quality. Sequential backward selection of features was used—starting at 16 feature group and in each step removing the feature that has the least impact on a cluster quality. For the first removed from the set feature, a rating of "16" is given, while for the last remaining feature, the highest rating "1" is given. Six experiments were performed using each cluster validity index separately. The summary of feature ratings is given in Table 6.5. The received ranking results show that the order of the features formed according to different indices does not completely coincide—different indexes prescribed not the same priority of the features. In the last column of the table, a common set of characteristics is presented by combining the results of all indices. Based on the last column, the first five features (highlighted in the Table) were chosen—**P3**, **P8**, **P12**, **P14**, and **P16**. They will be used to classify 2DEG image areas using a feed-forward multilayer artificial neural network.

It is important to acknowledge that here according to the selected set of characteristics, clusters are formed not by clustering methods but by hand-specified

(a)

ANN \ Expert	a	b	c
a	97, 8%	5, 1%	5, 6%
b	2, 2%	94, 9%	2, 3%
c	0%	0%	92, 1%

(b)

ANN \ Expert	a	b	c
a	98, 7%	7, 4%	5, 4%
b	1, 3%	92, 6%	1, 0%
c	0%	0%	93, 6%

Fig. 6.14 Results of 2DE image region classification into {a, b, c} classes: (**a**) training areas; (**b**) new areas

class information, while the quality of clusters is measured by those indices that are typically used for the optimal number of cluster selection during unsupervised classification.

Experiments on the use of ANN for Segmentation

Using a group of selected features, the three-layer feed-forward artificial neural network (ANN) was trained. The input and hidden layers had 10 and 3 neurons, respectively. Hidden layer neuron activation functions were of the tangent sigmoid type. Neural network was trained using the Levenberg-Marquardt algorithm that minimizes the quadratic error.

A set of training examples consisting of 230 gel image areas were collected from a single gel image and divided into three a–c classes. For neural network testing purposes, we used areas collected from the gel images not used for network training. Gel image area classification by ANN results is given in Fig. 6.14. ANN classification on testing data accuracy is 95%.

6.3.2 *Models of Protein Spots and Modeling*

After partitioning image into separate simple regions (initial image segmentation) where every region possibly contains single spot of protein, every region can be modelled. Protein spot modeling provides a parameterized description of spot using parameters of a specially selected model. After model fitting, protein spot in the target image can be replaced with its model. The objectives of the spot modeling can be as follows:

- Determine the contours of the spots.
- Separate overlapping spots.
- Parameterize protein spots.
- Reconstruct saturated or oversaturated spots.
- Generate realistic synthetic 2DE gel image.

6.3.2.1 Common Protein Spot Models

The simplest protein spot model is the circularly symmetric 2D Gaussian function (see (6.57)). Shape of this model changes the same way in all directions and is controlled by single σ parameter. So the model could be imagined as made using single Gaussian curve (referred to as M1G).

Model 6.1 (Circularly symmetric two-dimensional Gaussian model)

Circularly symmetric 2D Gaussian protein spot model:

$$\Psi_{\mathrm{M1G}}(x, y; \boldsymbol{\theta}_{\mathrm{M1G}}) = I^{\mathrm{B}} + I^{\mathrm{A}} \cdot \exp\left(-\frac{(x - x_{\mathrm{c}})^2 + (y - y_{\mathrm{c}})^2}{2\sigma^2}\right), \tag{6.57}$$

here (x_c, y_c) is coordinates of the protein spot model center; I^{B} is background intensity of the spot; I^{A} is peak intensity of the spot, excluding background; and σ is extent of the model (standard deviation of the Gaussian).

This model is used less frequently than the next Gaussian-based model, which could be imagined as made using two Gaussian curves (referred to as M2G). This is the most common model based on 2D Gaussian function which has additional degree of freedom. The second model can adapt its shape in two cardinal directions (along MW and pI axis) independently so it can adapt to protein spots which usually have different vertical and horizontal extents. This model demonstrates the symmetry along x and y axis.

Model 6.2 (Two-way adapting 2D Gaussian model)

2D Gaussian protein spot model, varying independently in two cardinal directions:

$$\Psi_{\mathrm{M2G}}(x, y; \boldsymbol{\theta}_{\mathrm{M2G}}) = I^{\mathrm{B}} + I^{\mathrm{A}} \cdot \exp\left(-\frac{(x - x_{\mathrm{c}})^2}{2\sigma_{\mathrm{x}}^2}\right) \cdot \exp\left(-\frac{(y - y_{\mathrm{c}})^2}{2\sigma_{\mathrm{y}}^2}\right), \tag{6.58}$$

here (x_c, y_c) is coordinates of the protein spot model center; I^{B} is background intensity of the spot; I^{A} is peak intensity of the spot, excluding background; and σ_{x} and σ_{y} are extent of the model in horizontal and vertical directions (standard deviation of the Gaussian).

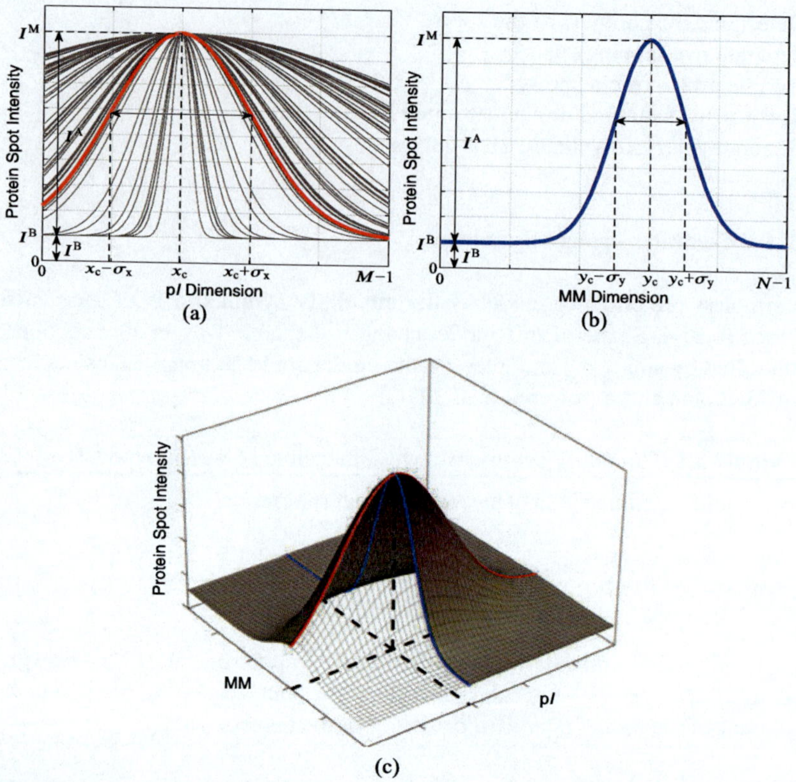

Fig. 6.15 Two-way adapting 2D Gaussian model (M2G). Horizontal (**a**) and vertical (**b**) cross sections of protein spot model and its 3D plot (**c**). Model's capacity to adapt its shape is shown by random sampling σ_x parameter and plotted as thin lines (**a**)

The two-way adapting 2D Gaussian model (M2G) (Fig. 6.15) can represent different protein diffusion on the molecular weight and isoelectric point axis. It is suitable for modeling most of the gel spots. But when protein spot exhibits intensity distortions like saturations or oversaturations (situations when top of the spot is flat or even concave), application of this model leaves high residual error. This model is suitable in cases when a result needs to be obtained quickly and with minimal requirements for shape representation accuracy (e.g., to replace spot with its model only to positively influence gel image segmentation algorithm). In cases when spots need to be reconstructed as accurately as possible, models representing asymmetries of the spots in three or more directions should be used.

Bettens et al. (1997) observed that when the protein concentration in the area is high, and the exposure time is long while imaging 2DE gel, effects of spot saturation may occur (a spot surface with a flat top). The Gaussian model can no longer accurately fit such a spot, so authors proposed to model spot shape by using a simplified diffusion process. In the case of 2DE, the medium in which the diffusion takes place is considered to be two-dimensional and anisotropic—there are two main cardinal directions in which diffusion occurs with different diffusion coefficients. It is also assumed that the material initially is distributed not in a single point but in a circle of radius θ_R. By removing the symmetric points from the equation and adding two additional position coordinates, and a parameter to estimate the background intensity, a protein spot "diffusion" model is obtained (referred to as MD).

Model 6.3 (Diffusion Model)

Diffusion model of protein spots, fitted to 2DE image region, whose pixel coordinates are x and y, is defined as follows:

$$\Psi_{MD}(x, y; \boldsymbol{\theta}_{MD}) = I^B + \frac{\theta_{C_0}}{2}\left[\operatorname{erf}\left(\frac{\theta_r + \theta_R}{2}\right) + \operatorname{erf}\left(\frac{\theta_r - \theta_R}{2}\right)\right] \tag{6.59a}$$

$$+ \frac{\theta_{C_0}}{\theta_R\sqrt{\pi}}\left[\exp\left(-\left(\frac{\theta_r + \theta_R}{2}\right)^2\right) + \exp\left(-\left(\frac{\theta_r - \theta_R}{2}\right)^2\right)\right],$$

here I^B is background intensity of the spot; θ_{C_0} is initial concentration of protein; θ_r is radius of protein circular diffusion area; θ_R is coefficient of diffusion defined as follows:

$$\theta_R = \left(\frac{(x - x_c)^2}{\theta_{Dx}} + \frac{(y - y_c)^2}{\theta_{Dy}}\right)^{-1/2}, \tag{6.59b}$$

here θ_{Dx} and θ_{Dy} are parameters, proportional to the diffusion in the corresponding directions, and x_c and y_c are coordinates of the protein spot model center.

The diffusion model (MD) (Fig. 6.16) can be used to represent protein spots with saturation distortions when the top of the spot is flat. The model can capture different diffusions in two cardinal directions like the M2G model, but additional parameters allow for this model to be more accurately applied to spots with flat peaks. The disadvantage of the diffusion model is that the spot shape variance is estimated in only two directions, and the application of the model due to the complexity and the need to select seven parameters requires more calculations. The diffusion model of protein spots is less suitable for modeling spots of irregular shapes—spots that have very low horizontal line reflection symmetry.

Rogers et al. (2003) presented a parametric model constructed as combination of statistics of spot shape variations and 2D Gaussian kernel. The model is derived from the observed spots rather than from the idealization of the spot formation process. It can distinguish common protein spots like other methods do and additionally groups of spots of a specific shape. Spot modeling is performed using a point distribution model. But this model is difficult to use for separation of overlapping spots.

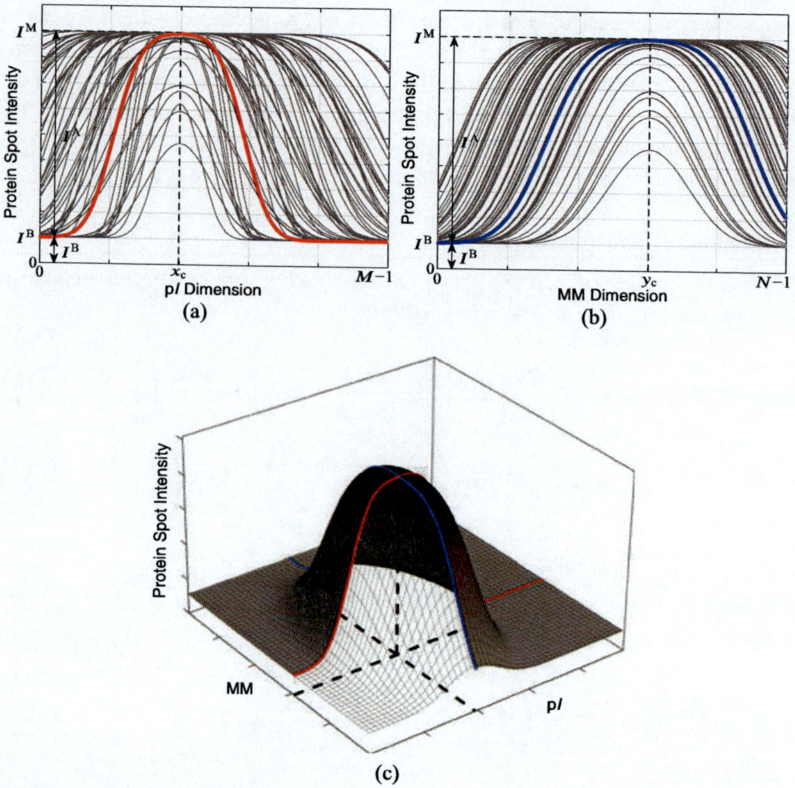

Fig. 6.16 The diffusion model (MD). Horizontal (**a**) and vertical (**b**) cross sections of protein spot model and its 3D plot (**c**). Model's capacity to adapt its shape is shown by random sampling θ_{Dx} and θ_r parameters and plotted as thin lines (**a**)

There are protein spots that have become asymmetrical in the vertical or horizontal or both directions. To deal with such spots, we need models that can adapt shape in 3 or 4 directions independently.

6.3.2.2 Need for Complex Protein Spot Models

In order to increase protein spot contour extraction precision (contours are used for calculation of relative volume of a protein), spots can be modeled, and their contours can be determined from the modeled spots. Models can also be used to separate overlapping spots by modeling them. But the variety of spot shapes created by asymmetrically shaped, saturated, and oversaturated spots requires powerful models to represent those non-standard spots.

Published sources describe cases that the most commonly used is anisotropic Gaussian model (M2G), followed by a model describing the diffusion process (MD). The diffusion model is more elastic than Gaussian, but it takes a much more time to optimize its parameters. In addition, the diffusion model has difficulty reproducing highly saturated spots with a flat surface. For these reasons, new models are presented and tested in this work:

- Gaussian-based complex models
- Spline-based (Π-shaped curve) models
- Bell-shaped function based models
- Sigmoid-based models

Structural elements for these models are 1D functions: Gaussian, spline (Π-shaped fuzzy logic membership function), bell-shaped fuzzy logic membership function, and sigmoid. In order to represent protein spot shape, these 1D functions (curves) are combined to create 2D functions (surfaces) that could be used as spot model.

Selection of appropriate protein spot model attracts attention to two main properties of the model—model's ability to represent diversity of protein spot shapes and the complexity of the model. The first criterion is tested by calculating the residual error between the fitted model and the real spot shape, and the second by the convergence rate. A simpler model with fewer parameters will have a simpler form of objective function, will converge faster, and will result in faster modeling process of spots. The representation power of the model decreases as the number of parameters decreases, so when making decision on a model selection, it is necessary to choose which criterion is more important.

6.3.2.3 Gaussian-Based Complex Models

To represent protein spot asymmetries along the molecular weight or isoelectric point axis of a 2DEG image, a protein spot model comprising combinations of different one-dimensional Gaussian functions may be constructed.

Simple symmetric 1D Gaussian function is defined as follows:

$$f_{\mathrm{G}}(x; \boldsymbol{\theta}_{\mathrm{G}}) = \exp\left(-\frac{(x - x_{\mathrm{c}})^2}{2\sigma_{\mathrm{x}}^2}\right), \tag{6.60}$$

simple asymmetric 1D Gaussian:

$$f_{2\mathrm{G}}(x; \boldsymbol{\theta}_{2\mathrm{G}}) = \varepsilon_{\mathrm{x}} \cdot \exp\left(-\frac{(x - x_{\mathrm{c}})^2}{2\sigma_{\mathrm{x1}}^2}\right) \tag{6.61}$$

$$+ (1 - \varepsilon_{\mathrm{x}}) \cdot \exp\left(-\frac{(x - x_{\mathrm{c}})^2}{2\sigma_{\mathrm{x2}}^2}\right),$$

and asymmetric 1D Gaussian with flat top formed by introducing center split:

$$f_{\mathrm{2Gft}}(x; \boldsymbol{\theta}_{\mathrm{2Gft}}) = \varepsilon_{\mathrm{x1}} \cdot \exp\left(-\frac{(x - (x_{\mathrm{c}} - d_{\mathrm{x}}))^2}{2\sigma_{\mathrm{x1}}^2}\right) \tag{6.62}$$

$$+ \varepsilon_{\mathrm{x2}} \cdot \exp\left(-\frac{(x - (x_{\mathrm{c}} + d_{\mathrm{x}}))^2}{2\sigma_{\mathrm{x2}}^2}\right) + (1 - \varepsilon_{\mathrm{x1}} - \varepsilon_{\mathrm{x2}}),$$

here x_{c} is coordinates of the Gaussian center and σ^2, σ_{x1}^2, σ_{x2}^2 is extent of the model in horizontal and vertical directions (standard deviation of the Gaussian); step functions forming the asymmetry of the model are:

$$\varepsilon_{\mathrm{x}} = \begin{cases} 1, & \text{if } x \le x_{\mathrm{c}}; \\ 0, & \text{otherwise}; \end{cases}$$

$$\varepsilon_{\mathrm{x1}} = \begin{cases} 1, & \text{if } x \le x_{\mathrm{c}} - d_{\mathrm{x}}; \\ 0, & \text{otherwise}; \end{cases} \quad \text{and} \quad \varepsilon_{\mathrm{x2}} = \begin{cases} 1, & \text{if } x \ge x_{\mathrm{c2}} + d_{\mathrm{x}}; \\ 0, & \text{otherwise}. \end{cases}$$

Model of Three Gaussian Functions

A mathematical model of the protein spot is constructed by combining symmetric (6.60) and asymmetric Gaussian functions (6.61) so that the model becomes asymmetric along the 2DE molecular weight axis. The resulting combination of Gaussian functions is appended with two additional parameters describing the background intensity of the spot I^{B} and the intensity of the spot without background (spot height) I^{A}.

Model 6.4 (Three Gaussian Functions Model)

Three Gaussian functions-based model of protein spots, used to represent 2DEG image region, whose pixel coordinates are x and y, is defined as follows:

$$\Psi_{\mathrm{M3G}}(x, y; \boldsymbol{\theta}_{\mathrm{M3G}}) = \tag{6.63}$$

$$I^{\mathrm{B}} + I^{\mathrm{A}} \left[\varepsilon_{\mathrm{y}} \cdot \exp\left(-\frac{(y - y_{\mathrm{c}})^2}{2\sigma_{\mathrm{y1}}^2}\right) + (1 - \varepsilon_{\mathrm{y}}) \cdot \exp\left(-\frac{(y - y_{\mathrm{c}})^2}{2\sigma_{\mathrm{y2}}^2}\right)\right] \cdot \exp\left(-\frac{(x - x_{\mathrm{c}})^2}{2\sigma_{\mathrm{x}}^2}\right),$$

here $\{\sigma_{x_1}, \sigma_{x_2}, \sigma_y\}$ is extent of the model in horizontal and vertical directions (standard deviation of the Gaussian); (x_c, y_c) is coordinates of the protein spot model center; I^B is background intensity of the spot; and I^A is peak intensity of the spot, excluding background.

The proposed M3G protein spot model (Fig. 6.17) is more flexible than the anisotropic bell-shaped model and is able to represent the asymmetry of the protein spot along the molecular weight axis. Saturated protein spots cannot be modeled using M3G because flat peaks cannot be obtained using one x_c and one y_c. Proposed three Gaussian functions-based model is parameterized by 7 variables.

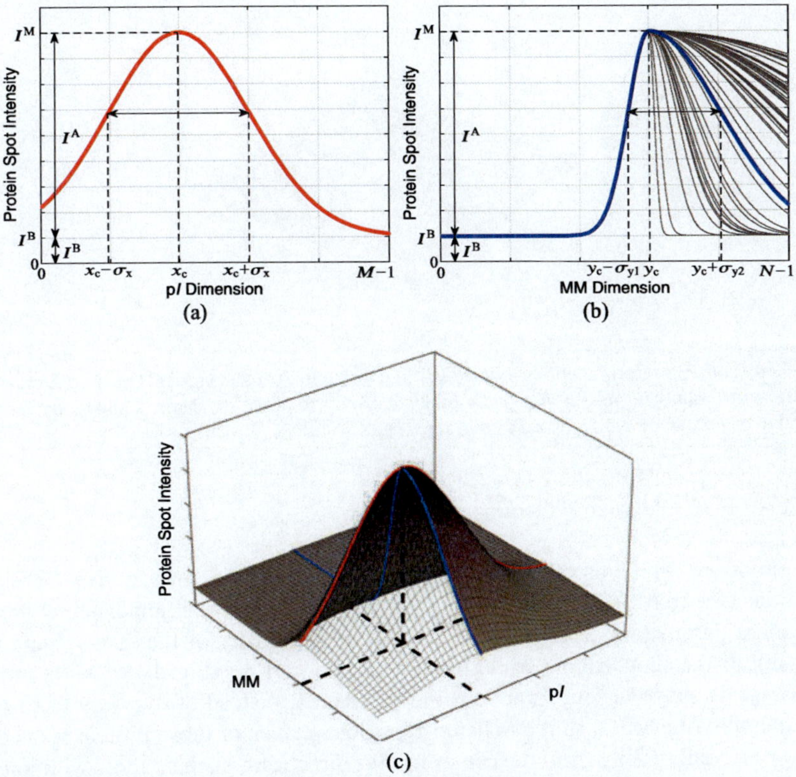

Fig. 6.17 Three Gaussian functions model (M3G). Horizontal (**a**) and vertical (**b**) cross sections of protein spot model and its 3D plot (**c**). Model's capacity to adapt its shape is shown by random sampling σ_{y2} parameter and plotted as thin lines (**b**)

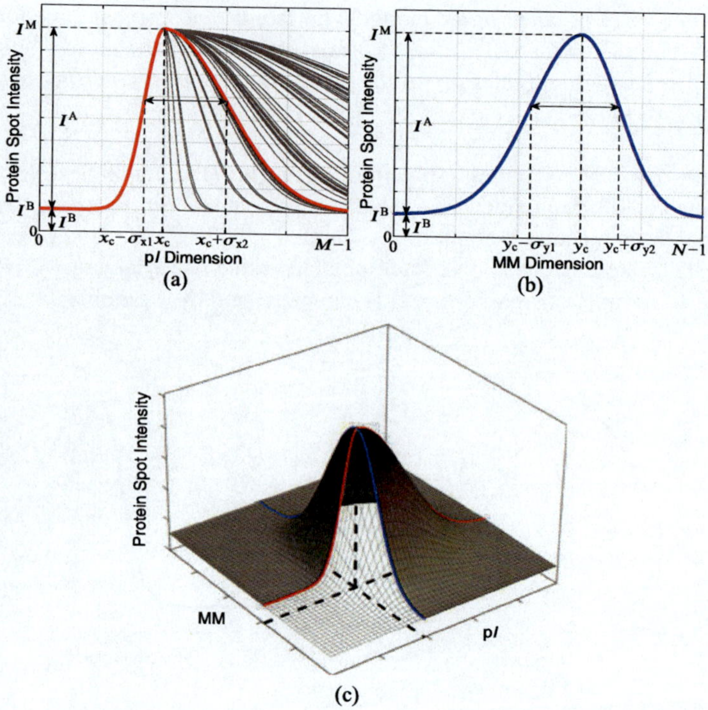

Fig. 6.18 Four Gaussian functions model (M4G). Horizontal (**a**) and vertical (**b**) cross sections of protein spot model and its 3D plot (**c**). Model's capacity to adapt its shape is shown by random sampling σ_{x2} parameter and plotted as thin lines (**a**)

Model of Four Gaussian Functions

The proposed M4G protein spot model consists of two asymmetric Gaussian functions (see (6.61)). The resulting combination of functions is appended with two additional parameters describing the background intensity of the spot I^B and the intensity of the spot without background (spot height) I^A (Fig. 6.18). M4G model can adapt its shape in four directions independently, instead of two-way like M4G or three-way like M3G, so it can better represent shapes of those protein spots that change unevenly going from the center in four directions, such as spots with longer tails or other irregularly shaped spots.

Model 6.5 (Four Gaussian Functions Model)

Four Gaussian functions-based model of protein spots, used to represent 2DEG image region, whose pixel coordinates are x and y, is defined as follows:

$$\Psi_{\mathrm{M4G}}(x, y; \boldsymbol{\theta}_{\mathrm{M4G}}) = I^{\mathrm{B}} + I^{\mathrm{A}} \cdot \left[\varepsilon_{\mathrm{x}} \exp\left(-\frac{(x-x_{\mathrm{c}})^2}{2\sigma_{\mathrm{x1}}^2}\right) + (1-\varepsilon_{\mathrm{x}}) \exp\left(-\frac{(x-x_{\mathrm{c}})^2}{2\sigma_{\mathrm{x2}}^2}\right)\right]$$

$$\times \left[\varepsilon_{\mathrm{y}} \exp\left(-\frac{(y-y_{\mathrm{c}})^2}{2\sigma_{\mathrm{y1}}^2}\right) + (1-\varepsilon_{\mathrm{y}}) \exp\left(-\frac{(y-y_{\mathrm{c}})^2}{2\sigma_{\mathrm{y2}}^2}\right)\right], \quad (6.64)$$

here $\{\sigma_{x_1}, \sigma_{x_2}, \sigma_{y_1}, \sigma_{y_2}\}$ *is extent of the model in horizontal and vertical directions (standard deviation of the Gaussian);* $(x_{\mathrm{c}}, y_{\mathrm{c}})$ *is coordinates of the protein spot model center;* I^{B} *is background intensity of the spot; and* I^{A} *is peak intensity of the spot, excluding background;*

$$\varepsilon_{\mathrm{x}} = \begin{cases} 1, & \textit{if } x \leq x_{\mathrm{c}}; \\ 0, & \textit{otherwise} \end{cases} \quad \text{and} \quad \varepsilon_{\mathrm{y}} = \begin{cases} 1, & \textit{if } y \leq y_{\mathrm{c}}; \\ 0, & \textit{otherwise}. \end{cases} \quad (6.65)$$

Using step functions ε_{x} and ε_{y}, and using different extents of the model in horizontal and vertical directions (standard deviations of the Gaussian) $(\sigma_{\mathrm{x1}}, \sigma_{\mathrm{x2}})$, $(\sigma_{\mathrm{y1}}, \sigma_{\mathrm{y2}})$ asymmetries of the model are formed along pI and MW axis (Fig. 6.18).

Although this model is flexible and allows the modeling of spots of different shape appearances, eight parameters need to be calculated to fit the model. The asymmetries of the protein spot along the pI axis in the 2DEG images are small, so selecting $\sigma_{x1} = \sigma_{2x}$ yields a saturated protein spot model M3G with a smaller number (7) of parameters. Although the M4G model is elastic on the sides, it cannot represent the flat top of the saturated spot.

Model of Four Gaussian Functions with Enabled Flat Top

The proposed M4GFT protein spot model consists of two asymmetric Gaussian functions with enabled flat top (see (6.62)).

Model 6.6 (Four Gaussian Functions with Flat Top Model)

Four Gaussian functions-based model of protein spots, used to represent 2DEG image region, whose pixel coordinates are x and y, is defined as follows:

$$\Psi_{\mathrm{M4Gft}}(x, y; \boldsymbol{\theta}_{\mathrm{M4Gft}}) = I^{\mathrm{B}} + I^{\mathrm{A}} \quad (6.66)$$

$$\times \left[\varepsilon_{\mathrm{x1}} \exp\left(-\frac{(x-(x_{\mathrm{c}}-d_{\mathrm{x}}))^2}{2\sigma_{\mathrm{x1}}^2}\right) + \varepsilon_{\mathrm{x2}} \exp\left(-\frac{(x-(x_{\mathrm{c}}+d_{\mathrm{x}}))^2}{2\sigma_{\mathrm{x2}}^2}\right) + (1-\varepsilon_{\mathrm{x1}}-\varepsilon_{\mathrm{x2}})\right]$$

$$\times \left[\varepsilon_{\mathrm{y1}} \exp\left(-\frac{(y-(y_{\mathrm{c}}-d_{\mathrm{y}}))^2}{2\sigma_{\mathrm{y1}}^2}\right) + \varepsilon_{\mathrm{y2}} \exp\left(-\frac{(y-(y_{\mathrm{c}}+d_{\mathrm{y}}))^2}{2\sigma_{\mathrm{y2}}^2}\right) + (1-\varepsilon_{\mathrm{y1}}-\varepsilon_{\mathrm{y2}})\right],$$

here $\{\sigma_{x_1}, \sigma_{x_2}, \sigma_{y_1}, \sigma_{y_2}\}$ *is extent of the model in horizontal and vertical directions (standard deviation of the Gaussian);* $(x_{\mathrm{c}}, y_{\mathrm{c}})$ *is coordinates of the protein spot* *model center; I^{B} is background intensity of the spot; and I^{A} is peak intensity of the spot, excluding background;*

$$\varepsilon_{x1} = \begin{cases} 1, & \text{if } x \le x_c - d_x; \\ 0, & \text{otherwise} \end{cases} \quad \text{and} \quad \varepsilon_{x2} = \begin{cases} 1, & \text{if } y \ge x_c + d_x; \\ 0, & \text{otherwise}. \end{cases} \tag{6.67}$$

$$\varepsilon_{y1} = \begin{cases} 1, & \text{if } x \le x_c - d_y; \\ 0, & \text{otherwise} \end{cases} \quad \text{and} \quad \varepsilon_{y2} = \begin{cases} 1, & \text{if } y \ge y_c + d_y; \\ 0, & \text{otherwise}. \end{cases} \tag{6.68}$$

Visualizations of the model (3D surface with vertical and horizontal cross sections) are shown in Fig. 6.19.

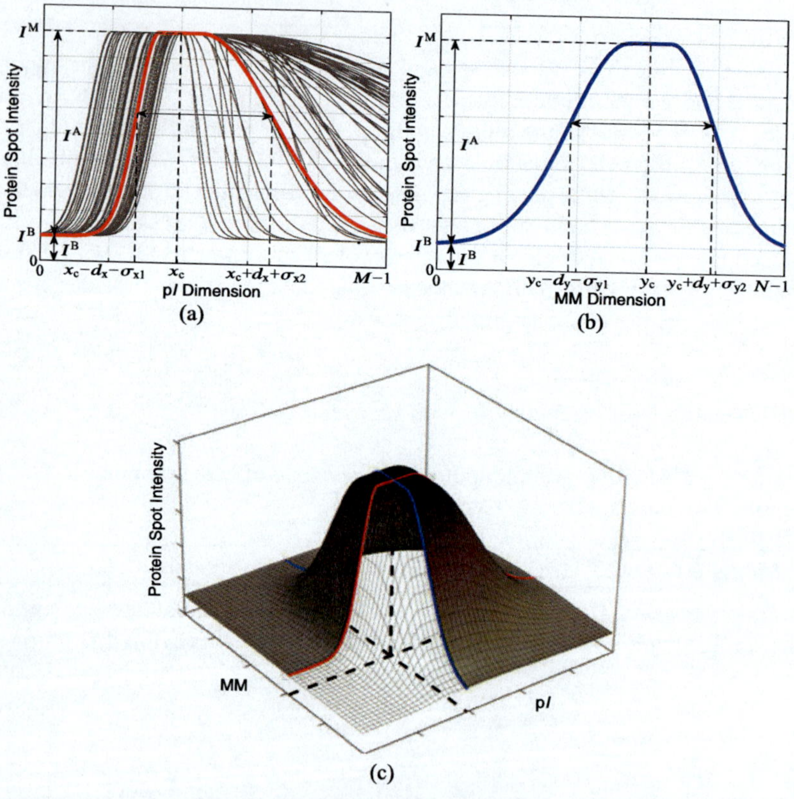

Fig. 6.19 Four Gaussian functions with flat top model (M4GFT). Horizontal (**a**) and vertical (**b**) cross sections of protein spot model and its 3D plot (**c**). Model's capacity to adapt its shape is shown by random sampling σ_{x2} and d_x parameters and plotted as thin lines (**a**)

6.3.2.4 Spline-Based (Π-Shaped) Models

This group of spot models is composed by combining two Π-shaped spline curves, which are composition of two S-shaped splines (functions typically are used as fuzzy logic membership functions (Jang et al. 1997)) which is defined as follows:

$$f_{2S}(x; \boldsymbol{\theta}_{2S}) = \tag{6.69}$$

$$\begin{cases} 0, & \text{if } x \le x_c - d_{x1} - d_{x2}; \\ 2\left(x - (x_c - d_{x1} - d_{x2})\right)^2 / d_{x2}^2, & \text{if } x_c - d_{x1} - d_{x2} < x \le x_c - d_{x1} - 0.5d_{x2}; \\ 1 - 2\left(x - (x_c - d_{x1})\right)^2 / d_{x2}^2, & \text{if } x_c - d_{x1} - 0.5d_{x2} < x \le x_c - d_{x1}; \\ 1, & \text{if } x_c - d_{x1} < x \le x_c + d_{x1}; \\ 1 - 2\left(x - (x_c + d_{x1})\right)^2 / d_{x3}^2, & \text{if } x_c + d_{x1} < x \le x_c + d_{x1} + 0.5d_{x3}; \\ 2\left(x - (x_c + d_{x1} + d_{x3})\right)^2 / d_{x3}^2, & \text{if } x_c + d_{x1} + 0.5d_{x3} < x \le x_c + d_{x1} + d_{x3}; \\ 0, & \text{if } x > x_c + d_{x1} + d_{x3}; \end{cases}$$

here $\{d_{x1}, d_{x2}, d_{x3}\}$ is parameters that determine the shape of the curve and x_c is the center of curve. The curve becomes symmetric if $d_{x2} = d_{x3}$.

Simple Π-Shaped Model

Simplified Π-shaped spline curve is achieved by constraining function (6.69) with $d_{x1} = 0$. The resulting combination of two constrained spline functions is appended with two additional parameters describing the background intensity of the spot I^B and the intensity of the spot without background (spot height) I^A. Proposed model is parameterized by 6 variables and is less capable to represent saturated spots.

Model 6.7 (Simple Π-Shaped Model)

Four spline functions-based model of protein spots, used to represent 2DEG image region, whose pixel coordinates are x and y, is defined as follows:

$$\Psi_{MP}(x, y; \boldsymbol{\theta}_{MP}) = I^B + I^A \cdot \Psi_{pI}(x) \cdot \Psi_{MM}(y), \tag{6.70a}$$

$$\Psi_{pI}(x) = \begin{cases} 0, & \text{if } x \le x_c - d_{x2}; \\ 2\left(x - (x_c - d_{x2})\right)^2 / d_{x2}^2, & \text{if } x_c - d_{x2} < x \le x_c - 0.5d_{x2}; \\ 1 - 2\left(x - (x_c)\right)^2 / d_{x2}^2, & \text{if } x_c - 0.5d_{x2} < x \le x_c; \\ 1, & \text{if } x_c < x \le x_c; \\ 1 - 2\left(x - (x_c)\right)^2 / d_{x3}^2, & \text{if } x_c < x \le x_c + 0.5d_{x3}; \\ 2\left(x - (x_c + d_{x3})\right)^2 / d_{x3}^2, & \text{if } x_c + 0.5d_{x3} < x \le x_c + d_{x3}; \\ 0, & \text{if } x > x_c + d_{x3}; \end{cases} \tag{6.70b}$$

$$\Psi_{\mathrm{MM}}(y) = \begin{cases} 0, & \text{if } y \le y_{\mathrm{c}} - d_{\mathrm{y2}}; \\ 2\left(y - (y_{\mathrm{c}} - d_{\mathrm{y2}})\right)^2/d_{\mathrm{y2}}^2, & \text{if } y_{\mathrm{c}} - d_{\mathrm{y2}} < y \le y_{\mathrm{c}} - 0.5d_{\mathrm{y2}}; \\ 1 - 2\left(y - (y_{\mathrm{c}})\right)^2/d_{\mathrm{y2}}^2, & \text{if } y_{\mathrm{c}} - 0.5d_{\mathrm{y2}} < y \le y_{\mathrm{c}}; \\ 1, & \text{if } y_{\mathrm{c}} < y \le y_{\mathrm{c}}; \\ 1 - 2\left(y - (y_{\mathrm{c}})\right)^2/d_{\mathrm{y3}}^2, & \text{if } y_{\mathrm{c}} < y \le y_{\mathrm{c}} + 0.5d_{\mathrm{y3}}; \\ 2\left(y - (y_{\mathrm{c}} + d_{\mathrm{y3}})\right)^2/d_{\mathrm{y3}}^2, & \text{if } y_{\mathrm{c}} + 0.5d_{\mathrm{y3}} < y \le y_{\mathrm{c}} + d_{\mathrm{y3}}; \\ 0, & \text{if } y > y_{\mathrm{c}} + d_{\mathrm{y3}}; \end{cases} \tag{6.70c}$$

here $\{d_{\mathrm{x2}}, d_{\mathrm{x3}}; d_{\mathrm{y2}}, d_{\mathrm{y3}}\}$ is parameters that determine the shape of the spot model; I^{B} is background intensity of the spot; and I^{A} is peak intensity of the spot, excluding background. x_{c} and y_{c} are centers of the spot.

Visualizations of the model (3D surface with vertical and horizontal cross sections) are shown in Fig. 6.20.

Π-Shaped Model With Flat Top

The resulting combination of spline functions is appended with two additional parameters describing the background intensity of the spot I^{B} and the intensity of the spot without background (spot height) I^{A}. Proposed model is parameterized by 8 variables.

Model 6.8 (Π-shape Model With Flat Top)

Four spline functions-based model of protein spots, used to represent 2DEG image region, whose pixel coordinates are x and y, is defined as follows:

$$\Psi_{\mathrm{MPft}}(x, y; \boldsymbol{\theta}_{\mathrm{MPft}}) = I^{\mathrm{B}} + I^{\mathrm{A}} \cdot \Psi_{\mathrm{pI}}(x) \cdot \Psi_{\mathrm{MM}}(y), \tag{6.71a}$$

$$\Psi_{\mathrm{pI}}(x) = \begin{cases} 0, & \text{if } x \le x_{\mathrm{c}} - d_{\mathrm{x1}} - d_{\mathrm{x2}}; \\ 2\left(\dfrac{x - (x_{\mathrm{c}} - d_{\mathrm{x1}} - d_{\mathrm{x2}})}{d_{\mathrm{x2}}}\right)^2, & \text{if } x_{\mathrm{c}} - d_{\mathrm{x1}} - d_{\mathrm{x2}} < x \le x_{\mathrm{c}} - d_{\mathrm{x1}} - 0.5d_{\mathrm{x2}}; \\ 1 - 2\left(\dfrac{x - (x_{\mathrm{c}} - d_{\mathrm{x1}})}{d_{\mathrm{x2}}}\right)^2, & \text{if } x_{\mathrm{c}} - d_{\mathrm{x1}} - 0.5d_{\mathrm{x2}} < x \le x_{\mathrm{c}} - d_{\mathrm{x1}}; \\ 1, & \text{if } x_{\mathrm{c}} - d_{\mathrm{x1}} < x \le x_{\mathrm{c}} + d_{\mathrm{x1}}; \\ 1 - 2\left(\dfrac{x - (x_{\mathrm{c}} + d_{\mathrm{x1}})}{d_{\mathrm{x3}}}\right)^2, & \text{if } x_{\mathrm{c}} + d_{\mathrm{x1}} < x \le x_{\mathrm{c}} + d_{\mathrm{x1}} + 0.5d_{\mathrm{x3}}; \\ 2\left(\dfrac{x - (x_{\mathrm{c}} + d_{\mathrm{x1}} + d_{\mathrm{x3}})}{d_{\mathrm{x3}}}\right)^2, & \text{if } x_{\mathrm{c}} + d_{\mathrm{x1}} + 0.5d_{\mathrm{x3}} < x \le x_{\mathrm{c}} + d_{\mathrm{x1}} + d_{\mathrm{x3}}; \\ 0, & \text{if } x > x_{\mathrm{c}} + d_{\mathrm{x1}} + d_{\mathrm{x3}}; \end{cases} \tag{6.71b}$$

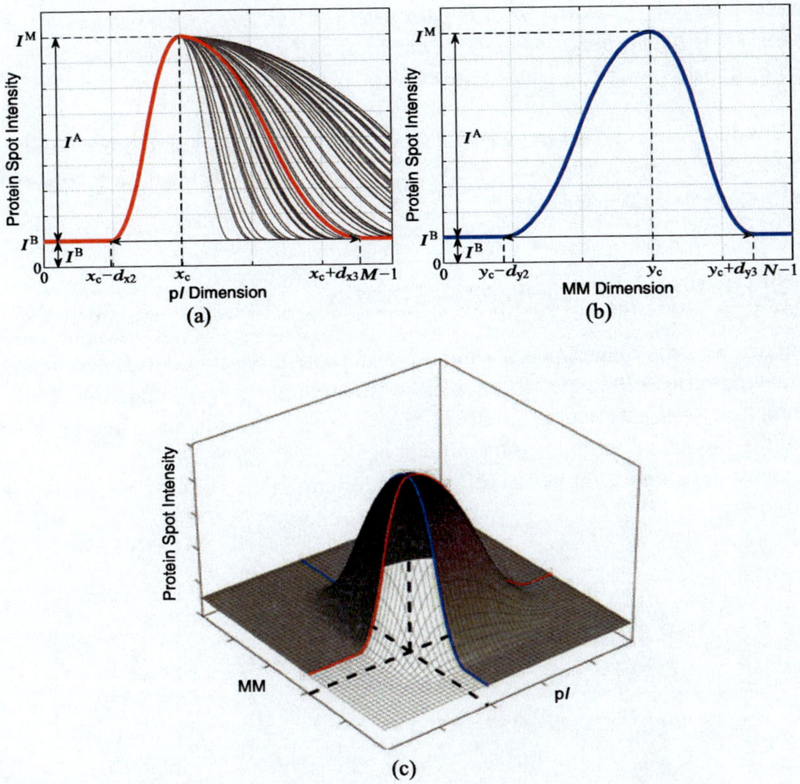

Fig. 6.20 Simple Π-shaped model (MP). Horizontal (**a**) and vertical (**b**) cross sections of protein spot model and its 3D plot (**c**). Model's capacity to adapt its shape is shown by random sampling d_{x3} parameter and plotted as thin lines (**a**)

$$\Psi_{\mathrm{MM}}(y) = \begin{cases} 0, & \text{if } y \le y_c - d_{y1} - d_{y2}; \\ 2\left(\dfrac{y-(y_c-d_{y1}-d_{y2})}{d_{y2}}\right)^2, & \text{if } y_c-d_{y1}-d_{y2} < y \le y_c-d_{y1}-0.5d_{y2}; \\ 1-2\left(\dfrac{y-(y_c-d_{y1})}{d_{y2}}\right)^2, & \text{if } y_c-d_{y1}-0.5d_{y2} < y \le y_c-d_{y1}; \\ 1, & \text{if } y_c-d_{y1} < y \le y_c+d_{y1}; \\ 1-2\left(\dfrac{y-(y_c+d_{y1})}{d_{y3}}\right)^2, & \text{if } y_c+d_{y1} < y \le y_c+d_{y1}+0.5d_{y3}; \\ 2\left(\dfrac{y-(y_c+d_{y1}+d_{y3})}{d_{y3}}\right)^2, & \text{if } y_c+d_{y1}+0.5d_{y3} < y \le y_c+d_{y1}+d_{y3}; \\ 0, & \text{if } y > y_c+d_{y1}+d_{y3}; \end{cases} \tag{6.71c}$$

here $\{d_{x1}, d_{x2}, d_{x3}; d_{y1}, d_{y2}, d_{y3}\}$ *is parameters that determine the shape of the spot model;* I^{B} *is background intensity of the spot; and* I^{A} *is peak intensity of the spot, excluding background.* x_c *and* y_c *are centers of the spot.*

Visualizations of the model (3D surface with vertical and horizontal cross sections) are shown in Fig. 6.21. This spot model is better situated for reconstruction of saturated spot models.

6.3.2.5 Bell-Shaped Function-Based Models

In addition to the Gaussian and spline-based protein spot models, new symmetric and asymmetric bell-shaped 2D models were applied for spot modeling. This kind of function was selected to create a protein spot representation model because changing the parameter θ_x^a results in a flat peak of the curve.

One-dimensional symmetric bell-shaped function (Jang et al. 1997) is expressed as follows:

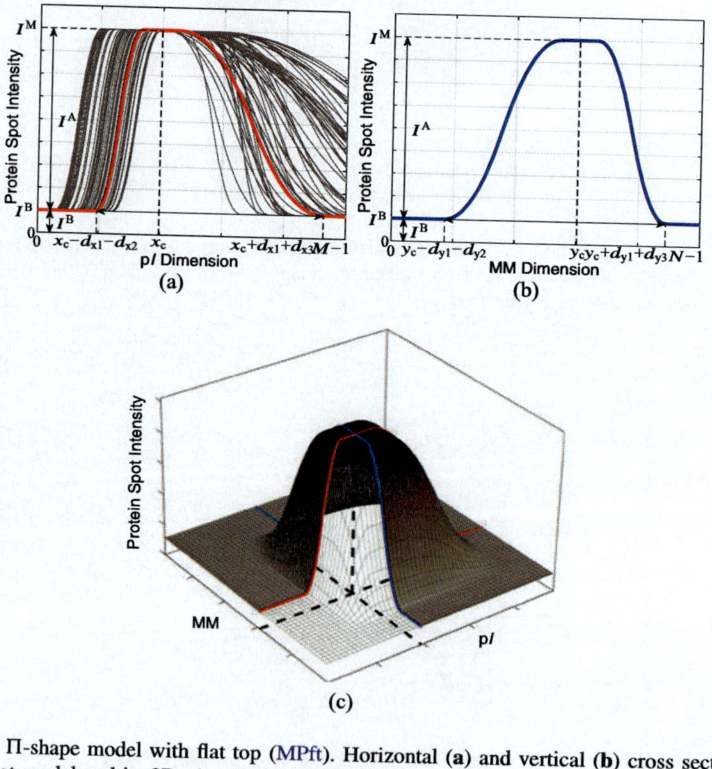

Fig. 6.21 Π-shape model with flat top (MPft). Horizontal (**a**) and vertical (**b**) cross sections of protein spot model and its 3D plot (**c**). Model's capacity to adapt its shape is shown by random sampling d_{x1} and d_{x3} parameters, and ploted as thin lines (**a**).

$$f_{\mathrm{B1}}(x) = \frac{1}{1 + \left|\frac{x - x_{\mathrm{c}}}{\theta_{\mathrm{x}}^{\mathrm{a}}}\right|^{2\theta_{\mathrm{x}}^{\mathrm{b}}}}, \tag{6.72}$$

here $\theta_{\mathrm{x}}^{\mathrm{a}}$ is distance to the center of the spot; $\theta_{\mathrm{x}}^{\mathrm{b}}$ is slope of model edges; and x_{c} is coordinate of the function center.

One-dimensional asymmetric bell-shaped function is expressed as follows:

$$f_{\mathrm{B2}}(x) = \varepsilon_{\mathrm{x}} \cdot \frac{1}{1 + \left|\frac{x - x_{\mathrm{c}}}{\theta_{\mathrm{x}}^{\mathrm{a}}}\right|^{2\theta_{\mathrm{x}}^{\mathrm{b1}}}} + (1 - \varepsilon_{\mathrm{x}}) \cdot \frac{1}{1 + \left|\frac{x - x_{\mathrm{c}}}{\theta_{\mathrm{x}}^{\mathrm{a}}}\right|^{2\theta_{\mathrm{x}}^{\mathrm{b2}}}}, \tag{6.73}$$

here $\theta_{\mathrm{x}}^{\mathrm{a}}$ is distance to the center of the curve; $\theta_{\mathrm{x}}^{\mathrm{b1}}$, $\theta_{\mathrm{x}}^{\mathrm{b2}}$ is slopes of the curve edges; x_{c} is coordinate of the curve center; step function forming the asymmetry of the curve is:

$$\varepsilon_{\mathrm{x}} = \begin{cases} 1, & \text{if } x \le x_{\mathrm{c}}; \\ 0, & \text{otherwise}. \end{cases}$$

The two-dimensional bell-shaped function can be composed in two ways: by combining two different symmetric or asymmetric 1D bell-shaped functions. Both combinations of functions can be used as mathematical model of protein spots, after adding parameters to define the background and intensity of the spot.

Two-Way Symmetric Bell-Shaped Model

This is a simpler version of the bell-shaped spot model which is composed using two symmetric different 1D functions. Model has 8 parameters.

Model 6.9 (Two-Way Symmetric Bell-Shaped Model)

Two-way symmetric bell-shaped function-based model of protein spots, used to represent 2DEG image region, whose pixel coordinates are x and y, is defined as follows:

$$\Psi_{\mathrm{MBs}}(x, y; \boldsymbol{\theta}_{\mathrm{MBs}}) = I^{\mathrm{B}} + I^{\mathrm{A}} \cdot \frac{1}{1 + \left|\frac{x - x_{\mathrm{c}}}{\theta_{\mathrm{x}}^{\mathrm{a}}}\right|^{2\theta_{\mathrm{x}}^{\mathrm{b}}}} \cdot \frac{1}{1 + \left|\frac{y - y_{\mathrm{c}}}{\theta_{\mathrm{y}}^{\mathrm{a}}}\right|^{2\theta_{\mathrm{y}}^{\mathrm{b}}}}, \tag{6.74a}$$

here I^{B} is background intensity of the spot; I^{A} is peak intensity of the spot, excluding background; $\theta_{\mathrm{x}}^{\mathrm{a}}$, $\theta_{\mathrm{y}}^{\mathrm{a}}$ is distance to the center of the spot; $(x_{\mathrm{c}}, y_{\mathrm{c}})$ is *coordinates of the protein spot model center; and θ_x^b, θ_y^b is the slopes of the model satisfying conditions:*

$$\theta_x^b < \frac{1}{2}x_c \quad \text{and} \quad \theta_y^b < \frac{1}{2}y_c\,; \tag{6.74b}$$

Visualizations of the model (3D surface with vertical and horizontal cross sections) are shown in Fig. 6.22.

Asymmetric Bell-Shaped Model

This is a more complex version of the bell-shaped spot model which is composed using two asymmetric different 1D functions. Model has 10 parameters.

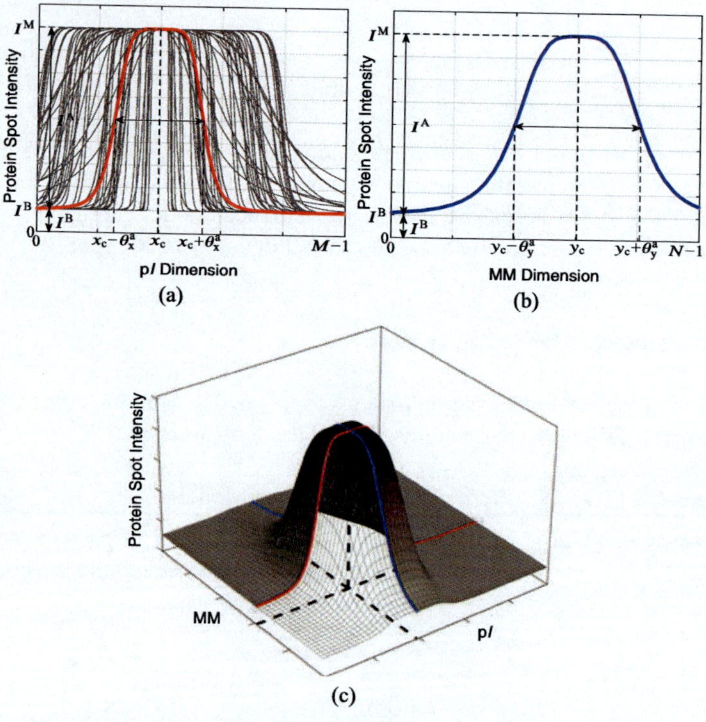

Fig. 6.22 Two-way symmetric bell-shaped model (MBS). Horizontal (**a**) and vertical (**b**) cross sections of protein spot model and its 3D plot (**c**). Model's capacity to adapt its shape is shown by random sampling θ_x^a and θ_x^b parameters and plotted as thin lines (**a**)

Model 6.10 (Asymmetric Bell-Shaped Model)

Asymmetric bell-shaped function-based model of protein spots, used to represent 2DEG image region, whose pixel coordinates are x and y, is defined as follows:

$$\Psi_{\mathrm{MBas}}\left(x, y; \boldsymbol{\theta}_{\mathrm{MBas}}\right) = I^{\mathrm{B}} + I^{\mathrm{A}} \varepsilon_{\mathrm{x}} \cdot \frac{1}{1+\left|\dfrac{x-x_{\mathrm{c}}}{\theta_{\mathrm{x}}^{\mathrm{a}}}\right|^{2\theta_{\mathrm{x}}^{\mathrm{b1}}}} + (1-\varepsilon_{\mathrm{x}}) \cdot \frac{1}{1+\left|\dfrac{x-x_{\mathrm{c}}}{\theta_{\mathrm{x}}^{\mathrm{a}}}\right|^{2\theta_{\mathrm{x}}^{\mathrm{b2}}}}$$

$$\times\, \varepsilon_{\mathrm{y}} \cdot \frac{1}{1+\left|\dfrac{y-y_{\mathrm{c}}}{\theta_{\mathrm{y}}^{\mathrm{a}}}\right|^{2\theta_{\mathrm{y}}^{\mathrm{b1}}}} + (1-\varepsilon_{\mathrm{y}}) \cdot \frac{1}{1+\left|\dfrac{y-y_{\mathrm{c}}}{\theta_{\mathrm{y}}^{\mathrm{a}}}\right|^{2\theta_{\mathrm{y}}^{\mathrm{b2}}}} \tag{6.75}$$

here I^{B} is background intensity of the spot; I^{A} is peak intensity of the spot, excluding background; $\theta_{\mathrm{x}}^{\mathrm{a}}$, $\theta_{\mathrm{y}}^{\mathrm{a}}$ is distance to the center of the spot; $(x_{\mathrm{c}}, y_{\mathrm{c}})$ is coordinates of the protein spot model center; and $\theta_{\mathrm{x}}^{\mathrm{b1}}, \theta_{\mathrm{x}}^{\mathrm{b2}}, \theta_{\mathrm{y}}^{\mathrm{b1}}, \theta_{\mathrm{y}}^{\mathrm{b2}}$ are the slopes of the model satisfying conditions:

$$\theta_{\mathrm{x}}^{\mathrm{b}} < \frac{1}{2} x_{\mathrm{c}} \quad \text{and} \quad \theta_{\mathrm{y}}^{\mathrm{b}} < \frac{1}{2} y_{\mathrm{c}}\,; \tag{6.76}$$

step functions forming the asymmetry of the model are:

$$\varepsilon_{\mathrm{x}} = \begin{cases} 1, & \text{if } x \le x_{\mathrm{c}} - d_{\mathrm{x}}; \\ 0, & \text{otherwise} \end{cases} \quad \text{and} \quad \varepsilon_{\mathrm{y}} = \begin{cases} 1, & \text{if } y \ge x_{\mathrm{c}} + d_{\mathrm{x}}; \\ 0, & \text{otherwise}. \end{cases}$$

Visualizations of the model (3D surface with vertical and horizontal cross sections) are shown in Fig. 6.23.

The protein spot model $\Psi_{\mathrm{MBs}}\left(x, y; \boldsymbol{\theta}_{\mathrm{MBs}}\right)$ is suitable for the approximation of round symmetric protein spots, and the second model $\Psi_{\mathrm{MBas}}(x, y; \boldsymbol{\theta}_{\mathrm{MBas}})$ for anisotropic ones. The anisotropic bell-shaped model estimates different protein dispersion in two orthogonal directions.

Using anisotropic bell-shaped function for modeling of 2DEG images results in eight model parameters for each protein spot (6 parameters are estimated using isotropic bell-shaped model). In practice, asymmetries of protein spots in the molecular weight axis occur in 2DEG images. Both presented protein spot models do not account for these asymmetries.

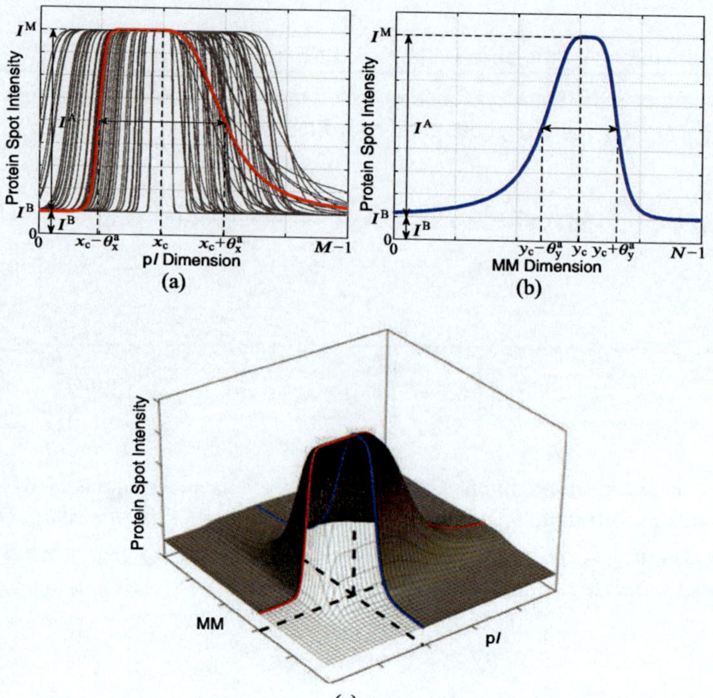

Fig. 6.23 Asymmetric bell-shaped model (MBAS). Horizontal (**a**) and vertical (**b**) cross sections of protein spot model and its 3D plot (**c**). Model's capacity to adapt its shape is shown by random sampling θ_x^a and θ_x^{b2} parameters and plotted as thin lines (**a**)

6.3.2.6 Sigmoid-Based Models

This group of spot models is built by combining two symmetric or asymmetric sigmoid-based curves. General sigmoid-based curve is a product of two sigmoids and is defined as follows:

$$f_S(x) = \frac{1}{1 + \exp\left(-\theta_x^{b1}\left(x - \left(x_c - \theta_x^a\right)\right)\right)} \cdot \frac{1}{1 + \exp\left(-\theta_x^{b2}\left(x - \left(x_c - \theta_x^a\right)\right)\right)}; \tag{6.77}$$

here θ_x^a is distance from the slope centers to the center of the curve; θ_x^{b1}, θ_x^{b2} is parameters that determine the steepness of the curve slopes; and x_c is coordinate of the function center. In a case when $\theta_x^{b1} = \theta_x^{b2}$, the curve becomes symmetric.

Two-Way Symmetric Sigmoid-Based Model

This is a simpler version of the sigmoid-based spot model which is composed using two symmetric different 1D functions. Model has 8 parameters.

Model 6.11 (Two-Way Symmetric Sigmoid-Based Model)

Two-way symmetric sigmoid function-based model of protein spots, used to represent 2DEG image region, whose pixel coordinates are x and y, is defined as follows:

$$\Psi_{\mathrm{MSs}}\left(x, y; \boldsymbol{\theta}_{\mathrm{MSs}}\right) = I^{\mathrm{B}} + I^{\mathrm{A}} \times \frac{1}{1+\exp\left(-\theta_{\mathrm{x}}^{\mathrm{b}}\left(x-\left(x_{\mathrm{c}}-\theta_{\mathrm{x}}^{\mathrm{a}}\right)\right)\right)} \cdot \frac{1}{1+\exp\left(\theta_{\mathrm{x}}^{\mathrm{b}}\left(x-\left(x_{\mathrm{c}}-\theta_{\mathrm{x}}^{\mathrm{a}}\right)\right)\right)} \times \frac{1}{1+\exp\left(-\theta_{\mathrm{y}}^{\mathrm{b}}\left(y-\left(y_{\mathrm{c}}-\theta_{\mathrm{y}}^{\mathrm{a}}\right)\right)\right)} \cdot \frac{1}{1+\exp\left(\theta_{\mathrm{y}}^{\mathrm{b}}\left(y-\left(y_{\mathrm{c}}-\theta_{\mathrm{y}}^{\mathrm{a}}\right)\right)\right)}; \tag{6.78}$$

here I^{B} is background intensity of the spot; I^{A} is peak intensity of the spot, excluding background; $\theta_{\mathrm{x}}^{\mathrm{a}}$, $\theta_{\mathrm{y}}^{\mathrm{a}}$ is distance from the slope centers to the center of the curve; $\theta_{\mathrm{x}}^{\mathrm{b}}$, $\theta_{\mathrm{y}}^{\mathrm{b}}$ is parameters that determine the steepness of the curve slopes; and x_{c} is coordinate of the function center.

Visualizations of the model (3D surface with vertical and horizontal cross sections) are shown in Fig. 6.24.

Asymmetric Sigmoid-Based Model

This is a more complex version of the sigmoid-based spot model which is composed using two asymmetric different 1D functions. Model has 10 parameters.

Model 6.12 (Asymmetric Sigmoid-Based Model)

Asymmetric sigmoid function-based model of protein spots, used to represent 2DEG image region, whose pixel coordinates are x and y, is defined as follows:

$$\Psi_{\mathrm{MSas}}\left(x, y; \boldsymbol{\theta}_{\mathrm{MSas}}\right) = I^{\mathrm{B}} + I^{\mathrm{A}} \times \frac{1}{1+\exp\left(-\theta_{\mathrm{x}}^{\mathrm{b1}}\left(x-\left(x_{\mathrm{c}}-\theta_{\mathrm{x}}^{\mathrm{a}}\right)\right)\right)} \cdot \frac{1}{1+\exp\left(\theta_{\mathrm{x}}^{\mathrm{b2}}\left(x-\left(x_{\mathrm{c}}-\theta_{\mathrm{x}}^{\mathrm{a}}\right)\right)\right)}$$

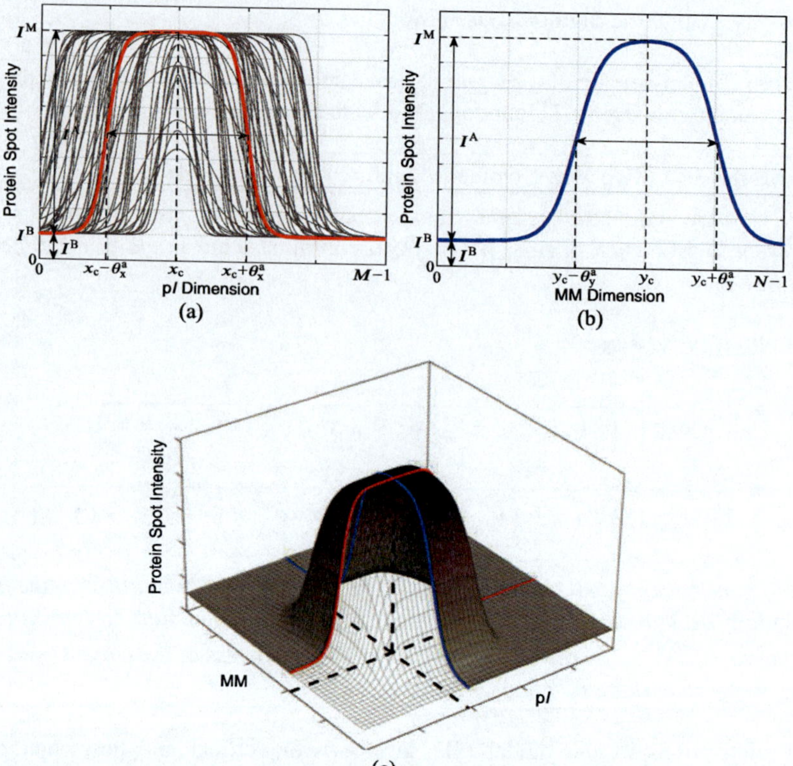

Fig. 6.24 Two-way symmetric sigmoid-based model (MSS). Horizontal (**a**) and vertical (**b**) cross sections of protein spot model and its 3D plot (**c**). Model's capacity to adapt its shape is shown by random sampling θ_x^a and θ_x^b parameters and plotted as thin lines (**a**)

$$\times \frac{1}{1+\exp\left(-\theta_y^{b1}\left(y-\left(y_c-\theta_y^a\right)\right)\right)} \cdot \frac{1}{1+\exp\left(\theta_y^{b2}\left(y-\left(y_c-\theta_y^a\right)\right)\right)}; \tag{6.79}$$

here I^B is background intensity of the spot; I^A is peak intensity of the spot, excluding background; θ_x^a, θ_y^a is distance from the slope centers to the center of the curve; θ_x^{b1}, θ_x^{b2}, θ_y^{b1}, θ_y^{b2} is parameters that determine the steepness of the curve slopes; and x_c is coordinate of the function center.

Visualizations of the model (3D surface with vertical and horizontal cross sections) are shown in Fig. 6.25.

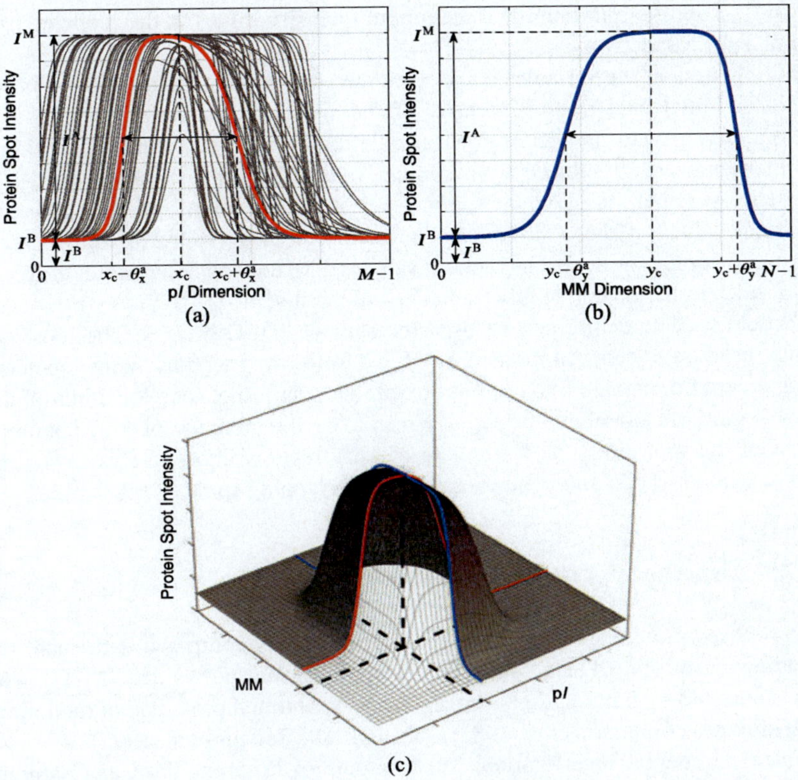

Fig. 6.25 Asymmetric sigmoid-based model (MSAS). Horizontal (**a**) and vertical (**b**) cross sections of protein spot model and its 3D plot (**c**). Model's capacity to adapt its shape is shown by random sampling θ_x^a and θ_x^{b2} parameters and plotted as thin lines (**a**)

6.3.2.7 Protein Spot Modeling

During modeling of protein spots, segment (region) of 2DEG image is approximated by a mathematical model. For constrained minimization problem, Matlab implementation of Trust-region-reflective least squares algorithm (Coleman and Li 1996, 1994) is used, which seeks to minimize the residual error E_r defined by the objective function:

$$E_r = \sum_{x,y \in S_{Gij}} \frac{\left(\Psi_t(x, y; \boldsymbol{\theta}_t) - I_G(x, y)\right)^2}{n_S\left(I_S^{\top} - I_S^{\perp}\right)^2}, \tag{6.80}$$

here $\Psi_t\,(x, y; \boldsymbol{\theta}_t)$ is model output (modeled spot), corresponding to the parameter vector $\boldsymbol{\theta}_t$; n_S is the number of points in the modeled region S_{Gij} of image; and

$I_S^{\perp}$ and $I_S^{\top}$ are the minimum and maximum intensity values in the modeled image region S_{Gij}, respectively.

Initial guess of the variables is based on the current image region which is being modeled. Initial background intensity of the spot and peak intensity of the spot without background are determined as minimum value of the image region $I_S^{\perp}$ and maximum value reduced by minimum value of the image region $(I_S^{\top} - I_S^{\perp})$, respectively. Other variables that define spot shape are initialized taking averaged historical values of these variables. The main stopping criteria of the solver are maximum number of iterations (default value is 400) and lower bound on the change in the value of the objective function $\Delta^{\perp} E_r$ (default value is 10^{-6}).

Protein spot modeling can be used to segment 2DEG images. The most commonly used segmentation procedure is as follows: detect as many centers as possible; split the image into regions containing single spot (one local minimum in image region); fit a mathematical spot model to each region; use obtained parameter values of the spot model as features to classify regions into two classes—regions with or without spots; and identify and split overlapping spots.

6.3.2.8 Experimental Comparison of Spot Models

The proposed protein spot models were experimentally compared with each other, including anisotropic 2D Gaussian (M2G) and diffusion spot models (M2G), which were suggested by other authors for regular and saturated protein spot modeling in 2DEG images (Anderson et al. 1981; Garrels 1989; Bettens et al. 1997).

For experimental investigation, 12, 864 spots of different sizes and saturation level were extracted from a set of scanned images of two-dimensional electrophoresis gels. The collection of image patches, each preferably containing single spot, was divided into three groups:

(a) Low-intensity spots with $I^{A} < 100$, in total 6532 image
(b) Medium-intensity spots with $100 \leq I^{A} < 180$, in total 4853 image patches;
(c) High-intensity and saturated protein spots with $I^{A} \geq 180$, in total 1479 image patches;

Representative spot samples from each group were visualized in 3D and presented in Fig. 6.26.

Results of experimental comparison of the models are presented as model fitting accuracy (Fig. 6.27), relative time of model fitting (Fig. 6.28), and speed-accuracy tradeoff (Fig. 6.29).

Residual errors in Fig. 6.27 uncover how different spot models are able to fit various protein spots. Image patches containing spots were grouped according to spot intensity into three categories to better reveal spot models' power to aproximate protein spots of different sizes and appearance. The lowest residual error in all three groups was obtained using asymmetric bell-shaped model (MBAS). Slightly higher errors were made using four Gaussian functions with flat top model (M4GFT), asymmetric sigmoid-based model (MSAS), and four Gaussian

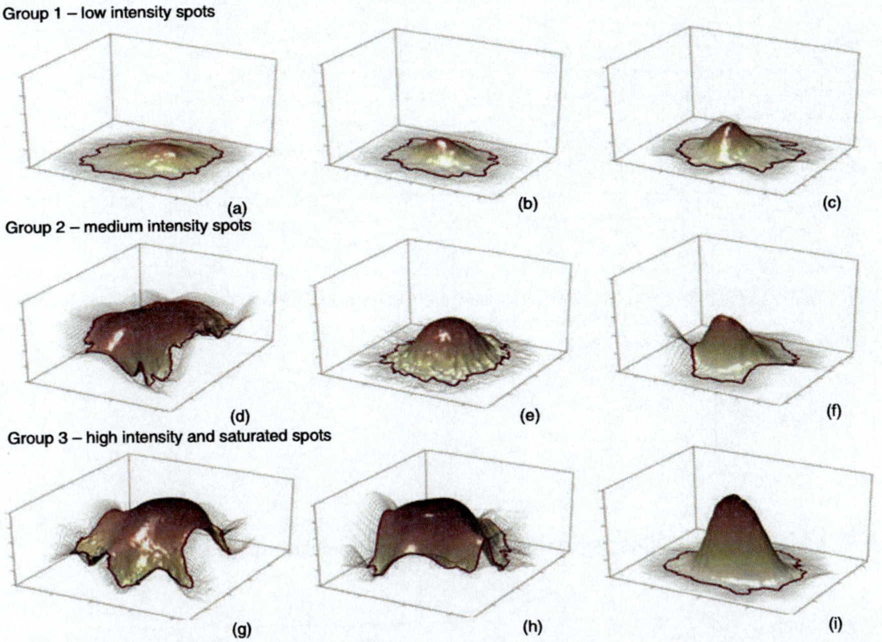

Fig. 6.26 3D visualization of representative protein spot samples from three groups: low-intensity spots **(a)**–**(c)**; medium-intensity spots **(d)**–**(f)**; high-intensity and saturated spots **(g)**–**(i)**

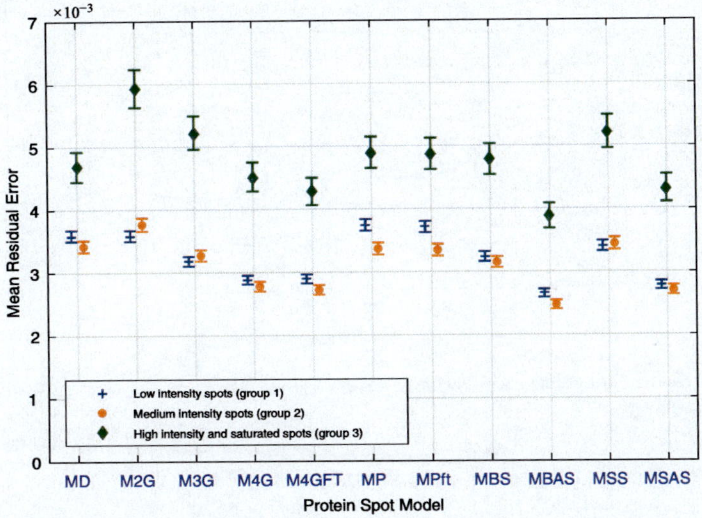

Fig. 6.27 Model fitting accuracy expressed as residual errors. Error bars indicate 95% confidence intervals

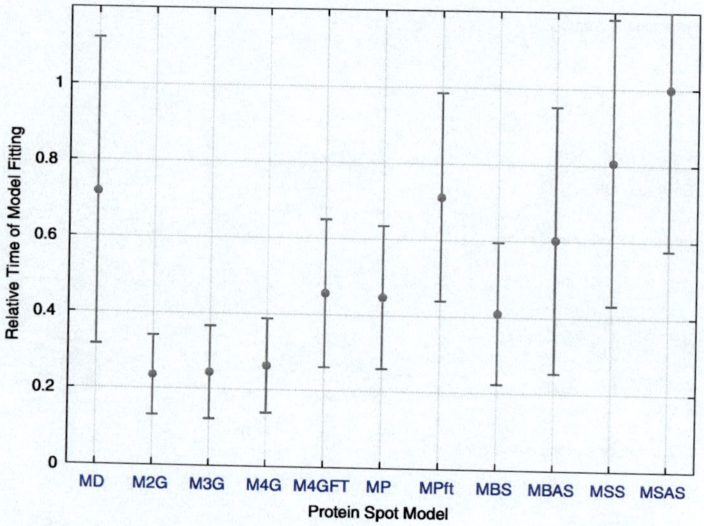

Fig. 6.28 Spot model fitting relative duration. Error bars indicate standard deviations

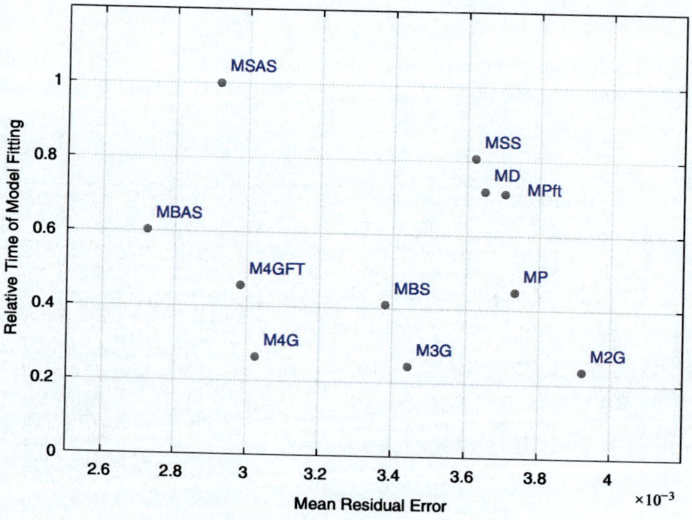

Fig. 6.29 Model fitting accuracy (residual errors) versus relative time of model fitting (speed-accuracy tradeoff)

functions model (M4G). The highest errors were demonstrated by the two-way adapting (anisotropic) 2D Gaussian model (M2G). Models were behaving similarly while approximating spots of low intensity (group 1) and medium intensity (group 2). Approximation error of high-intensity and saturated protein spots was higher for all models, but trends of accuracies of models are very similar in all three groups.

Timing experiments (Fig. 6.28) showed that the shortest relative duration of model fitting is achieved if using Gaussian-based models: two-way adapting (anisotropic) 2D Gaussian model (M2G), three Gaussian functions model (M3G), and four Gaussian functions model (M4G). Four Gaussian functions with flat top model (M4GFT) is next to two-way symmetric bell-shaped model (MBS) and simple Π-shaped model (MP). The slowest fitting was while using asymmetric sigmoid-based model (MSAS).

Speed-accuracy tradeoff plot (Fig. 6.29) reveals that three models are on the lower-left front: asymmetric bell-shaped model (MBAS), four Gaussian functions model (M4G), and four Gaussian functions with flat top model (M4GFT). The difference in residual error between M4G and M4GFT is much smaller than model fitting time; therefore M4G model may be preferable. In summary, if more accurate and slower model is needed, the asymmetric bell-shaped model (MBAS) should be selected. For speed over accuracy, four Gaussian functions model (M4G) should be used.

6.3.3 Protein Spot Parametrization

Spot quantity, isoelectric point (pI), and molecular weight (MW) are important parameters to estimate during parametrization of protein spots. By having the centers of the protein spots, after elimination of the vertical and horizontal geometric distortions of the gels, we can determine the parameters of the proteins pI and MW. Having spot contours, we will be able to estimate relative amounts of proteins and their changes. Later these data are processed by statistical methods to assess their reliability.

6.3.3.1 Estimation of Protein pI and MW Values

Isoelectric points of investigated proteins are calculated by linearly interpolating between known pI values of protein spots that were transferred from the standard gel. If immobilized pH gradient (IPG) strips are used, the pH calibration curve is provided by the manufacturer, and it can be used to calculate pI values of proteins lying between proteins of known pI.

Molecular weights (MW) of proteins are determined by comparing their mobilities with those of several marker proteins (standards) of known molecular weight. By measuring the migration distance of each protein (standards and unknowns) from the top of the resolving gel, the relative mobility $\nu^{(r)}$ of each protein can be calculated. Linear relationship between decimal logarithm of molecular weight and relative mobility of proteins provides means to compute MW of unknown proteins. Relative mobility of the protein is the distance migrated by the r-th protein from the top of the gel $y^{(r)}$ compared to height of the gel $N_{\mathrm{y}}^{\mathrm{A}}$:

$$\nu^{(r)} = \frac{y^{(r)}}{N_{\mathrm{y}}^{\mathrm{A}}}. \tag{6.81}$$

Example of protein MW determination and the influence of vertical geometrical distortions are presented in Fig. 6.5.

6.3.3.2 Estimation of Protein Quantity Changes

Data on maximum spot intensity or total spot intensity (spot volume) can be used in the analysis of changes in protein expression. These assumptions will be correct when the maximum intensity value of the protein spot is in the center (the spot is convex when visualized 3D as hill) and if the background variation is compensated. Maximum intensity of the spot, also called optical density, is the value of intensity at the center of the spot:

$$I_{\mathrm{OD}}^{(r)} \triangleq I\left(x_0^{(r)}, y_0^{(r)}\right) \approx \max_{(x,y)\in \boldsymbol{S}^{(r)}} I(x, y), \tag{6.82}$$

here $\boldsymbol{S}^{(r)}$ is region of r-th protein spot and $\left(x_0^{(r)}, y_0^{(r)}\right)$ is coordinates of spot center.

Spot quantity can be calculated using the original spot or its mathematical model approximation. In both cases, the total intensity is calculated inside the spot region (inside contour area of segmented spot). Determination of such contours was objective of gel image segmentation stage. Spot quantity $V^{(r)}$ is defined as the total intensity of a spot in a gel image and corresponds to the amount of protein in the actual spot on the gel. The total intensity of an object is the sum of the intensities of all the pixels that make up the object:

$$V^{(r)} = \sum_{(x,y)\in \boldsymbol{S}^{(r)}} I(x, y). \tag{6.83}$$

The quantity of each protein spot in a gel $V^{(r)}$ is divided by the total quantity of all the protein spots in that gel in order to obtain the final normalized quantity of the given spot:

$$\widehat{V^{(r)}} = \frac{V^{(r)}}{\sum_r V^{(r)}}. \tag{6.84}$$

Process of normalization compensates for non-expression-related variations in spot total intensity. The motivation for this normalization is that the total protein amount is constant in all samples.

Normalized quantities are used to calculate the ratio of spot quantity between images:

$$\widehat{\Delta V_{G1\to 2}^{(r)}} = \begin{cases} \widehat{V_{G2}^{(r)}} \Big/ \widehat{V_{G1}^{(r)}}, & \text{if } \widehat{V_{G2}^{(r)}} > \widehat{V_{G1}^{(r)}}; \\ -\widehat{V_{G1}^{(r)}} \Big/ \widehat{V_{G2}^{(r)}}, & \text{otherwise}, \end{cases} \quad (6.85)$$

here $\widehat{V_{G1}^{(r)}}, \widehat{V_{G2}^{(r)}}$ is normalized quantity of the spot from the first and the second gels, respectively. One can notice that $\widehat{\Delta V_{G1\to 2}^{(r)}} > 1$ when spot from the second gel is bigger (spot increased going from 1 to 2 gel) and $\widehat{\Delta V_{G1\to 2}^{(r)}} < -1$ when the spot from the first gel is bigger.

These quantity ratios cannot directly provide exact information about the change of expression of proteins as the changes may have occur due to existence of variances in biological and technical parts of experiment. The quantitative estimates presented here should be processed by statistical means to test the hypotheses. Statistical analysis may be applied after replicates of identical experiments are performed, where the number and order of experiments are determined by the experimental design methods. Additionally it is common practice to select only ratios that are larger than selected threshold—protein spots with a change of quantity larger than two-fold are expected as changed expression.

Section Generalization

1. Designed and implemented automatic 2DEG image segmentation method, based on multilayer perceptron, minimizes the loss of wanted information, and the amount of errors therefore enables to analyze and evaluate larger amounts of data without increase of the number of experiments:
   - Image regions are classified into three classes ("spot exists in the region," "only part of the spot exists in the region," and "there is no spot in the region") at 95% accuracy.
   - Computationally expensive oversegmentation analysis and correction are performed only on image regions of class "only part of the spot exists in the region" so the required oversegmentation postprocessings are reduced by 20–30%.
   - Allows to specify classification results of individual 2DEG image regions according to image region classification of aligned images and to correct significant errors: lost image regions with protein spots and erroneously merged regions.

2. New models were proposed and investigated for parametrization of regular and saturated protein spots:
   - Gaussian-based complex models
   - Spline-based (Π-shaped curve) models
   - Bell-shaped function-based models
   - Sigmoid-based models
3. For accuracy over speed, asymmetric bell-shaped model (MBAS) should be selected. For speed over accuracy, four Gaussian functions model (M4G) should be used.

References

Ahmad N, Zhang J, Brown PJ, James D, Birch JR, Racher AJ, Smales CM (2006) On the statistical analysis of the GS-NS0 cell proteome: imputation, clustering and variability testing. Biochim. Biophys. Acta, Proteins Proteomics 1764(7):1179–1187

Aittokallio T, Salmi J, Nyman TA, Nevalainen OS (2005) Geometrical distortions in two-dimensional gels: applicable correction methods. J Chromatogr B Anal Technol Biomed Life Sci 815(1–2):25–37

Alterovitz G, Liu J, Chow J, Ramoni MF (2006) Automation, parallelism, and robotics for proteomics. Proteomics 6(14):4016–4022

Anderson NL, Taylor J, Scandora AE, Coulter BP, Anderson NG (1981) The TYCHO system for computer-analysis of two-dimensional gel-electrophoresis patterns. Clin Chem 27(11):1807–1820

Becher B, Knofel AK, Peters J (2006) Time-based analysis of silver-stained proteins in acrylamide gels. Electrophoresis 27(10):1867–1873

Bettens E, Scheunders P, Vandyck D, Moens L, Vanosta P (1997) Computer analysis of two-dimensional electrophoresis gels: a new segmentation and modeling algorithm. Electrophoresis 18(5):792–798

Biron DG, Brun C, Lefevre T, Lebarbenchon C, Loxdale HD, Chevenet F, Brizard JP, Thomas F (2006) The pitfalls of proteomics experiments without the correct use of bioinformatics tools. Proteomics 6(20):5577–5596

Bookstein FL (1989) Principal warps: Thin-plate splines and the decomposition of deformations. IEEE Trans. Pattern Anal. Mach. Intell. 11(6):567–585

Brauner JM, Groemer TW, Stroebel A, Grosse-Holz S, Oberstein T, Wiltfang J, Kornhuber J, Maler JM (2014) Spot quantification in two dimensional gel electrophoresis image analysis: comparison of different approaches and presentation of a novel compound fitting algorithm. BMC Bioinf 15(1):181

Canny J (1986) A computational approach to edge-detection. IEEE Trans Pattern Anal Mach Intell 8(6):679–698

Coleman TF, Li Y (1994) On the convergence of reflective newton methods for large-scale nonlinear minimization subject to bounds. Math Program 67(2):189–224

Coleman TF, Li Y (1996) An interior trust region approach for nonlinear minimization subject to bounds. SIAM J Optim 6(2):418–445

Corzett TH, Fodor IK, Choi MW, Walsworth VL, Chromy BA, Turteltaub KW, Mccutchen-Maloney SL (2006) Statistical analysis of the experimental variation in the proteomic characterization of human plasma by two-dimensional difference gel electrophoresis. J Proteome Res 5(10):2611–2619

de Jesus JR, Guimarães IC, Arruda MAZ (2019) Quantifying proteins at microgram levels integrating gel electrophoresis and smartphone technology. J Proteomics 198:45–49

Dowsey AW, Yang GZ (2008) The future of large-scale collaborative proteomics. Proc IEEE 96(8):1292–1309

Dowsey AW, Dunn MJ, Yang GZ (2008) Automated image alignment for 2D gel electrophoresis in a high-throughput proteomics pipeline. Bioinformatics 24(7):950–957

Dowsey AW, English JA, Lisacek F, Morris JS, Yang GZ, Dunn MJ (2010) Image analysis tools and emerging algorithms for expression proteomics. Proteomics 10(23):4226–4257

Dzemyda G, Kurasova O, Žilinskas J (2008) Daugiamačių duomenų vizualizavimo metodai. Matematikos ir informatikos institutas, Vilnius

Eravci M, Fuxius S, Broedel O, Weist S, Eravci S, Mansmann U, Schluter H, Tiemann J, Baumgartner A (2007) Improved comparative proteome analysis based on two-dimensional gel electrophoresis. Proteomics 7(4):513–523

Fernandez-Lozano C, Seoane JA, Gestal M, Gaunt TR, Dorado J, Pazos A, Campbell C (2016) Texture analysis in gel electrophoresis images using an integrative kernel-based approach. Sci. Rep. 6:19256

Fodor IK, Nelson DO, Alegria-Hartman M, Robbins K, Langlois RG, Turteltaub KW, Corzett TH, Mccutchen-Maloney SL (2005) Statistical challenges in the analysis of two-dimensional difference gel electrophoresis experiments using decyder (tm). Bioinformatics 21(19):3733–3740

Garrels JI (1989) The quest system for quantative analysis of two-dimensional gels. J. Biol. Chem. 264(9):5269–5282

Glasbey CA, Mardia KV (1998) A review of image-warping methods. J. Appl. Stat. 25(2):155–171

Goez MM, Torres-Madroñero MC, Röthlisberger S, Delgado-Trejos E (2018) Preprocessing of 2-dimensional gel electrophoresis images applied to proteomic analysis: a review. Genomics Proteomics Bioinf 16(1):63–72

Gonzalez RC, Woods RE, Eddins SL (2003) Digital image processing using MATLAB(R). Prentice Hall, Englewood Cliffs

Gustafsson JS, Blomberg A, Rudemo M (2002) Warping two-dimensional electrophoresis gel images to correct for geometric distortions of the spot pattern. Electrophoresis 23(11):1731–1744

Hunsucker SW, Duncan MW (2006) Is protein overlap in two-dimensional gels a serious practical problem? Proteomics 6(5):1374–1375

Jang JSR, Sun CT, Mizutani E (1997) Neuro-fuzzy and soft computing: a computational approach to learning and machine intelligence. In MATLAB curriculum series. Prentice Hall, Englewood Cliffs

Jegelevicius D, Lukosevicius A, Paunksnis A, Barzdziukas V (2002) Application of data mining. technique for diagnosis of posterior uveal melanoma. Informatica 13(4):455–464

Johansson B (2004) Low level operations and learning in computer vision. Phd. Linköping University, Linköping

Kasperavičius P (1997) Patentologija. Baltic ECO, Vilnius

Keller Y, Averbuch A (2006) Multisensor image registration via implicit similarity. IEEE Trans Pattern Anal Mach Intell 28(5):794–801

Kim YI, Cho JY (2019) Gel-based proteomics in disease research: is it still valuable? Biochim Biophys Acta (BBA)-Proteins Proteomics 1867(1):9–16

Kirvaitis R (1999) Loginės schemos. Vilnius, Enciklopedija

Kohlrausch J, Rohr K, Stiehl H (2005) A new class of elastic body splines for nonrigid registration of medical images. J Math Imaging Vision 23(3):253–280

Kostopoulou E, Zacharia E, Maroulis D (2014) An effective approach for detection and segmentation of protein spots on 2-D gel images. IEEE J Biomed Health Inf 18(1):67–76

Kostopoulou E, Katsigiannis S, Maroulis D (2015) 2d-gel spot detection and segmentation based on modified image-aware grow-cut and regional intensity information. Comput Methods Programs Biomed 122(1):26–39

Kovesi PD (2020) MATLAB and Octave functions for computer vision and image processing. http://www.peterkovesi.com/matlabfns/

Krogh M, Fernandez C, Teilum M, Bengtsson S, James P (2007) A probabilistic treatment of the missing spot problem in 2D gel electrophoresis experiments. J Proteome Res 6(8):3335–3343

Laptik R, Navakauskas D (2005) Application of artificial neural networks for the recognition of car number plates. Elektronika ir elektrotechnika 8(64):27–31. in Lithuanian
Laptik R, Navakauskas D (2007) Application of Ant Colony Optimization for image segmentation. Elektronika ir elektrotechnika 8(80):13–18
Laptik R, Navakauskas D (2009) MAX-MIN Ant System in image processing. Elektronika ir elektrotechnika 1(89):21–24
Lo SL, You T, Lin Q, Joshi SB, Chung MCM, Hew CL (2006) SPLASH: systematic proteomics laboratory analysis and storage hub. Proteomics 6(6):1758–1769
Lowe DG (2004) Distinctive image features from scale-invariant keypoints. Int J Comput Vision 60(2):91–110
Loy G, Zelinsky A (2003) Fast radial symmetry for detecting points of interest. IEEE Trans Pattern Anal Mach Intell 25(8):959–973
Luhn S, Berth M, Hecker M, Bernhardt J (2003) Using standard positions and image fusion to create proteome maps from collections of two-dimensional gel electrophoresis images. Proteomics 3(7):1117–1127
Mahnke RC, Corzett TH, Mccutchen-Maloney SL, Chromy BA (2006) An integrated proteomic workflow for two-dimensional differential gel electrophoresis and robotic spot picking. J Proteome Res 5(9):2093–2097
Marczyk M (2017) Mixture modeling of 2-d gel electrophoresis spots enhances the performance of spot detection. IEEE Trans. Nanobiosci. 16(2):91–99
Mateika D, Martavicius R (2008) Large image formation using harris-plessey corner detection algorithm. Elektronika ir elektrotechnika 5(85):21–24
Matuzevičius D (2010a) Automatinė dvimatės elektroforezės gelių su mažiausiais geometriniais iškraipymais atranka. Mokslas—Lietuvos Ateitis Elektronika ir Elektrotechnika 2(1):9–13
Matuzevičius D (2010b) Dvimatės elektroforezės gelių vaizdų analizė taikant intelektualiuosius metodus. Vilniaus Gedimino Technikos Universitetas, Daktaro Disertacija
Matuzevičius D, Navakauskas D (2005) Investigation of segmentation methods for proteomics. Elektronika ir elektrotechnika 7(63):66–70. in Lithuanian
Matuzevičius D, Navakauskas D (2008) Feature selection for segmentation of 2-D electrophoresis gel images. In: Proceedings of the 11th International Biennial Baltic Electronics Conference, BEC 2008, Tallinn, Estonia, pp 341–344
Matuzevičius D, Navakauskas D (2010) Comparison of distance measures according to suitability for 2D electrophoresis image registration using synthetic image data and SOFN. In: Romaniuk RS, Kulpa KS (eds) Proceedings of SPIE, Photonics applications in astronomy, communications, industry, and high-energy physics experiments, vol 7745. SPIE, Bellingham, p CID 7745 16
Matuzevičius D, Serackis A, Navakauskas D (2007) Mathematical models of oversaturated protein spots. Elektronika ir elektrotechnika 1(73):63–68
Matuzevicius D, Zurauskas E, Navakauskiene R, Navakauskas D (2008) Improved proteomic characterization of human myocardium and heart conduction system by using computational methods. Biologija 4:283–289
Matuzevičius D, Serackis A, Navakauskas D (2010) Application of K-Means and MLP in the automation of matching of 2DE gel images. In: Lecture notes in computer science: proceedings of the 20th international conference on artificial neural networks, ICANN 2010, vol 1. Springer, Thessaloniki, pp 541–550
Miller MD, Acey RA, Lee LYT, Edwards AJ (2001) Digital imaging considerations for gel electrophoresis analysis systems. Electrophoresis 22(5):791–800
Millioni R, Puricelli L, Sbrignadello S, Iori E, Murphy E, Tessari P (2012) Operator-and software-related post-experimental variability and source of error in 2-DE analysis. Amino Acids 42(5):1583–1590
Miura K (2001) Imaging and detection technologies for image analysis in electrophoresis. Electrophoresis 22(5):801–813
Miura K (2003) Imaging technologies for the detection of multiple stains in proteomics. Proteomics 3(7):1097–1108

Moche M, Albrecht D, Maaß S, Hecker M, Westermeier R, Büttner K (2013) The new horizon in 2D electrophoresis: New technology to increase resolution and sensitivity. Electrophoresis 34(11):1510–1518

Moritz B, Meyer HE (2003) Approaches for the quantification of protein concentration ratios. Proteomics 3(11):2208–2220

Morris JS, Clark BN, Gutstein HB (2008) Pinnacle: a fast, automatic and accurate method for detecting and quantifying protein spots in 2-dimensional gel electrophoresis data. Bioinformatics 24(4):529–536

Morris JS, Clark BN, Wei W, Gutstein HB (2009) Evaluating the performance of new approaches to spot quantification and differential expression in 2-dimensional gel electrophoresis studies. J Proteome Res 9(1):595–604

Natale M, Maresca B, Abrescia P, Bucci EM (2011) Image analysis workflow for 2-D electrophoresis gels based on ImageJ. Proteomics Insights 4:37–49

Navakauskas D (2005) Grid computing for proteomics. In: Notes of seminar at the lithuanian academy of sciences

Navakauskiene R, Treigyte G, Borutinskaite VV, Matuzevicius D, Navakauskas D, Magnusson KE (2012) Alpha-dystrobrevin and its associated proteins in human promyelocytic leukemia cells induced to apoptosis. J Proteomics 75(11):3291–3303. https://doi.org/10.1016/j.jprot.2012.03.041

Navakauskiene R, Borutinskaite VV, Treigyte G, Savickiene J, Matuzevicius D, Navakauskas D, Magnusson KE (2014) Epigenetic changes during hematopoietic cell granulocytic differentiation—comparative analysis of primary CD34+cells, KG1 myeloid cells and mature neutrophils. BMC Cell Biol 15:4. https://doi.org/10.1186/1471-2121-15-4

Nhek S, Tessema B, Indahl UG, Martens H, Mosleth EF (2015) 2d electrophoresis image segmentation within a pixel-based framework. Chemom Intell Lab Syst 141:33–46

Nickelson L, Asmontas S, Martavicius R, Engelson V (2006) Microwave pulse propagation inside a 3D heart model. Elektronika ir Elektrotechnika 8(72):5–10

Oliveira BM, Coorssen JR, Martins-de Souza D (2014) 2DE: The phoenix of proteomics. J Proteomics 104:140–150

Penney GP, Weese J, Little JA, Desmedt P, Hill DLG, Hawkes DJ (1998) A comparison of similarity measures for use in 2-D-3-D medical image registration. IEEE Trans Med Imaging 17(4):586–595

Pivoriūnas A, Surovas A, Borutinskaitė V, Matuzevičius D, Treigytė G, Savickienė J, Tunaitis V, Aldonytė R, Jarmalavičiūtė A, Suriakaitė K, Liutkevičius E, Venalis A, Navakauskas D, Navakauskienė R, Magnusson KE (2010) Proteomic analysis of stromal cells derived from the dental pulp of human exfoliated deciduous teeth. Stem Cells Dev 19(7):1081–1093

Pleisner KP, Hoffmann F, Kriegel K, Wenk C, Wegner S, Sahlstrom A, Oswald H, Alt H, Fleck E (1999) New algorithmic approaches to protein spot detection and pattern matching in two-dimensional electrophoresis gel databases. Electrophoresis 20(4–5):755–765

Pomastowski P, Buszewski B (2014) Two-dimensional gel electrophoresis in the light of new developments. TrAC, Trends Anal Chem 53:167–177

Quadroni M, James P (1999) Proteomics and automation. Electrophoresis 20(4–5):664–677

Rabilloud T, Chevallet M, Luche S, Lelong C (2010) Two-dimensional gel electrophoresis in proteomics: past, present and future. J Proteomics 73(11):2064–2077

Rashwan S, Sarhan A, Faheem MT, Youssef BA (2015) Fuzzy watershed segmentation algorithm: an enhanced algorithm for 2d gel electrophoresis image segmentation. Int J Data Min Bioinf 12(3):275–293

Rodriguez A, Fernandez-Lozano C, Dorado J, Rabuñal JR (2014) Two-dimensional gel electrophoresis image registration using block-matching techniques and deformation models. Anal Biochem 454:53–59

Rogers M, Graham J, Tonge RP (2003) Statistical models of shape for the analysis of protein spots in two-dimensional electrophoresis gel images. Proteomics 3(6):887–896

Savelonas MA, Mylona EA, Maroulis D (2012) Unsupervised 2d gel electrophoresis image segmentation based on active contours. Pattern Recognit 45(2):720–731

Sengar RS, Upadhyay AK, Singh M, Gadre VM (2016) Analysis of 2d-gel images for detection of protein spots using a novel non-separable wavelet based method. Biomed Signal Process Control 25:62–75

Serackis A (2008) Vaizdo rekonstravimo technologijos baltymų pėdsakams parametrizuoti. daktaro disertacija. Vilniaus Gedimino Technikos Universitetas, Vilniaus

Serackis A, Navakauskas D (2008) Reconstruction of overlapped protein spots using RBF networks. Elektronika ir Elektrotechnika 1(81):83–88

Serackis A, Matuzevičius D, Navakauskas D (2006) Reconstruction of protein spots using DSP modules. In: Proceedings of 29th international conference on fundamentals of electrotechnics and circuit theory, IC-SPETO 2006, vol 2. Gliwice-Ustron, Poland, pp 573–576

Serackis A, Matuzevičius D, Navakauskas D (2010) 2DE gel image preprocessing using self-organizing maps. In: Romaniuk RS, Kulpa KS (eds) Proceedings of SPIE Photonics applications in astronomy, communications, industry, and high-energy physics experiments. SPIE, Washington, vol 7745, p CID 7745 1N

Simutis R (1997) Exploratory analysis of biochemical processes using hybrid modeling methods. In: Komorowski J, Zytkow J (eds) Principles of data mining and knowledge discovery. Lecture notes in artificial intelligence, vol 1263. Springer, Berlin, pp 200–210

Smilansky Z (2001) Automatic registration for images of two-dimensional protein gels. Electrophoresis 22(9):1616–1626

Sonka M, Hlavac V, Boyle R (2007) Image processing, analysis, and machine vision. Thomson-Engineering, London

Thornbury JR (1994) Clinical efficacy of diagnostic-imaging—love it or leave it. Am J Roentgenol 162(1):1–8

Treigys P, Saltenis V, Dzemyda G, Barzdziukas V, Paunksnis A (2008) Automated optic nerve disc parameterization. Informatica 19(3):403–420

Treigytė G, Zaikova I, Matuzevičius D, Čeksterytė V, Dabkevičienė G, Kurtinaitienė B, Navakauskienė R (2014) Comparative proteomic analysis of pollen of trifolium pratense, t. alexandrinum and t. repens. Zemdirbyste-Agriculture 101(4):453–460

Valledor L, Jorrín J (2011) Back to the basics: maximizing the information obtained by quantitative two dimensional gel electrophoresis analyses by an appropriate experimental design and statistical analyses. J Proteomics 74(1):1–18

Vedaldi A, Fulkerson B (2008) VLFeat: an open and portable library of computer vision algorithms. http://www.vlfeat.org/

Veeser S, Dunn MJ, Yang GZ (2001) Multiresolution image registration for two-dimensional gel electrophoresis. Proteomics 1(7):856–870

Vijayendran C, Burgerneister S, Friehs K, Niehaus K, Flaschel E (2007) 2DBase: 2D-PAGE database of escherichia coli. Biochem Biophys Res Commun 363:822–827

Vlahou A (2008) Clinical proteomics: methods and protocols, vol 428. Springer, Berlin

Wiener N (1965) Cybernetics: or the control and communication in the animal and the machine, 2nd edn. MIT Press, Cambridge

Wu Y, Zhang L (2011) Comparison of two academic software packages for analyzing two-dimensional gel images. J Bioinf Comput Biol 9(06):775–794

Zagorchev L, Goshtasby A (2006) A comparative study of transformation functions for nonrigid image registration. IEEE Trans Image Process 15(3):529–538

Zech H, Echtermeyer C, Wöhlbrand L, Blasius B, Rabus R (2011) Biological versus technical variability in 2-D DIGE experiments with environmental bacteria. Proteomics 11(16):3380–3389

Zhang B, VerBerkmoes NC, Langston MA, Uberbacher E, Hettich RL, Samatova NF (2006) Detecting differential and correlated protein expression in label-free shotgun proteomics. J Proteome Res 5(11):2909–2918. https://doi.org/10.1021/pr0600273

Zhang L, Wen Q, Mao HP, Luo N, Rong R, Fan JJ, Yu XQ (2013) Developing a reproducible method for the high-resolution separation of peritoneal dialysate proteins on 2-D gels. Protein Expression Purif 89(2):196–202

Part III
Protein Localization in Leukemia

Chapter 7
Protein Translocation into the Nucleus during Leukemia Cell Differentiation

Various vital cellular processes involve the translocation of proteins between various compartments of the cell, including between the cytoplasm and the nucleus. Certain short amino acid sequences in a protein, classed as signal sequences or signal peptides, can direct its localization, even if translocation also happens when these signal sequences do not exist. Proteins translocation into the nucleus could proceed actively and by diffusion (usually it is characteristic for the proteins with molecular weight less than 40 kDa). Movements between the cytoplasm and nucleus occur through nuclear pore complexes (NPCs) embedded in the nuclear membrane. These macromolecular structures consist of several nucleoporins. The complex of nucleoporins allow both passive transport and as well interacting with importin, exportin, and other molecules that are necessary in order to transport the protein in different cellular processes. Transport of different proteins is conducted either dependently or independently on transport receptors. As well as assisting to the nucleocytoplasmic transport, nucleoporins also play a substantive role in cell differentiation, potentially by their direct gene interaction (Kabachinski and Schwartz 2015; Khan et al. 2020).

Protein translocation into the nucleus can occur during or after protein translation. Many transcription factors enter the nucleus after modification in the cytoplasm or undergo modification after the translocation. We have developed the method for the assessment of protein translocation from the cytosol into the nucleus (Kulyte et al. 2001), identified transcription factors translocated into the nucleus during granulocytic differentiation of leukemic cells, demonstrated that some of identified transcription factors undergo tyrosine phosphorylation in leukemic cells exposed to all-trans retinoic acid (Navakauskiene et al. 2003) and even dystrobrevin-γ (a splice isoform of dystrobrevin-α) was altered by tyrosine phosphorylation soon after ATRA induced granulocytic differentiation of acute promyelocytic leukemia cells (Kulyte et al. 2002).

R. Navakauskienė et al., *Epigenetics and Proteomics of Leukemia*,
https://doi.org/10.1007/978-3-030-68708-3_7

7.1 Method for Protein Translocation Assay from Cytosol into the Nucleus

Here we introduce the methods for the assessment of protein translocation into the nucleus and tyrosine phosphorylation (Kulyte et al. 2001) produced with permission from Biotechniques as agreed by Future Science Ltd.

Phosphotyrosine signaling is essential in cell regulation—from cell receptor activation, through stimulation of signal networks and nuclear targeting, to concluding cellular responses. Clearly, it would be of great value to know when and where proteins are tyrosine phosphorylated or dephosphorylated. We developed a novel, nonradioactive approach to elucidate the temporal and spatial characteristics of protein tyrosine phosphorylation and nuclear targeting. Furthermore, we demonstrated the feasibility of applying the technique to analyze phosphotyrosine signaling and transport in proliferating human promyelocytic leukemia (HL-60) cells. The method can be applied to identify whether protein tyrosine phosphorylations and dephosphorylations happen in the cytosol or the nucleus and to determine whether such modifications are related with nuclear traffic. Acute promyelocytic leukemia (HL-60) cells were used as the experimental model. Biotinylated cytosolic proteins obtained from donor cells were used to locate nuclear transport in permeabilized cells of recipient. Later, 2D gel electrophoresis was applied to fractionate the nuclear and cytosolic proteins obtained from the recipient cells that were afterwards blotted onto polyvinylidene difluoride membranes. The membranes were developed with streptavidin and then reprobed with anti-phosphotyrosine antibodies. The main benefit of the protocol is that it is simple to perform, and reproducible results are gained by overlaying the patterns of biotinylated and/or tyrosine-phosphorylated proteins. Besides, several hundred nuclear and cytosolic proteins can be analyzed simultaneously. Thus, the comparison of 2D gel electrophoresis maps of biotinylated and tyrosine-phosphorylated proteins allows evaluate the involvement of trafficking of the latter proteins in cell signaling. Below we are presenting comprehensive methodological description to monitor the nucleocytoplasmic translocation of proteins and their modifications—phosphorylation and dephosphorylation.

The cytosolic fraction of proliferating HL-60 cells was prepared principally as described by Adam et al. (1990). First, proteins in the fraction were biotinylated by exposure to 5 mM N-hydroxysuccinimide biotin for 30 min at 4 °C. Later, unbound biotin was removed by subjecting the cytosolic fraction to extensive dialysis against nuclear import buffer (20 mM HEPES, pH 7.3, 110 mM potassium acetate, 2 mM magnesium acetate, 5 mM sodium acetate, 5 mM sodium fluoride, 1 mM EGTA, 2 mM DTT, 1 mM orthovanadate, and 1 mg/mL each aprotinin, pepstatin, and leupeptin).

The digitonin permeabilization of HL-60 cells and assaying nuclear import was performed as followed. The cells were harvested by low speed centrifugation at 1500 × g and then washed twice with ice-cold PBS (pH 7.3) and once with import buffer. The cell pellet was then gently resuspended to 5 × 105 cells/mL in ice-cold nuclear import buffer comprising digitonin which concentration was

30 mg/mL (diluted from a 20 mg/mL stock solution in dimethyl sulfoxide). The cells were afterwards placed on ice for 5 min, and the reaction mixture was gently inverted once a minute. After incubation, the cells were centrifuged at 1500 × g for 7 min at 4 °C and then washed at at a minimum of three times with ice-cold import buffer in order to remove digitonin and prevent nuclear membranes permeabilization. Freshly digitonin permeabilized HL-60 cells were applied in the nuclear import experiments. The permeabilized cells were mixed with cytosolic fraction and complete nuclear import buffer comprising an energy-regenerating system (1 mM ATP, 5 mM creatine phosphate, and 20 U/mL creatine phosphokinase). Nuclear import was allowed to proceed for 30 min at 30 °C while being gently mixed by hand. The integrity of the nuclear membrane and protein transport function was determined by optical sectioning with a confocal laser-scanning microscope (Sarastro 2000; Molecular Dynamics, Sunnyvale, CA, USA). Probes—the polyoma virus structural protein, VP1, which carries a nuclear localization sequence (NLS), and BSA, which lacks NLS, were labeled with fluorescein isothiocyanate (FITC). Cytosolic proteins that were biotinylated and subsequently marked with fluorescent streptavidin (streptavidin-ALEXA) constituted the third type of probe. When using the fluoresceinated proteins VP1 and BSA, nuclei were counterstained with propidium iodide. In experiments with the streptavidin probed proteins, DNA was marked with SYTOX®-Green nucleic acid stain. Isolated nuclei were microscopically pure, and imported proteins were found in the nuclei but not attached to it. We also confirmed that the import of cytosolic proteins occurred in the presence, but not in the absence, of ATP.

Subsequently the isolation and analysis of nuclear and cytosolic proteins was performed. Cytosol and nuclei were isolated as specified by Antalis and Godbolt (1991), with several modifications. In summary, cells were collected, washed twice in PBS (pH 7.5), resuspended to 3×10^7 cells/mL in solution A (10 mM NaCl, 10 mM Tris-HCl, pH 7.5, 3 mM $MgCl_2$, 0.05% Nonidet P-40, 5 mM sodium fluoride, 1 mM PMSF, 1 mg/mL aprotinin, pepstatin, and leupeptin, and 1 mM sodium orthovanadate), and then allowed to swell for 15 min at 0 °C. Afterwards, the cell suspension was shaken vigorously by hand and right away mixed in a 1:1 ratio (v/v) with solution B (i.e., solution A supplemented with 0.6 M sucrose B). The cell homogenates were then centrifuged at 1500 × g for 5 min. The supernatant, corresponding to the cytosolic fraction, was clarified by centrifuging at 1500 × g for 15 min and then analyzed or frozen at −76 °C. Pelleted nuclei were washed twice by centrifuging at 1500 × g for 5 min, first with a 1 : 1 mixture of solutions A and B and then three times with solution C (i.e., a 1 : 1 (v/v) mixture of solutions A and B without Nonidet P-40). The nuclei were afterwards analyzed for purity and integrity under a light microscope and then presented for further analysis or frozen at −76 °C. Total nuclear proteins were prepared by mixing approximately 5×10^7 nuclei/mL in lysis solution (8 M urea, 4% CHAPS, 1% DTT, 0.8% 2-D pharmalytes (pH 3–10), and 1 mM PMSF) and then centrifuging at 1500 × g for 30 min 2D gel electrophoresis, including isoelectric focusing and gel electrophoresis (SDS-PAGE), was performed. The proteins were subsequently blotted on Immobilon polyvinylidene difluoride (PVDF) membrane and probed

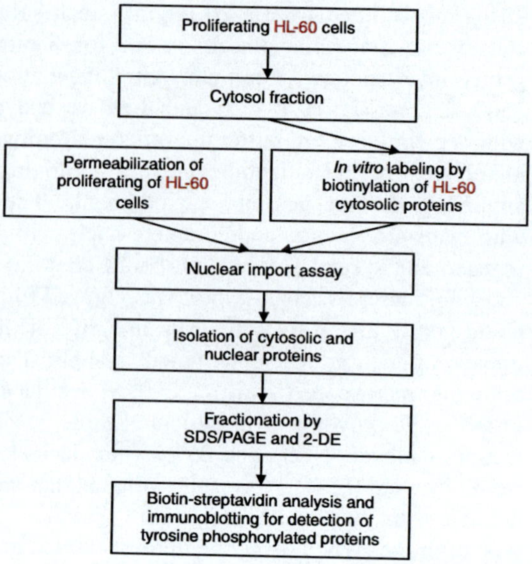

Fig. 7.1 The steps of methodology representing a powerful toll for studying protein translocation into the nucleus

with antiphosphotyrosine antibody or streptavidin-horseradish peroxidase (HRP). Enhanced chemiluminescence (ECL) was used to detect immunoreactive bands. Figure 7.1 represents the general scheme of described methodology.

By using the experimental strategy described in Fig. 7.1 we examined the location of proteins showing different degrees of tyrosine phosphorylation in proliferating HL-60 cells. More than 400 cytosolic and 300 nuclear peptides were detected by silver staining; both the electrophoretic patterns and the relative amounts of the proteins were highly reproducible (data not shown).

Figure 7.2 illustrates the characteristic tyrosine phosphorylation of representative segments (30–100 kDa and pI 4–8.5) of the total number of cytoplasmic and nuclear protein 2D gel electrophoresis maps. To determine whether a certain tyrosine phosphorylated cytosolic or nuclear protein is biotinylated, two ECL films obtained from the same PVDF membrane were probed with anti-phosphotyrosine antibodies and streptavidin, respectively, and then merged. In this way, five different cases of protein modifications could be detected: (i) biotinylation and tyrosine phosphorylation in both the cytoplasm and the nucleus (arrows in Fig. 7.2); (ii) biotinylation, but not tyrosine phosphorylation in the cytoplasm, and both biotinylation and tyrosine phosphorylation in the nucleus (circles in Fig. 7.2); (iii) biotinylation and tyrosine phosphorylation in the cytoplasm, but only a biotinylation signal in the nucleus (triangles in Fig. 7.2); (iv) no biotinylation at all, and tyrosine phosphorylation only in the nuclei of recipient cells in which nuclear import had been allowed to proceed (squares in Fig. 7.2); and (v) biotinylation and tyrosine phosphorylation only in the cytosol (diamonds in Fig. 7.2). The conclusions we drew from these observations are summarized in Table 7.1, and the patterns of protein tyrosine phosphorylation revealed by 2D gel electrophoresis are shown in Fig. 7.2.

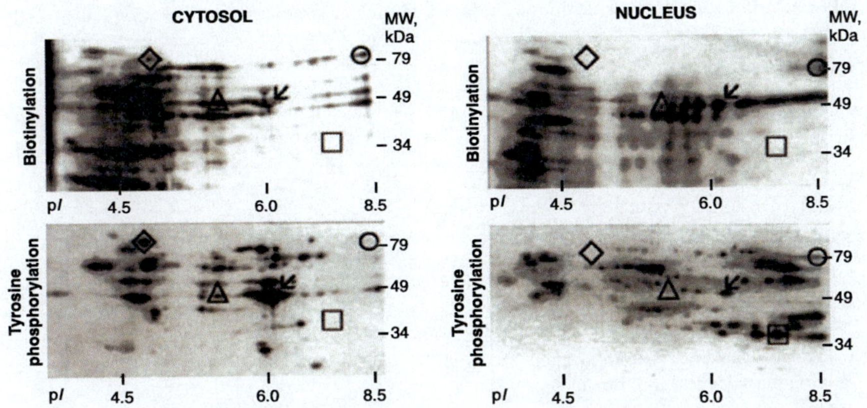

Fig. 7.2 2D gel electrophoresis (2DE) patterns of biotinylated and tyrosine-phosphorylated cytosolic (CYTOSOL) and nuclear (NUCLEUS) proteins isolated from proliferating recipient HL-60 cells after nuclear import. Digitonin-permeabilized recipient HL-60 cells devoid of their own cytoplasm were incubated with donor cytoplasm containing biotinylated cytosolic proteins. After allowing nuclear import, cytosolic and nuclear proteins were isolated, fractionated by 2DE, and analyzed for streptavidin-HRP (biotinylation) or reactivity to anti-phosphotyrosine antibodies (tyrosine phosphorylation). Symbols: arrows, proteins tyrosine phosphorylated in the cytosol and then translocated into the nucleus; circles, proteins translocated into the nucleus and then tyrosine phosphorylated; triangles, proteins that were tyrosine phosphorylated in the cytoplasm and then dephosphorylated before or after translocation into the nucleus; squares, tyrosine-phosphorylated proteins residing permanently in the nucleus; diamonds, tyrosine-phosphorylated proteins residing permanently in the cytoplasm. According Kulyte et al. (2001), reproduced with permission from Biotechniques as agreed by Future Science Ltd.

Concerning the spots denoted by triangles in Fig. 7.2, there are other even more weakly biotinylated protein spots with detectable tyrosine phosphorylation in the nucleus fraction. Moreover, when the measurement time was extended 3- to 5-fold, still no signal was detected. We are therefore convinced that the absence of tyrosine phosphorylation is due to dephosphorylation and not to an effect of insufficient sensitivity of detection. In the fourth case, the proteins apparently resided permanently in the nucleus because they were not found to be (but theoretically may well have been) biotinylated in the cytoplasmic fraction. An explanation for this might be that newly translated nuclear proteins are rapidly translocated from the cytoplasm, and, therefore, the relative amount of nascent nuclear proteins in the cytoplasmic fraction at the moment of extraction is very small. Moreover, it is possible that proteins from disassembled nuclei of mitotic cells contaminate the cytoplasmic fraction. However, such nuclear proteins should not interfere significantly with our method because less than 5% of the cells in a proliferating, asynchronous HL-60 population are mitotic (Savickiene et al. 1999). In summary, our method represents a powerful new tool for studying temporal and spatial aspects of protein tyrosine phosphorylation and translocation of the proteins to the nucleus in different types of cells and physiological states.

Table 7.1 Location of tyrosine phosphorylation of nuclear proteins in proliferating HL-60 cells

Cell Compartment of Protein Tyrosine Phosphorylation*	Symbol	Cytosol		Nucleus	
		Biotinylation	pY	Biotinylation	pY
1. Proteins tyrosine phosphorylated in the cytosol and subsequently translocated to the nucleus	↙	+	+	+	+
2. Proteins translocated into the nucleus and then tyrosine phosphorylated	○	+	−	+	+
3. Proteins tyrosine phosphorylated in the cytoplasm and then dephosphorylated before or after translocation into the nucleus	△	+	+	+	−
4. Tyrosine-phosphorylated proteins residing permanently in the nucleus	□	−	−	−	+
5. Tyrosine-phosphorylated proteins residing permanently in the cytoplasm	◊	+	+	−	−

According Kulyte et al. (2001), reproduced with permission from Biotechniques as agreed by Future Science Ltd.

*– Cytosolic proteins from donor cells were biotinylated and assessed after nuclear import into penneabilized recipient ceels. The nuclear import was allowed to proceed for 30 min, and the cytosolic and nuclear proteins in the recipient cells were subsequently fractionated by 2-D gel electrophoresis. The proteins were then transferred to PVDF membranes and probed with streptavidin (Biotinylation) and anti-phosphotyrosine (pY) antibodies, respectively.

7.2 Translocation of Transcription Regulators into the Nucleus

Signal transduction from the cell surface to the nucleus in response to extracellular stimuli controlled by a group of integrated signaling events, involving initiation of protein kinase cascades (Hunter 2000). There are two ways to transmit protein phosphorylation signals to the nuclear genome: activated protein kinases may be transfered from the cytoplasm into the nucleus, where they phosphorylate target transcription factors; and/or transcription factors may be kept in a latent state in the cytoplasm and enter into the nucleus after phosphorylation by cytoplasmic protein kinases (Bagnato et al. 2020). The first kind of signaling prevails, because most of transcription factors are nuclear proteins that reside in the cytoplasm only once while being synthesized. For example, upon activation, the MAP kinases ERK1 and ERK2 are transferred from the cytoplasm into the nucleus (Maik-Rachline et al. 2019), where they phosphorylate several substrates. Also, when the haematopoietic cells are treated with agents stimulating differentiation or apoptosis, certain isoforms of protein kinase C (PKC) as a result move into the nucleus (Poli et al. 2018).

Transferring of extracellular signals into the nucleus is well portrayed through NFκB, which is retained in the cytoplasm of unstimulated cells by interaction with the IκB inhibitor, and may be transferred into the nucleus only after phosphorylation (Chen and Greene 2004; Huang and Hung 2013). Likewise, a wide range of growth factors and cytokines elicit signaling cascades that stimulate rapid relocalization of STATs to the nucleus (Reich and Liu 2006). It seems that certain phosphorylations of transcription factors are involved in all systems responsible for cell differentiation and proliferation regulation (Hunter and Karin 1992; Karin and Hunter 1995), hence it is crucial to determine tyrosine phosphorylated proteins that take a role in the signaling pathways of differentiation and evaluate how they are related to proteins expressed by ATRA activated genes. We were looking for alernative tyrosine phosphorylated, ATRA induced transcription regulators that are transferred into the nucleus during commitment of HL-60 cells to granulocytic differentiation.

Using the approach described above (Kulyte et al. 2001) we examined the tyrosine phosphorylation and cytoplasm-to-nucleus transference of proteins in proliferating and differentiating HL-60 cell. We demonstrated that the proteins transferred into the nucleus had been phosphorylated either in the cytoplasm (before transference) or in the nucleus itself (principally in the nucleoplasm or when weakly bound to chromatin).

In order to conduct nuclear import assay permeabilized proliferating HL-60 cells were incubated with biotinylated cytoplasmic proteins that had been isolated from HL-60 cells which were proliferating or had differentiated for 0.5, 1, 6, 18, 24, 48, 72 h (data not shown) and 96 h (Fig. 7.3). Biotinylated cytoplasmic proteins ("total cytoplasm" in Fig. 7.3) of proliferating and differentiating HL-60 cells (respectively designated a and b in Fig. 7.3) were fractionated by SDS gel electrophoresis, evaluated for biotinylation and tyrosine phosphorylation, and afterwards applied for nuclear import assay. The qualitative electrophoretic patterns of these proteins were fairly similar in spite of differentiation stage, therefore it may be argued that the patterns created by nuclear-translocated proteins in the lanes designated a and b in Fig. 7.3 represent proliferating control cells and cells differentiated for 96 h.

The fractions of non-histone chromosomal, nucleoplasmic, and residual nuclear proteins (respectively designated "chromatin", "nucleoplasm", and "residual nucleus" in Fig. 7.3) were isolated from the nuclei of permeabilized proliferating HL-60 cells after the nuclear import assay and were afterwards subjected to SDS gel electrophoresis. From that point the translocation of proteins (biotinylated ones) into the nucleus and tyrosine phosphorylation were evaluated. Greatest difference can be noticed between proteins from proliferating and differentiated cells (a and b lanes in Fig. 7.3) targeted to the same nuclear protein fraction and between the presence of the same type of proteins in the individual nuclear fractions ("chromatin", "nucleoplasm", and "residual proteins" in Fig. 7.3). Several of long-term tyrosine phosphorylated proteins were related with the nuclear matrix, demonstrating that these proteins are responsible for formation and maintenance of the neutrophil phenotype. The results of SDS electrophoresis motivated us to analyze proteins that are increasingly transfered into the nucleus during progressing differentiation to evaluate whether several of them are normally responsible for terminating the cell cycle and for ensuring survival of the cell. In order to identify the nuclear translocated and tyrosine phosphorylated proteins in proliferating and differentiating HL-60 cells, the nuclear proteins of the experimental cells were

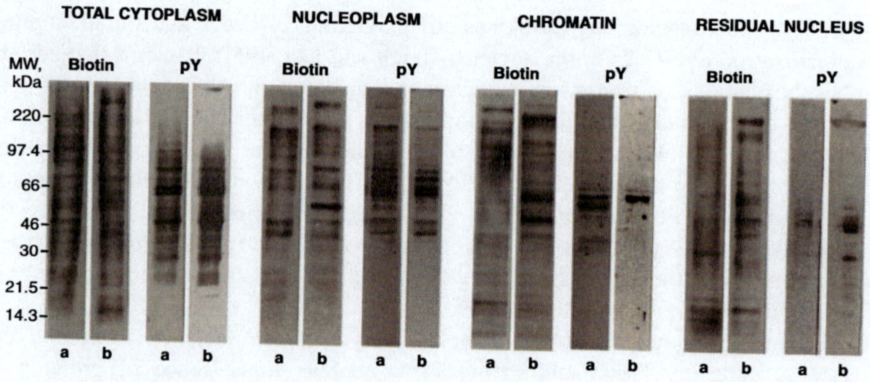

Fig. 7.3 Electrophoretic patterns and tyrosine phosphorylation of proteins initially isolated from the cytoplasm of proliferating or differentiating HL-60 cells and then allowed to transfer into the nucleus of other proliferating HL-60 cells. The isolated proteins originated from proliferating (a) and differentiating (b) HL-60 cells. In the experiments, we evaluated HL-60 cells for biotinylated cytoplasmic proteins (TOTAL CYTOPLASM), and for nuclear proteins that were obtained (after the nuclear import assay) from isolated nuclei with 0.15 M NaCl (NUCLEOPLASM), 0.35 M NaCl (CHROMATIN), or 2% SDS, 0.05 mM Tris (pH 6.8), and 50 mM DTT (RESIDUAL NUCLEUS). Migration of the molecular size-marker proteins is presented to the left (kDa values). Cytosolic proteins were isolated from proliferating and differentiating HL-60 cells and afterwards biotinylated. The nuclear import assay was conducted by incubating permeabilized control (proliferating) HL-60 cells with the biotinylated cytosolic proteins for 30 min at 30 °C. Later, the nuclei were isolated, and proteins were obtained with 0.15 M, or 0.35 M NaCl, or with 2% SDS, 0.05 mM Tris (pH 6.8), and 50 mM DTT and then fractionated by SDS-PAGE on a 7–15% acrylamide gel gradient and translocated onto Immobilon™ PVDF membrane. The membrane was analyzed for protein biotinylation with Streptavidin-HRP (Biotin), reprobed with an anti-phosphotyrosine antibody (IgG2bk), and afterwards developed with an enhanced chemiluminescence detection system. According Navakauskiene et al. (2003), License No 1025075-1

isolated and separated by 2DE, stained by silver or they were translocated onto PVDF membranes and detected either by the anti-phosphotyrosine antibodies, or by antibodies against transcription factors evaluated by us. The nuclear proteins that are characteristic of proliferating cells are presented in Fig. 7.4A, and proteins that accumulated in the nuclei after 18 h of differentiation are presented in Fig. 7.4B. 2DE patterns of nuclear proteins (Fig. 7.4) allow to argue that most of proteins in the nucleus of control and differentiating cells vary according to their amount to a lesser extent than according to tyrosine phosphorylation.

Tyrosine phosphorylation of nuclear proteins in the cells varied within 18 h of inducing differentiation, in comparison to what we noticed in proliferating HL-60 cells (Fig. 7.4; PY). Additionally for determination of changed nuclear translocation and tyrosine phosphorylation of separate proteins, we applied monoclonal antibodies against C/EBPbeta, c-Myb, STAT3, STAT5a, STAT5b, NFκB p50, and NFκB p65 to disclose the positions of these proteins in the 2DE membranes (Fig. 7.4, Silver). After comparison of the 2DE membranes with the aforementioned antibodies as well as anti-phosphotyrosine antibodies, we could identify

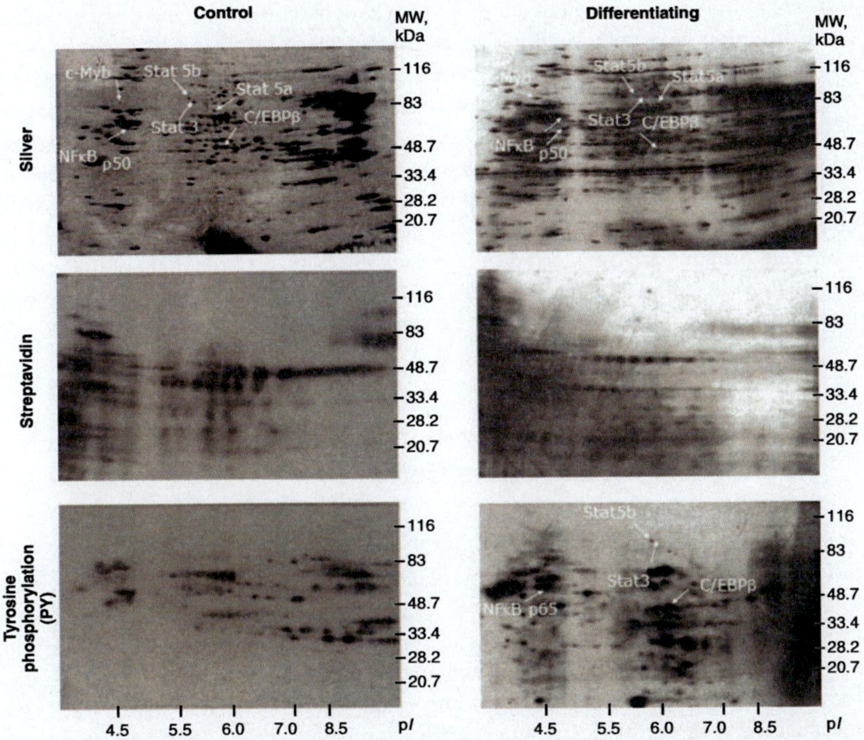

Fig. 7.4 Identification of proteins translocated into the nuclei of proliferating and differentiating HL-60 cells. Nuclear proteins were isolated followed by two-dimensional gel electrophoresis (2-DE) protein fractionation and transfer onto an Immobilon™ PVDF membrane, and were subsequently analyzed with an anti-phosphotyrosine antibody (IgG2bk, PY) and antibodies to C/EBPbeta, c-Myb, STAT3, STAT 5a, STAT 5b, NFκB p50, and NFκB p65. An enhanced chemiluminescence detection system was applied to visualize the protein spots. Migration of molecular size standards is shown by the kDa values to the left. According Navakauskiene et al. (2003), License No 1025075-1

which of the analyzed transcription factors were tyrosine-phosphorylated. We found that none of the examined transcription factors were tyrosine-phosphorylated in proliferating (control) HL-60 cells (Fig. 7.4A, PY). On the other hand, after differentiation was stimulated, STAT3, STAT5b, C/EBPbeta and NFκB p65 were tyrosine-phosphorylated. The tyrosine phosphorylation and nuclear translocation of STATs, C/EBPbeta, and NFκB after the onset of differentiation allow to argue that these proteins play a role in the mechanisms of granulocytic lineage commitment, despite the fact that they act through various intracellular signaling pathways.

Expression of transcription factors necessary for lineage commitment of differentiating cells (C/EBPbeta and c-myb) and for survival of differentiated cells (STATs and NFκB) was analyzed in the HL-60 cell line. Differentiatsion was stimulated by treating the cells with all-trans retinoic acid. c-Myb expression in the nucleus restored at the precommitment stage (18 h) what concurred with the highest nuclear level of C/EBPbeta, which allows to argue that these transcription factors have a combinatorial interaction in granulocytic signalling pathway. Expression of STAT5a and STAT5b changed during differentiation, although no important variations were seen in STAT3 levels. Increased cytosolic level of NFκB p65 during pre-commitment and commitment stages of granulocytic differentiation corresponded with augmentation of the STAT5a protein level what could prove their possible cooperation during granulocytic-lineage commitment of HL-60 cells. Our results show that proteins are transferred into the nuclei of HL-60 cells that are undergoing differentiation induced by treatment with all-trans retinoic acid, and that this nuclear import may play a role in the signal transduction that is required for differentiation.

7.3 Dystrobrevin-γ Translocation into the Nucleus of Acute Promyelocytic Leukemia Cells

We first demonstrated that human dystrobrevin-γ is expressed in cells of the promyelocytic HL-60 line, and biochemically it exists in two cell compartments—the cytosol and the nucleus (Kulyte et al. 2002). To analyze variations in the nuclear protein expression pattern and specific tyrosine phosphorylation in proliferating HL-60 cells and HL-60 cells that had undergone ATRA induced differentiation for 30 min, total nuclear proteins were fractionated by 2DE, silver stained or investigated with antibodies against tyrosine phosphorylation. Taking into account the tyrosine phosphorylation pattern of the transferred proteins (see Chap. 5, Fig. 5.6), we determined the largest variation for proteins with an acidic pI. The total amount of proteins did not change substantially after the brief ATRA stimulated differentiation. Despite this, we noticed a difference between the differentiating (ATRA 30 min) and the proliferating HL-60 cells in regard to tyrosine phosphorylated proteins detected in the nuclei. Several of these proteins appeared in the nucleus early in differentiation (30 min ATRA treatment) and remained tyrosine-phosphorylated during the precommitment stage of differentiation (up to 24 h). Two groups of tyrosine-phosphorylated proteins could be determined: those that were tyrosine-phosphorylated in the nuclei of both proliferating and ATRA differentiated cells; those that were not tyrosine-phosphorylated in the nuclei of proliferating cells but were phosphorylated in the nuclei of cells exposed to ATRA for 30 min. By applying the mass spectrometry analysis we found out that dystrobrevin-γ undergoes tyrosine phosphorylation soon (30 min) after the onset of ATRA induced differentiation and transfers into the nucleus (Kulyte et al. 2002). These results further support a signaling role for the peripheral DAPC complex (dystrophin-associated protein complex) (Fig. 7.5).

Little is known about the dystrobrevin-γ role in nuclear dystrophin complex, and it remains unclear whether it has structural and/or signaling properties. Based on the characteristics of the tyrosine phosphorylation and distribution of dystrobrevin

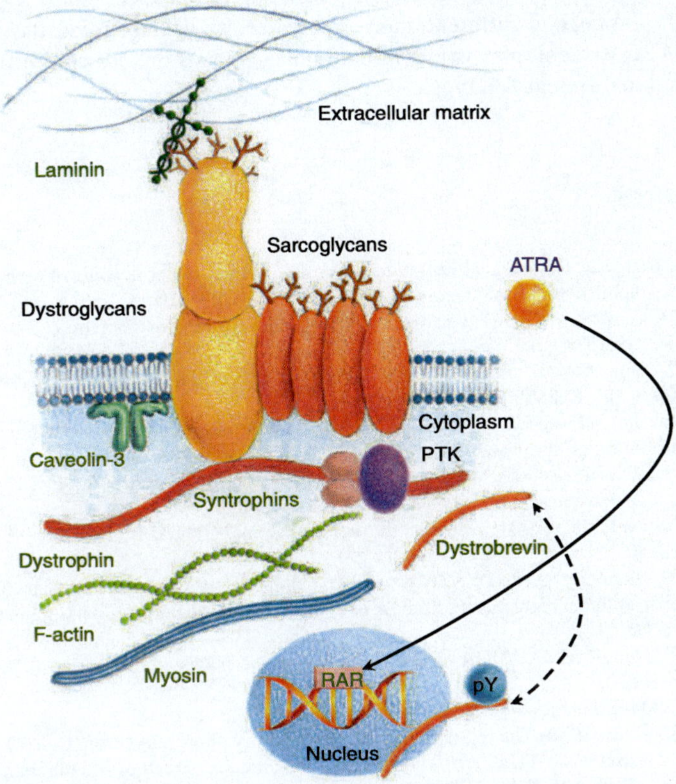

Fig. 7.5 The proposed model showing possible involvement of dystrobrevin-γ in dystroglycan-sarcoglycan complex

isoforms it was shown (Balasubramanian et al. 1998) that phosphotyrosine containing tail isoforms of dystrobrevin are the specialized components of the dystroglycan complex which render the complex sensitive to regulation by tyrosine kinases. It was examined by González et al. (2000) the subcellular distribution of the dystrophin Dp1 various splicing isoforms and conclusions were made that it identifies the nuclear or cytoplasmic distribution of Dp1. According to Blake et al. (1999) there Dp1 and dystrobrevin are related, but Dp1 is not able to act as a functional alternative for full-length dystrophin. Quite recently, Martínez-Vieyra et al. (2018) examined cellular distribution of dystrobrevin-α and its splice isoforms in human promyelocytic leukemia HL-60 cells and in the HL-60 cells induced to neutrophil-like cells with DMSO. Their results suggested that dystrobrevin-α isoform 1, which is 94 kDa, expression diminished when the cells reach their mature forms. The research group also demonstrated that dystrobrevin-α is crucial to lobulate and shape the nuclei of neutrophils induced from HL-60 cells. Their results have shown a significant relationship between the nucleoskeleton and cytoskeleton, which is disrupted by dystrobrevin-α depletion, translating into nuclear-shape abnormalities during the process of differentiation. These results demonstrated the probability that a DAPC-like complex exists in the nucleus but its precise role still should be discovered in different cell types.

References

Adam SA, Marr RS, Gerace L (1990) Nuclear protein import in permeabilized mammalian cells requires soluble cytoplasmic factors. J Cell Biol 111(3):807–816

Antalis TM, Godbolt D (1991) Isolation of intact nuclei from hematopoietic cell types. Nucleic Acids Res 19(15):4301

Bagnato G, Leopizzi M, Urciuoli E, Peruzzi B (2020) Nuclear functions of the tyrosine kinase Src. Int J Mol Sci 21(8):2675

Balasubramanian S, Fung ET, Huganir RL (1998) Characterization of the tyrosine phosphorylation and distribution of dystrobrevin isoforms. FEBS Lett 432(3):133–140

Blake DJ, Hawkes R, Benson MA, Beesley PW (1999) Different dystrophin-like complexes are expressed in neurons and glia. J Cell Biol 147(3):645–658

Chen LF, Greene WC (2004) Shaping the nuclear action of NF-κB. Nat Rev Mol Cell Biol 5(5):392–401

González E, Montanez C, Ray PN, Howard PL, Garcia-Sierra F, Mornet D, Cisneros B (2000) Alternative splicing regulates the nuclear or cytoplasmic localization of dystrophin dp71. FEBS Lett 482(3):209–214

Huang WC, Hung MC (2013) Beyond NF-κB activation: nuclear functions of IκB kinase α. J Biomed Sci 20(1):3

Hunter T (2000) Signaling-2000 and beyond. Cell 100(1):113–127

Hunter T, Karin M (1992) The regulation of transcription by phosphorylation. Cell 70(3):375–387

Kabachinski G, Schwartz TU (2015) The nuclear pore complex–structure and function at a glance. J Cell Sci 128(3):423–429

Karin M, Hunter T (1995) Transcriptional control by protein phosphorylation: signal transmission from the cell surface to the nucleus. Curr Biol 5(7):747–757

Khan AU, Qu R, Ouyang J, Dai J (2020) Role of nucleoporins and transport receptors in cell differentiation. Front Physiol 11:1–12. https://doi.org/10.3389/fphys.2020.00239

Kulyte A, Navakauskiene R, Treigyte G, Gineitis A, Magnusson KE (2001) Parallel assessment of tyrosine phosphorylation and nuclear targeting of proteins. Biotechniques 31(3):510+

Kulyte A, Navakauskiene R, Treigyte G, Gineitis A, Bergman T, Magnusson KE (2002) Characterization of human alpha-dystrobrevin isoforms in HL-60 human promyelocytic leukemia cells undergoing granulocytic differentiation. Mol Biol Cell 13(12):4195–4205. https://doi.org/10.1091/mbc.E02-03-0128

Maik-Rachline G, Hacohen-Lev-Ran A, Seger R (2019) Nuclear ERK: mechanism of translocation, substrates, and role in cancer. Int J Mol Sci 20(5):1194

Martínez-Vieyra I, Pacheco-Tapia G, Reyes-López C, Méndez-Méndez JV, Cerecedo D (2018) Role of α-Dystrobrevin in the differentiation process of HL-60 cells. Exp Cell Res 370(2):591–600

Navakauskiene R, Kulyte A, Treigyte G, Gineitis A, Magnusson KE (2003) Translocation of transcription regulators into the nucleus during granulocyte commitment of HL-60 cells. Biochem. Cell Biol. Biochimie et Biologie Cellulaire 81(4):285–295. https://doi.org/10.1139/O03-055

Poli A, Ratti S, Finelli C, Mongiorgi S, Clissa C, Lonetti A, Cappellini A, Catozzi A, Barraco M, Suh PG, et al. (2018) Nuclear translocation of PKC-α is associated with cell cycle arrest and erythroid differentiation in myelodysplastic syndromes (MDSs). FASEB J 32(2):681–692

Reich NC, Liu L (2006) Tracking STAT nuclear traffic. Nat Rev Immunol 6(8):602–612
Savickiene J, Gineitis A, Stigbrand T (1999) Modulation of apoptosis of proliferating and differentiating HL-60 cells by protein kinase inhibitors: suppression of PKC or PKA differently affects cell differentiation and apoptosis. Cell Death Differ 6(7):698–709. https://doi.org/10.1038/sj.cdd.4400541

Chapter 8
Protein Visualization in Leukemia Cells

The visualization of protein complexes in living cells allows to investigate the interaction of proteins in their usual environment and to identify their subcellular localization. Subcellular localization is crucial to protein function and has been suggested as a mean to accomplish functional diversity and, concurrently, economize on protein design and synthesis (Butler and Overall 2009). Subcellular localization defines the access of proteins to interacting partners and the post-translational modification machinery and allows the integration of proteins into functional biological networks. Aberrantly localized proteins have been related to various human diseases, for instance, Alzheimer's disease, cancer, kidney stones, and others.

It was found that epigenetic regulation has an impact on gene expression, and recent research shows that aberrant DNA methylation patterning and histone modifications can have a function in leukemogenesis. For the purpose of accentuation of the co-operation of epigenetic mechanisms acting during the latter process it is necessary to identify their potential as biomarkers of granulocytic differentiation as well as their cellular localization during differentiation processes. In the Chapter 8 we will present our research displaying the visualization of proteins, their isoforms and modifications during hematopoietic cell granulocytic differentiation as well as leukemia cell differentiation and apoptosis processes (Kulyte et al. 2002; Borutinskaite et al. 2011; Navakauskiene et al. 2012, 2014).

R. Navakauskienė et al., *Epigenetics and Proteomics of Leukemia*,
https://doi.org/10.1007/978-3-030-68708-3_8

8.1 Comparative Analysis of Histones during Granulocytic Differentiation

Hematopoietic stem cells particularly exhibit self-renewal and differentiation into mature distinct hematopoietic lineages. The latest state-of-art technologies have led to the determination of molecular, cellular, and functional heterogeneity of the HSC population (Laurenti and Göttgens 2018; Haas et al. 2018).

CD34 is a marker of hematopoietic stem cells and hematopoietic progenitor cells as well as non-hematopoietic cells like embryonic fibroblasts and vascular endothelial progenitors (Sidney et al. 2014). Clinically, CD34 is related with the selection and enrichment of hematopoietic stem cells for bone marrow transplants. Human hematopoietic progenitor cells are CD34 positive (CD34+). These cells obtained from healthy human blood, KG1 cells representing obstructed differentiation at an initial stage of hematopoietic development, and mature human neutrophils may consequently be applied epigenetic surveys. CD34+ cells create a valuable model system where progression from quiescent to cycling to differentiated states can be associated with the alterations in chromatin rearrangements. Alterations in histones H3 and H4 modifications being related to chromatin activation, i.e., H3K4me3, H3K9ac, H3K9ac/S10p and H4hyperAc, and reactivation of methylation-silenced genes could be obvious in hematopoietic primary CD34+ cells, KG1 cells, and mature neutrophils. We employed 1D, 2D gel fractionation histones, their qvariants and alterations of modifications as well as computational analyses of confocal images to analyze such alterations of histone modifications in these cell populations (Navakauskiene et al. 2014).

In order to analyze active chromatin formation and histone modifications during granulocytic differentiation we investigated the modification status of H3K4me3, H3K9ac, H3K9ac/S10p and H4hyperAc histones in human hematopoietic progenitor CD34+ cells, which were untreated or treated with phenylbutyrate as a HDAC inhibitor or RG108 as a non-nucleoside DNMT inhibitor in human myeloid leukemia KG1 cells, and mature human neutrophils. In our study we visualized histones in different ways (1 and 2 dimentional electrophoresis and confocal microscopy, Figs. 8.1, 8.2, 8.3, 8.4, 8.5, and 8.6). Core histones H2A, H2B, H3, and H4 wrap DNA and make an impact on chromatin condensation levels due to both histone and DNA modifica-tions. The chromatin structure is crucial in gene regulation during cell development, proliferation, differentiation and apoptosis, and core histones as well as linker histone H1 variants could be significant factors for the maintenance of stem cell pluripotency, DNA condensation, and gene expression regulation (Godde and Ura 2009; Lord et al. 1990; Terme et al. 2011; Pan and Fan 2016). In fact, some of histone variants, i.e., H3.3, H2A.Z, H2A.X, and macro H2A have indivi-dual roles when regulation of chromatin structure occurs (Barrero et al. 2013; Wang et al. 2018). In our study we investigated core histones and linker histone H1 distributions in hematopoietic CD34+ stem cells, control myeloid leukemia KG1 cells, KG1 cells stimulated to granulocytic differentiation by applying DNMT inhibitor RG108 and HDAC inhibitor phenylbutyrate and mature human neutrophils (NF). Isolated histones were fractionated in an AUT (0.9 M acetic acid,

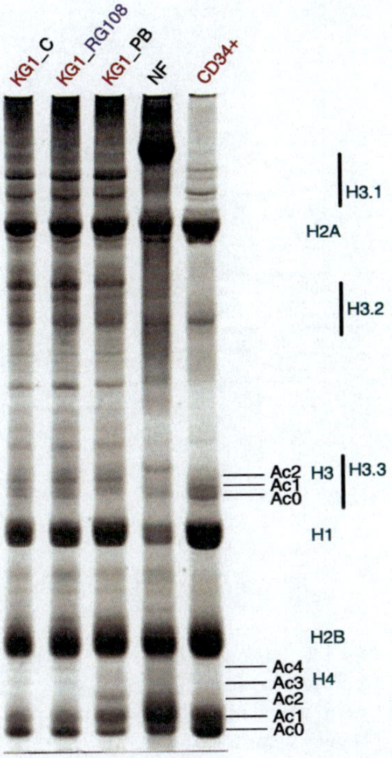

Fig. 8.1 Histone distribution in hematopoietic cells. Histones were isolated from untreated (KG1_C), and treated for 24 h with 25 μM RG108 (KG1_RG) or 6 h with 4 mM PB (KG1_PB) KG1 cells, human mature neutrophils (NF) and hematopoietic CD34+ stem cells. Isolated histones were fractionated in an AUT system based on the section. Representative images from one of three experiments showing similar results are presented. According Navakauskiene et al. (2014)

6 M urea, 4 mM Triton X-100, 15% polyacrylamide) system and stained (Fig. 8.1) or after fractionation sub-fractionated with SDS/PAGE (Fig. 8.2). We showed (Fig. 8.1) that the expression of linker histone H1 lowered during differentiation and mature human neutrophils have small amount of it. Several research groups have shown that H1 variants are varyingly expressed during cell differentiation: the pluripotent cells (ES and iPS4F1) have lower levels of the histone variant H1.0 and higher levels of the H1.3 and H1.5 variants, while the others, i.e., H1.2 and H1.4, have not shown any important changes (Terme et al. 2011; Di Liegro et al. 2018). Histone H3 is a significant epigenetic target due to its various modification states (Li and Zeng 2019). In our study (Navakauskiene et al. 2014) we demonstrated that H3.1 and H3.2 slightly dropped in CD34+ and NF in comparison to KG1 cells represented differentiating hematopoietic cells (Figs. 8.1 and 8.2). The level of histone H3 variant H3.3 lowered only in mature human neut-rophils (Figs. 8.1 and 8.2), when an expression of active gene was reduced. It has been shown that H3.3 containing nucleosomes are enriched in an active chromatin (Ahmad and Henikoff 2002). Jin and Felsenfeld (2006) have shown that H3.3 may have a direct impact on activation of the chicken folate receptor and β-globin genes. We have shown (Figs. 8.1 and 8.2) that histones H2A and H2B did not display apparent variances in all types of investigated cells. However, the acetylation of histones H3 and H4 was striking in KG1 cells stimulated to granu-locytic differentiation by HDAC inhibitor phenylbutyrate (PB) and ATRA (Merzvinskyte et al. 2006; Navakauskiene et al.

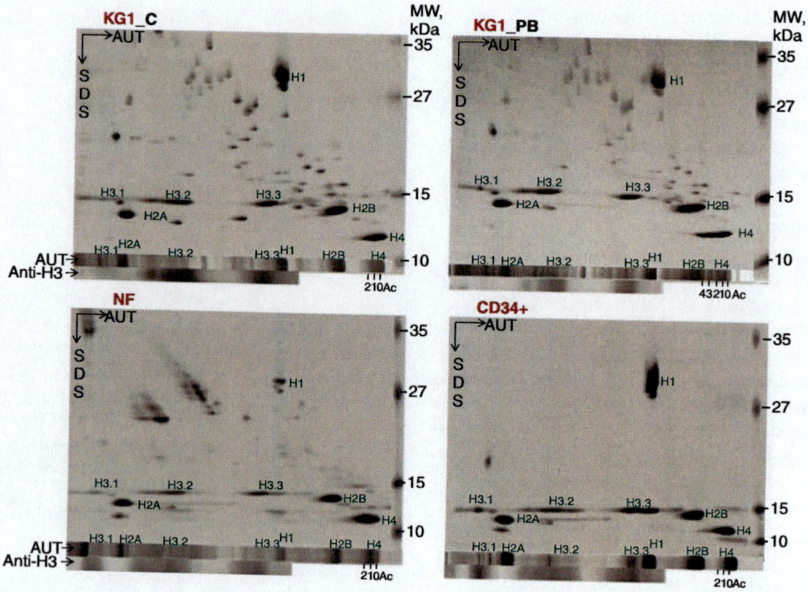

Fig. 8.2 Histone variants and their modifications during hematopoietic cell development. Histones were isolated from hematopoietic CD34+ stem cells, myeloid leukemia untreated (KG1_C), and treated with 4 mM PB for 48 h (KG1_PB) KG1 cells, and human mature neutrophils (NF). Isolated histones were fractionated in the AUT system (AUT) and then evaluated with antibodies against total histone H3 (Anti-H3) or AUT strips were fractionated in SDS/PAGE (SDS). Histones (H1, H2A, H2B, H4), their variants (H3.1, H3.2, H3.3) and histone H4 acetylation forms (Ac 0, 1, 2, 3, 4) are respectively marked in the images. Representative images from one of three experiments showing similar results are presented. According Navakauskiene et al. (2014)

2014), but not with DNMT inhibitor RG108 (Fig. 8.1). Our results allow to argue that core histones and their variants as well as the linker histone H1 distribute va-riously during granulocytic differentiation of hematopoietic cells and that their distribution ref-lects the differentiation status of hematopoietic cells.

The confocal microscopy images of the fluorescence intensity were evaluated by using computational methods of histones H3K4me3 (Fig. 8.3), H3K9ac (Fig. 8.4), H3K9ac/S10p (Fig. 8.5) and H4hyperAc (Fig. 8.6) as well as ratios of the median values of the fluorescent intensities. The hematopoietic progenitor CD34+ cells and neutrophils demonstrated very similar histone modification levels (Figs. 8.4, 8.5, and 8.6), except for H3K4me3; the latter exists in transcriptionally active chromatin and in neutrophils its level was declined (Fig. 8.3). Furthermore, it was lower in control KG1 cells (Fig. 8.3) in comparison with CD34+ cells. H3K9me3 deregulation in AML is associated preferentially with a decline of the modifications in core promoter regions. Müller-Tidow et al. (2010) showed that a decline in H3K4me3 levels at CREs (cAMP Response Element) was related with higher CRE-driven promoter activity *in vivo* in AML blasts. Besides, there are extensive alterations of H3K9me3 levels at gene promoters in AML and suggestions that inhibitors of H3K9 methylation can be applied as therapeutic agents (Müller-Tidow

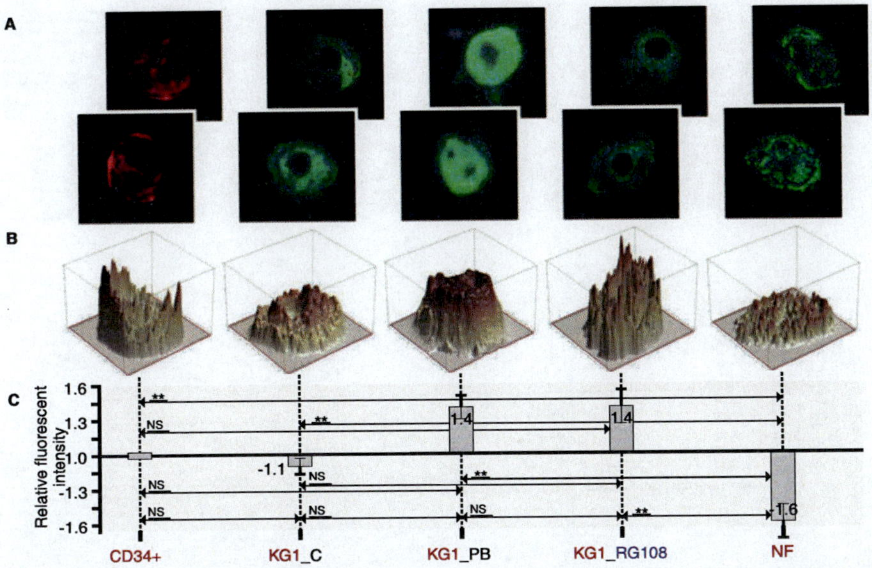

Fig. 8.3 Histone H3K4me3 modification levels in hematopoietic cells. (**A**) Representative confocal images of fluorescence intensity of histone H3K4me3 in CD34+ cells (CD34+), KG1 control (KG1_C), KG1 cells treated for 48 h with 4 mM phenylbutyrate (KG1_PB) or 25 μM RG108 (KG1_RG) and human mature neutrophils (NF) from two of three experiments showing similar results; (**B**) 3D profile of fluorescent intensity of corresponding bottom image from (**A**); (**C**) Ratios of median values of fluorescent intensities are demonstrated. Total fluorescence intensity of each cell category (CD34+, KG1_C, KG1_PB, KG1_RG, NF) was applied for computation of median values. Ratios represent fold-change (increased intensity—positive fold-change; decreased intensity—negative fold-change) of KG1_C, KG1_PB, KG1_RG, NF compared to CD34+. The Wilcoxon rank—sum test was used for statistical analysis: $P < 0.05$ (∗); $P < 0.01$ (∗∗); NS—no significant change. The bars represent fold enrichment of the modified histones relative to the control (CD34+ cells). Data is the mean ±SD from three independent experiments. According Navakauskiene et al. (2014)

et al. 2010; Monaghan et al. 2019). Paul et al. (2010) noticed that reactivation of p15INK4b expression in AML cell lines and patient blasts applying for treatment 5-aza-2'-deoxycytidine (decitabine) and trichostatin A (TSA) increased H3K4me3 and maintained H3K27me3 enrichment at p15INK4b. These data show that AML cells with p15INK4b DNA methylation have an altered histone methylation pattern in comparison with unmethylated samples and that these alterations are reversible by epigenetic drugs.

In general our findings demonstrate the importance of histones variants and modifications for the hematopoietic cell development as well as granulocytic differentiation. These changes could be valuable for the leukemia induced granulocytic differentiation.

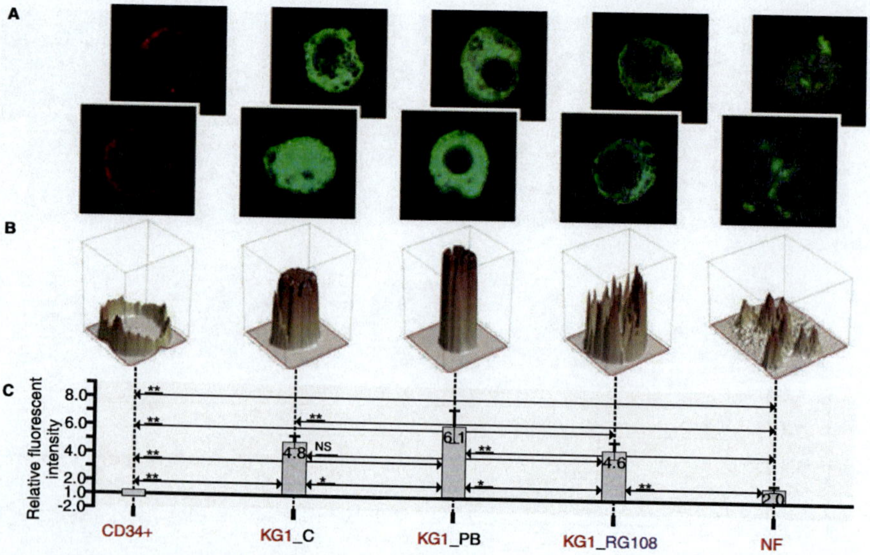

Fig. 8.4 Histone H3K9ac modifications during hematopoiesis. (**A**) Representative confocal images of fluorescence intensity of histone H3K9ac in CD34+ cells (CD34+), KG1 control (KG1_C), KG1 cells treated for 48 h with 4 mM phenylbutyrate (KG1_PB) or 25 μM RG108 (KG1_RG) and human mature neutrophils (NF) from two of three experiments showing similar results; (**B**) 3D profile of fluorescent intensity of corresponding bottom image from (**A**); (**C**) Ratios of median values of fluorescent intensities are shown. Total fluorescence intensity of each cell category (CD34+, KG1_C, KG1_PB, KG1_RG, NF) was applied for computation of median values. Ratios represent fold-change (increased intensity—positive fold-change; decreased intensity—negative fold-change) of KG1_C, KG1_PB, KG1_RG, NF compared to CD34+. The Wilcoxon rank—sum test was used for statistical analysis: $P < 0.05$ (∗); $P < 0.01$ (∗∗); NS—no significant change. The bars represent fold enrichment of the modified histones relative to the control (CD34+ cells). Data is the mean ±SD from three independent experiments. According Navakauskiene et al. (2014)

8.2 Visualization of Dystrobrevin-α in Leukemia Cells

In the previous chapters (Chaps. 5 and 7) we described the role of dystrobrevin-α during leukemia cell proliferation, differentiation and apoptosis. It is known that this protein is one of the components of the dystrophin-associated protein complex at the sarcolemma. dystrobrevin protein belongs to the dystrophin protein family (Blake et al. 1996; Gingras et al. 2016).

The exact role of dystrobrevin is unidentified yet, but some authors proposed that dystrobrevin can play a role in signaling processes. It was demonstrated that dystrophin protein is a substrate for a few kinases, like CaM kinase, casein kinase II, and protein kinase c (Senter et al. 1995; Bruno 2014).

In our studies we demonstrated that dystrobrevin-α isoforms were expressed in APL (NB4, HL-60) cells. The expression level and distribution of dystrobrevin-

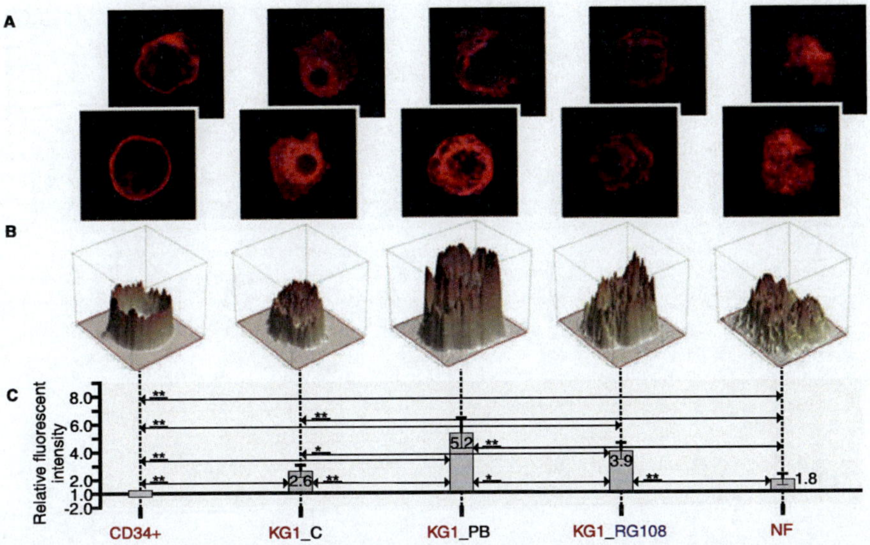

Fig. 8.5 Histone H3K9ac/S10p modifications during hematopoiesis. (**A**) Representative confocal images of fluorescence intensity of histone H3K9ac/S10p in CD34+ cells (CD34+), KG1 control (KG1_C), KG1 cells treated for 48 h with 4 mM phenylbutyrate (KG1_PB) or 25 μM RG108 (KG1_RG) and human mature neutrophils (NF) from two of three experiments showing similar results; (**B**) 3D profile of fluorescent intensity of corresponding bottom image from (**A**); (**C**) Ratios of median values of fluorescent intensities are presented. Total fluorescence intensity of each cell category (CD34+, KG1_C, KG1_PB, KG1_RG, NF) was used for computation of median values. Ratios represent fold-change (increased intensity—positive fold-change; decreased intensity—negative fold-change) of KG1_C, KG1_PB, KG1_RG, NF compared to CD34+. The Wilcoxon rank—sum test was used for statistical analysis: $P < 0.05$ (∗); $P < 0.01$ (∗∗); NS—no significant change. The bars represent fold enrichment of the modified histones relative to the control (CD34+ cells). Data is the mean ±SD from three independent experiments. According Navakauskiene et al. (2014)

α isoforms vary in separate cell fractions: cytoplasm, nucleus, hydrophobic cell compartments (Kulyte et al. 2002; Borutinskaite et al. 2011). We noticed that in proliferating and differentiating NB4 cells only the dystrobrevin-γ isoform went through tyrosine phosphorylation (Kulyte et al. 2002). Our results support an observation, that the phosphorylation state of dystrobrevin-γ in the nucleus is altered during differentiation process after promyelocytic leukemia cells were treated with ATRA.

In order to identify the localization of human dystrobrevin-α during granulocytic differentiation, we examined its distribution in HL-60 cells during three main stages of differentiation and in neutrophils obtained from peripheral blood after labeling with antibody-dystrobrevin and detection with confocal microscope (Kulyte et al. 2002). We observed that the distribution of dystrobrevin-α altered in the HL-60 cells during the separate stages of differentiation, seen as distinct staining alternating between the plasma membrane, the cytoplasm, and the nucleus (Fig. 8.7).

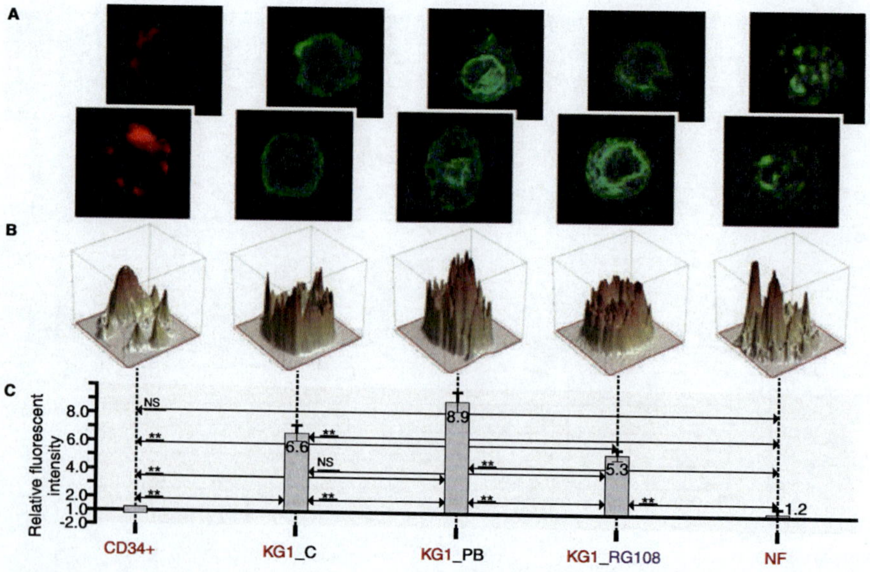

Fig. 8.6 Histone H4 hyperAc modification during hematopoiesis. (**A**) Representative confocal images of fluorescence intensity of histone H4 hyperAc in CD34+ cells (CD34+), KG1 control (KG1_C), KG1 cells treated for 48 h with 4 mM phenylbutyrate (KG1_PB) or 25 μM RG108 (KG1_RG) and human mature neutrophils (NF) from two of three experiments showing similar results; (**B**) 3D profile of fluorescent intensity of corresponding bottom image from (**A**); (**C**) Ratios of median values of fluorescent intensities are presented. Total fluorescence intensity of each cell category (CD34+, KG1_C, KG1_PB, KG1_RG, NF) was used for computation of median values. Ratios represent fold-change (increased intensity—positive fold-change; decreased intensity—negative fold-change) of KG1_C, KG1_PB, KG1_RG, NF compared to CD34+. Wilcoxon rank sum test was used for statistical analysis: $P < 0.05$ (∗); $P < 0.01$ (∗∗); NS—no significant change. The bars represent fold enrichment of the modified histones relative to the control (CD34+ cells). Data is the mean ±SD from three independent experiments. According Navakauskiene et al. (2014)

It seemed that protein dispersed throughout the whole cell volume, especially in the proliferating cells. It looked as though during the commitment stage dystrobrevin-α was more concentrated close to the membrane of the cell, despite the fact that the pattern was still diffuse (Fig. 8.7). Nevertheless, after 96 h of differentiation (i.e., in granulocyte-like cells), it was possible to notice clearly the clusters at the membranes of cells and nuclei. Interestingly, we inspected obvious differences in dystrobrevin immunostaining between the isolated normal neutrophils and the HL-60 cells undergoing induced differentiation *in vitro* (Fig. 8.7): both types of cells showed staining adjacent to the cell and nuclear membrane, but only the neutrophils demonstrated especially strong dystrobrevin immunoreactivity in the nuclei.

In our next study we confirmed the dystrobrevin-α localization in NB4 cells (Borutinskaite et al. 2011). We showed that dystrobrevin-α is primarily localized in the cytoplasm adjacent to the plasma membrane of proliferating NB4 cells.

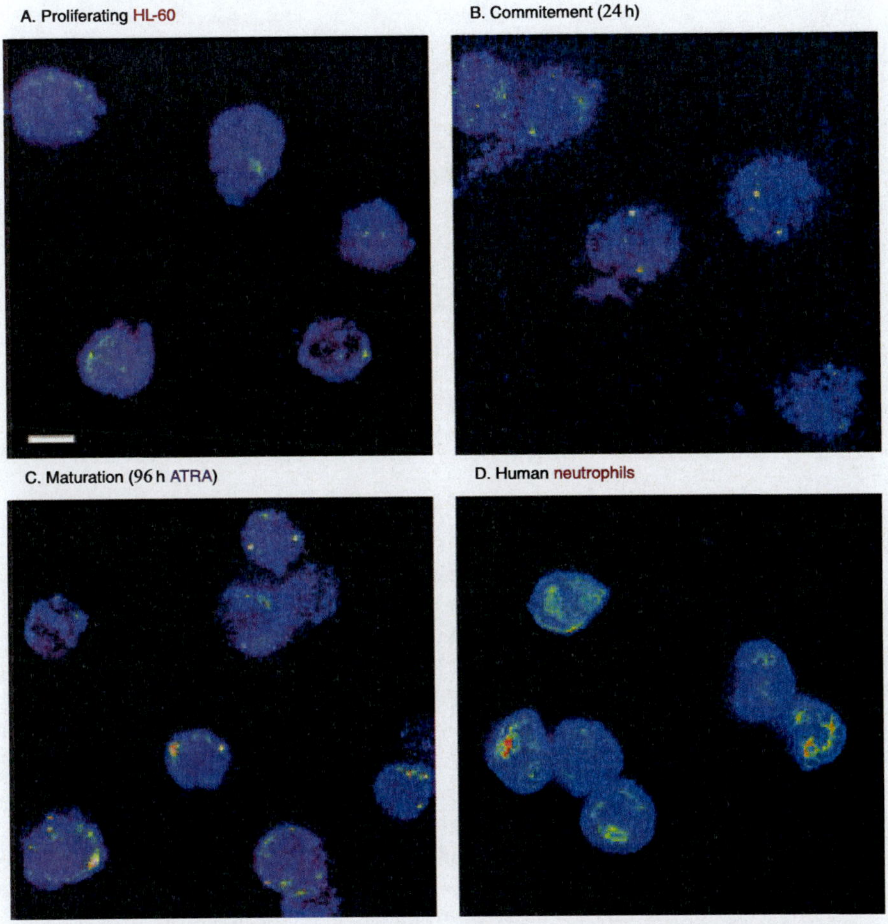

Fig. 8.7 Distribution of human dystrobrevin in HL-60 cells undergoing granulocytic differentiation and in human neutrophils. The confocal images present the localization of human dystrobrevin in proliferating HL-60 cells (**A**), HL-60 cells at the commitment stage of differentiation (**B**), after 24 h treatment with ATRA, mature granulocytes (**C**), after 96 h ATRA), and peripheral neutrophils isolated from human blood (**D**). All cells were fixed, labeled and the images were obtained under identical conditions. To visualize fluorescence from entire cells, some areas in the images were allowed to saturate. The color scale is shown on the left. Scale bar, 5 μm. According Kulyte et al. (2002), License No 1025167-10

After the treatment of NB4 cells with ATRA, cells underwent differentiation into granulocyte-like cells and dystrobrevin-α was identified more in the nucleus. Our data demonstrated that dystrobrevin-α changed its sub-cellular localization during the granulocytic differentiation of NB4 cells (Fig. 8.8)—after 96 h of ATRA treatment dystrobrevin-α localization in the cell became more or less the same as in human neutrophils.

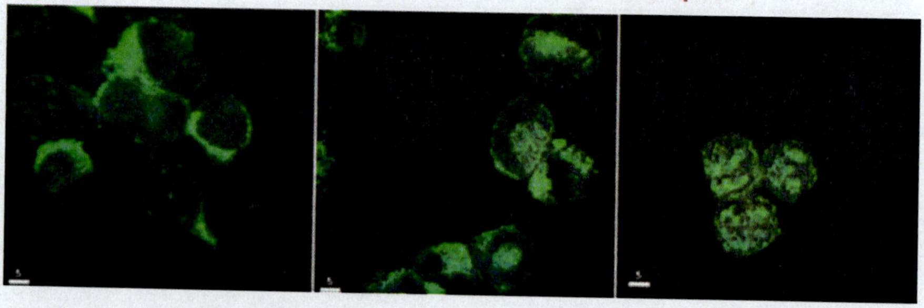

Fig. 8.8 Distribution of human dystrobrevin-α in NB4 cells. Confocal images demonstrating the localization of human dystrobrevin-α in proliferating (control), induced to granulocytic differentiation with 1 μM ATRA NB4 cells, and in human neutrophils. Bar: 5 μm. According Borutinskaite et al. (2011), License No 4796590532046

Immunostaining of mouse and ratskeletal muscle with dystrobrevin-α-specific antibodies determined that the protein is localized close to the sarcolemma and there are higher concentrations at the neuromuscular junctions, which indicates that it has a functional task at both sites (Blake et al. 1996; Grady et al. 2000; Gingras et al. 2016). Additionally, in our study, dystrobrevin-α was related to the nuclear membrane in k cells (Kulyte et al. 2002). While evaluating the difference in fluorescence between the cytoplasm and the nucleus, it can be suggested that dystrobrevin accumulates both on the cell membrane and in the nucleus (Borutinskaite et al. 2011; Kulyte et al. 2002). Since dystrobrevin is widely located along the cell and nuclear membranes and is clustered in differentiating HL-60 cells, this protein may also have a significant role in human neutrophils. We noticed the increase in dystrobrevin-α around/in the nuclei of motile human neutrophils, and it is certainly intriguing.

By applying the method of mass spectrometry we determined the proteins that co-immunoprecipitated with dystrobrevin. We identified two cytoskeletal proteins related to dystrobrevin-α in the granulocyte-like cells after HL-60 cell induced differentiation. These proteins were defined as human actin and myosin light chain (Kulyte et al. 2002). To investigate sites of co-localization of dystrobrevin-α and actin, we stained normal human neutrophils and granulocyte-like cells differentiated *in vitro* with the antibody against dystrobrevin-α and the fluorescein-5-isothiocyanate (FITC) labeled secondary antibody (Fig. 8.9A–C, green channel). The actin network was labeled with Alexa 594-phalloidin (Fig. 8.9, red channel). Nonetheless, a high-resolution microscopic analysis of the distribution of cytoskeletal proteins is obstructed by the small size of HL-60 and neutrophil cells. In any case, dystrobrevin-α was partially enriched at the outer envelope of the nuclei in both the human neutrophils and the mature granulocytes (Fig. 8.9B and C). The subcellular allocation of actin and dystrobrevin-α was alike and presented co-localization at the membrane, coupled with immunolabeling of dystrobrevin

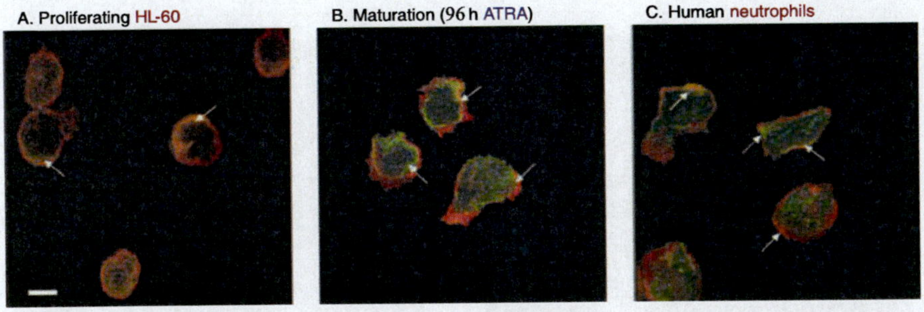

Fig. 8.9 Double-labeling of human dystrobrevin and F-actin in HL-60 cells undergoing granulocytic differentiation and in neutrophils from adult humans. The confocal images show colocalization of human dystrobrevin and F-actin in proliferating HL-60 cells (**A**), mature granulocytes (**B**), 96 h with ATRA, and human neutrophils (**C**). Cells were treated and visualized as described in the legend of Fig. 8.1. F-actin was stained with Alexa 594-phalloidin. Sections were ~0.6 μm in thickness, and the images were obtained under identical conditions. Scale bar, 5 μm. According Kulyte et al. (2002), License No 1025167-10

around the nucleus (arrows in Fig. 8.9). In B and C, there are several regions that demonstrate separate green and red fluorescence, implying that there are membrane domains where the proteins are adjacent to each other but not co-localizing. Some faint nuclear labeling for dystrobrevin was also noticed in proliferating HL-60 cells. Together, these findings verify previous observations and show that both *in vitro* and *in vivo* dystrobrevin-α interacts with actin.

We also identified proteins interacting with dystrobrevin-α in human acute promyelocytic leukemia NB4 cells. The proteins were determined during MALDI-TOF-MS and ESI-MS-MS analysis (Borutinskaite et al. 2011). All determined proteins can be separated into groups according to their function: (1) structural proteins like actin, tubulin, desmin, stathmin that have a significant role in cell growth and/or maintenance; (2) signaling proteins STAG1, RIBA and (3) other proteins like chaperones, those can participate in protein metabolism, signal transduction, and gene transcription regulation.

We confirmed co-localization of dystrobrevin-α with actin and HSP90 in proliferating NB4 cells (Fig. 8.10) (Borutinskaite et al. 2011). The indirect interdependency between dystrobrevin and F-actin via dystrophin is now acknowledged. The DAPC has primarily two types of functions, that is, it has mechanical and signaling tasks. Mechanical models involve the actin–dystrophin–dystroglycan–laminin axis, which implies a mechanical relation between the cytoskeleton and the extracellular matrix (Roberts 2001; Murphy and Ohlendieck 2016). The signaling role is determinded based on the fact that occurrence of tyrosine phosphorylation of dystrobrevin is specific to the stage of differentiation.

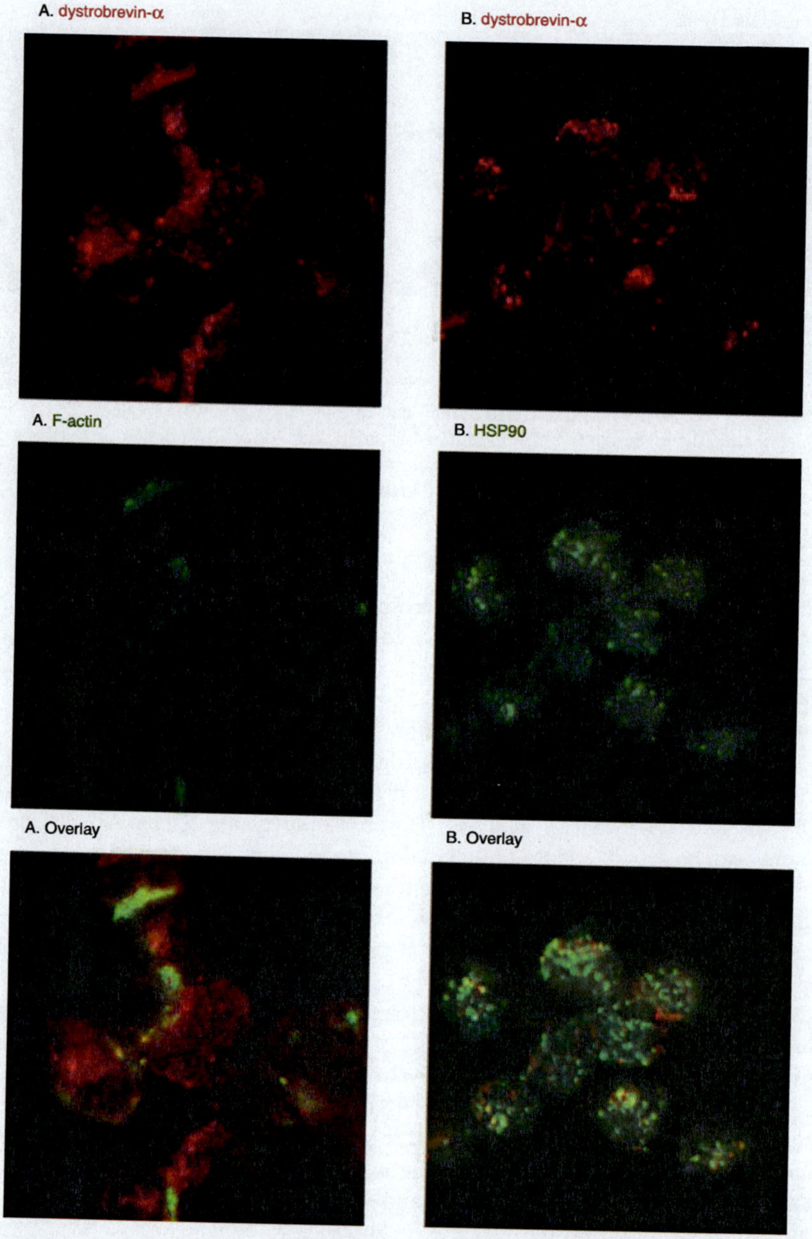

Fig. 8.10 Dual immunofluorescence labeling of human dystrobrevin-α and actin filaments (F-actin) (**A**) or dystrobrevin-α and HSP90 proteins (**B**) in proliferating NB4 cells. Cells were fixed and incubated with Alexa 488-phalloidin or anti-mouse HSP90 antibody (visualized with Alexa 488-conjugated goat anti-mouse antibodies), and then incubated with anti-rabbit dystrobrevin-α antibody (visualized with Alexa 563-conjugated goat anti-rabbit antibodies). The cells display clear co-localization between dystrobrevin-α and F-actin (**A**) or dystrobrevin-α and HSP90 (**B**) proteins at cytoplasm. Bar: 5 μm. According Borutinskaite et al. (2011), License No 4796590532046

8.3 Dystrobrevin-α Localization in Leukemia Cells Induced to Apoptosis

To describe the localization of human dystrobrevin-α and its splice isoforms in the process of apoptosis the human acute promyelocytic leukemia HL-60 cells were labeled with anti-dystrobrevin-α antibodies at separate stages of apoptosis induction, i.e., at 0.5, 3, 6, 9, 18 h with 68 μM etoposide (Et) alone or plus 25 μM Z-VAD(OH)-FMK. Figure 8.11 presents dystrobrevin-α localization in proliferating HL-60 cells (A), cells were treated for 9 h with 68 μM etoposide alone (B), or with 68 μM etoposide plus 25 μM Z-VAD(OH)-FMK (C). We discovered that allocation of dystrobrevin-α and its splice isoforms altered during apoptosis. In the proliferating cells, the protein exhibited throughout the whole cell, it had higher concentration in the places where cells attached to each other (Fig. 8.11A). After stimulation of apoptosis, when approximately 70% of cell population was apoptotic, it was possible to observe the clusters of dystrobrevin-α and its splice isoforms in the cytoplasm of the cells (Fig. 8.11B). Besides, we noticed similarities in the

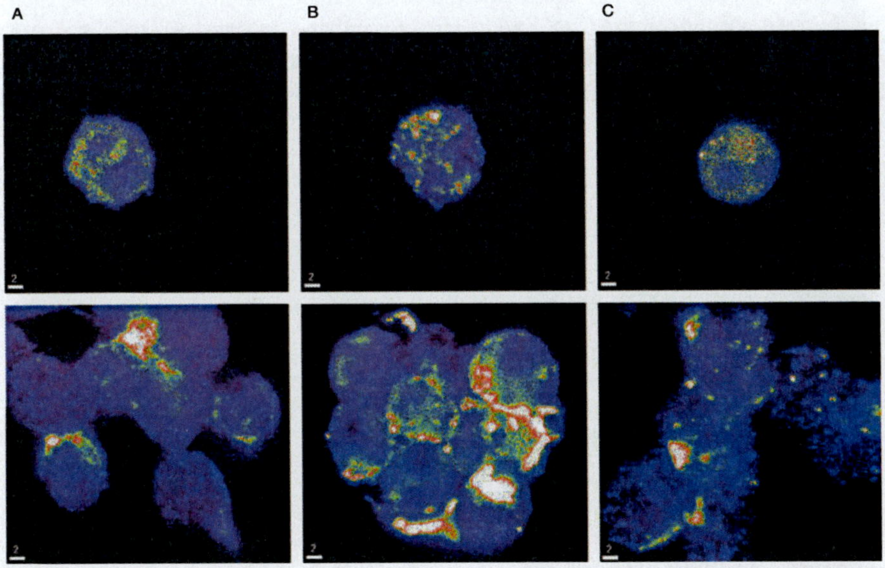

Fig. 8.11 Cellular distribution of human dystrobrevin in control HL-60 cells and in cells induced to apoptosis. The confocal images show the localization of human dystrobrevin-α in proliferating control cells (**A**), induced to apoptosis with 68 μM etoposide (**B**), or treated with both 68 μM etoposide and 25 μM Z-VAD(OH)-FMK (**C**). All cells were fixed and labeled with primary antibodies against dystrobrevin-α then incubated with secondary Alexa 488-coupled antibodies. Thereafter, the cells were allowed to adhere on microscope slides coated with poly-L-lysine in a wet chamber, mounted in mounting media and images were obtained by using Sarastro 2000 confocal laser scanning microscope (Molecular Dynamics, Sunnyvale, CA) under identical conditions. Bar: 2 μM. According Navakauskiene et al. (2012), License No 4797021401553

dystrobrevin-α immunostaining of the proliferating cells and those treated both with etoposide and Z-VAD(OH)-FMK (Fig. 8.11C). Here, dystrobrevin-α and its splice isoforms seemed to be in all cell compartments but with higher concentration in the cytoplasm. Casually, results alike were achieved for the other promyelocytic cell line NB4.

Confocal microscopy of control and apoptotic HL-60 cells demonstrated that dystrobrevin-α had a higher concentration not only in the cytoplasm little while later than the treatment of cells with etoposide was applied, but also in areas of cell-cell contacts. The presence of the caspase inhibitor Z-VAD(OH)-FMK suspended this accumulation, and allocation is similar to that in the control cells (Fig. 8.11). We thus made an assumption that this could reflect distinct functions of dystrobrevin-α and that etoposide modulated the affinity of dystrobrevin-α for other proteins being affected by or regulating the apoptosis process.

Previously, we have determined dystrobrevin-α relation with actin and myosin light chain both in human promyelocytic leukemia cells and human neutrophils (Kulyte et al. 2002). Accidentally, we have preliminary evidence that the silencing of dystrobrevin-α, has a major impact on the structure of F-actin cytoskeleton in adherent HeLa cells (unpublished data). Since there is an evidence that actin cytoskeleton is involved in regulation of many signaling processes, such as receptor affinity, apoptosis, cell cycle control and others, our results could present new information on how dystrobrevin-α potentially participates in signal transduction in myeloid or phagocyte cells during cell differentiation and apoptosis progression.

References

Ahmad K, Henikoff S (2002) The histone variant H3.3 marks active chromatin by replication-independent nucleosome assembly. Mol Cell 9(6):1191–1200

Barrero MJ, Sese B, Marti M, Belmonte JCI (2013) Macro histone variants are critical for the differentiation of human pluripotent cells. J Biol Chem 288(22):16110–16116

Blake DJ, Nawrotzki R, Peters MF, Froehner SC, Davies KE (1996) Isoform diversity of dystrobrevin, the murine 87-kda postsynaptic protein. J Biol Chem 271(13):7802–7810

Borutinskaite VV, Magnusson KE, Navakauskiene R (2011) α-Dystrobrevin distribution and association with other proteins in human promyelocytic NB4 cells treated for granulocytic differentiation. Mol Biol Rep 38(5):3001–3011. https://doi.org/10.1007/s11033-010-9965-9

Bruno C (2014) Dystrophin complex functions as a scaffold for signalling proteins. Biochim Biophys Acta (BBA)-Biomembr 1838(2):635–642

Butler GS, Overall CM (2009) Proteomic identification of multitasking proteins in unexpected locations complicates drug targeting. Nat Rev Drug Discovery 8(12):935–948

Di Liegro CM, Schiera G, Di Liegro I (2018) H1.0 linker histone as an epigenetic regulator of cell proliferation and differentiation. Genes 9(6):310

Gingras J, Gawor M, Bernadzki KM, Grady RM, Hallock P, Glass DJ, Sanes JR, Proszynski TJ (2016) α-Dystrobrevin-1 recruits Grb2 and α-catulin to organize neurotransmitter receptors at the neuromuscular junction. J Cell Sci 129(5):898–911

Godde JS, Ura K (2009) Dynamic alterations of linker histone variants during development. Int J Dev Biol 53(2–3):215–224. https://doi.org/10.1387/ijdb.082644jg

Grady RM, Zhou H, Cunningham JM, Henry MD, Campbell KP, Sanes JR (2000) Maturation and maintenance of the neuromuscular synapse: genetic evidence for roles of the dystrophin–glycoprotein complex. Neuron 25(2):279–293

Haas S, Trumpp A, Milsom MD (2018) Causes and consequences of hematopoietic stem cell heterogeneity. Cell Stem Cell 22(5):627–638

Jin C, Felsenfeld G (2006) Distribution of histone H3.3 in hematopoietic cell lineages. Proc Natl Acad Sci 103(3):574–579

Kulyte A, Navakauskiene R, Treigyte G, Gineitis A, Bergman T, Magnusson KE (2002) Characterization of human alpha-dystrobrevin isoforms in HL-60 human promyelocytic leukemia cells undergoing granulocytic differentiation. Mol Biol Cell 13(12):4195–4205. https://doi.org/10.1091/mbc.E02-03-0128

Laurenti E, Göttgens B (2018) From haematopoietic stem cells to complex differentiation landscapes. Nature 553(7689):418–426

Li D, Zeng Z (2019) Epigenetic regulation of histone H3 in the process of hepatocellular tumorigenesis. Biosci. Rep. 39(8):1–9

Lord KA, Hoffman-Liebermann B, Liebermann DA (1990) Complexity of the immediate early response of myeloid cells to terminal differentiation and growth arrest includes ICAM-1, Jun-B and histone variants. Oncogene 5(3):387–396

Merzvinskyte R, Treigyte G, Savickiene J, Magnusson KE, Navakauskiene R (2006) Effects of histone deacetylase inhibitors, sodium phenyl butyrate and vitamin B3, in combination with retinoic acid on granulocytic differentiation of human promyelocytic leukemia HL-60 cells. In: Diederich M (ed) Signal Transduction Pathways, PT B: Stress Signaling and Transcriptional Control, Annals of the New York Academy of Sciences, vol 1091, pp 356–367. https://doi.org/10.1196/annals.1378.080. Cell Signaling World 2006 Conference, Luxembourg, Luxembourg, Jan 25–28, 2006

Monaghan L, Massett ME, Bunschoten RP, Hoose A, Pirvan PA, Liskamp RMJ, Jørgensen HG, Huang X (2019) The emerging role of H3K9me3 as a potential therapeutic target in acute myeloid leukaemia. Front Oncol 9:705

Müller-Tidow C, Klein HU, Hascher A, Isken F, Tickenbrock L, Thoennissen N, Agrawal-Singh S, Tschanter P, Disselhoff C, Wang Y, et al. (2010) Profiling of histone H3 lysine 9 trimethylation levels predicts transcription factor activity and survival in acute myeloid leukemia. Blood J Am Soc Hematol 116(18):3564–3571

Murphy S, Ohlendieck K (2016) The biochemical and mass spectrometric profiling of the dystrophin complexome from skeletal muscle. Comput Struct Biotechnol J 14:20–27

Navakauskiene R, Treigyte G, Borutinskaite VV, Matuzevicius D, Navakauskas D, Magnusson KE (2012) Alpha-dystrobrevin and its associated proteins in human promyelocytic leukemia cells induced to apoptosis. J Proteomics 75(11):3291–3303. https://doi.org/10.1016/j.jprot.2012.03.041

Navakauskiene R, Borutinskaite VV, Treigyte G, Savickiene J, Matuzevicius D, Navakauskas D, Magnusson KE (2014) Epigenetic changes during hematopoietic cell granulocytic differentiation—comparative analysis of primary CD34+cells, KG1 myeloid cells and mature neutrophils. BMC Cell Biol 15:4. https://doi.org/10.1186/1471-2121-15-4

Pan C, Fan Y (2016) Role of H1 linker histones in mammalian development and stem cell differentiation. Biochim Biophys Acta (BBA) Gene Regul Mech 1859(3):496–509

Paul TA, Bies J, Small D, Wolff L (2010) Signatures of polycomb repression and reduced H3K4 trimethylation are associated with p15INK4b DNA methylation in AML. Blood J Am Soc Hematol 115(15):3098–3108

Roberts RG (2001) Dystrophins and dystrobrevins. Genome Biol 2(4):reviews3006–1

Senter L, Ceoldo S, Petrusa MM, Salviati G (1995) Phosphorylation of dystrophin—effects on actin-binding. Biochem Biophys Res Commun 206(1):57–63. https://doi.org/10.1006/bbrc.1995.1009

Sidney LE, Branch MJ, Dunphy SE, Dua HS, Hopkinson A (2014) Concise review: evidence for CD34 as a common marker for diverse progenitors. Stem cells 32(6):1380–1389

Terme JM, Sesé B, Millán-Ariño L, Mayor R, Belmonte JCI, Barrero MJ, Jordan A (2011) Histone H1 variants are differentially expressed and incorporated into chromatin during differentiation and reprogramming to pluripotency. J Biol Chem 286(41):35347–35357

Wang Y, Long H, Yu J, Dong L, Wassef M, Zhuo B, Li X, Zhao J, Wang M, Liu C, et al. (2018) Histone variants H2A.Z and H3.3 coordinately regulate PRC2-dependent H3K27me3 deposition and gene expression regulation in mES cells. BMC Biol 16(1):1–18

Chapter 9
Computational Methods for Protein Localization Analysis

Protein localization and quantitative evaluation are the most common tasks of the experiments seeking to answer important biological questions (Hawkes and Spence 2019; Markaki and Harz 2017; Chiarini-Garcia and Melo 2011; Kubitscheck 2017). Quantitative fluorescence microscopy techniques are popular tools used for solving these tasks (Jerome and Price 2018; Seruca et al. 2017; Waters and Wittmann 2014; Rebollo and Bosch 2019). It is impossible to get quantitative results from fluorescence microscopy digital images without using computational tools. There are quite a number of open source and commercial software tools specifically designed for microscopy data analysis (Bajcsy et al. 2018; Eliceiri et al. 2012; Li et al. 2013; Usaj et al. 2016). Software packages exist as specialized collections of image processing and analysis algorithms, e.g., `CellProfiler` (McQuin et al. 2018; Lamprecht et al. 2007; Carpenter et al. 2006; Kamentsky et al. 2011; Jones et al. 2009), `ImageJ` (Rueden et al. 2017; Schneider et al. 2012), `Fiji` (Schindelin et al. 2012), `Icy` (De Chaumont et al. 2012), `BioImageXD` (Kankaanpää et al. 2012), `Ilastik` (Berg et al. 2019; Sommer et al. 2011), `MATtrack` (Courtney et al. 2015), `CellSegm` (Hodneland et al. 2013), `CellCognition` (Held et al. 2010), `TMARKER` (Schüffler et al. 2013), `MorphoLibJ` (Legland et al. 2016), `Cellomics`, `MetaXpress` or as general purpose collections, e.g., `OpenCV` (Kaehler and Bradski 2016; Domínguez et al. 2017), `ITK` (Johnson et al. 2015a,b), `VTK` (Schroeder et al. 2006), `Scikit-Image` (Van der Walt et al. 2014), `ImgLib2` (Pietzsch et al. 2012), `TrackMate` (Tinevez et al. 2017), `MATLAB`. Generally separate algorithms need to be combined into customized workflow designed for specific biological problem at hand. For the more common problems to deal, almost-ready-to-use solutions may be offered by software packages as workflow templates. But in either case the user will have to make more or less manual adjustments to prepare required workflow for microscopy data analysis.

In cases when suitable workflow for user's microscopy visual data analysis exists, then minimal required adjustments could include initial tuning of various parameters of the image preprocessing and object segmentation algorithms, e.g., the typical size

R. Navakauskienė et al., *Epigenetics and Proteomics of Leukemia*,
https://doi.org/10.1007/978-3-030-68708-3_9

of the objects, noise reduction settings, thresholding parameters. The user has to manually optimize such parameters to achieve the correct segmentation of objects in images (Miura and Sladoje 2020; Markaki and Harz 2017). Considering that images of various conditions may be present, it may be tedious work to find parameters that lead to satisfactory results of the image analysis. We seek to reduce mandatory initial user involvement that is needed before starting the analysis of a new portion of fluorescence microscopy images.

In this chapter, we review the key steps of automated fluorescence image analysis workflow—image preparation with the focus on image preprocessing, object segmentation, which includes nucleus and cell segmentation followed by the postprocessing, and data analysis (fluorescence intensity quantification, differential analysis, statistical data analysis, and data visualization). We highlight the strengths and weaknesses of the popular image processing and analysis algorithms used as realizations of specific workflow steps, expose decisions and parameter selections that still must be done manually, and suggest possible solutions on how to automate them. We are seeking to increase the automation level of the workflow and thereby minimize user involvement by:

- automatically detecting optimal initial parameters by exploiting regularities of the objects in the images;
- selecting or designing image analysis algorithms and their configurations such that would lead to the adaptability to the particular data set;
- introducing automated detection of the output data inconsistencies that may be indicators of existing errors in the analysis results, and bringing user attention to review and manage such situations.

Using the described strategy, we have developed the tool (workflow) for automated analysis of fluorescence confocal images and employed it to evaluate histone modifications changes in cell populations (Navakauskiene et al. 2014).

9.1 Overview of Typical Workflow of Automated Fluorescence Image Analysis

In this Section, we have a look at the typical workflow of automated fluorescence image analysis—basic building blocks (or key steps) needed to create automated fluorescence image analysis workflow (Fig. 9.1). These key steps are image preparation (channel separation of acquired images, image preprocessing), image analysis (object segmentation, postprocessing), and data analysis (quantification and differential analysis, statistical analysis, visualization) (Lundberg and Borner 2019; Hegde et al. 2018; Ljosa and Carpenter 2009; Furtado and Henry 2002; Sajjad et al. 2016; Meijering et al. 2016; Hussain et al. 2018). We also highlight challenges of automatic analysis that arise in separate image analysis stages (Markaki and Harz 2017; Nandy et al. 2012; Cao et al. 2018; Elen and Turan 2018; Rebollo and

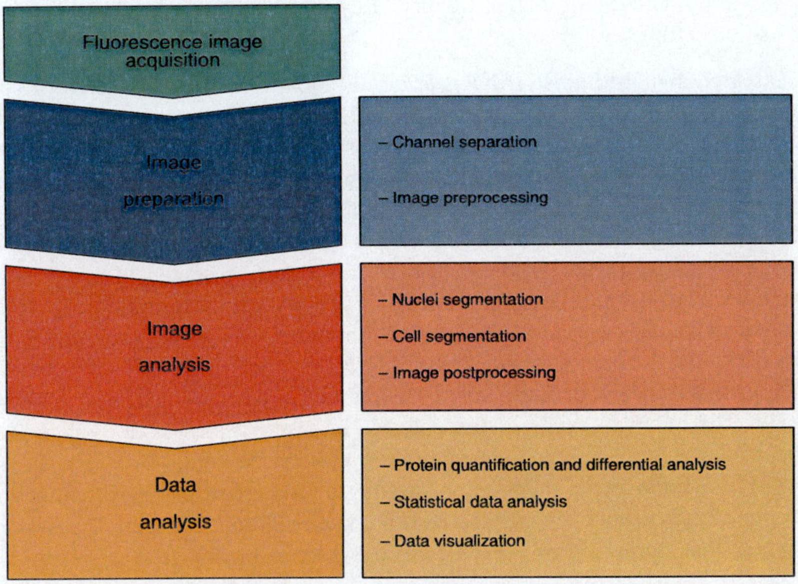

Fig. 9.1 Outline of a typical workflow of automated fluorescence image analysis

Bosch 2019). We expose these challenges by presenting experimental results of the most common image processing algorithms used as building blocks for microscopy data analysis workflow. Development of the image analysis algorithms requires solving tasks posed by variations of cell morphology, noise, and other acquisition artifacts (uneven sample illumination, fluorescence cross-talk) (Qiao et al. 2012; Schnipper et al. 2017; Li et al. 2013; Haidekker 2010; Paulik et al. 2017).

9.1.1 Fluorescence Image Acquisition

The first step of fluorescence microscopy-based studies is fluorescence image acquisition (Fig. 9.1A) (Jerome and Price 2018; Hawkes and Spence 2019; Rebollo and Bosch 2019). This step can be attributed to image analysis workflow because its proper setup is crucial for acquiring reliable findings and biologically meaningful results (North 2006; Seruca et al. 2017; Lawlor 2019; Gomes-Alves et al. 2018). Various issues should be considered to correctly setup imaging system (Waters 2009; Phoulady et al. 2017; Model and Burkhardt 2001; Yang et al. 2018). The user should optimize image acquisition parameters like magnification, the number of images per sample, exposure times, resolution (pixel size), camera binning, gain, required channels (Usaj et al. 2016; Rebollo and Bosch 2019; Markaki and Harz 2017; Sluder and Wolf 2013). Carefully adjusted acquisition conditions will increase the amount of collected biologically meaningful and relevant information by providing images of better quality. Key notes about adjustment of image acquisition settings:

- Increased signal-to-noise (SNR) ratio will allow to reduce the need for subsequent image preprocessing operations and to simplify object segmentation task. SNR may be increased by increasing exposure times and laser power, but it must be done with caution to manage possible photobleaching and photodamage. SNR also may be increased by camera binning at the expense of lower resolution.
- Selection of the optimal pixel size and magnification should be made considering the scale of the smallest objects of interest presented in the biological sample. Pixel should be small enough to preserve the resolution of the microscope (Markaki and Harz 2017). It is preferable to set the pixel size 3 times smaller than the object we want to measure.
- Optimal utilization of the camera's dynamic range may be set by tuning the gain. The maximal range of the possible intensity values in the image should be achieved while preventing saturation effects. Separate biological samples from the same experiment may require different optimal gain settings, but compromises should be made to keep imaging conditions constant. It is important to note that increasing gain also boosts the noise in the image.

9.1.2 Image Preparation

During the image preparation, image channel splitting (Fig. 9.2) and image preprocessing may be performed. Typical preprocessing procedures may include noise reduction, background removal, illumination correction, contrast enhancement, intensity normalization (Markaki and Harz 2017; Schwarzfischer et al. 2011; Rittscher et al. 2008; Rebollo and Bosch 2019). In this Subsection, we will take a quick look at channel splitting and image preprocessing tasks and show application results of several popular preprocessing algorithms.

Channel Splitting

In a multi-color fluorescence image, separate channels convey different biological information. Typically one of the channels serves as a nuclear marker (contains intensity information of labeled nuclei). Other channels may contain a fluorescence intensity profile of the other proteins of interest labeled using a different fluorescent dye (Fig. 9.2). These intensity profiles are related by spatial location, but the objects' appearance in these channels may be very different, so it may need different approaches of image segmentation to extract semantic information (localization and boundaries of separate objects). Channel splitting needs to be executed prior to subsequent automated image processing and analysis. When seeking to automate image analysis workflow, channel splitting may be at least partly automated by using algorithmic detection of nuclear marker channel as it typically distinguishes itself by having a more uniform morphology of objects compared to other channels. Automated selection of the nuclei channel is also useful when manual channel

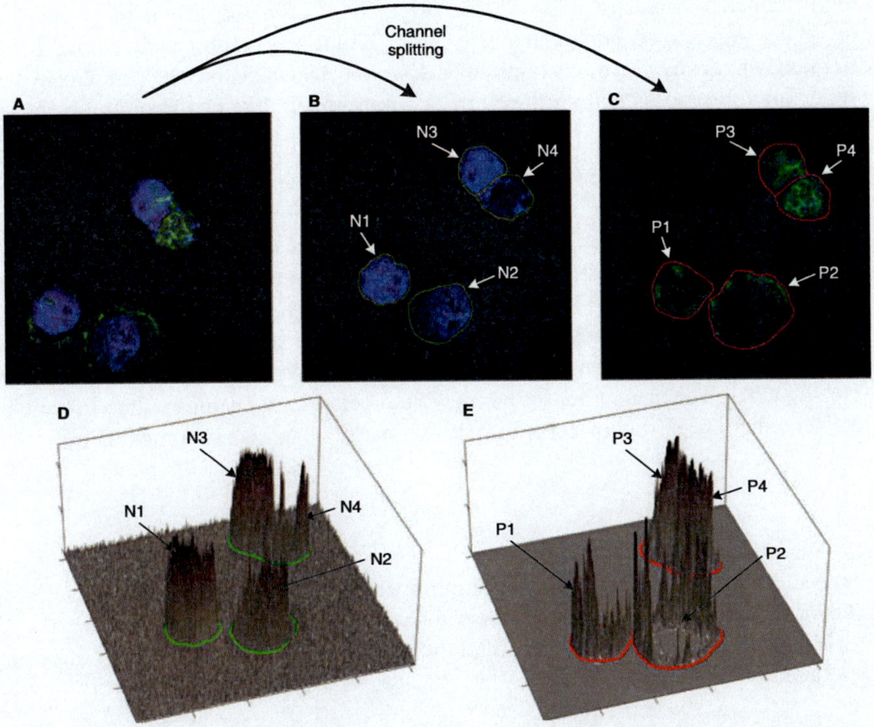

Fig. 9.2 Performing channel separation and visualization of manually segmented regions of nuclei and its surroundings. Original image (**A**), extracted nuclei channel (**B**) and a channel of target biomolecule (**C**). Fluorescence intensity visualized as a surface in (**D**) and (**E**) better reveals existing noise

splitting is performed because it allows detection of possible human errors by comparing results provided by human and the algorithm.

Image Preprocessing

Objectives of image preprocessing may be related to the improvement of image segmentation results or general enhancement of image visual quality and attractiveness. Depending on these objectives, different image preprocessing algorithms would be applied. Preferable image modifications for visualization purposes are related to the human visual system—images with higher contrast of objects are more attractive and easier to perceive. Existing high-frequency noise for the human eye is easy to deal with. Therefore, local contrast enhancement methods like unsharp masking may be useful. Deconvolution methods also can be used to enhance images by removing blur (Pawley 2006; Wallace et al. 2001).

Image preprocessing in order to improve segmentation outcome is oriented to suppressing or removing spatial frequency components different than those that make up the objects we want to extract. Typically we want to attenuate high-frequency intensity noise (noise originating from the acquisition process, or object appearance nonhomogeneity from biological nature) and remove low-frequency intensity variations (uneven background). Such distortions complicate the segmentation process and produce errors during pixel classification into object or background.

Common methods to attenuate high-frequency noise are spatial linear low-pass filters (such as the mean or the Gaussian filters) or non-linear noise reduction techniques that preserve object boundaries (median, anisotropic diffusion, bilateral, non-local means filters) (Haidekker 2010; Gonzalez and Woods 2018; Bankman 2008; Russ and Neal 2017). In general, linear filtering is being implemented using convolution. Kernels of mean and Gaussian filters are separable therefore filtering may be speeded-up by using two simpler kernels ($1 \times N_y$ and $N_x \times 1$) instead of single ($N_x \times N_y$). The mean filter further may be accelerated by using recursive formula implementation. Kernel of 2D circularly symmetric Gaussian is computed by:

$$K_G(x, y) = \frac{1}{2\pi\sigma^2} \exp\left(-\frac{x^2 + y^2}{2\sigma^2}\right), \tag{9.1}$$

here x, y—the distances from the origin in the horizontal and vertical axis; σ—standard deviation of the Gaussian distribution.

Image filtering using a mean filter in general case may be done by moving average filter:

$$I^o(x, y) = \frac{1}{N_x \cdot N_y} \sum_{(i,j)\in S_{xy}} I(i, j), \tag{9.2}$$

here S_{xy}—filter window of size $N_x \times N_y$ pixels.

A major drawback of low pass frequency filters is that they also blur boundaries of the objects together with noise attenuation. Blurred boundaries (i.e., reduced contrast) increase the uncertainty of object boundary detection. Non-linear filters may be used to reduce noise and preserve contrast (object boundaries). A commonly used non-linear filter is the median filter, which replaces each pixel by the median value in its neighborhood, which is defined by sliding window size:

$$I^o(x, y) = \underset{(i,j)\in S_{xy}}{\text{median}}\, I(i, j), \tag{9.3}$$

here S_{xy}—filter window of size $N_x \times N_y$ pixels.

Other examples of non-linear edge-preserving filters used for noise reduction in images are bilateral filter (Tomasi and Manduchi 1998), anisotropic diffusion (Perona and Malik 1990), Non-Local Means filter (Buades et al. 2005). The bilateral filter replaces each pixel's intensity with a weighted average of intensity values from its neighborhood, where weights depend not only on spatial distances but also on pixel intensity differences. Anisotropic diffusion is a locally adaptive filter which is realized as an approximation of the generalized diffusion equation by iteratively applying a set of computations until a sufficient degree of smoothing is obtained. Non-local means filter measures the similarity of the target pixel to the rest of the pixels and uses similarities to compute the weighted mean. Usage of such adaptive filters has additional costs of an increased number of parameters that must be tuned before applying the filter.

Requirements on filter selection and its optimal configuration originate from the segmentation algorithm's requirements for the input image (how the image has to appear) in order that the segmentation algorithm could produce maximally accurate segmentation results. For example, thresholding-based algorithms require high contrast and homogeneous objects to produce binary mask having regions that are solid and with precise boundaries. Watershed-based segmentation algorithms require that every object of interest in the image appeared as a dark spot in the light background and contained only one local minimum. Otherwise, more or less postprocessing operations will be needed to correct segmentation results (oversegmentation, undersegmentation, holes, spurious/single pixels). The noise reduction effects of various filtering methods (Gaussian, median, bilateral, anisotropic diffusion, non-local means filters) on segmentation by thresholding outcomes are presented in Fig. 9.3.

Background removal and uneven illumination correction have a dual purpose. One of the purposes is to improve image segmentation, the same as of previously described image smoothing techniques. The second one is to prepare an image for fluorescence intensity quantification by removing the background signal from objects' intensity. Varying background intensity is most likely caused due to imaging specifics—due to the instrument setup and imaging parameters, uneven illumination, or autofluorescence.

The most commonly used method for separating particular size objects from slowly varying background intensity is morphological top-hat transform. Mathematical morphology is a technique for processing and analysis of geometrical structures and is based on two basic operations—erosion and dilation. Grayscale erosion and dilation work like sliding window operations, which replaces each pixel by the minimum or maximum value in its neighborhood, which is defined by the structuring element $\boldsymbol{D}$. Top-hat transform is the difference between the input image and its morphological opening:

$$\boldsymbol{I}^{\mathrm{o}} = \boldsymbol{I} - \boldsymbol{I} \circ \boldsymbol{D}\,. \tag{9.4}$$

The opening operation (denoted by $\circ$) is the erosion (denoted by $\ominus$) followed by the dilation (denoted by $\oplus$) using the particular structuring element $\boldsymbol{D}$, which should have a shape of similar morphology but larger by size compared to the objects in the image we want to leave (nuclei, cells), and smaller in size compared to objects we seek to remove (varying background):

$$\boldsymbol{I} \circ \boldsymbol{D} = (\boldsymbol{I} \ominus \boldsymbol{D}) \oplus \boldsymbol{D}\,, \tag{9.5}$$

here $\boldsymbol{D}$—disk-shaped structuring element.

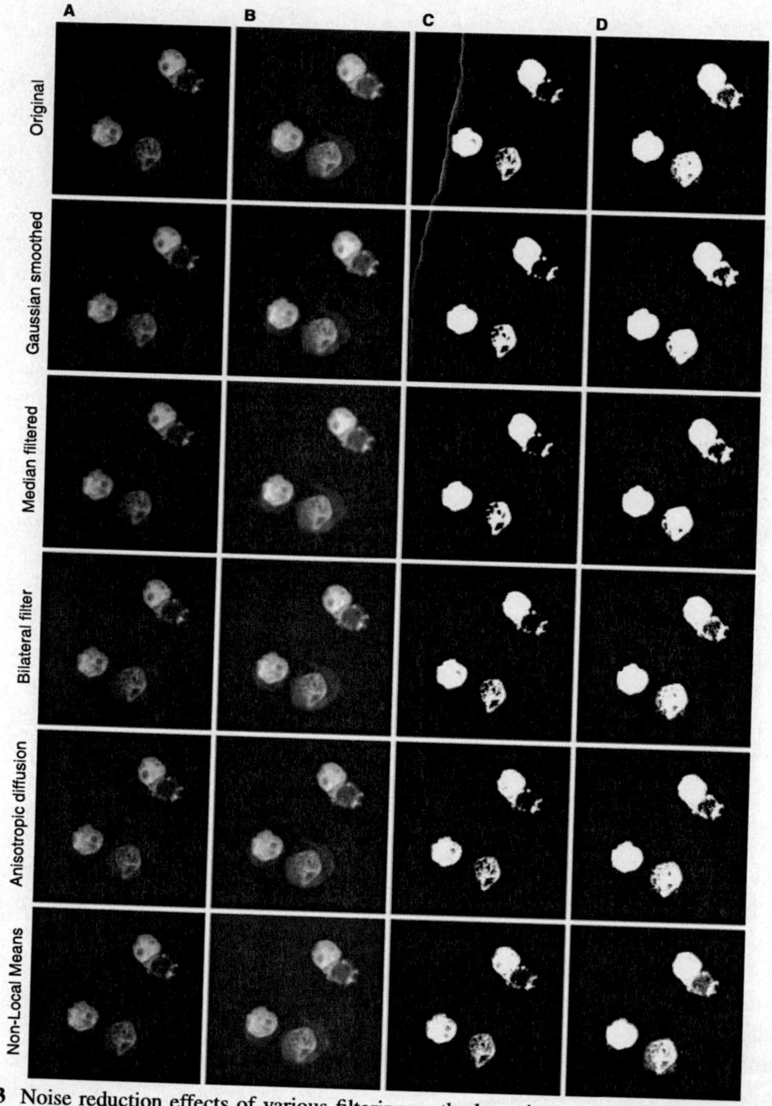

Fig. 9.3 Noise reduction effects of various filtering methods on image segmentation by thresholding. The array of images is composed as follows: column A contains images displayed using linear intensity mapping; column B contains images displayed using gamma intensity mapping to enhance variations in low intensities range; in columns C and D are shown the segmentation results of images from column A using two different thresholds. Rows hold images processed by different filters: original image, Gaussian smoothed, median filtered, using the bilateral filter, using anisotropic diffusion, Non-Local Means filtered. Segmentation of Gaussian and median filtered images gave binary masks that have the most solid objects and with least free/detached pixels compared to the rest results

9.1.3 *Image Analysis*

The image analysis stage, during which the object segmentation is performed, is the critical step of automated fluorescence image analysis. The purpose of image analysis is to extract information about the locations of individual objects and delineate their boundaries. The objects we want to segment may be nuclei, whole cells, or locations of some protein of interest. The extracted information about segmented objects is typically encoded in a labeled image, in which pixel values represent object labels. Created labeled images can be used for further measurements of objects to collect geometrical and/or photometric features.

9.1.3.1 Nuclei Segmentation

The goal of nuclei segmentation is to generate a labeled image-mask that defines regions of separate nuclei. Such a mask may serve as a region of interest (RoI) to define whole regions of nuclei, as a precursor for regions of nuclei boundaries, or as a seed area for whole cell segmentation.

The most common image segmentation methods used for nuclei detection and segmentation are:

- thresholding (Otsu 1979; Kapur et al. 1985; Xue and Zhang 2012);
- Watershed transform (Vincent and Soille 1991; Soille 2013);
- local image filters (Wang and Cao 2019; Gudla et al. 2008);
- deformable models (Sethian 1999; Kass et al. 1988);
- mathematical morphology based methods (Soille 2013; Fouad et al. 2016; Vincent 1993).

Image thresholding can be successfully applied for object segmentation when there is a high contrast between object and background, and objects are not touching (Hegde et al. 2019; Ranefall and Wahlby 2016; Andrade et al. 2019). In case when there are objects of different intensities or objects are touching, segmentation results are not good enough to be used as masks for cell parameterization. Besides, there exists a difficult task of selecting the right threshold. Well known threshold selection methods, like Otsu (1979), may not provide optimal value for the current problem. Struggles of thresholding to segment objects are shown in Fig. 9.4. Image preprocessing before applying thresholding may improve the results (Fig. 9.5), but then additional decisions on the preprocessing algorithm and its parameters selection will be required.

In order to separate touching cells or nuclei, other methods that take into account spatial relations between pixels are required (Prinyakupt and Pluempitiwiriyawej 2015). One of such methods is Watershed transformation, which isolates local minima in separate image regions forming separate catchment basins (Vincent and Soille 1991; Soille 2013). The 2D and 3D visualization of the Watershed algorithm in action is presented in Fig. 9.6. Due to the nature of the algorithm, Watershed

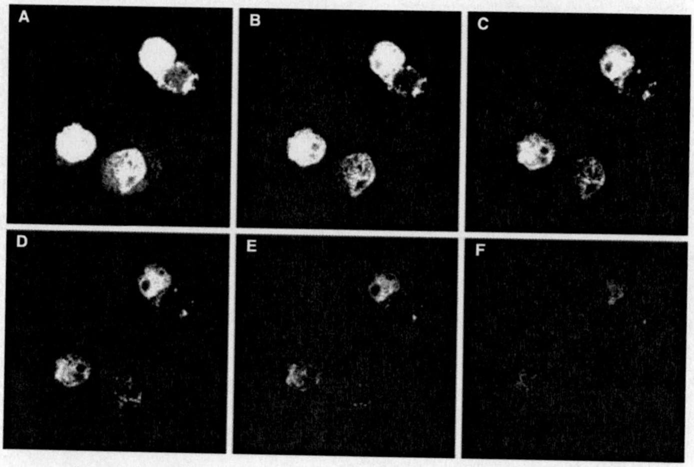

Fig. 9.4 The problem of selecting the right threshold. It is illustrated by examples of cell nuclei segmentation by thresholding at increasing threshold levels. The threshold in case (B) is the same as given by the Otsu method

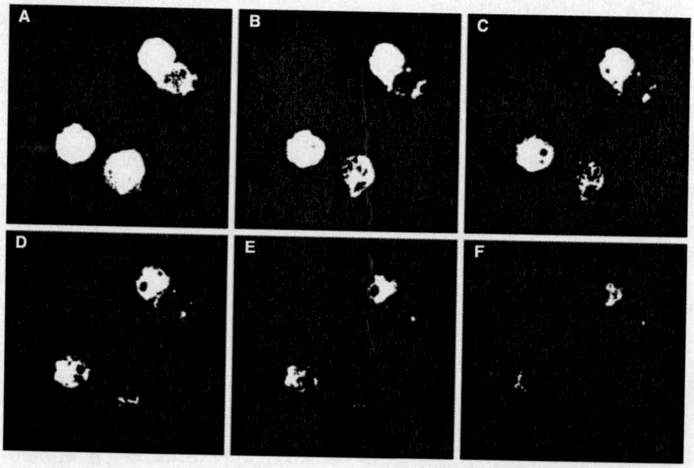

Fig. 9.5 Cell nuclei segmentation by thresholding at increasing threshold levels after the image was smoothed using a median filter of size 5×5. The threshold in case (B) is the same as given by the Otsu method

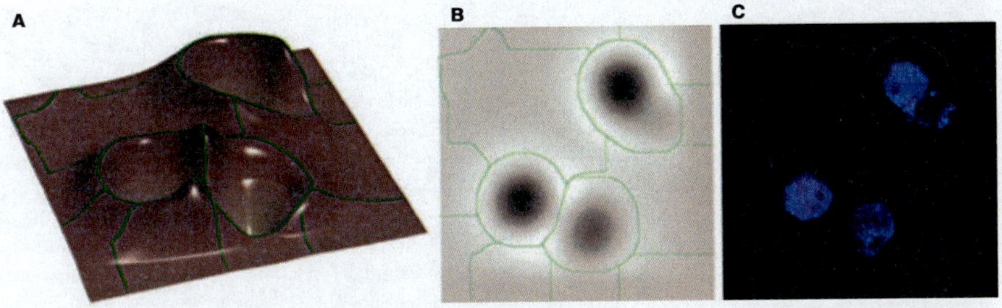

Fig. 9.6 Visualization of Watershed segmentation principle in 3D (**A**) and 2D (**B**). Green Watershed contours are the result of segmentation of intensity map (**B**), which is an outcome of nuclei image (**C**) preprocessing by Laplacian of Gaussian filter

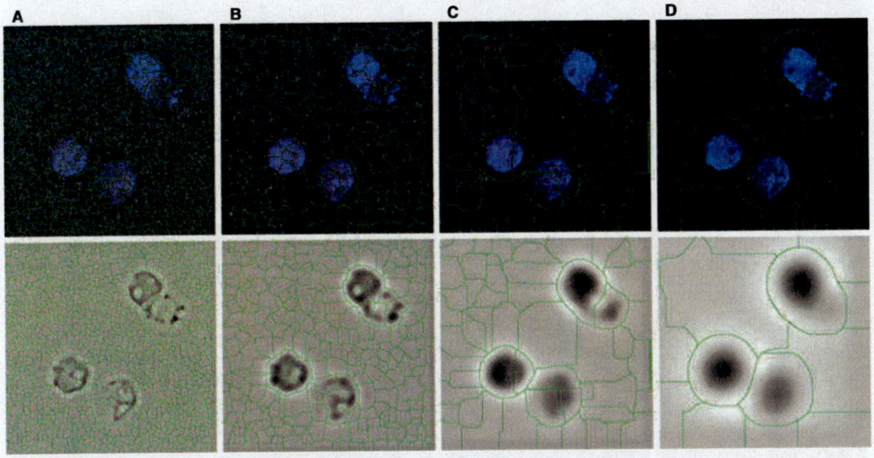

Fig. 9.7 The problems of Watershed noise sensitivity and selection of image preprocessing type to balance between oversegmentation and undersegmentation. They are illustrated by examples of Watershed segmentation of images that were preprocessed by Laplacian of Gaussian filters with increasing variance (from (**A**) to (**D**)) (in the first row—original images, in the second—preprocessed)

segmentation is very sensitive to the noise that creates false local minima, and the algorithm treats them the same way as the local minima corresponding to the object region. Elimination of false local minima or postprocessing of oversegmentations is required to employ the Watershed algorithm as an object segmenter. Results of the Watershed segmentation algorithm applied to differently preprocessed images are presented in Fig. 9.7.

Local image filters try to amplify some specific shape features of the object and exploit these extracted features to detect object centroids and contours (Bashar et al. 2012; Li et al. 2007; McCullough et al. 2008; Nandy et al. 2009; Choudhry 2016). A simple example of this kind of algorithms is edge detection. The difficulties of edge detection to form connected and closed contours are shown in Fig. 9.8. Deformable

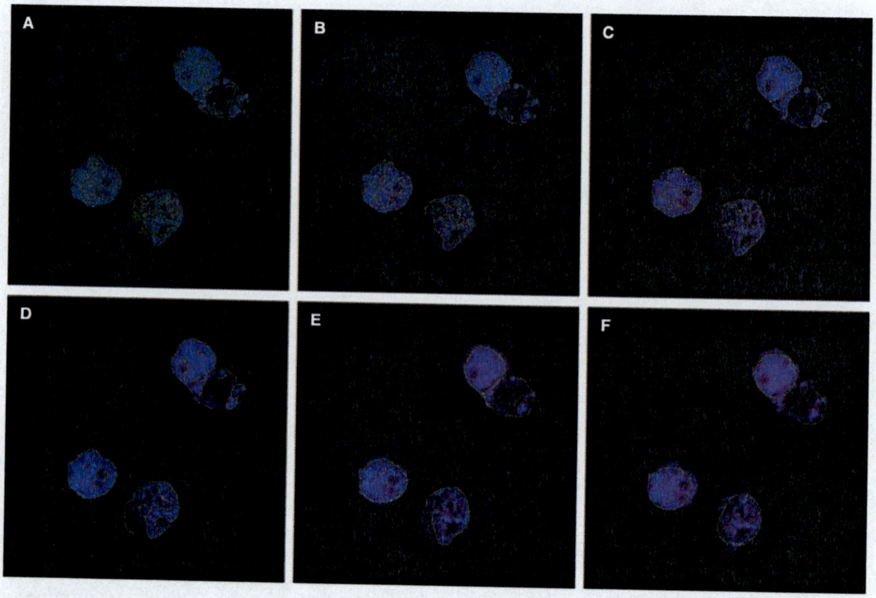

Fig. 9.8 Examples of edge detection-based nuclei segmentation using differently smoothed images (resulted contours are superimposed on original images). The algorithm fails to extract the solid and closed boundaries of the objects

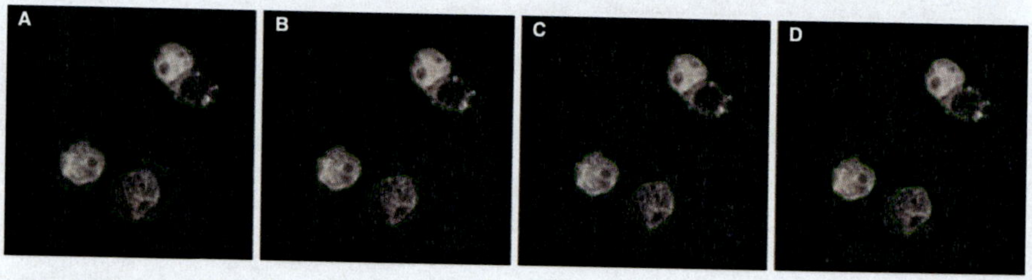

Fig. 9.9 Examples of segmentation results by fast marching method using different thresholds. Optimal threshold values for individual objects are different

models are based on level sets (Sethian 1999; Kass et al. 1988). This group of methods typically require some form of initial contour from which to start the evolution of the contour around the object (nucleus, cell) we want to delineate. Some algorithms may require to provide additional parameter values for the algorithm. Segmentation results using fast marching (Fig. 9.9) and active contours (Fig. 9.10) show that it is a difficult task to select a set of parameters that provides good segmentation results for all the objects in the image.

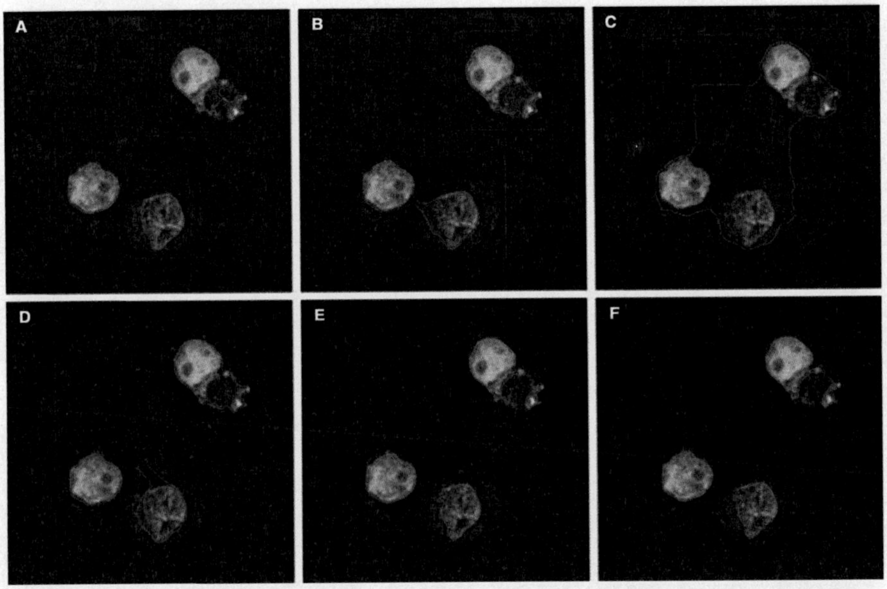

Fig. 9.10 The problem of parameter selection for segmentation using active contours. Different initializations (**A**)–(**C**) and different iteration counts (**D**)–(**F**) provide a set of divergent results

9.1.3.2 Cell Segmentation

Cell segmentation is a detection task of fluorescent intensities in other channels than the nucleus and linking these intensities to a particular cell defined by nuclei. The segmentation process may be driven by using information about the nucleus location as a seed point to grow the area of the cell. Each pixel outside the nuclei that shows some minimal required intensity above the background can be treated as belonging to one of the cells and, at the same time, to one of the nuclei. So the segmentation of the cell may be treated as the task of assigning outer pixels (that possess some minimal fluorescent intensity) to one of the nuclei. Pixels without any signal are harder to assign to one of the candidate cells or background using only intensity information which is absent. The damage to the measurements would remain minimal if the no-signal pixels were classified incorrectly in cases when we are interested only in total fluorescent intensity present in cell area but not in mean intensity or cell size. A class of "no-signal pixel" may be inferred using additional knowledge about cell convexity and/or uniformity (continuity).

9.1.3.3 Image Postprocessing

The final stage of image analysis is image postprocessing. Postprocessing may be required if the output data of nuclei and cell segmentation stages contained minor errors that need to be corrected before starting data analysis or if modifications of segmented object representation are needed. At this stage, remaining errors that should be dealt with are oversegmentations (needs segment merging), undersegmentations (needs segment splitting), holes in the objects (filling holes), touching objects, fake objects (remove artifacts), geometrical distortions of object shapes. Transformations of the object representation may be changing object region masks to object contours, object skeletonization.

Additionally, at this stage, manual error correction may be performed. Erroneous segmentation results may be presented to the user for review, and later manual corrections through an interactive interface may be applied.

9.1.4 Data analysis

The objective of the quantitative fluorescence image analysis is to extract certain quantitative features that would allow us to discover biologically meaningful information about the biological experiment in progress (Waters and Wittmann 2014; Seruca et al. 2017; Ouyang et al. 2019; Antony et al. 2013). Quantification of fluorescence intensity and differential analysis, the execution of statistical data analysis, and data visualization to present measurements are finalizing quantitative fluorescence image analysis steps.

Localization or quantity of fluorescence in the images may be of interest and described by measuring image pixel intensity values in the defined regions, providing quantitative features. These regions are identified during the object segmentation stage. Although various processed images were created for image segmentation, intensity measurements should be performed on the background-corrected original images (Waters 2009; Ljosa and Carpenter 2009). In order to make accurate measurements of the fluorescence intensities, the background must be corrected (Waters and Wittmann 2014). After feature extraction, statistical data analysis should be performed to evaluate differences (or their absence) and their statistical significance between experimental group samples (Dukes and Sullivan 2007; Krzywinski and Altman 2014a). These differences may be related to the biological processes which are under investigation. Distribution of data from quantitative imaging experiments needs to be determined (normal or not) and appropriate statistics selected to analyze the data (Usaj et al. 2016). More frequently data are not normally distributed (Krzywinski and Altman 2013a). Data should be visualized with included selected statistics, and the type of statistics used reported (Krzywinski and Altman 2013b). It is recommended to choose such data visualization type that would present as much information as possible and reveal underlying data distribution (Krzywinski and Altman 2014b; Collins and Huett 2018).

Subsection Generalization

1. The main steps of automated fluorescence image analysis workflow are:
   - fluorescence image acquisition;
   - image preparation;
   - image analysis;
   - data analysis.
2. The most common image segmentation methods used for nuclei detection and segmentation are:
   - Otsu thresholding;
   - Watershed transformation;
   - deformable models;
   - mathematical morphology operations;
   - local image filters.

9.2 Automatic Protein Localization in Fluorescence Images of Blood Cells

We aim to develop an automated fluorescence image analysis tool that would allow us to minimize the amount of required manual operations during the initial tuning of various parameters of the segmentation algorithm as well as during the inspection of segmentation results.

We have identified several bottlenecks of our fluorescence image analysis workflow. These bottlenecks occur due to the demand for intense manual intervention of the laboratory scientist. Providing automation solutions to deal with these bottlenecks would be the most beneficial in order to increase experiment flow speed. Given that we already have a functional system from the previous fluorescence image analysis experiments, the most significant bottlenecks of adapting the system for the new experiment would be as follows (and possible approaches to remove them by using automation solutions):

B1. The need to manually adjust parameters of segmentation algorithms (including parameters of image preprocessing and postprocessing stages). Approaches to manage it:

- Reduce the number of parameters that must be adjusted manually by designing image analysis algorithms and their configurations such that would lead to the adaptability to the particular data set.
- Automatically detect optimal values of required parameters by exploiting regularities of the objects in the images.

B2. The amount of labor and time required to inspect and correct object segmentation results (detected boundaries of cell nuclei and other cell areas of interest). Approaches to manage it:

- Automatically detect inconsistencies in output data that may be indicators of existing errors in the analysis results, and bringing user attention to review and manage such situations.
- Provide user-friendly interface tools for error correction.

B3. Other preparatory tasks for the image analysis experiment—image channel separation, assigning images to experimental groups. Approaches to manage it:

- Suggest for the user probable nuclei channel in the images.
- Automatically assign images to distinct experimental groups according to the directories structure where image files are located.

In the following subsections, we propose possible solutions to overcome listed bottlenecks using the provided approaches. A prerequisite for successful automatic analysis, in this case, is the availability of as complete as possible image set (all images from all experimental groups). Automatization is enabled based on the expectation that all the images from all experimental groups are available for the image analysis system. Some steps need some statistics derived from a group of images to initialize its parameters. Providing a single or tiny pack of images for analysis may lead to incorrect predictions about image contents (i.e., an incorrect guess of nuclei channel, morphological nuclei groups by size and shape).

9.2.1 Fluorescence Image Acquisition Setup

Here we describe the configuration of a confocal imaging system for fluorescence image acquisition. Confocal imaging was performed using Bio-Rad Radiance 2100 and Radiance 2000MP laser scanning systems (Carl Zeiss AG, Germany). Images were taken in sequence after inserting the signal enhancing lenses by activating channel 1 (blue, not used): Mai-Tai laser (815 nm), with dichroic beam splitter 500DCLPXR, blocking filter BGG22 and emission filter D488/10; channel 2 (green): Argon laser (488 nm), no blocking filter and emission filter HQ545/40; and channel 3 (red): Argon laser (488 and 514 nm), no blocking filter and the emission filter E600 LP. The Radiance scan heads are attached to the Nikon Eclipse TE2000U (Tokyo, Japan) optical microscope. The microscope is equipped with PlanApo DicH x60 oil immersion objective (NA 1.40).

9.2.2 Image Preparation

In the previous Sect. 9.1 we described image preparation as the initial stage of image analysis during which the image preprocessing and image channel separation is performed. Other configuration tasks of the experiment, like the assignment of the images to the experimental groups, may also be attributed to the current stage. At the beginning of the current Section, we listed the identified bottlenecks of microscopy image analysis workflow we want to solve. Bottlenecks related to the image preparation stage are as follows: manual adjustment of parameters of the image preprocessing algorithms (B1), separation of image channels, and distribution of collected images into experimental groups (B3).

According to the sequence of operations of the workflow, distribution of collected microscopy images into experimental groups, and decomposition of images into individual images according to the color channels are the earliest but the least labor-intensive. Images were assigned to distinct experimental groups using the information of the directory structure where the image files were located. Names of experimental groups may be initialized from the folder names and edited by the user if needed.

Channel separation is required because individual channels of fluorescence images convey different biological information. One of the image channels hold intensity information of visualized nuclei, and the rest of the channels (signal channels) may contain fluorescence intensity information of the labeled biomolecules under investigation. Nuclei channel may be recognized by some common appearance features, as it typically contains more fluorescence intensity and less intensity variations between images from different experimental groups. The channel separation algorithm may provide its initial guess about the nuclei channel, and the user should review and confirm or correct it. If there is only a single kind of biomolecules under investigation, the other, non-nuclei, channel containing fluorescence data can automatically be treated as the signal channel, and the last channel would be left aside. In case there are two differently labeled biomolecules of interest, then a manual assignment of both channels to the signal channels will be required.

Manual adjustment of image preprocessing algorithm parameters is a challenging and responsible task. The purpose of image preprocessing is an improvement of image segmentation results—to suppress or remove such image properties that mislead the segmentation process. Typically the user has to adjust parameters of preprocessing, segmentation, and postprocessing algorithms in tandem. It means that we get a search space that grows exponentially with the number of parameters. Additionally, the user may be encountering a situation where no proper combination of parameters exists. A review of several popular preprocessing algorithms that can be found in literature and software packages is presented in the previous Section, where we experimentally exposed shortcomings of those algorithms and difficulties in readapting them when experimental conditions change. We achieved a reduction of the number of parameters that should be adjusted manually by designing image analysis algorithms that can adapt to the particular data set and do not need image preprocessing. The development of adaptive segmentation algorithms is described in the next Subsection.

9.2.3 Nuclei Detection and Segmentation

Our approach to creating an automatic algorithm for fluorescence image analysis that does not need its parameter tuning due to change of nuclei morphology and/or image acquisition settings, or some other factors, is based on these concepts:

- scale-space representation of images to achieve multiscale processing. We use Gaussian and Laplacian pyramids;
- automatic estimation of nuclei average size using multiscale blob detection and using average size to initialize other segmentation algorithms;
- marker guided Watershed transform for extraction of super-pixels that would be used in the subsequent analysis;
- region split-merge approach to post-process initial image partitioning into regions (super-pixels);
- MSER (Maximally Stable Extremal Regions) detector to generate resolution independent region of interest (RoI) proposals of images;
- usage of multiple independent sub-algorithms for nuclei segmentation and employing voting strategy in order to generate a unified hypothesis about nuclei regions and increase the reliability of segmentation results.

Nuclei segmentation is a critical step for successful fluorescent image analysis. This step's segmentation results are used as a seed point to segment the rest parts of the cell. To create an algorithm that would give maximally reliable detection results, we combined three independent detection and segmentation algorithms. The results of these separate algorithms later are merged into a single hypothesis using a voting approach.

All algorithms use some kind of multiscale image analysis. This approach is necessary to be able to detect nuclei of various sizes and shapes. The first independent branch (algorithm) is based on Gaussian and Laplacian pyramid (see Fig. 9.11). The evolution of local maxima (centroid areas) between scales is analyzed to detect centroids that optimally describe nuclei locations and to get estimate of nuclei average size. The second branch is using Watershed transform to fragmentize images of the Laplacian pyramid (see Fig. 9.12). This operation forms super-pixels that later are merged into larger blobs spanning the area of nuclei (Fig. 9.13). The third branch is based on maximally stable extremal regions (MSER)—the method used for blob detection in images (see Fig. 9.14).

Before running individual branches of the algorithm, for efficiency, we pre-compute some intermediate images as they will be reused in several steps of the main algorithm. Precomputing includes the construction of Gaussian and Laplacian pyramids. Gaussian pyramid (see Fig. 9.11) is a set of images where the lowest level

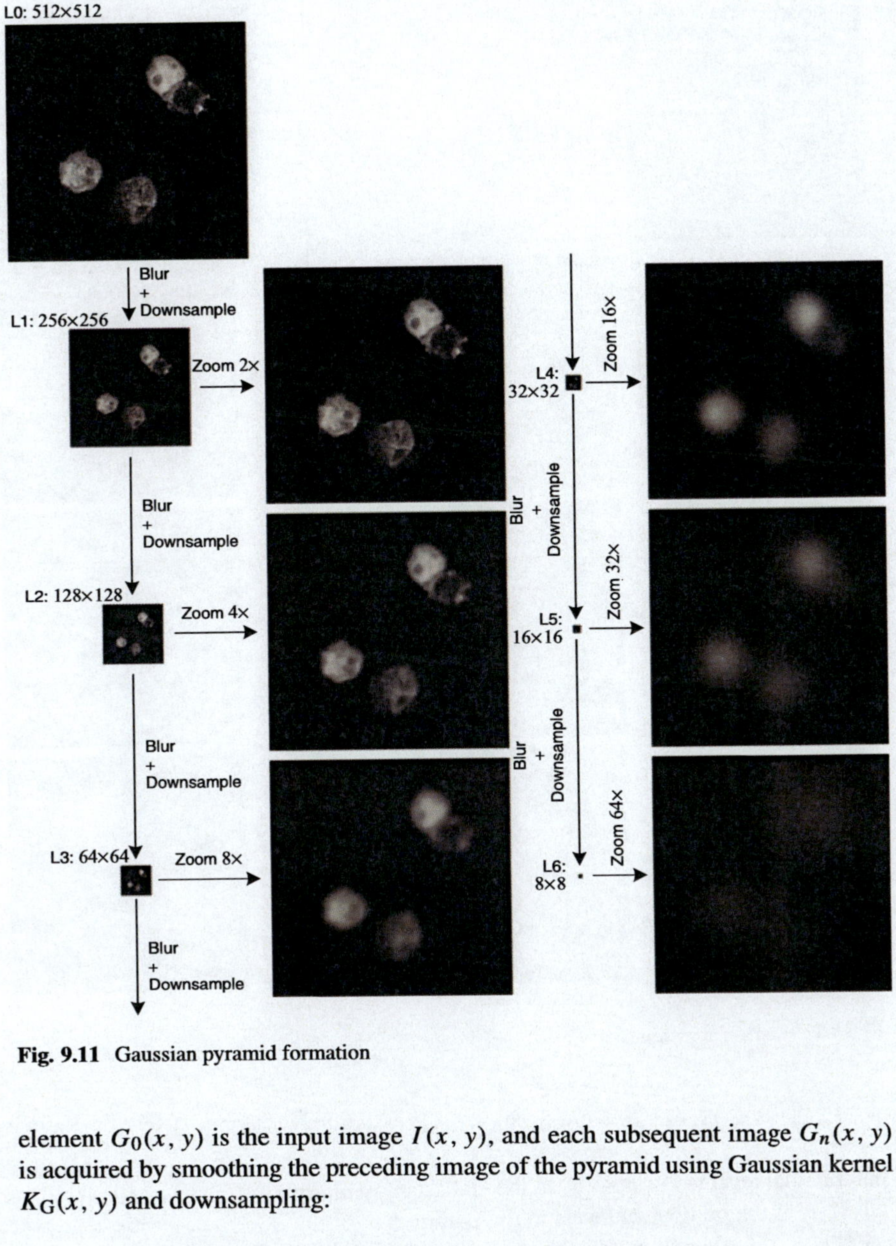

Fig. 9.11 Gaussian pyramid formation

element $G_0(x, y)$ is the input image $I(x, y)$, and each subsequent image $G_n(x, y)$ is acquired by smoothing the preceding image of the pyramid using Gaussian kernel $K_G(x, y)$ and downsampling:

$$G_{n+1}(x, y) = \text{downsample}\,(G_n(x, y) * K_G(x, y)) \; . \tag{9.6}$$

Laplacian pyramid (see Fig. 9.12) is a set of bandpass images $L_n(x, y)$, which are the differences of adjacent images of the Gaussian pyramid:

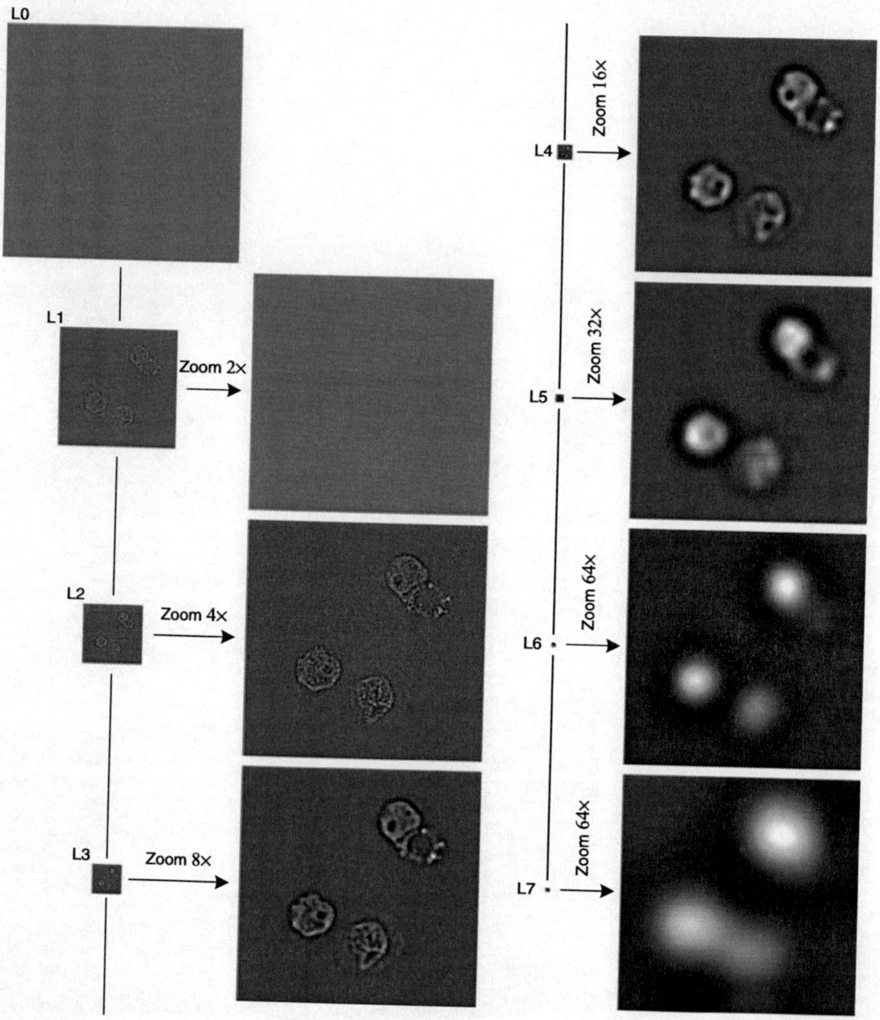

Fig. 9.12 Laplacian pyramid

$$L_n(x, y) = G_n(x, y) - \text{upsample}\,(G_{n+1}(x, y)) \quad \forall n \in [0, N-2], \tag{9.7}$$

but the final level is the same as in the Gaussian pyramid: $L_{N\text{-}1}(x, y) = G_{N\text{-}1}(x, y)$, where N is the number of levels in the pyramid.

The need to manually adjust parameters of segmentation algorithms was one of the bottlenecks (B1) of the microscopy image analysis workflow we aim to solve. To address this bottleneck, one of the suggested approaches was to automatically detect optimal values of required parameters by exploiting regularities of the objects in the images. If we want to discover some regularities across the batch of microscope

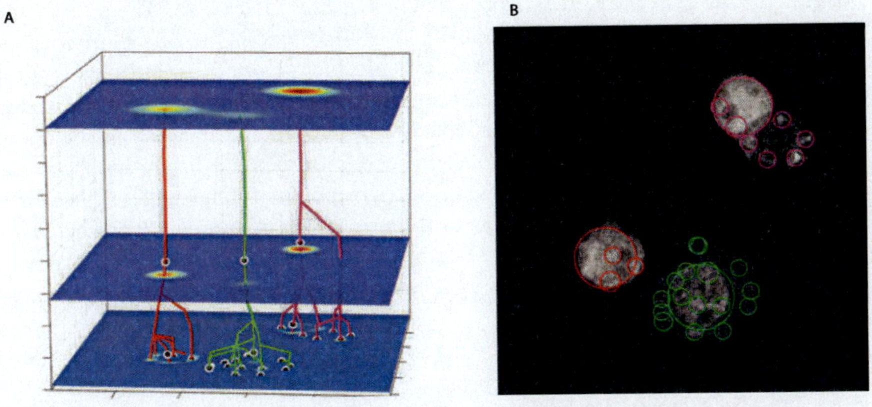

Fig. 9.13 Multiscale blob detection with automatic scale selection. Detected local extrema locations in scale space marked with white circles (**A**) and corresponding decoded locations of detected blob structures drawn as colored circles (**B**). Paths in (**A**) show the evolution of local spatial extrema as scale changes, and the color of the line encodes the clustering result of the extrema points to indicate dependencies of points to different nuclei. Finally, only the extrema points with the largest scale from each path are kept for estimation of nuclei size statistics

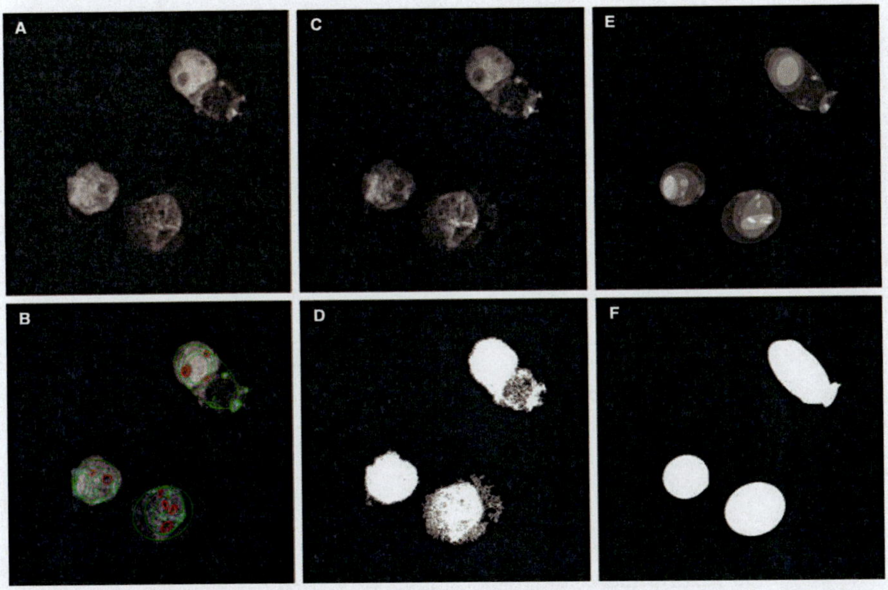

Fig. 9.14 Application of Maximally Stable Extremal Regions (MSER) algorithm for blob detection in images. Original image (A) is processed by MSER and detected stable regions fitted with ellipses (B) where green ellipses are for bright blobs and red—for dark blobs. Image reconstructed from a subset of detected regions (C) and the binary mask (D) after thresholding reconstructed image at the lowest threshold. In (E) is the image reconstructed by voting from the ellipses (B), and its thresholded version at the lowest threshold (F) to be used as a mask of nuclei regions

images, in the first place, it is worth looking at the nuclei size. The size of the nuclei would be the least scattered attribute of the object. One of the approaches to detect unknown scales of the objects is to use ideas from the multiscale interest point detectors—the scale-normalized Laplacian operator for blob detection. Using this method it is possible to simultaneously estimate spatial location and scale of dark and bright blobs. Nuclei in the fluorescence images are bright blobs, so the Laplacian operator ∇^2 will give strong negative responses for the nuclei whose sizes (represented by radius r) are similar to the size of the Gaussian (represented by standard deviation of the Gaussian, σ) kernel $K_{\mathrm{G}}(x, y; \sigma)$, that was used to smooth the image $I(x, y)$ before applying the Laplacian operator. The smoothed image:

$$I^{\mathrm{L}}(x, y; \sigma) = I(x, y) * K_{\mathrm{G}}(x, y; \sigma) \tag{9.8}$$

is a scale space representation at the scale σ. The relationship between the blob radius r and the Gaussian size σ is: $r = \sqrt{2}\sigma$. To find the scale (and location) where the scale-normalized Laplacian of Gaussian (LoG) $\sigma^2\nabla^2 K_{\mathrm{G}}(x, y; \sigma)$ would give strong negative responses for the bright nuclei, we have to apply this filter to the image at the range of scales producing a 3D volume $\nabla^2 I^{\mathrm{L}}(x, y; \sigma)$ and detecting in it scale-space local minima. Scaled-normalized LoG filters may be approximated by the Difference of Gaussian (DoG) filters, whose scales differ by the factor k:

$$\sigma^2\nabla^2 K_{\mathrm{G}}(x, y; \sigma) \approx \frac{1}{k-1}\left(K_{\mathrm{G}}(x, y; k\sigma) - K_{\mathrm{G}}(x, y; \sigma)\right), \tag{9.9}$$

so the images from the Gaussian pyramid may be used for efficient computation of convolution of image $I(x, y)$ with DoG filter bank by performing simple image subtractions. It is the same operation as the construction of Laplacian pyramid, just extended to have additional intermediate scales.

To reduce false-positive locations of the nuclei centers, the scale-space local minima points are pruned by clustering according to dependence to the same object and selecting only points with the largest scale. Clustering is performed by detecting local minima only in space and for all scales, and by tracking the evolution of these local minima as the scale changes (visualization presented in Fig. 9.13). After that, local minima on the paths and in the scale are detected, and only the points with the largest scale from each path are kept. Deriving scale statistics from collected local point scales provides estimates of the nuclei size in the images.

Another proposed approach to overcome the bottleneck of manual adjustments of segmentation algorithm parameters is to reduce the number of parameters that must be adjusted manually by designing image analysis algorithms and their configurations that would lead to the adaptability to the particular data set. A combination of multiscale image analysis methods (utilization of Gaussian and Laplacian pyramids in analysis, using maximally stable extremal regions algorithm) and parameter-free Watershed transformation enables to achieve a reduction of parameters.

Analysis of the images from Gaussian and Laplacian pyramids in a similar way like in nuclei size estimation provides more reliable locations of nuclei centers, which are used for marker controlled Watershed segmentation. Applying Watershed transformation on images of Laplacian pyramid gives a multiscale segmentation of nuclei image. The Watershed regions from multiple scales are used for analysis in the last step of the segmentation algorithm, where voting is performed to generate unified a hypothesis. Maximally stable extremal regions (MSER) algorithm is another method for blob detection in images. MSERs are such image regions (connected components) that are stable (nearly the same) through a range of thresholds (Fig. 9.14). A resulting mask (Fig. 9.14F) is used in the last step of the segmentation algorithm, where voting is performed.

9.2.4 Cell Segmentation

The task of cell segmentation involves detecting regions where the fluorescent signal of interest should be quantified. This signal is related to biological processes that we want to analyze. These can be cases where we want to discriminate localization of proteins between nuclear and cytoplasmic areas. In other cases, knowing the area of the whole cell is sufficient and just care to separate it from the area of other cells. Nuclear regions of cells were determined during previous steps. In this processing stage, information about nuclei regions is used to delineate the whole cell's regions. Previously detected nuclei serve as seed points to expand the regions of interest to the extents of the cells.

Region growing approach is used to expand nuclei regions until the probable edge of the cell is reached. This process can be seen as the assignment of outer (perinuclear) pixels to one of the nuclei. Because some pixels may belong to a background class (do not belong to any cell), therefore, pixel should possess some minimal fluorescence intensity above background intensity in order to be classified as belonging to one of the nuclei/cells. Region growing may be completed more efficiently and with more consistency if classification is performed not on simple pixels individually but on pixel groups (super-pixels). Super-pixel grid is a Watershed transform of the image that is a result of adding all the channels of the initial image and saturating the result. Super-pixels are assigned to a particular cell class related to the nucleus using the lowest geodesic distance criteria. A geodesic distances map is generated using the image of combined channels and the regions from the nuclei mask as seed areas.

9.2.5 Image Postprocessing

In our microscopy image analysis workflow, the image postprocessing step was dedicated to the semi-manual inspection of segmentation results. The inspection involves the detection and correction of possible segmentation errors. Correction may be done by interactively providing corrections to segmentation masks of the objects (nuclei and/or cells), or by choosing to objectively exclude erroneous segments because they fall out of the target group distribution. The amount of labor and time required to inspect object segmentation results was one of the bottlenecks of the microscopy image analysis workflow we aim to solve (B2). To overcome the bottleneck, the following approaches were suggested: automatically detect inconsistencies in output data that may be indicators of existing errors in the analysis results, bring user's attention to the picked out situations to check and manage them, and provide user-friendly graphical interface tools for error correction.

The sorts and the sources of the segmentation errors:

- algorithm failure to split highly overlapped nuclei;
- oversegmented nuclei;
- non-cell related artifacts in images that were segmented and classified as nuclei/cells;
- particular cell phenotypes that are not related to the biological experiment in progress and should be filtered out.

The strategy to make the inspection process more efficient is to prioritize segmentation results that must be reviewed. Arranging inspection tasks in this way would allow limiting the review process just to some subset of images that must get the most attention and consequently reduce the total workload. The strategy's implementation is achieved by using specialized graphical interfaces for algorithm-assisted error detection and for entering the corrections. Essential parts of the graphical interface are summarized presentation of segmentation results and interactive tools for corrections input.

Concepts of graphical presentation of segmentation results for algorithm-assisted error detection:

- automatically select the outlier nuclei candidates based on geometric features (area, circularity, solidity) and photometric features (total fluorescence intensity of the nucleus), and present them sorted by the distance from the mean in descending order;
- automatically select the outlier cell candidates based on photometric features (total fluorescence intensity of the cell), and present them sorted according to the distance from the mean in descending order;
- display scatterplots of nuclei/cells geometric and photometric features; visualize interclass and intraclass variations of the features; clicking the data point displays segmentation results of the associated object.

Design solutions on the user interface to enter corrections:

- display original nuclei or cells image with the overlayed super-pixel grid. Super-pixels are homogeneous regions of the image, smaller than the objects, so the super-pixel grid appears as an oversegmented image. Typically there exists a

subset of super-pixels that overlaps with the object region with a high IoU (intersection over union) score; therefore, it is possible to construct an object mask by selecting a group of super-pixels;
- select nuclei or cells layer to activate; start correcting from the nuclei layer;
- perform manual correction of the selected object mask (object selections come from the algorithm-assisted error detection step). For more details, refer to the list of user interface elements for the manual correction of the selected object mask;
- rerun automatic cell segmentation after correction of nucleus segmentation and, if required, continue to the cell segmentation correction.

User interface elements for the manual correction of the selected object mask:

- For nucleus segmentation, if the nucleus layer is selected:
  - mouse click super-pixel to assign it to the nearest nucleus; shift-click to classify as background; ctrl-click to assign it to the additional instance of the nucleus (if needed to correct undersegmentation case);
  - drag-and-drop super-pixel to the area of the required nucleus to assign it to that class;
  - use lasso selection tool—selecting super-pixels outside nucleus together with the nucleus regions (full or parts), assigns super-pixels to the nucleus class.
- For cell segmentation, if cell layer is activated (or select particular cell layer, If exist more that one):
  - click super-pixel to assign it to the nearest cell; shift-click to classify as background;
  - drag-and-drop super-pixel to the area of the required cell to assign it to that class;
  - use selection tool—selecting super-pixels outside cell together with the cell regions (full or parts), assigns super-pixels to cell class.

9.2.6 *Data Analysis and Visualization*

After the isolation of individual cells, the total fluorescence intensity of each cell can be obtained. Protein quantity in the particular area of the cell is related to the total fluorescence intensity in a segmented region of the corresponding cell. The total intensity of an object is the sum of the intensities of all the pixels that make up the object:

$$V^{(r)} = \sum_{(x,y)\in S^{(r)}} I(x, y). \tag{9.10}$$

here $S^{(r)}$ – segmented region of r-th cell; x, y – pixel coordinates.

Application of designed system, and acquired fluorescence image analysis results are presented in Fig. 8.3, 8.4, 8.5, and 8.6. After quantifying all cells, median values of total fluorescence intensities of each cell grouped by class were computed. The ratios between median values represent fold-change in protein expression. An increase of fluorescence intensity yields a positive fold-change and a decrease—a negative fold-change. Wilcoxon rank-sum test was used to evaluate statistically significant changes.

If data provided by the fluorescence image analysis are not normally distributed, the Wilcoxon rank-sum test may be used as a nonparametric alternative to the two-sample t-test for independent samples. The Wilcoxon rank-sum test allows a hypothesis test of the equality of medians of two samples.

References

Andrade AR, Vogado LHS, de M S Veras R, Silva RRV, Araujo FHD, Medeiros FNS (2019) Recent computational methods for white blood cell nuclei segmentation: a comparative study. Comput Methods Programs Biomed 173:1–14

Antony PPMA, Trefois C, Stojanovic A, Baumuratov AS, Kozak K (2013) Light microscopy applications in systems biology: opportunities and challenges. Cell Commun Signaling 11(1):24

Bajcsy P, Chalfoun J, Simon M (2018) Web microanalysis of big image data. Springer, New York

Bankman I (2008) Handbook of medical image processing and analysis, 2nd edn. Elsevier, Amsterdam

Bashar MK, Komatsu K, Fujimori T, Kobayashi TJ (2012) Automatic extraction of nuclei centroids of mouse embryonic cells from fluorescence microscopy images. Plos One 7(5):e35550

Berg S, Kutra D, Kroeger T, Straehle CN, Kausler BX, Haubold C, Schiegg M, Ales J, Beier T, Rudy M, Eren K, Cervantes JI, Xu B, Beuttenmueller F, Wolny A, Zhang C, Koethe U, Hamprecht FA, Kreshuk A (2019) Ilastik: interactive machine learning for (bio)image analysis. Nat Methods 16(12): 1226–1232. https://doi.org/10.1038/s41592-019-0582-9

Buades A, Coll B, Morel JM (2005) A review of image denoising algorithms, with a new one. Multiscale Model. Simul. 4(2):490–530

Cao H, Liu H, Song E (2018) A novel algorithm for segmentation of leukocytes in peripheral blood. Biomed Signal Process Control 45:10–21

Carpenter AE, Jones TR, Lamprecht MR, Clarke C, Kang IH, Friman O, Guertin DA, Chang JH, Lindquist RA, Moffat J, et al. (2006) Cellprofiler: image analysis software for identifying and quantifying cell phenotypes. Genome Biol. 7(10):R100

Chiarini-Garcia H, Melo RCN (2011) Light Microscopy. In: Methods and Protocols. Springer, New York

Choudhry P (2016) High-throughput method for automated colony and cell counting by digital image analysis based on edge detection. PloS one 11(2):e0148469. https://doi.org/10.1371/journal.pone.0148469

Collins A, Huett A (2018) A multi-phenotypic imaging screen to identify bacterial effectors by exogenous expression in a HeLa cell line. Sci Data 5(1):1–12. https://doi.org/10.1038/sdata.2018.81

Courtney J, Woods E, Scholz D, Hall WW, Gautier VW (2015) MATtrack: a matlab-based quantitative image analysis platform for investigating real-time photo-converted fluorescent signals in live cells. PloS one 10(10):e0140209

De Chaumont F, Dallongeville S, Chenouard N, Hervé N, Pop S, Provoost T, Meas-Yedid V, Pankajakshan P, Lecomte T, Le Montagner Y, et al. (2012) Icy: an open bioimage informatics platform for extended reproducible research. Nat Methods 9(7):690–696

Domínguez C, Heras J, Pascual V (2017) IJ-OpenCV: Combining ImageJ and OpenCV for processing images in biomedicine. Comput Biol Med 84:189–194

Dukes KA, Sullivan LM (2007) A review of basic biostatistics. Evaluating Techniques in Biochemical Research Cell Press, Cambridge, MA, pp 50–56

Elen A, Turan MK (2018) A new approach for fully automated segmentation of peripheral blood smears. Int J Adv Appl Sci 5(1):81–93

Eliceiri KW, Berthold MR, Goldberg IG, Ibáñez L, Manjunath BS, Martone ME, Murphy RF, Peng H, Plant AL, Roysam B, et al. (2012) Biological imaging software tools. Nat Methods 9(7):697

Fouad S, Landini G, Randell D, Galton A (2016) Morphological separation of clustered nuclei in histological images. In: International Conference on Image Analysis and Recognition. Springer, New York, pp 599–607

Furtado A, Henry R (2002) Measurement of green fluorescent protein concentration in single cells by image analysis. Anal Biochem 310(1):84–92

Gomes-Alves AG, Maia AF, Cruz T, Castro H, Tomas AM (2018) Development of an automated image analysis protocol for quantification of intracellular forms of Leishmania spp. PloS one 13(8):e0201747. https://doi.org/10.1371/journal.pone.0201747

Gonzalez RC, Woods RE (2018) Digital Image Processing, 4th edn. Prentice Hall, Englewood Cliffs

Gudla PR, Nandy K, Collins J, Meaburn KJ, Misteli T, Lockett SJ (2008) A high-throughput system for segmenting nuclei using multiscale techniques. Cytometry Part A J Int Soc Anal Cytol 73(5):451–466

Haidekker M (2010) Advanced biomedical image analysis. Wiley, New York

Hawkes PW, Spence JCH (2019) Springer handbook of microscopy. Springer, New York

Hegde RB, Prasad K, Hebbar H, Sandhya I (2018) Peripheral blood smear analysis using image processing approach for diagnostic purposes: A review. Biocybern Biomed Eng 38(3):467–480

Hegde RB, Prasad K, Hebbar H, Singh BMK (2019) Development of a robust algorithm for detection of nuclei of white blood cells in peripheral blood smear images. Multimedia Tools Appl 78(13):17879–17898

Held M, Schmitz MHA, Fischer B, Walter T, Neumann B, Olma MH, Peter M, Ellenberg J, Gerlich DW (2010) Cellcognition: time-resolved phenotype annotation in high-throughput live cell imaging. Nat Methods 7(9):747–754

Hodneland E, Kögel T, Frei DM, Gerdes HH, Lundervold A (2013) Cellsegm-a MATLAB toolbox for high-throughput 3D cell segmentation. Source Code Biol Med 8(1):16

Hussain A, Ghosh S, Kalkhoran SB, Hausenloy DJ, Hanssen E, Rajagopal V (2018) An automated workflow for segmenting single adult cardiac cells from large-volume serial block-face scanning electron microscopy data. J Struct Biol 202(3):275–285

Jerome WG, Price RL (2018) Basic confocal microscopy. Springer, New York

Johnson HJ, McCormick MM, Ibanez L (2015a) The ITK software guide book 1: introduction and development guidelines, vol 1. Kitware, Inc., New York

Johnson HJ, McCormick MM, Ibanez L (2015b) The ITK software guide book 2: design and functionality, vol 2. Kitware, Inc., New York

Jones TR, Carpenter AE, Lamprecht MR, Moffat J, Silver SJ, Grenier JK, Castoreno AB, Eggert US, Root DE, Golland P, et al. (2009) Scoring diverse cellular morphologies in image-based screens with iterative feedback and machine learning. Proc Natl Acad Sci 106(6):1826–1831

Kaehler A, Bradski G (2016) Learning OpenCV 3: computer vision in C++ with the OpenCV library. O'Reilly Media, Inc., New York

Kamentsky L, Jones TR, Fraser A, Bray MA, Logan DJ, Madden KL, Ljosa V, Rueden C, Eliceiri KW, Carpenter AE (2011) Improved structure, function and compatibility for cellprofiler: modular high-throughput image analysis software. Bioinformatics 27(8):1179–1180

Kankaanpää P, Paavolainen L, Tiitta S, Karjalainen M, Päivärinne J, Nieminen J, Marjomäki V, Heino J, White DJ (2012) Bioimagexd: an open, general-purpose and high-throughput image-processing platform. Nat Methods 9(7):683–689

Kapur JN, Sahoo PK, Wong AKC (1985) A new method for gray-level picture thresholding using the entropy of the histogram. Comput Vision Graphics Image Proc 29(3):273–285

Kass M, Witkin A, Terzopoulos D (1988) Snakes: active contour models. Int J Comput Vision 1(4):321–331

Krzywinski M, Altman N (2013a) Points of significance: importance of being uncertain. Nat Methods 10(9):809–810

Krzywinski M, Altman N (2013b) Points of significance: significance, p values and t-tests. Nat Methods 10(11):1041–1042

Krzywinski M, Altman N (2014a) Points of significance: comparing samples–part i. Nat Methods 11(4):215–216

Krzywinski M, Altman N (2014b) Points of significance: visualizing samples with box plots. Nat Methods 11(2):119–120

Kubitscheck U (2017) Fluorescence microscopy: from principles to biological applications. Wiley, New York

Lamprecht MR, Sabatini DM, Carpenter AE (2007) Cellprofiler: free, versatile software for automated biological image analysis. Biotechniques 42(1):71–75

Lawlor D (2019) Introduction to light microscopy: tips and tricks for beginners. Springer, Berlin

Legland D, Arganda-Carreras I, Andrey P (2016) Morpholibj: integrated library and plugins for mathematical morphology with imagej. Bioinformatics 32(22):3532–3534

Li G, Liu T, Nie J, Guo L, Malicki J, Mara A, Holley SA, Xia W, Wong STC (2007) Detection of blob objects in microscopic zebrafish images based on gradient vector diffusion. Cytometry Part A J Int Soc Anal Cytol 71(10):835–845

Li F, Yin Z, Jin G, Zhao H, Wong STC (2013) Bioimage informatics for systems pharmacology. PLoS Comput Biol 9(4):e1003043

Ljosa V, Carpenter AE (2009) Introduction to the quantitative analysis of two-dimensional fluorescence microscopy images for cell-based screening. PLoS Comput Biol 5(12):e1000603

Lundberg E, Borner GH (2019) Spatial proteomics: a powerful discovery tool for cell biology. Nat. Rev. Mol. Cell Biol. 20(5):285–302

Markaki Y, Harz H (2017) Light microscopy. In: Methods and protocols. Springer, New York

McCullough DP, Gudla PR, Harris BS, Collins JA, Meaburn KJ, Nakaya MA, Yamaguchi TP, Misteli T, Lockett SJ (2008) Segmentation of whole cells and cell nuclei from 3-d optical microscope images using dynamic programming. IEEE Trans. Med. Imaging 27(5):723–734

McQuin C, Goodman A, Chernyshev V, Kamentsky L, Cimini BA, Karhohs KW, Doan M, Ding L, Rafelski SM, Thirstrup D, et al. (2018) Cellprofiler 3.0: Next-generation image processing for biology. PLoS Biol 16(7):e2005970

Meijering E, Carpenter AE, Peng H, Hamprecht FA, Olivo-Marin JC (2016) Imagining the future of bioimage analysis. Nat Biotechnol 34(12):1250

Miura K, Sladoje N (2020) Bioimage data analysis workflows. Springer, Berlin

Model MA, Burkhardt JK (2001) A standard for calibration and shading correction of a fluorescence microscope. Cytometry J Int Soc Anal Cytol 44(4):309–316

Nandy K, Gudla PR, Meaburn KJ, Misteli T, Lockett SJ (2009) Automatic nuclei segmentation and spatial fish analysis for cancer detection. In: Proceedings of the 2009 annual international conference of the IEEE engineering in medicine and biology society. IEEE, New York, pp 6718–6721

Nandy K, Gudla PR, Amundsen R, Meaburn KJ, Misteli T, Lockett SJ (2012) Automatic segmentation and supervised learning-based selection of nuclei in cancer tissue images. Cytometry Part A 81(9):743–754

Navakauskiene R, Borutinskaite VV, Treigyte G, Savickiene J, Matuzevicius D, Navakauskas D, Magnusson KE (2014) Epigenetic changes during hematopoietic cell granulocytic differentiation—comparative analysis of primary CD34+cells, KG1 myeloid cells and mature neutrophils. BMC Cell Biol 15:4. https://doi.org/10.1186/1471-2121-15-4

North AJ (2006) Seeing is believing? a beginners' guide to practical pitfalls in image acquisition. J Cell Biol 172(1):9–18

Otsu N (1979) A threshold selection method from gray-level histograms. IEEE Trans Syst Man Cybern 9(1):62–66

Ouyang W, Winsnes CF, Hjelmare M, Cesnik AJ, Åkesson L, Xu H, Sullivan DP, Dai S, Lan J, Jinmo P, et al. (2019) Analysis of the human protein atlas image classification competition. Nat Methods 16(12):1254–1261

Paulik R, Micsik T, Kiszler G, Kaszal P, Szekely J, Paulik N, Varhalmi E, Premusz V, Krenacs T, Molnar B (2017) An optimized image analysis algorithm for detecting nuclear signals in digital whole slides for histopathology. Cytometry Part A 91A(6, SI):595–608. https://doi.org/10.1002/cyto.a.23124

Pawley J (2006) Handbook of biological confocal microscopy, vol 236. Springer, New York

Perona P, Malik J (1990) Scale-space and edge detection using anisotropic diffusion. IEEE Trans. Pattern Anal. Mach. Intell. 12(7):629–639

Phoulady HA, Goldgof D, Hall LO, Mouton PR (2017) A framework for nucleus and overlapping cytoplasm segmentation in cervical cytology extended depth of field and volume images. Comput Med Imaging Graphics 59:38–49

Pietzsch T, Preibisch S, Tomančák P, Saalfeld S (2012) ImgLib2–generic image processing in Java. Bioinformatics 28(22):3009–3011

Prinyakupt J, Pluempitiwiriyawej C (2015) Segmentation of white blood cells and comparison of cell morphology by linear and naïve bayes classifiers. Biomed Eng Online 14(1):63

Qiao G, Zong G, Sun M, Wang J (2012) Automatic neutrophil nucleus lobe counting based on graph representation of region skeleton. Cytometry Part A 81(9):734–742

Ranefall P, Wahlby C (2016) Global gray-level thresholding based on object size. Cytometry Part A 89A(4):385–390. https://doi.org/10.1002/cyto.a.22806

Rebollo E, Bosch M (2019) Computer optimized microscopy: methods and protocols. Springer, New York

Rittscher J, Machiraju R, Wong STC (2008) Microscopic image analysis for life science applications. Artech House, New York

Rueden CT, Schindelin J, Hiner MC, DeZonia BE, Walter AE, Arena ET, Eliceiri KW (2017) Imagej2: imagej for the next generation of scientific image data. BMC Bioinf 18(1):529

Russ JC, Neal FB (2017) The image processing handbook, 7th edn. CRC Press, West Palm Beach

Sajjad M, Khan S, Jan Z, Muhammad K, Moon H, Kwak JT, Rho S, Baik SW, Mehmood I (2016) Leukocytes classification and segmentation in microscopic blood smear: a resource-aware healthcare service in smart cities. IEEE Access 5:3475–3489

Schindelin J, Arganda-Carreras I, Frise E, Kaynig V, Longair M, Pietzsch T, Preibisch S, Rueden C, Saalfeld S, Schmid B, et al. (2012) Fiji: an open-source platform for biological-image analysis. Nat Methods 9(7):676–682

Schneider CA, Rasband WS, Eliceiri KW (2012) NIH image to imagej: 25 years of image analysis. Nat Methods 9(7):671–675

Schnipper N, Stassen HH, Kallinich T, Sperling K, Hoffmann K (2017) Image analysis of neutrophil nuclear morphology: learning about phenotypic range and its reliable analysis from patients with pelger-huët-anomaly and treated with colchicine. Cytometry Part B Clin Cytometry 92(6):541–549

Schroeder W, Martin K, Lorensen B (2006) The visualization toolkit, 4th edn. Kitware, New York

Schüffler PJ, Fuchs TJ, Ong CS, Wild PJ, Rupp NJ, Buhmann JM (2013) TMARKER: a free software toolkit for histopathological cell counting and staining estimation. J Pathol Inf 4(2):2

Schwarzfischer M, Marr C, Krumsiek J, Hoppe P, Schroeder T, Theis FJ (2011) Efficient fluorescence image normalization for time lapse movies. In: Proc. Microscopic Image Analysis with Applications in Biology. MIAAB

Seruca R, Suri JS, Sanches JM (2017) Fluorescence imaging and biological quantification. CRC Press, New York

Sethian JA (1999) Level set methods and fast marching methods: evolving interfaces in computational geometry, fluid mechanics, computer vision, and materials science, vol 3. Cambridge University, Cambridge

Sluder G, Wolf DE (2013) Digital microscopy. Academic Press, New York

Soille P (2013) Morphological image analysis: principles and applications. Springer, New York

Sommer C, Straehle CN, Köthe U, Hamprecht FA (2011) Ilastik: interactive learning and segmentation toolkit. In: Proceedings of the 8th IEEE international symposium on biomedical imaging (ISBI 2011), pp 230–233. https://doi.org/10.1109/ISBI.2011.5872394, 1
Tinevez JY, Perry N, Schindelin J, Hoopes GM, Reynolds GD, Laplantine E, Bednarek SY, Shorte SL, Eliceiri KW (2017) TrackMate: an open and extensible platform for single-particle tracking. Methods 115:80–90
Tomasi C, Manduchi R (1998) Bilateral filtering for gray and color images. In: Sixth International Conference on Computer Vision (IEEE Cat. No. 98CH36271). IEEE, New York, pp 839–846
Usaj MM, Styles EB, Verster AJ, Friesen H, Boone C, Andrews BJ (2016) High-content screening for quantitative cell biology. Trends Cell Biol. 26(8):598–611
Van der Walt S, Schönberger JL, Nunez-Iglesias J, Boulogne F, Warner JD, Yager N, Gouillart E, Yu T (2014) scikit-image: image processing in Python. Peer J 2:e453
Vincent L (1993) Morphological grayscale reconstruction in image analysis: applications and efficient algorithms. IEEE Trans Image Process 2(2):176–201
Vincent L, Soille P (1991) Watersheds in digital spaces: an efficient algorithm based on immersion simulations. IEEE Trans. Pattern Anal. Mach. Intell. (6):583–598
Wallace W, Schaefer LH, Swedlow JR (2001) A workingperson's guide to deconvolution in light microscopy. Biotechniques 31(5):1076–1097
Wang Y, Cao Y (2019) Leukocyte nucleus segmentation method based on enhancing the saliency of saturation component. J. Algorithms Comput. Technol. 13:1748302619845783
Waters JC (2009) Accuracy and precision in quantitative fluorescence microscopy
Waters J, Wittmann T (2014) Quantitative imaging in cell biology. Academic Press, New York
Xue JH, Zhang YJ (2012) Ridler and calvard's, kittler and illingworth's and otsu's methods for image thresholding. Pattern Recognit Lett 33(6):793–797
Yang SJ, Berndl M, Ando DM, Barch M, Narayanaswamy A, Christiansen E, Hoyer S, Roat C, Hung J, Rueden CT, Shankar A, Finkbeiner S, Nelson P (2018) Assessing microscope image focus quality with deep learning. BMC Bioinf 19(1):1–9. https://doi.org/10.1186/s12859-018-2087-4

Future Therapy in Leukemia

Acute myeloid leukemia is a type of leukemia which distinguishing feature is the invasion of bone marrow, blood, and other tissues by rapidly proliferating, undifferentiated cells of the blood system—blasts. Acute myeloid leukemia was incurable 50 years ago, but now the disease is overcome by 35–40% of adult patients younger than 60 years and 5–15% of patients older than 60 years (Döhner et al. 2015). Patients foremost are treated with chemotherapy. The standard combination of chemotherapy drugs is called 7 + 3, i.e., 7 day continuous cytarabine infusion and daunorubicin/idarubicin for 1 to 3 days (Dombret and Gardin 2016). If treatment is stopped prematurely, relapse will inevitably occur. Various drug combinations are often used to treat relapse, also many patients underwent allogenic HSC transplantation during the treatment.

About 2/3 of AML patients achieve remission after standard chemotherapy, but patients experience many side effects during treatment, such as fatigue, decreased resistance to infections, anemia, bleeding, nausea, hair loss, changes in hormone levels in the body (www.cancerresearchuk.org). 25–50% of recoverees relapse within 5 years of stopping treatment. Chemotherapy is ineffective in 1/3 of patients. In order to improve the effectiveness of treatment and reduce side effects, new treatments are constantly being sought.

Genomic analysis reveals complex genomic alterations and gene expression changes in cancer diseases. Detection of chromosomal abnormalities and gene mutations help to categorize patients into risk groups and select the most successful treatment method. For example, FMS-like tyrosine kinase 3 (FLT3) plays an important role in hematopoietic cell survival, proliferation, and differentiation. Its mutation is the most frequent genetic alteration associated with a poor prognosis in AML patients. Therefore, FLT3 is a promising molecular target for the treatment of AML patients with FLT3 mutations. There are two major types of FLT3 mutations: internal tandem duplication mutations (FLT3-ITD) and point mutations or deletion in the tyrosine kinase domain (FLT3-TKD). The tyrosine kinase inhibitors (TKI), including tandutinib, sunitinib, midostaurin, lestaurtinib and sorafenib, were sub-

R. Navakauskienė et al., *Epigenetics and Proteomics of Leukemia*,
https://doi.org/10.1007/978-3-030-68708-3

jected to clinical trials for evaluating efficacy and safety (De Kouchkovsky and Abdul-Hay 2016; Kiyoi et al. 2020).

Approximately 20% of AML patients have an isocitrate dehydrogenases (IDH1 or IDH2) mutation. IDH enzymes are involved in multiple metabolic and epigenetic cellular processes. The IDH1 or IDH2 mutations lead to DNA hypermethylation, aberrant gene expression, cell proliferation, and abnormal differentiation. Several novel therapies specifically targeting mutant IDH had been developed (Ragon and DiNardo 2017; Montalban-Bravo and DiNardo 2018). However, the treatment of leukemia with characteristic chromosomal aberrations are challenging, as patients may be or become resistant to standard therapy during treatment and may relapse after remission. Therefore, there is a necessity to develop new therapies.

In certain cases of cancer, especially in acute promyelocytic leukemia, differentiation therapy is applied. The differentiation therapy is induced with all-trans retinoic acid. Studies have been performed with other nuclear hormone ligands, such as vitamin D compounds and PPARγ agonists, or with other agents, such as hematopoietic cytokines or compounds acting on the epigenetic landscape, but they have not been successful. Therefore, ATRA is still successfully used for differentiation therapy of leukemia, especially APL. However, in some cases (relapse or resistance for repeated ATRA treatment) with ATRA, it is rational to use combination therapy with epigenetic modifiers, e.g., the inhibitors of histone methyltransferases and/or deacetylases as well as inhibitors of DNA methyltransferases are used together with all-trans retinoic acid. These studies are extensively described in the monograph. During such treatments cancer cells are induced to differentiate and re-enter the normal way of development. In the monograph we overviewed the efficiency of combined epigenetic therapy with ATRA for leukemia and demonstrated what the impact of such combined treatment is promising for leukemia treatment—the differentiation of leukemia cells is induced, leukemia cell growth is inhibited, and the promotion of apoptosis increases.

The genome defines what we have and do not have—it is our genetic information. The epigenome, meanwhile, ensures the stable propagation of genetic information and the proper functioning of cells. Therefore, epigenetic modifications (histone, DNA modification) are an integral part of normal cell development. Epigenetic modifications and their location result in the restructuring of chromatin and related proteins, thereby affecting cellular processes such as cell cycle, differentiation, DNA repair, growth arrest, and cell death. Thus, the changes in the epigenome are enormous in causing changes in chromatin remodeling and consequently in disease pathogenesis. Recent therapies are based on genetic testing, the application of targeted therapies and epigenetic combined treatments. Combined epigenetic therapy is particularly important in cancer and leukemia. Genetic aberrations are accompanied by epigenetic changes, therefore epigenetic therapy, especially in combination with other therapies, is thought to be an important and powerful treatment for cancer (Navakauskiene 2017; Ahuja et al. 2016). Unregulated activity of DNMTs, HDACs, and some HMTs is often observed, which may lead to inhibition of the expression of genes that inhibit cancer. Inhibitors of these epigenetic modulators could potentially restore the expression of cancer inhibitory genes by inducing chromatin remodeling (Kwon and Shin 2011). Many processes in the cell, for instance, intensified proliferation or blockade of differentiation, are affected by epigenetic changes (Dawson and Kouzarides 2012; Jones et al. 2016). The importance of epigenetic changes in carcinogenesis is also indicated by the fact that a relatively large proportion of mutations characteristic of leukemic diseases are detected in the genes important for epigenetic regulation (Kaushansky and Zhan 2018). For example, the naturally occurring polyphenol EGCG (epigallocatechin-3-gallate) has the ability to restrain DNMT1 and HDAC activity as well as induce the apoptosis (Khan et al. 2015; Moradzadeh et al. 2018; Borutinskaitė et al. 2018; Vitkeviciene et al. 2018).

The goal of epigenetic therapy is to induce such epigenetic changes in the cell that would either help to restore the impaired differentiation process or, like in traditional chemotherapy, cause cell death. However, high doses of drugs are used to cause cancer cell death, which can have serious side effects on the body (Lee and Lee 2019). When cells are treated with therapeutic agents, the cells undergo apoptosis or therapy-induced cellular aging, depending on the extent of the damage caused. The latter is defined as irreversible cell cycle arrest (Provinciali et al. 2013). In general, cell aging is caused by less toxic doses of chemical agents (Chang et al. 1999), so it is thought to be safer and cause fewer side effects than induction of apoptosis (Lee and Lee 2019). Although induction of apoptosis is thought to give the better outcomes in cancer treatment, the capability of chemical agent to induce cell aging is nevertheless a benefit (Nardella et al. 2011). The ability of some chemical agents inducing the cancer cell aging could also be evaluated for the leukemia treatment.

Metabolic reprogramming is a new feature characteristic of cancer. Presently main directions of the AML treatment are emphasized—combined therapy (management of cellular processes such as differentiation, apoptosis, epigenetic regulation) of leukemia and metabolic vulnerability for precise treatment of refractory leukemia (Choi et al. 2020; Burnett and Stone 2020; Richard-Carpentier and DiNardo 2019; Farge et al. 2017; Vitkeviciene et al. 2019; Vitkevičienė et al. 2019; Visnjic et al. 2019). For the leukemia therapy could be also used the drugs that are approved for other diseases like malaria, diabetes (Vitkevičienė et al. 2019; Yuan et al. 2020; Biondani and Peyron 2018).

Therefore, the focus has recently been on the search for new oncological drugs targeting to mitochondrial activity. To date, significant efforts have been directed to the therapeutic application of glycolysis, whereas mitochondrial oxidative phosphorylation (OXPHOS) has not been extensively studied. Only in recent years very promising studies emerged. A study on IACS-010759, a small molecule inhibitor in the mitochondrial electron transport chain, was recently published (Molina et al. 2018). It was found that treatment with IACS-010759 potentially inhibits cancer cell proliferation and induces apoptosis in OXPHOS-based models of acute myeloid leukemia. IACS-010759 is presently being assessed in clinical trials phase 1 with recurrent/refractory AML and solid tumors (Molina et al. 2018). Our recent study (Vitkevičienė et al. 2019) demonstrated that oxidative phosphorylation inhibition by using only oxidative phosphorylation inhibitors atovaquone and metformin or them in different combinations with cytosine analogue cytarabine and apoptosis inducer venetoclax prompts anticancerous alterations in therapy-resistant-acute myeloid leukemia cells of patients. Therefore, we suggest that oxidative phosphorylation inhibitors might have a significant additional role in combinatory treatment of leukemia.

The biggest challenge in the treatment of leukemia is Relapsed/Refractory AML. Patients who have not achieved complete remission after two cycles of induction chemotherapy are usually diagnosed as having refractory leukemia. Patients who have reached remission but after some time leukemia blasts again increased are diagnosed as relapsed leukemia. The treatment of Relapsed/Refractory AML cases is a challenge for the decades. According to www.LLS.org the following drugs are FDA approved for Relapsed/Refractory AML: enasidenib (Idhifa), gemtuzumab ozogamicin (Mylotarg), ivosidenib (Tibsovo), gilteritinib (Xospata).

Relapsed/Refractory AML patients are treated in clinical trials based on their previous treatment history. Targeted sequencing of genes that frequently mutate AML is used to identify active mutations, such as FLT3 or IDH1/IDH2. Inhibitors based on identified gene mutations are used in therapy. In the absence of active mutations, then conventional agents are combined with other new therapies, such as monoclonal antibodies and other target drugs, various combinations of chemotherapy are also used, and allogeneic HSC transplantation is performed (Bose et al. 2017; Zucenka et al. 2020).

Following the example of recent therapies and our extended analysis the combination of new epigenetic inhibitors—HDACis, HMTis, DNMTis, the tyrosine kinase inhibitors, the oxidative phosphorylation inhibitors together with chemotherapy would also affect cell proliferation in patients with leukemia and the initial data give hope that pharmacological chromatin remodeling will allow to create the new strategies for future therapy of leukemia.

References

Ahuja N, Sharma AR, Baylin SB (2016) Epigenetic therapeutics: a new weapon in the war against cancer. Annu Rev Med 67:73–89

Biondani G, Peyron JF (2018) Metformin, an anti-diabetic drug to target leukemia. Front Endocrinol 9:446

Borutinskaitė V, Virkšaitė A, Gudelytė G, Navakauskienė R (2018) Green tea polyphenol EGCG causes anti-cancerous epigenetic modulations in acute promyelocytic leukemia cells. Leuk Lymphoma 59(2):469–478. https://doi.org/10.1080/10428194.2017.1339881

Bose P, Vachhani P, Cortes JE (2017) Treatment of relapsed/refractory acute myeloid leukemia. Curr Treat Options Oncol 18(3):17

Burnett A, Stone R (2020) AML: new drugs but new challenges. Clin Lymphoma Myeloma Leuk 20(6):341–350

Chang BD, Broude EV, Dokmanovic M, Zhu H, Ruth A, Xuan Y, Kandel ES, Lausch E, Christov K, Roninson IB (1999) A senescence-like phenotype distinguishes tumor cells that undergo terminal proliferation arrest after exposure to anticancer agents. Cancer Res 59(15):3761–3767

Choi JH, Bogenberger JM, Tibes R (2020) Targeting apoptosis in acute myeloid leukemia: current status and future directions of BCL-2 inhibition with venetoclax and beyond. Target Oncol 15:147–162

Dawson MA, Kouzarides T (2012) Cancer epigenetics: from mechanism to therapy. Cell 150(1):12–27. https://doi.org/10.1016/j.cell.2012.06.013

De Kouchkovsky I, Abdul-Hay M (2016) Acute myeloid leukemia: a comprehensive review and 2016 update. Blood Cancer J 6. https://doi.org/10.1038/bcj.2016.50

Döhner H, Weisdorf DJ, Bloomfield CD (2015) Acute myeloid leukemia. New England J Med 373(12):1136–1152

Dombret H, Gardin C (2016) An update of current treatments for adult acute myeloid leukemia. Blood J Amer Society Hematol 127(1):53–61

Farge T, Saland E, de Toni F, Aroua N, Hosseini M, Perry R, Bosc C, Sugita M, Stuani L, Fraisse M (2017) Chemotherapy-resistant human acute myeloid leukemia cells are not enriched for leukemic stem cells but require oxidative metabolism. Cancer Discov 7(7):716–735

Jones PA, Issa JPJ, Baylin S (2016) Targeting the cancer epigenome for therapy. Nat Rev Genet 17(10):630

Kaushansky K, Zhan H (2018) The regulation of normal and neoplastic hematopoiesis is dependent on microenvironmental cells. Adv Biol Regul 69:11–15

Khan MA, Hussain A, Sundaram MK, Alalami U, Gunasekera D, Ramesh L, Hamza A, Quraishi U (2015) (-)-Epigallocatechin-3-gallate reverses the expression of various tumor-suppressor genes by inhibiting DNA methyltransferases and histone deacetylases in human cervical cancer cells. Oncol Rep 33(4):1976–1984

Kiyoi H, Kawashima N, Ishikawa Y (2020) FLT3 mutations in acute myeloid leukemia: therapeutic paradigm beyond inhibitor development. Cancer Sci. https://doi.org/10.1111/cas.14274

Kwon MJ, Shin YK (2011) Epigenetic regulation of cancer-associated genes in ovarian cancer. Int J Mol Sci 12(2):983–1008

Lee S, Lee JS (2019) Cellular senescence: a promising strategy for cancer therapy. BMB Rep 52(1):35

Molina JR, Sun Y, Protopopova M, Gera S, Bandi M, Bristow C, McAfoos T, Morlacchi P, Ackroyd J, Agip ANA (2018) An inhibitor of oxidative phosphorylation exploits cancer vulnerability. Nat Med 24(7):1036–1046

Montalban-Bravo G, DiNardo CD (2018) The role of IDH mutations in acute myeloid leukemia. Future Oncol 14(10):979–993

Moradzadeh M, Roustazadeh A, Tabarraei A, Erfanian S, Sahebkar A (2018) Epigallocatechin-3-gallate enhances differentiation of acute promyelocytic leukemia cells via inhibition of PML-RARα and HDAC1. Phytother Res 32(3):471–479

Nardella C, Clohessy JG, Alimonti A, Pandolfi PP (2011) Pro-senescence therapy for cancer treatment. Nat Rev Cancer 11(7):503–511

Navakauskiene R (2017) Combination epigenetic therapy. In: Tollefsbol T (ed) Handbook of epigenetics: the new molecular and medical genetics, 2nd edn. Academic Press, Cambridge; Elsevier, Amsterdam, chap 41, pp 623–632

Provinciali M, Cardelli M, Marchegiani F, Pierpaoli E (2013) Impact of cellular senescence in aging and cancer. Curr Pharm Des 19(9):1699–1709

Ragon BK, DiNardo CD (2017) Targeting IDH1 and IDH2 mutations in acute myeloid leukemia. Curr Hematol Malig Rep 12(6):537–546

Richard-Carpentier G, DiNardo CD (2019) Single-agent and combination biologics in acute myeloid leukemia. Hematology Am Soc Hematol Educ Program 2019(1):548–556

Visnjic D, Dembitz V, Lalic H (2019) The role of AMPK/mTOR modulators in the therapy of acute myeloid leukemia. Curr Med Chem 26(12):2208–2229

Vitkeviciene A, Baksiene S, Borutinskaite V, Navakauskiene R (2018) Epigallocatechin-3-gallate and BIX-01294 have different impact on epigenetics and senescence modulation in acute and chronic myeloid leukemia cells. Eur J Pharmacol 838:32–40. https://doi.org/10.1016/j.ejphar.2018.09.005

Vitkeviciene A, Skiauteryte G, Zucenka A, Stoskus M, Gineikiene E, Borutinskaite V, Griskevicius L, Navakauskiene R (2019) HDAC and HMT inhibitors in combination with conventional therapy: a novel treatment option for acute promyelocytic leukemia. J Oncol 2019:11. https://doi.org/10.1155/2019/6179573

Vitkevičienė A, Janulis V, Žučenka A, Borutinskaitė V, Kaupinis A, Valius M, Griškevičius L, Navakauskienė R (2019) Oxidative phosphorylation inhibition induces anticancerous changes in therapy-resistant–acute myeloid leukemia patient cells. Mol Carcinog 58(11):2008–2016

Yuan F, Cheng C, Xiao F, Liu H, Cao S, Zhou G (2020) Inhibition of mTORC1/P70S6K pathway by Metformin synergistically sensitizes acute myeloid leukemia to Ara-C. Life Sci 243:117276

Zucenka A, Pileckyte R, Trociukas I, Peceliunas V, Vaitekenaite V, Maneikis K, Davainis L, Zvirblis T, Stoskus M, Gineikiene E, et al. (2020) Outcomes of relapsed or refractory acute myeloid leukemia patients failing venetoclax based salvage therapies. Eur J Haematol 1–9

Appendix A
Clinical Trials of Drugs for Leukemia

Drugs for Chronic Lymphocytic Leukemia

Table A.1 Clinical trials of drugs for Chronic Lymphocytic Leukemia: Chemical Taxonomy provided by Classyfire; ATC Codes-code provided by Anatomical Therapeutic Chemical Classification System

Drug name	ATC code	Chemical taxonomy	Description	Clinical trials
Acalabrutinib	L01XE51	–	An inhibitor of Bruton's tyrosine kinase (BTK). Has antineoplastic activity.	NCT04115631 NCT02972840 NCT02180711 NCT03580928 NCT03527147 etc.
Alemtuzumab	L04AA34	–	A recombinant DNA-derived monoclonal antibody against CD52.	NCT02766465 NCT01659606 NCT03989466 NCT02626715 NCT02038478 etc.
Arzerra (Ofatumumab)	L01XC10	–	A monoclonal antibody against the B cell CD20 with potential antineoplastic activity.	NCT03136146 NCT02877303 NCT02199184 NCT02049515
Bendamustine Hydrochloride	L01AA09	Benzimidazoles	The hydrochloride salt of bendamustine. Has antimetabolite activities.	NCT04115631 NCT03295240 NCT03113422 NCT03834688 etc.

(continued)

R. Navakauskienė et al., *Epigenetics and Proteomics of Leukemia*,
https://doi.org/10.1007/978-3-030-68708-3

Table A.1 (continued)

Drug name	ATC code	Chemical taxonomy	Description	Clinical trials
Bendeka (Bendamustine Hydrochloride)	–	–	See Bendamustine Hydrochloride.	–
Calquence (Acalabrutinib)	–	–	See Acalabrutinib.	–
Campath (Alemtuzumab)	–	–	See Alemtuzumab.	–
Chlorambucil	L01AA02	Organonitrogen compounds	An antineoplastic aromatic nitrogen mustard	–
Copiktra (Duvelisib)	–	Isoquinolines and derivatives	An inhibitor of phosphoinositide-3 kinase (PI3K). Has immunomodulating and antineoplastic activities.	NCT03372057 NCT02783625 NCT04331119 etc.
Cyclophosphamide	L01AA01	Organonitrogen compounds	An alkylating agent. Has antineoplastic and immunosuppressive activities.	NCT02488967 NCT03220022 NCT03012100 NCT04119336
Dexamethasone	R01AD53 D07XB05 R01AD03 S01CB01 S02CA06	Steroids and steroid derivatives	A synthetic adrenal corticosteroid. Has anti-inflammatory properties.	NCT03937635 NCT03150693 NCT03914625 NCT02553460 NCT03556332
Duvelisib	–	Isoquinolines and derivatives	An inhibitor of phosphoinositide-3 kinase (PI3K) with immunomodulating and antineoplastic activities	NCT03372057 NCT02783625 NCT03534323 NCT03892044 etc.
Fludarabine Phosphate	L01BB05	Purine nucleosides	The phosphate salt of vidarabine (ara-A). Has antineoplastic activity.	NCT03813147 NCT03105336 NCT03164057 NCT00919503 etc.
Gazyva (Obinutuzumab)	L01XC15	Carboxylic Acids and Derivatives	A monoclonal antibody against CD20.	NCT03701282 NCT03737981 NCT03269669 NCT02498951 etc.
Ibrutinib	L01XE27	Benzene and substituted derivatives	An inhibitor of Bruton's tyrosine kinase (BTK) with potential antineoplastic activity.	NCT03701282 NCT03737981 NCT02443077 NCT03220022
Idelalisib	L01XX47	Imidazopyrimidines	A phosphoinositide-3 kinase (PI3K) inhibitor with immunomodulating and antineoplastic activities.	NCT03133221 NCT02332980 NCT03878524 NCT02536300 NCT02457598

(continued)

Table A.1 (continued)

Drug name	ATC code	Chemical taxonomy	Description	Clinical trials
Imbruvica (Ibrutinib)	–	–	See Ibrutinib.	–
Leukeran (Chlorambucil)	–	–	See Chlorambucil.	–
Mechlorethamine Hydrochloride	–	–	The hydrochloride salt of mechlorethamine with antineoplastic and immunosuppressive activities.	–
Mustargen (Mechlorethamin Hydrochloride)	L01AA05	Organonitrogen compounds	See Mechlorethamine Hydrochloride.	–
Obinutuzumab	L01XC15	Carboxylic Acids and Derivatives	An antibody against CD20 with potential antineoplastic activity.	NCT03701282 NCT03737981 NCT03269669 etc.
Ofatumumab	L01XC10	Carboxylic Acids and Derivatives	An antibody directed against the B cell CD20 cell surface antigen with potential antineoplastic activity.	NCT03136146 NCT02877303 NCT02199184 NCT02049515
Prednisone	A07EA03 H02AB07	Steroids and steroid derivatives	A synthetic glucocorticoid with anti-inflammatory and immunomodulating properties.	NCT03419234 NCT03984448 NCT03269669 NCT03914625 NCT03574571 NCT02723994
Rituxan (Rituximab)	L01XC02	Carboxylic Acids and Derivatives	An antibody directed against the CD20 antigen.	NCT03984448 NCT02900976 NCT03220022 NCT03267433 NCT04212013 NCT03919175 NCT03467867 NCT03864419 NCT03719131 etc.
Rituxan Hycela (Rituximab and Hyaluronidase Human)	See Rituxan (Rituximab).	See Rituxan (Rituximab).	A combination of rituximab and the enzyme hyaluronidase, with antineoplastic activity.	NCT03267433 NCT04212013 NCT03919175 NCT03467867 NCT03864419 NCT03719131
Rituximab	–	–	See Rituxan (Rituximab).	–
Rituximab and Hyaluronidase Human	–	–	See Rituxan Hycela (Rituximab and Hyaluronidase Human).	–
Treanda (Bendamustine Hydrochloride)	–	–	See Bendamustine Hydrochloride.	–

(continued)

Table A.1 (continued)

Drug name	ATC code	Chemical taxonomy	Description	Clinical trials
Truxima (Rituximab)	–	–	See Rituxan (Rituximab).	–
Venclexta (Venetoclax)	–	–	See Venclexta (Venetoclax).	–
Venetoclax	–	–	See Venclexta (Venetoclax).	–
Zydelig (Idelalisib)	–	–	See Idelalisib.	–

Drugs for Chronic Myeloid Leukemia

Table A.2 Clinical trials of drugs for Acute Lymphoblastic Leukemia Chronic Myeloid Leukemia: Chemical Taxonomy provided by Classyfire; ATC Codes-code provided by Anatomical Therapeutic Chemical Classification System

Drug name	ATC code	Chemical taxonomy	Description	Clinical trials
Bosulif (Bosutinib)	L01XE14	Quinolines and derivatives	A quinolone derivative and dual kinase inhibitor. Has potential antineoplastic activity.	NCT03654768 NCT03516279 NCT03610971 NCT04329325 NCT03854903 etc.
Bosutinib	–	–	See Bosulif (Bosutinib).	–
Busulfan	L01AB01	Organic sulfonic acids and derivatives	A derivative of dimethane-sulfonate with antineoplastic and cytotoxic properties.	NCT03126916 NCT03613532 NCT03303950 etc.
Busulfex (Busulfan)	–	–	See Busulfan.	–
Cyclophosphamide	–	–	–	–
Cytarabine	–	–	–	–
Dasatinib	–	–	–	–
Dexamethasone	–	–	–	–
Gleevec (Imatinib Mesylate)	–	–	–	–
Hydrea (Hydroxyurea)	L01XX05	Carboximidic acids and derivatives	A monohydroxyl-substituted urea (hydroxycarbamate) antimetabolite.	NCT03165734 NCT03165734 NCT03944915 NCT02626715 NCT01050855
Hydroxyurea	–	–	See Hydrea (Hydroxyurea).	–
Iclusig (Ponatinib Hydrochloride)	–	–	–	–

(continued)

Table A.2 (continued)

Drug name	ATC code	Chemical taxonomy	Description	Clinical trials
Imatinib Mesylate	–	–	–	–
Mechlorethamine Hydrochloride	–	–	–	–
Mustargen (Mechlorethamine Hydrochloride)	L01AA05	Organonitrogen compounds	The hydrochloride salt of mechlorethamine with antineoplastic and immunosuppressive activities.	–
Myleran (Busulfan)	–	–	See Busulfan.	–
Nilotinib	L01XE08	Benzene and substituted derivatives	Bcr-Abl tyrosine kinase inhibitor with antineoplastic activity.	NCT03654768 NCT03516279 NCT04205903
Omacetaxine Mepesuccinate	L01XX40	Cephalotaxus alkaloids	A plant alkaloid homoharringtonine isolated from the evergreen tree Cephalotaxus with potential antineoplastic activity.	NCT03564873
Ponatinib Hydrochloride	–	–	–	–
Sprycel (Dasatinib)	–	–	–	–
Synribo (Omacetaxine Mepesuccinate)	–	–	See Omacetaxine Mepesuccinate.	–
Tasigna (Nilotinib)	–	–	See Nilotinib.	–

Drugs for Acute Lymphocytic Leukemia

Table A.3 Clinical trials of drugs for Acute Lymphocytic Leukemia: Chemical Taxonomy provided by Classyfire; ATC Codes-code provided by Anatomical Therapeutic Chemical Classification System

Drug name	ATC code	Chemical taxonomy	Description	Clinical trials
Arranon (Nelarabine)	L01BB07	Purine nucleosides	arabinonucleoside antimetabolite. Has antineoplastic activity.	NCT03020030 NCT03117751 NCT02763384 NCT03328104
Asparaginase Erwinia chrysanthemi	–	–	Enzyme isolated from the bacterium (E. carotovora).	NCT03914625 NCT02521493 NCT04293562 NCT02723994

(continued)

Table A.3 (continued)

Drug name	ATC code	Chemical taxonomy	Description	Clinical trials
Asparlas (Calaspargase Pegol-mknl)	–	–	Conjugated L-asparaginase II (E. coli) and succinimidyl carbonate monomethoxypolyethylene glycol (SC-PEG). Complex has antineoplastic activity.	–
Besponsa (Inotuzumab Ozogamicin)	L01XC26	–	A CD22-targeted cytotoxic immunoconjugate. Has potential antineoplastic activity.	NCT03150693 NCT03959085 NCT03739814 NCT03991884
Blinatumomab	L01XC19	Carboxylic Acids and Derivatives	An anti-CD19/anti-CD3 bispecific monoclonal antibody. Has antineoplastic activities.	NCT03914625 NCT03739814 NCT02143414 NCT03117751 NCT02879695
Blincyto (Blinatumomab)	–	–	See Blinatumomab.	–
Calaspargase Pegol-mknl	–	–	See Asparlas (Calaspargase Pegol-mknl).	–
Cerubidine (Daunorubicin Hydrochloride)	L01DB02 L01XY01	Anthracyclines	An anthracycline antineoplastic antibiotic.	NCT03150693 NCT03959085 NCT03959085 NCT04326439
Clofarabine	L01BB06	Purine nucleosides	A purine nucleoside analog. Has antineoplastic activity.	NCT04220008 NCT01119066 NCT01701986 NCT02135874 NCT02053545 NCT01701986 NCT04220008
Clolar (Clofarabine)	–	–	See Clofarabine.	–
Cyclophosphamide	L01AA01	Organonitrogen compounds	An alkylating agent. Has antineoplastic and immunosuppressive activities.	NCT02488967 NCT03220022 NCT03012100 NCT04119336
Cytarabine	L01BC01 L01XY01	Pyrimidine nucleosides	An antimetabolite analogue of cytidine.	NCT04115631 NCT03150693 NCT03959085 NCT03007147 NCT03701308 NCT03504410 NCT03303339

(continued)

Table A.3 (continued)

Drug name	ATC code	Chemical taxonomy	Description	Clinical trials
Dasatinib	L01XE06	Benzene and substituted derivatives	A synthetic small-molecule inhibitor of SRC-family protein-tyrosine kinases.	NCT02465060 NCT03654768 NCT03516279 NCT02143414 NCT03020030 NCT03595917
Daunorubicin Hydrochloride	L01DB02 L01XY01	Anthracyclines	An anthracycline antineoplastic antibiotic.	NCT03150693 NCT03959085 NCT03007147 NCT02521493 NCT03164057
Dexamethasone	R01AD53 D07XB05 R01AD03 S01CB01 S02CA06	Steroids and steroid derivatives	A synthetic adrenal corticosteroid. Has anti-inflammatory properties.	NCT03937635 NCT03150693 NCT03914625 NCT02553460 NCT03556332
Doxorubicin Hydrochloride	L01DB01	Anthracyclines	An anthracycline antibiotic with antineoplastic activity.	NCT02488967 NCT03984448 NCT03907488 NCT03914625 NCT03959085
Erwinaze (Asparaginase Erwinia Chrysanthemi)	–	–	An enzyme isolated from the bacterium (E. carotovora).	NCT03914625 NCT02521493 NCT04293562 NCT02723994 NCT03164057
Gleevec (Imatinib Mesylate) or Imatinib Mesylate	L01XE01	Benzene and substituted derivatives	A tyrosine kinase inhibitor with antineoplastic activity.	NCT03516279 NCT03007147 NCT03023046 NCT02043587
Iclusig (Ponatinib Hydrochloride)	L01XE24	Benzene and substituted derivatives	A receptor tyrosine kinase (RTK) inhibitor. Has antiangiogenic and antineoplastic activities.	NCT03838692 NCT03576547 NCT02272998
Inotuzumab Ozogamicin	L01XC26	Carboxylic Acids and Derivatives	A CD22-targeted cytotoxic immunoconjugate with potential antineoplastic activity.	NCT03150693 NCT03959085 NCT03739814 NCT03677596 NCT03991884 NCT03962465 NCT02311998 NCT01371630 NCT03851081
Kymriah (Tis-agenlecleucel)	–	–	–	NCT03570892 NCT03876769 NCT03610724 NCT03630159 NCT04134117 NCT04225676

(continued)

Table A.3 (continued)

Drug name	ATC code	Chemical taxonomy	Description	Clinical trials
Marqibo (Vincristine Sulfate Liposome)	L01CA02	Vinca alkaloids	A vincristine sulfate with potential antineoplastic activity.	NCT02879643 NCT03136146 NCT03851081
Mercaptopurine	L01BB02	Imidazopyrimidines	A thiopurine-derivative antimetabolite with antineoplastic and immunosuppressive activities.	NCT03150693 NCT03914625 NCT03959085 NCT03007147 NCT02723994
Methotrexate	L04AX03 L01BA01	Carboxylic acids and derivatives	Folate derivative with antineoplastic and immunosuppressant activities.	NCT03150693 NCT03914625 NCT03959085
Nelarabine	L01BB07	Purine nucleosides	Purine nucleoside analog with antineoplastic activity.	NCT03020030 NCT03117751 NCT02763384 NCT03328104 NCT03808610 NCT00501826
Oncaspar (Pegaspargase)	L01XX24	Carboxylic Acids and Derivatives	A complex of polyethylene glycol conjugated with L-asparaginase.	NCT03914625 NCT03959085 NCT03007147 NCT02553460 NCT03384654 NCT02723994
Pegaspargase	–	–	See Oncaspar (Pegaspargase).	–
Ponatinib Hydrochloride	–	–	See Iclusig (Ponatinib Hydrochloride).	–
Prednisone	A07EA03 H02AB07	Steroids and steroid derivatives.	A synthetic glucocorticoid with anti-inflammatory and immunomodulating properties.	NCT03419234 NCT03984448 NCT03269669 NCT03914625 NCT03574571 NCT02723994
Purinethol (Mercaptopurine)	–	–	See Mercaptopurine.	–
Purixan (Mercaptopurine)	–	–	See Mercaptopurine.	–
Rubidomycin (Daunorubicin Hydrochloride)	–	–	See Daunorubicin Hydrochloride.	–

(continued)

Table A.3 (continued)

Drug name	ATC code	Chemical taxonomy	Description	Clinical trials
Sprycel (Dasatinib)	–	–	See Dasatinib.	–
Tisagenlecleucel	–	–	Autologous T lymphocytes.	NCT03570892 NCT03876769 NCT03610724 NCT03630159 NCT01853631 NCT01853631 NCT02906371 etc.
Trexall (Methotrexate)	–	–	See Methotrexate.	–
Vincristine Sulfate	L01CA02	Vinca alkaloids	A natural alkaloid isolated from the plant Vinca rosea Linn with antimitotic and antineoplastic activities.	NCT03984448 NCT03150693 NCT03914625 NCT03959085 NCT03007147

Drugs for Acute Myeloid Leukemia

Table A.4 Clinical trials of drugs for Acute Myeloid Leukemia: Chemical Taxonomy provided by Classyfire; ATC Codes-code provided by Anatomical Therapeutic Chemical Classification System

Drug name	ATC code	Chemical taxonomy	Description	Clinical trials
Arsenic Trioxide	L01XX27	Metalloid organides	An arsenic compound with antineoplastic activity.	NCT01409161 NCT02699723 NCT01835288
Azacitidine	L01BC07	Organooxygen compounds	A pyrimidine nucleoside analogue of cytidine. Has antineoplastic activity.	NCT03813147 NCT03248479 NCT03013998 NCT03164057 NCT02719574 NCT02752035 etc.
Cerubidine (Daunorubicin Hydrochloride)	L01DB02 L01XY01	Anthracyclines	An anthracycline antineoplastic antibiotic.	NCT03150693 NCT03959085 NCT03959085 NCT04326439
Cyclophosphamid	L01AA01	Organonitroger compounds	An alkylating agent. Has antineoplastic and immunosuppressive activities.	NCT02488967 NCT03220022 NCT03012100 NCT04119336

(continued)

Table A.4 (continued)

Drug name	ATC code	Chemical taxonomy	Description	Clinical trials
Cytarabine	L01BC01 L01XY01	Pyrimidine nucleosides	An antimetabolite analogue of cytidine.	NCT04115631 NCT03150693 NCT03959085 NCT03007147 NCT03701308 NCT03504410 NCT03303339
Daunorubicin Hydrochloride	–	–	See Cerubidine (Daunorubicin Hydrochloride).	–
Daunorubicin Hydrochloride and Cytarabine Liposome	–	–	A combination of the antineoplastic agents cytarabine and daunorubicin.	NCT04293562 NCT04231851 NCT03555955 NCT04075747
Daurismo (Glasdegib Maleate)	L01XX63	Benzene and substituted derivatives	The maleate salt form of glasdegib, an orally bioavailable small molecule. Has antineoplastic activities.	NCT04231851 NCT03416179
Dexamethasone	R01AD53 D07XB05 R01AD03 S01CB01 S02CA06	Steroids and steroid derivatives	An adrenal corticosteroid. Has anti-inflammatory properties.	NCT03937635 NCT03150693 NCT03914625 NCT02553460 NCT03556332
Doxorubicin Hydrochloride	L01DB01	Anthracyclines	An anthracycline antibiotic with antineoplastic activity.	NCT02488967 NCT03984448 NCT03907488 NCT03914625 NCT03959085
Enasidenib Mesylate	L01XX59	Triazines	The mesylate salt form of enasidenib with potential antineoplastic activity.	NCT04203316 NCT03732703 NCT03825796 NCT03728335 NCT03683433
Gemtuzumab Ozogamicin	L01XC05	Carboxylic Acids and Derivatives	An anti-CD33 monoclonal antibody.	NCT04293562 NCT03904251 NCT01409161 NCT04070768 NCT03900949 etc.
Gilteritinib Fumarate	L01XE54	Piperidines	The fumarate salt form of gilteritinib. Has antineoplastic activities.	NCT04293562 NCT03730012
Glasdegib Maleate	–	–	See Daurismo (Glasdegib Maleate).	–
Idamycin PFS (Idarubicin Hydrochloride)	L01DB06	Anthracyclines	The hydrochloride salt of the anthracycline antineoplastic antibiotic idarubicin.	NCT03164057 NCT03194932 NCT02665143

(continued)

Table A.4 (continued)

Drug name	ATC code	Chemical taxonomy	Description	Clinical trials
Idarubicin Hydrochloride	–	–	See Idamycin PFS (Idarubicin Hydrochloride).	–
Idhifa (Enasidenib Mesylate)	–	–	See Enasidenib Mesylate.	–
Ivosidenib	L01XX62	Carboxylic acids and derivatives	An inhibitor of isocitrate dehydrogenase type 1 (IDH1). Has antineoplastic activity.	NCT04195555 NCT03013998 NCT02074839 NCT03564821 NCT03173248 etc.
Midostaurin	L01XE39	Indoles and derivatives	An indolocarbazole multikinase inhibitor with potential antiangiogenic and antineoplastic activities.	NCT03258931 NCT03836209 NCT04075747 NCT02115295 NCT03591510
Mitoxantrone Hydrochloride	L01DB07	Anthracenes	The hydrochloride salt of an anthracenedione antibiotic with antineoplastic activity.	NCT02521493 NCT04293562 NCT02553460 NCT03983824 etc.
Mylotarg (Gemtuzumab Ozogamicin)	–	–	See Gemtuzumab Ozogamicin.	–
Onureg (Azacitidine)	–	–	See Azacitidine.	–
Rubidomycin (Daunorubicin Hydrochloride)	–	–	See Cerubidine (Daunorubicin Hydrochloride).	–
Rydapt (Midostaurin)	–	–	See Midostaurin.	–
Tabloid (Thioguanine)	L01BB03	Imidazopyrimidines	A guanosine analogue antimetabolite.	NCT03150693 NCT03914625 NCT03959085 NCT03007147 NCT01920737 etc.
Thioguanine	–	–	See Tabloid (Thioguanine).	–
Tibsovo (Ivosidenib)	–	–	See Ivosidenib.	–
Trisenox (Arsenic Trioxide)	–	–	See Arsenic Trioxide.	–
Venclexta (Venetoclax)	L01XX52	Diazinanes	An inhibitor of the anti-apoptotic protein Bcl-2, with potential antineoplastic activity.	NCT03701282 NCT03737981
Venetoclax	–	–	See Venclexta (Venetoclax).	–

(continued)

Table A.4 (continued)

Drug name	ATC code	Chemical taxonomy	Description	Clinical trials
Vincristine Sulfate	–	–	–	–
Vyxeos (Daunorubicin Hydrochloride and Cytarabine Liposome)	–	–	See Daunorubicin Hydrochloride and Cytarabine Liposome.	–
Xospata (Gilteritinib Fumarate)	–	–	See Gilteritinib Fumarate.	–

Appendix B
Cell Lines Used in *in vitro* Study

Acute promyelocytic leukemia (APL) is a subtype of acute myeloid leukemia (AML) when block the differentiation process at the promyelocytic stage. Usually acute promyelocytic leukemia has specific t(15;17) chromosomal translocation that leads to the formation of a chimeric gene that binds to the PML and RARα genes.

AML cell lines investigated in the study are shortly described and presented in the Table B.1.

HL-60 cell line is derived from a peripheral blood of APL patient and provides a unique model system for *in vitro* studies of cellular and molecular events associated with proliferation and differentiation into granulocyte/monocyte/macrophage lineage leukemic cells and normal cells. Most cells have various cell surface antigens specific for immature myeloid cells (Birnie 1988). Cytogenetic analysis of HL-60 cells is an evidence of multiple karyotypic abnormalities, containing monosomy, trisomy and tetrasomy, and various chromosomal translocations (Donti et al. 1991; Wolman et al. 1985). HL-60 is p53 null and exceptionally sensitive to diverse apoptotic stimuli including damage of DNA (Kim et al. 2007). HL-60 cells have high degree of c-myc as well (Mangano et al. 1998). HL-60 cells are lacking t(15;17) and fusion PML-RARα. In this cell line, differentiation and apoptosis are two independent processes. HL-60 proliferates continuously in suspension culture (RPMI media) with serum (10% FBS) and antibiotic chemicals. The doubling time of HL-60 cells is about 36–48 hours. We showed that around 2–7% of HL-60 control cells have CD11b on their cell surface. Spontaneous differentiation into mature granulocytes can be provoked by compounds such as dimethyl sulfoxide (DMSO), or all-trans retinoic acid. Other compounds, e.g., 1,25-dihydroxyvitamin D3, GM-CSF, and 12-O-tetradecanoylphorbol-13-acetate (TPA), can provoke HL-60 to differentiate into monocytic, eosinophil, and macrophage-like phenotypes, accordingly (Mangelsdorf et al. 1984; Olsson et al. 1983).

K562 cell line, a human Caucasian chronic myelogenous leukemia cell line, derived from pleural effusion of 53-year-old female with CML in terminal blast crisis. The K562 blasts are multipotential, hematopoietic malignant cells that spon-

R. Navakauskienė et al., *Epigenetics and Proteomics of Leukemia*,
https://doi.org/10.1007/978-3-030-68708-3

Table B.1 Description of AML cell lines presented in the study. Culture conditions—Complete growth medium: RPMI or DMEM; Atmosphere: air, 95% and carbon dioxide (CO_2), 5%; Temperature: 37 °C

Cell lines	Derivation morphology	Molecular genetics & Cytogenetics
HL-60 cells AML FAB-M2 (now referred to as AML with maturation (WHO))	Peripheral blood leukocytes were obtained by leukopheresis from a 36-year-old Caucasian female with acute promyelocytic leukemia. 90–95% cells are myeloblastic/promyelocitic, large, blast-like, rounded nuclei (2–4 nucleoli), azurophilic granules. CD3-, CD4+, CD13+, CD14-, CD15+, CD19-, CD33+, CD34-, HLA-DR-.	Myc+ + + (Collins & Groudine, 1982). p53 gene on chromosome 17pl3 (Isobe et al., 1986) has been largely deleted (Wolf & Rotter, 1985), one allele of the GM-CSF gene on chromosome 5q21-q23 is rearranged and partly deleted (Huebner et al., 1985). Cytogenetics: human flat-moded hypotetraploid karyotype with hypodiploid sideline and 1.5% polyploidy—82–88<4n>XX, −X, −X, −8, −8, −16, −17, −17, +18, +22, +2mar, ins(1;8)(p?31;q24hsr)×2, der(5)t(5;17)(q11;q11)×2, add(6)(q27)×2, der(9)del(9)(p13)t(9;14)(q?22;q?22)×2, der(14)t(9;14)(q?22;q?22)×2, der(16)t(16;17)(q22;q22)×1–2, add(18)(q21)—sideline with: −2, −5, −15, del(11)(q23.1q23.2)—c-myc amplicons present in der(1) and in both markers.
KG1	Bone marrow of a 59-year-old man with erythroleukemia that developed into acute myeloid leukemia (AML) at relapse mostly round, single cells in suspension. CD3-, CD13+, CD14-, CD15+, CD19-, CD33+, CD34+, HLA-DR+.	Expression of fusion gene FGFR1OP2-FGFR1 (OP2-FGFR1). Cytogenetics: human hypodiploid karyotype with 4.5% polyploidy—45(42–47)<2n>X/XY, −4, +8, +8, −12, −17, −20, +2mar, der(5;17)(q10;q10)del(5)(q?11q?13), dup(7)(q12q33), del(7)(q22q35), i(8q)×2, der(8)t(6;8)(p11;q22), der(8)t(8;12)(p11;q13), der(11)t(1;11)(q13-21;p11-p13), der(16)t(?12;16)(?p13;q13/21)—resembles published karyotype.
K562	Pleural effusion of a 53-year-old woman with chronic myeloid leukemia (CML) in blast crisis round large, single cells in suspension CD3-, CD14-, CD15+, CD19-, CD33+, CD71+, CD235a+	Expression of fusion gene BCR-ABL1Cytogenetics: human hypotriploid karyotype without sharp mode— 61–68<3n>XX, −X, −3, +7, −13, −18, +3mar, del(9)(p11/13), der(14)t(14;?)(p11;?), der(17)t(17;?)(p11/13;?), der(?18)t(15;?18)(q21;?q12), del(X)(p22)—two markers appear from FISH to have arisen from Ph.

(continued)

Table B.1 (continued)

Cell lines	Derivation morphology	Molecular genetics & Cytogenetics
NB4 APL = AML FAB M3	Bone marrow of a 23-year-old woman with acute promyelocytic leukemia in second relapse. CD3-, CD4+, CD11b-, CD13+, CD14-, CD15+, CD19-, CD33+, CD34-, CD3+, HLA-DR-.	The (15;17) PML-RARα fusion gene. Cytogenetics: human hypertriploid karyotype with 3% polyploidy—78(71–81)<3n>XX, −X, +2, +6, +7, +7, +11, +12, +13, +14, +17, −19, +20, +4mar, der(8)t(8;?)(q24;?), der(11)t(11;?)(?→::11p15→11q22.1::11 q13→22.1:), der(12)t(12;?)(p11;?), 14p+, t(15;17)(q22;q11–12.1), der(19)t(10;19)(q21.1;p13.3)×2—identity confirmed.
THP-1	Peripheral blood of a 1-year-old boy with acute monocytic leukemia (AML) at relapse round, single cells in suspension, partly in clusters. CD3-, CD4+, CD13+, CD15+, CD19-, CD34-, cyCD68+, HLA-DR+.	Expression of fusion gene KMT2A-MLLT3 (MLL-MLLT3; MLL-AF9). Cytogenetics: human near-tetraploid karyotype—94(88–96)<4n>XY/XXY, −Y, +1, +3, +6, +6, −8, −13, −19, −22, −22, +2mar, add(1)(p11), del(1)(q42.2), i(2q), del(6)(p21)×2–4, i(7p), der(9)t(9;11)(p22;q23)i(9)(p10)×2, der(11)t(9;11)(p22;q23)×2, add(12)(q24)×1–2, der(13)t(8;13)(p11;p12), add(?18)(q21)—carries t(9;11) associated with AML M5

taneously differentiate into recognizable progenitors of the granulocyte, erythrocyte, and monocytic series (Lozzio and Lozzio 1975; Andersson et al. 1979).

KG1 cell line was derived by H. P. Koeffler and D. W. Golde (Koeffler et al. 1980). A bone marrow aspirate was derived from a 59-year-old Caucasian male with erythroleukemia that developed into acute myelogenous leukemia. KG1 myeloblasts can differentiate into macrophages in the presence of phorbol esters. The KG1 cell line consists of myeloblasts and promyelocytes mainly. An exceptional feature of the KG1 cells is their nearly complete dependence on colony-stimulating factor for proliferation in soft-gel culture.

NB4 cells, the very first isolated human APL line, with the typical t(15;17) chromosomal translocation, that results in the expression of the PML-RARα fusion protein (Mozziconacci et al. 2002). This cell line was isolated from patient's bone marrow with morphologically, cytogenetically, and clinically representative APL after a second exposure to ATRA. PML-RARα enhances leukemogenesis of APL. It obstructs the differentiation and boosts the survival of myeloid precursor cells (Fagioli et al. 1994). ATRA, 9-cis all-trans retinoic acid, and RARα selective agonists prompt growth inhibition, apoptosis, and granulocytic differentiation. In this cell line, in contrast to HL-60 cells, RXR agonist activity is very low. To induce apoptosis it is enough to activate PML-RARα and/or RARα. Exposure of ATRA in NB4 cells is known to induce an increase in the expression of genes encoding procaspase-1, -7, -8, and -9. Apoptosis initiated by ATRA and other differentiation-inducing retinoids is accompanied by protein PARP cleavage. PARP (poly [ADP-ribose] polymerase) is a substrate for caspase-3, -6, and -7. Assessment of ATRA and 9-cis ATRA impact on cytochrome c release from mitochondria into the cytosol revealed that 9-cis ATRA releases more cytochrome c into the cytosol than ATRA. After exposure to both ATRA and 9-cis ATRA NB4, decreased levels of BCL-2 and elevated BAX were detected. Thus, the apoptogenic activity of retinoids in APL involves the following processes: alteration of the BCL-2/BAX ratio, release of cytochrome c from mitochondria, and activation of caspases. Retinoid-dependent degradation of anti-apoptotic PML-RARα (at least in NB4 cells) may be a determining factor in the initiation of all these processes (Gianni et al. 2000).

THP-1 cell line was derived from the peripheral blood of a 1-year-old male AML patient (Tsuchiya et al. 1980). This cell line has been extensively used in research of immune responses while cells are not just in the monocyte state but as well in the macrophage-like state. THP-1 cells in the monocyte state may be differentiated into a macrophage-like phenotype using either phorbol-12-myristate-13-acetate (PMA), 1α, 25-dihydroxyvitamin D3 (vD3) or macrophage colony-stimulating factor (M-CSF) (Chanput et al. 2014). The typical doubling time in THP-1 monocytes is approximately from 35 to 50 h.

References

Andersson LC, Jokinen M, Gahmberg CG (1979) Induction of erythroid differentiation in the human leukemia cell line K562. Nature 278:364–365

Birnie GD (1988) The HL60 cell line: a model system for studying human myeloid cell differentiation. Br J Cancer Suppl 9:41–45

Chanput W, Mes JJ, Wichers HJ (2014) THP-1 cell line: an in vitro cell model for immune modulation approach. Int Immunopharmacol 23(1):37–45. https://doi.org/10.1016/j.intimp.2014.08.002

Donti E, Lanfrancone L, Pelicci PG, Birnie GD, Dallafavera R (1991) Loss of amplification and appearance of a novel translocation site of the C-MYC oncogene in HL-60 leukemia-cells. Cancer Genet Cytogenet 56(1):57–64. https://doi.org/10.1016/0165-4608(91)90362-X

Fagioli M, Grignani F, Ferrucci PF, ALcalay M, Mencarelli A, Nicoletti I, Grignani F, Pelicci PG (1994) Effect of the acute promyelocytic leukemia PML/RAR-ALPHA protein on differentiation and survival of myeloid precursors. Leukemia 8(1):S7–S11

Gianni M, Ponzanelli I, Mologni L, Reichert U, Rambaldi A, Terao M, Garattini E (2000) Retinoid-dependent growth inhibition, differentiation and apoptosis in acute promyelocytic leukemia cells. Expression and activation of caspases. Cell Death Differ 7(5):447–460. https://doi.org/10.1038/sj.cdd.4400673

Kim EK, Kwon KB, Han MJ, Song MY, Lee JH, Ko YS, Shin BC, Yu J, Lee YR, Ryu DG, Park JW, Park BH (2007) Induction of G1 arrest and apoptosis by Scutellaria barbata in the human promyelocytic leukemia HL-60 cell line. Int J Mol Med 20(1):123–128

Koeffler HP, Billing R, Lusis AJ, Sparkes R, Golde DW (1980) An undifferentiated variant derived from the human acute myelogenous leukemia cell line (KG-1). Blood 56(2):265–273

Lozzio CB, Lozzio BB (1975) Human chronic myelogenous leukemia cell-line with positive Philadelphia chromosome. Blood 45(3):321–334. https://doi.org/10.1182/blood-2016-08-736025

Mangano R, Piddini E, Carramusa L, Duhig T, Feo S, Fried M (1998) Chimeric amplicons containing the c-myc gene in HL60 cells. Oncogene 17(21):2771–2777. https://doi.org/10.1038/sj.onc.1202434

Mangelsdorf DJ, Koeffler HP, Donaldson CA, Pike JW, Haussler MR (1984) 1,25-Dihydroxyvitamin D3-induced differentiation in a human promyelocytic leukemia cell line (HL-60): receptor-mediated maturation to macrophage-like cells. J Cell Biol 98(2):391–398

Mozziconacci MJ, Rosenauer A, Restouin A, Fanelli M, Shao WL, Fernandez F, Toiron Y, Viscardi J, Gambacorti-Passerini C, Miller WH, Lafage-Pochitaloff M (2002) Molecular cytogenetics of the acute promyelocytic leukemia-derived cell line NB4 and of four all-trans retinoic acid-resistant subclones. Genes Chromosomes Cancer 35(3):261–270. https://doi.org/10.1002/gcc.10117

Olsson IL, Breitman TR, Sarngadharan MG, C GR (1983) Mechanisms for induction of differentiation in the human promyelocytic cell line HL-60. Haematol Blood Transfus 28:384–389

Tsuchiya S, Yamabe M, Yamaguchi Y, Kobayashi Y, Konno T, Tada K (1980) Establishment and characterization of a human acute monocytic leukemia cell line (THP-1). Int J Cancer 26(2):171–176. https://doi.org/10.1002/ijc.2910260208

Wolman SR, Lanfrancone L, Dalla-Favera R, Ripley S, Henderson AS (1985) Oncogene mobility in a human leukemia line HL-60. Cancer Genet Cytogenet 17(2):133–141. https://doi.org/10.1016/0165-4608(85)90023-8

Appendix C
Chemical Agents Used in the Studies

Table C.1 Chemical agents used *in vitro*/*ex vivo*/*in vivo* studies[a]

Agents	Formula	Structure
3-DU, 3-Deazauridine	$C_{10}H_{13}NO_6$	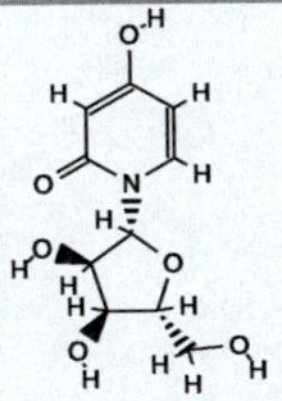
6-MMPR, 6-Methylthiopurine riboside	$C_{11}H_{14}N_4O_4S$	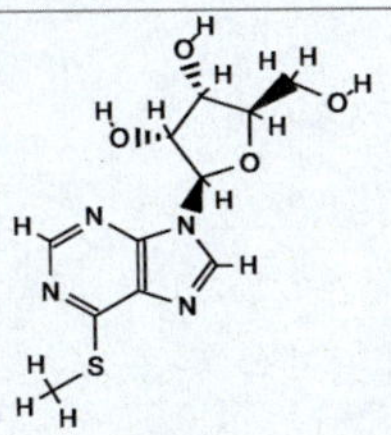
6-TG, 6-Thioguanine	$C_5H_5N_5S$	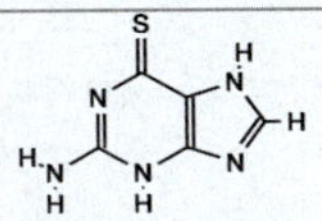
ATRA, all-trans retinoic acid	$C_{20}H_{28}O_2$	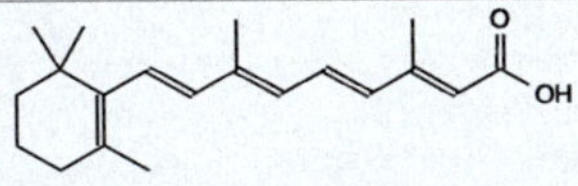

(continued)

R. Navakauskienė et al., *Epigenetics and Proteomics of Leukemia*,
https://doi.org/10.1007/978-3-030-68708-3

Table C.1 (continued)

Agents	Formula	Structure
Belinostat	$C_{15}H_{14}N_2O_4S$	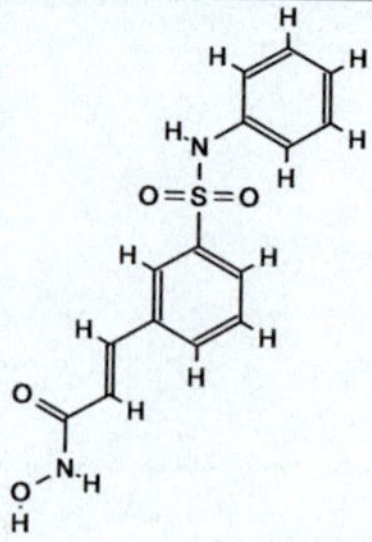
BIX-01294, N-(1-benzylpiperidin-4-yl)-6,7-dimethoxy-2-(4-methyl-1,4-diazepan-1-yl)quinazolin-4-amine	$C_{28}H_{38}N_6O_2$	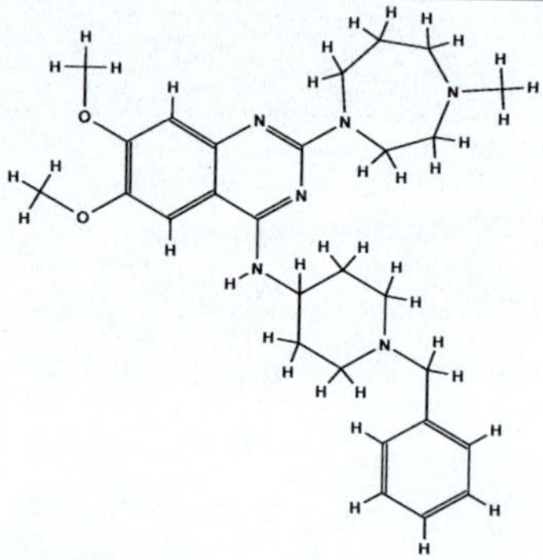
Cytarabine, cytosine arabinoside (AraC)	$C_9H_{13}N_3O_5$	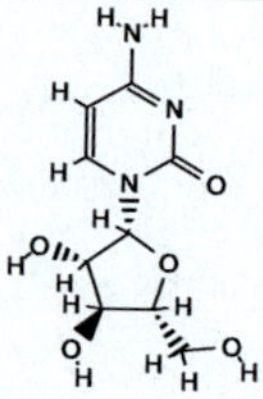
Daunorubicin	$C_{27}H_{29}NO_{10}$	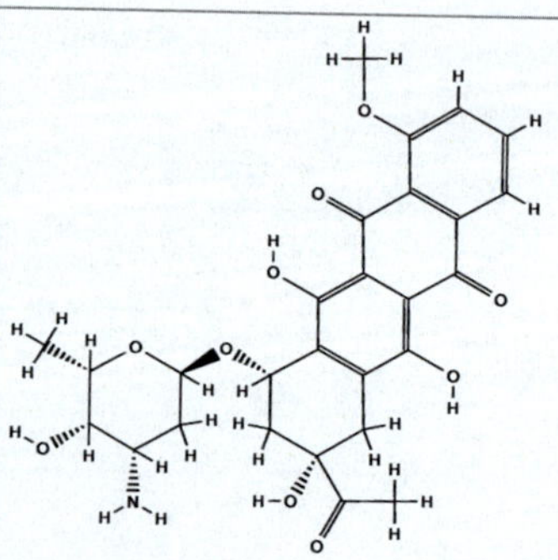

(continued)

Table C.1 (continued)

Agents	Formula	Structure
dbcAMP, dibutyryl cyclic AMP	$C_{18}H_{24}N_5O_8P$	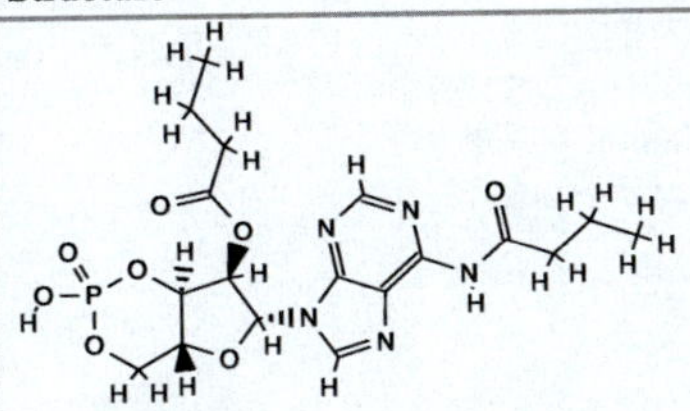
DZNep, 3-Deazaneplanocin A	$C_{12}H_{14}N_4O_3$	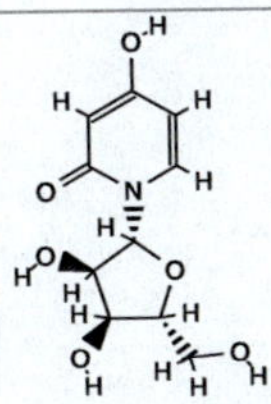
EGCG, (-)-Epigallocatechin gallate	$C_{22}H_{18}O_{11}$	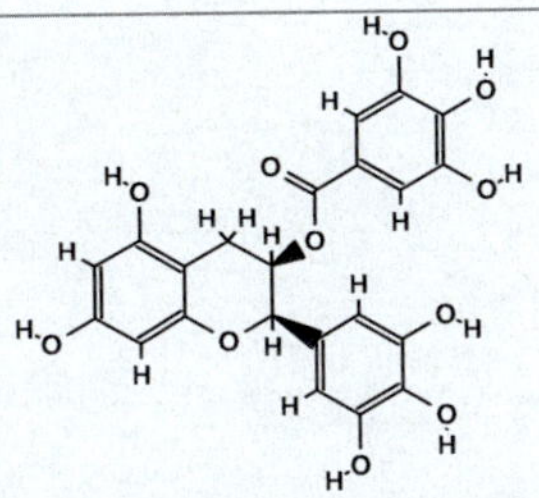
Etoposide	$C_{29}H_{32}O_1$	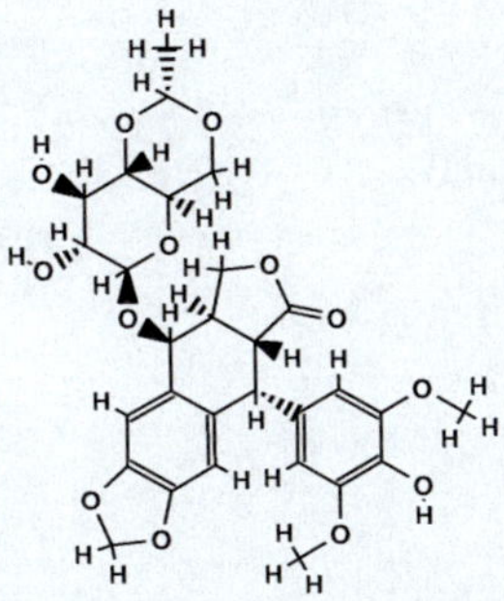
FK228, Romidepsin	$C_{24}H_{36}N_4O_6S_2$	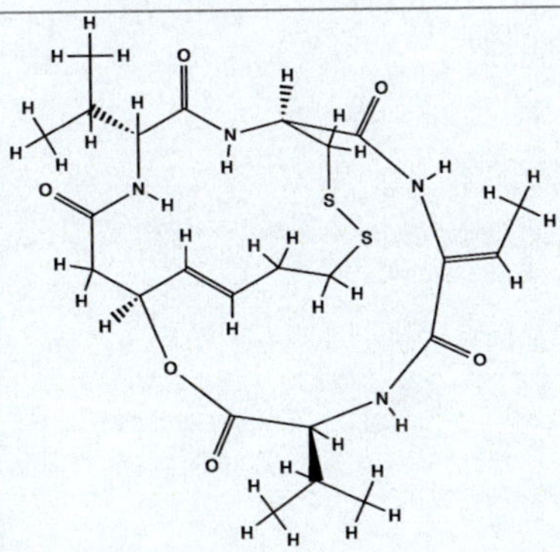

(continued)

Table C.1 (continued)

Agents	Formula	Structure
H-89, N-[2-[[3-(4-Bromophenyl)-2-propenyl]amino]ethyl]-5-isoquinolinesulfonamide	$C_{20}H_{20}BrN_3O_2S$	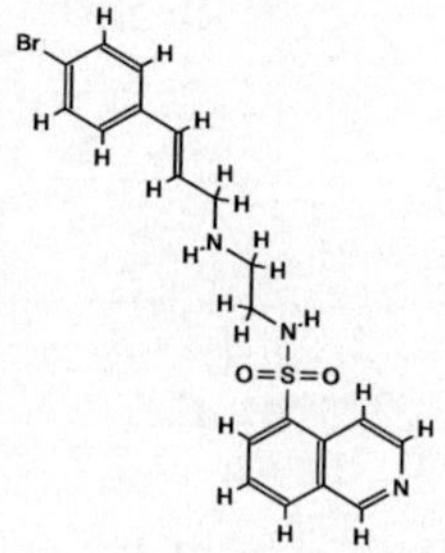
Idarubicin	$C_{26}H_{27}NO_9$	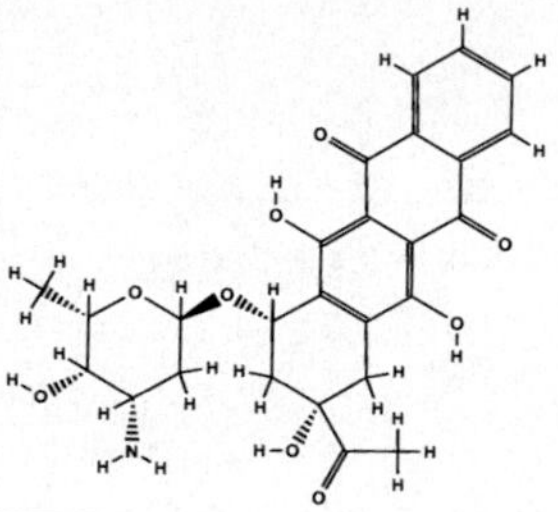
Metformin	$C_4H_{11}N_5$	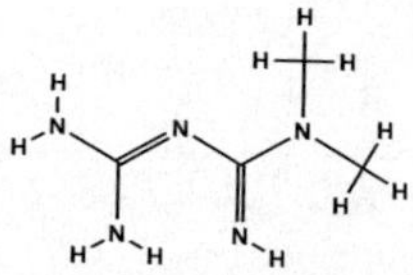
Procaine, Vitamin H3, Novocaine	$C_{13}H_{20}N_2O_2$	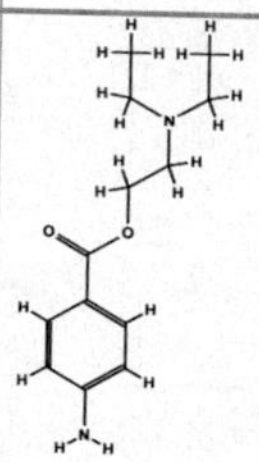
RG108, N-Phthalyl-L-tryptophan	$C_{19}H_{14}N_2O_4$	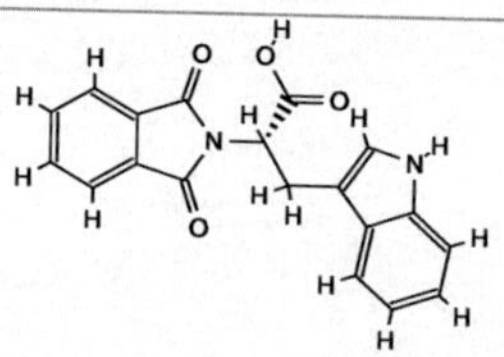

(continued)

Table C.1 (continued)

Agents	Formula	Structure
Sodium butyrate	$C_4H_7NaO_2$	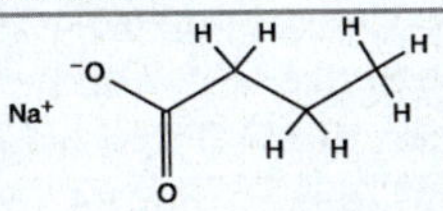
TSA, Trichostatin A	$C_{17}H_{22}N_2O_3$	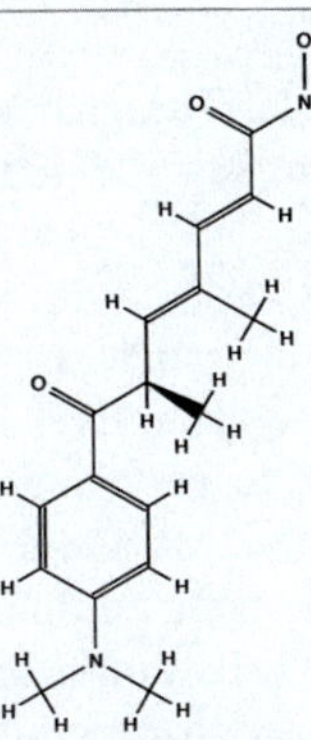
Z-VAD(OH)-FMK, Z-Val-Ala-Asp Fluoromethyl ketone	$C_{21}H_{28}FN_3O_7$	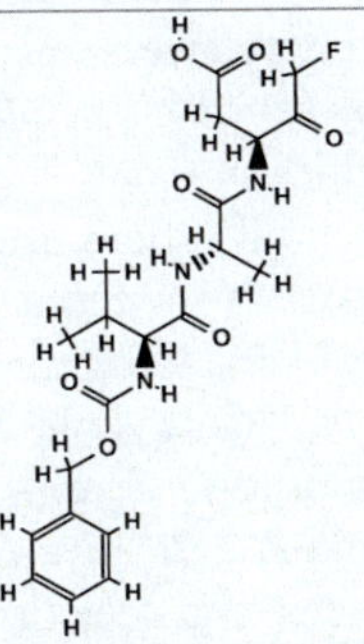
Zebularine	$C_9H_{12}N_2O_5$	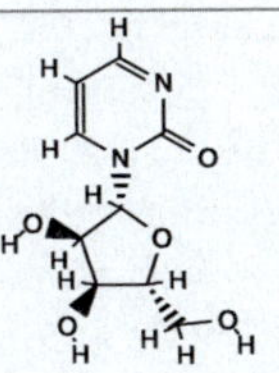

[a] Data taken from https://www.ebi.ac.uk/chembl

3-DU (DAU, deazauridine) is an investigational chemical compound. Deazauridine a nucleoside analogue, which blocks growth of tumor cells in culture and and *in vivo* (Brockman et al. 1975). It is converted intracellularly to 3-deazauridine triphosphate, which competitively inhibits cytidine triphosphate synthetase thereby inhibiting biosynthesis of the nucleic acid cytidine 5′-triphosphate (CTP). 3-DU obstructs the growth of L1210 leukemia cells if a concentration of 6 μM is applied and dose-dependently reduces mortality in a mouse model of leukemia. It also enhances the incorporation of decitabine into DNA in HL-60 myeloid and MOLT-3 lymphoid leukemia cells when used at a concentration of 20 μM. 3-DU (100 or

150 mg/kg), when combined with decitabine, reduces mortality in L1210 leukemia's mouse model (McPartland et al. 1974).

6-MMPR (6-Methylthiopurine riboside, Methylthioinosine, DB02896) is an experimental compound, a purine derivative which has antineoplastic and antiangiogenic properties. 6-MMPR inhibits amidophosphoribosyltransferase, the first committed step in de novo purine synthesis, and blocks fibroblast growth factor-2 (FGF2)-induced cell.

6-TG (6-thioguanine, Thioguanine) is FDA approved for consolidation treatment of acute non-lymphocytic leukemias. 6-TG is a purine analogue that is applied to treat the acute and chronic myelogenous leukemias (Munshi et al. 2014). Thioguanine therapy is related with inferior, typically transient and asymptomatic elevations in serum aminotransferase levels and has also been related with uncommon cases of cholestatic acute liver injury and chronic liver injury, which lead to portal hypertension due to nodular regenerative hyperplasia. Nucleotides of 6-TG are integrated in both the DNA and the RNA by phosphodiester linkages, and some investigations have demonstrated that integration of such false bases contributes to the cytotoxicity of 6-TG. Its tumor inhibitory properties may be caused by one or more of its impacts on feedback inhibition of de novo purine synthesis; inhibition of purine nucleotide interconversions; or incorporation into the DNA and RNA. The general outcome of its action is a sequential blockade of the utilization and synthesis of the purine nucleotides.

ATRA (all-trans retinoic acid, tretinoin (DB00755)) is a drug used to treat an acne and is FDA approved for treatment of APL with other agents like daunorubicin, arsenic trioxide, idarubicin,etc. Tretinoin, derived from vitamin A, is necessary for normal cell growth and embryonic development. An excess of tretinoin can be teratogenic. It is also used to treat psoriasis, acne vulgaris, and several other skin diseases. Other vitamin A derivative (isotretinoin) used to treat the severe acne and some forms of skin, head, and neck cancer, and other conditions. Also many clinical trials are still going with ATRA (NCT03878524; NCT02712905; NCT02339740; NCT02688140; NCT01987297, etc.).

Antileukemic effects of ATRA first were mentioned around 1980, when was demonstrated that ATRA can induce differentiation of HL-60 promyeloblasts into mature neutrophils. It was shown that 1 μM ATRA induce leukemic cells' differentiation up to 95% with a concomitant loss of their ability to proliferate. After 10 years of the first reports on clinical use of ATRA in APL patients were reported and complete remission rates exceed 80% to 90% in APL patients. The results of clinical trials showed that chemotherapy (daunorubicin together with cytarabine) combination with ATRA was more significant than chemotherapy alone and longer follow-up has confirmed. So, in 1995 initial FDA approval was made for ATRA, and in 2004 ATRA was used for standard therapy to initiate remission in APL patients or APL patients relapsed from anthracycline chemotherapy, or for whom anthracycline-based chemotherapy is contraindicated.

Belinostat (PXD101) is FDA approved drug (Beleodaq), indicated to treat the patients with relapsed or refractory peripheral T-cell lymphoma (PTCL) with manageable safety profile. Belinostat is a chemical compound that inhibits HDAC class

I and II enzymes activities. Belinostat Clinical trials are still going on for different diseases like lymphoma, epithelial cancer, etc. (NCT02137759; NCT03432741; NCT00993616; NCT0377292; etc.). It is a potential alternative treatment for patients who did not experience adequate response to first-line drugs for PTCL. Patients with baseline thrombocytopenia may get belinostat therapy. Research shows that this HDAC inhibitor is able to stop the proliferation of cancer cells and induce their apoptosis (Khan et al. 2008; Havas et al. 2016). Belinostat *in vitro* and clinical studies evaluating the impact of this chemical on carcinoma and other hematologic or rigid tumors (Damaskos et al. 2016; Puvvada et al. 2016; Campbell and Thomas 2017). The applicability of belinostat in chemotherapy with carboplatin is also being investigated. It is noteworthy that, when evaluating the effects on solid tumors, belinostat was used in combination with conventional chemotherapeutic agents (Thomas et al. 2014; Vitfell-Rasmussen et al. 2016). Belinostat has also been used as monotherapy in the study of effects on newly diagnosed AML. However, when used alone, belinostat was not highly effective. Notwithstanding, belinostat plus the proteasome inhibitor bortezomib caused apoptosis of AML and ALL (acute lymphocytic leukemia) cells and primary blasts, but it is non-toxic to normal CD34 (+) cells (Dai et al. 2011). Clinical trials have also been conducted to assess the impact of the combination of belinostat I and bortezomib on recurrent/refractory AML and myelodysplastic syndrome (Clinical Trial Identification in ClinicalTrials.gov Database NCT01075425).

BIX-01294 is an investigational chemical compound. BIX-01294 is a compound with low molecular weight which selectively acts as a specific inhibitor for EHMT1/GLP and EHMT2/G9A. It was the first selective inhibitor of any lysine methyltransferase (KMT) inhibitor. Lysine methyltransferase inhibitor reduces H3K9me2 levels, inhibits H3K36 methylation by oncoproteins NSD1, NSD2, and NSD3. Studies show that BIX-01294 effectively inhibits growth of cancer cells (Cho et al. 2011; Cui et al. 2015; Ren et al. 2015). The decrease in EHMT protein levels is associated with chromosomal instability and telomere lesions leading to cellular senescence, making EHMT2 an attractive target for cancer therapy (Kramer 2016).

Clinical trials are still going on with combination of idarubicin and other agents like fludarabine, quizartinib, and glycosylated G-CSF "FLAG-IDA" (NCT04112589), and others (NCT03164057, NCT03860844, NCT03250338).

Cytarabine (DB00987) is FDA approved drug for the treatment of blast phase of chronic myelocytic leukemia, acute non-lymphocytic leukemia, and acute lymphocytic leukemia. Cytarabine a cytosine analogue and antineoplastic agent mostly used to treat the acute leukemia. Cytarabine is related with a low rate of transient serum enzyme and bilirubin elevations during therapy, but has only seldom been implicated in cases of clinically apparent acute liver injury with jaundice. Cytarabine is indicated to treat the non-Hodgkin's lymphoma in children, Hodgkin's lymphomas, and myelodysplastic syndrome (Maddocks 2018). Cytarabine is an antineoplastic antimetabolite intended for the therapy of various forms of leukemia, e.g., acute myelogenous leukemia and meningeal leukemia. Antimetabolites masquerade as purine or pyrimidine—which become the building blocks of DNA. They became an obstacle which does not allow these substances to be integrated into

DNA during the S cell cycle phase, by stopping normal development and division. Cytarabine is metabolized intracellularly into its active triphosphate form (cytosine arabinoside triphosphate). This metabolite then causes a damage to DNA by multiple mechanisms, including the inhibition of DNA repair through an effect on beta-DNA polymerase, inhibition of alpha-DNA polymerase, and incorporation into DNA. The last mechanism is probably the most significant. Although the mechanism of action is not entirely investigated, it seems that cytarabine acts through the inhibition of DNA polymerase. A limited, but important, incorporation of cytarabine into both DNA and RNA has also been estimated.

Daunorubicin (daunomycin, Cerubidine/Vyxeos, DB00694) is a FDA approved chemotherapy medication, used for treatment of cancer, especially of AML, acute lymphocytic leukemia (ALL), chronic myelogenous leukemia (CML), and Kaposi's sarcoma. Daunorubicin and doxorubicin are anthracycline antibiotics that may intercalate between base pairs and hinder DNA synthesis. Anthracycline are the class of drugs (antibiotics) used in cancer chemotherapy, isolated from fungi (Streptomyces peucetius, Streptomyces caesius) with anti-proliferative effects. Anthracyclines act via inhibition of topoisomerase I and II, which are crucial for DNA replication, inducing cellular defects like mitochondrial perturbations, oxidative stress, lipid and proteotoxic stress. Anthracyclines lead to cell death and to irreversible tissue damage. Anthracyclines were introduced into clinical oncology in 1960s and still continue as major cancer treatment components. Different clinical trials (NCT03416179; NCT03701308; NCT02038777; NCT04109066, etc.) are performed.

dbcAMP (dibutyryl cyclic AMP sodium salt, Bucladesine, DB13242) is an experimental drug, a cell-permeable cAMP equivalent that triggers cAMP dependent protein kinase (PKA) or the cAMP/PKA signaling pathway. In some countries like Japan Bucladesine is registered as an experimental drug and is used as cardiac stimulant. dbcAMP is applied as cell signaling modulator and a morphological differentiation inducer in such cells as Schwann cells. It is used in numerous neural differentiation studies. It was shown that monoblast lines such as U937 can develop locomotor capacity only if such agents as dbcAMP or cytokines are added, due to their ability to drive differentiation into monocyte morphology. Bucladesine is a cyclic nucleotide analogue which penetrates cell membrane and imitates the action of endogenous cAMP. It acts as a phosphodiesterase inhibitor (Azami et al. 2010).

DZNep (3-Deazaneplanocin A) is an investigational chemical compound. DZNep is an inhibitor of adenosylhomocysteine hydrolase and histone methyltransferases capable of inhibiting trimethylation of histone H3 lysine 27 and histone H4 lysine 20. DZNep is a derivative of the natural antibiotic neplanocin-A, in which ribose is replaced by a cyclopentenyl ring. This HMTI activates the cell cycle regulators p16INK4a, p17KIP1, and p27Kip1 expression, arresting the cell cycle during the G1 cell cycle phase and thereby provoke apoptosis (Nakagawa et al. 2014; Takashina et al. 2016). Studies have demonstrated that DZNep induces apoptosis exclusively in cancerous cells and not in healthy cells (Tan et al. 2007). In addition, DZNep has been shown to interact with HDAC and DNMT (Zhou et al. 2011).

EGCG (epigallocatechin-3-gallate, DB12116) is an investigational chemical compound. EGCG is a phenolic antioxidant found in various plants, for example, in green tea. Green tea is the most common and extremely popular drink in the world. It has been shown to protect against cancer using animal models. Depending on brew time and temperature, it is possible that it will be 100–200 mg EGCG in one cup of green tea. EGCG is a potential antioxidant that can play an important part in treating conditions such as cancer. It is found only in green tea, but not in black tea, because exposure to hot or boiling water significantly reduces EGCG content and high temperatures are used in the production of black tea. It has been shown that EGCG, along with other flavonoids, may be significant in the treatment of brain, prostate, uterine, or bladder cancers. There is an assumption that it may play a greater role when treating the hormone-related cancers such as breast or prostate cancer (Stuart et al. 2006). Clinical trials with EGCG are underway (NCT02891538; NCT03928847; NCT02580279; NCT03622762; NCT02577393, etc.).

EGCG exhibits the ability to impact different epigenetic mechanisms such as DNA methyltransferase (DNMT) inhibition, histone acetyltransferase (HAT) inhibition, histone modifications via histone deacetylase (HDAC), or non-coding RNA expression (Daniel and Tollefsbol 2015). EGCG is methylated by catechol-O-methyltransferases to form MeEGCG and DiMeEGCG. EGCG is also a potential inhibitor of catechol-O-methyltransferase activity. Catechol-O-methyltransferase and DNA methyltransferase belong to the same family of methyltransferases, both enzymes having a similar core structure and active center. Thus, EGCG can also inhibit DNMT by binding to the active site of the enzyme. Hypermethylation of CpG islands in promoter regions is a significant mechanism for silencing the expression of numerous important genes involved in suppression of cancer. Studies have shown that EGCG inhibits DNMT activity and causes CpG demethylation and reactivation of silenced methylated genes (Khan et al. 2015).

EGCG inhibits HDAC activity and increases histone acetylation levels in cell lines of skin, prostate, and breast cancer. EGCG has been shown to reduce HDAC1, HDAC2, and HDAC3 iRNA expression. EGCG had a stronger inhibitory effect on DNA methylation compared to catechin. The presence of certain concentrations of magnesium ions greatly increased the impact of EGCG on the inhibition of DNMT1 (Lee et al. 2005). It was estimated that EGCG decreases the global DNA methylation level in A431 cells in a dose-dependent manner (Gu et al. 2009). EGCG reduced levels of 5-methylcytosine, DNMT1, DNTM3a and DNMT3b information RNA and protein levels, DNA methyltransferase activity. EGCG also weakened histone deacetylase activity and raised levels of acetylated lysine 9 and 14 in histone H3 and acetylated lysine 5, 12, 16 in histone H4, but lowered histone H3 lysine 9 methylation. These histone modifications are significant for transcriptional suppression in many genes that are important in cancer treatment (Nandakumar et al. 2011). It was also estimated that EGCG inhibits the growth of the renal cell carcinoma cell line 786-0 and promotes their apoptosis (Gu et al. 2009).

Etoposide (DB00773) is FDA approved drug for treatment of refractory testicular tumors, small cell lung cancer, other malignancies such as lymphoma, non-lymphocytic leukemia, and glioblastoma multiform. Etoposide a topoisomerase inhibitor and present a type of chemotherapy, also known as Toposar, Vepesid, Etopophos. It is used for treatment of a number of different types of cancer. The mechanism of action of etoposide is as a topoisomerase inhibitor (Baldwin and Osheroff 2005). Etoposide is a semisynthetic derivative of podophyllotoxin, a substance obtained from the mandrake root Podophyllum peltatum. Due to potent antineoplastic properties, etoposide binds to and inhibits topoisomerase II and its function in ligating cleaved DNA molecules that leads to the accumulation of single- or double-strand DNA breaks, the inhibition of DNA replication and transcription, and apoptotic cell death. Etoposide acts principally in the G2 and S phases of the cell cycle. It is used as a therapy in such malignancies as lymphoma, non-lymphocytic leukemia, and glioblastoma multiform (name of drug is Vepesid). FDA approved clinical trials: NCT00334815; NCT02003222; NCT02443077; NCT03811002; NCT00453154.

FK228 (depsipeptide, Romidepsin, DB06176) is a natural prodrug that inhibits histone deacetylases (Furumai et al. 2002). It was FDA approved on November 5th 2009 as Romidepsin (trade name Istodax). Romidepsin, previously known by the research codes FK-228, FR-901228, and NSC-630176, is a histone deacetylase inhibitor used to treat cutaneous T-cell lymphoma (CTCL) in patients who have received at a minimum one systemic therapy earlier. Romidepsin becomes active once taken up into the cell. The active metabolite has a free thiol group, which interacts with zinc ions in the active site of class 1 and 2 HDAC enzymes. This leads to inhibition of enzymatic activity. Some tumors have overexpressed HDACs and down-regulated/mutated histone acetyltransferases. This disbalance of HDAC relative to histone acetyltransferase can cause a decline in regulatory genes that lead to tumorigenesis. Inhibition of HDAC by FK228 may restore normal gene expression in cancer cells and lead to cell cycle arrest and apoptosis. Clinical trials for solid tumors as combination of standard therapy and Romidepsin were performed (NCT01537744).

H-89 is an investigational chemical compound. H-89 is a PKA inhibitor which competitively inhibits the phosphorylation of serine and threonine residues of this enzyme (Hidaka et al. 1990). This chemical compound has been often investigated to research PKA signaling pathways in neuronal. The possible involvement of protein kinases in apoptotic cell death caused by neurotoxic compounds was estimated in earlier studies (Lochner and Moolman 2006).

Idarubicin (IDA, drug Idamycin, DB01177) is FDA approved for myeloid leukemia (AML) in adults. Idarubicin an anthracycline antileukemic agent that inhibits DNA and RNA synthesis by intervening between base pairs. Idarubicin is metabolized primarily to idarubicinol, which accumulates in plasma. Idarubicin has monophasic clearance and 15 hours of a half-life cycle of elimination. The 2 major anthracyclines used for induction therapy of AML are idarubicin and daunorubicin. Doses of daunorubicin for treatment are 45 to 90 mg/m^2. The most common dose for IDA is 12 mg/m^2. Several studies have shown that patients initially treated with high doses of daunorubicin, i.e., >60 mg/m^2 were more likely to achieve complete remission and had a reduced risk of relapse compared to patients receiving the standard dose of daunorubicin. In 2015, a study comparing the efficacy of high-dose daunorubicin with standard-dose idarubicin for induction of AML was performed (Li et al. 2015). It was demonstrated that IDA in comparison to other anthracyclines extends entire survival and disease-free survival, increases complete remission rate, and lowers relapse rate, although rises the risks of death on induction therapy and grade 3/4 mucositis (a kind of painful inflammation and ulceration of mucous membranes lining the digestive tract). No difference in other different grade 3/4 adverse events was found. No statistically significant difference was found between the two regimens in achieving complete remission and survival (Owattanapanich et al. 2019). Because all common chemotherapeutic agents are harmful to all cells of the body, not just cancerous ones, the aim is always to use a drug that is less cytotoxic and requires a lower dose to achieve the same effect, so IDA could be better than daunorubicin.

Metformin is a medicine (drug name Glumetza, DB00331) a first FDA approved agent to treat the type 2 diabetes. It can be used alone or combined with thiazolidinedione, sulfonylureas, or other hypoglycemic agents. Metformin is extensively used antidiabetic molecule, which created great interest over the last decade as a possible anticancer molecule. Metformin inhibits mitochondrial respiration, activates the AMPK (5′ adenosine monophosphate-activated protein kinase) tumor suppressor pathway, promotes catabolic energy-saving responses, and blocks anabolic processes related with abnormal cell proliferation. Metformin is used to treat polycystic ovary syndrome (PCOS), although such treatment is not officially approved. Metformin also reduces hyper glycaemia and increases insulin-stimulated glucose uptake in liver, muscle, and fat. Lowering the level of insulin and insulin-like growth factor-1 (IGF-1) obstructs cancer growth promotion. Cancer cells are known to modify metabolism to meet their strong biochemical needs related to their proliferation. Metabolic dependence (e.g., glucose dependence) develops in cancer cells and may be attractive to new therapeutic strategies (Biondani and Peyron 2018). Nowadays clinical trials for treatment of other diseases than diabetes is going on for head and neck squamous cell carcinoma treatment (NCT03618654), multiple sclerosis (NCT04121468), rectal cancer (NCT03053544), etc.

Procaine (DB00721)—the FDA approved drug used for local anesthesia. Like the structurally related procainamide (also used in clinical practice to treat the arrhythmia), it was estimated that this substance has DNA methylation inhibitory properties. In addition, the cells treated with this preparation are inhibited in the M phase and their total number is significantly reduced. This indicates that cancer cell proliferation and malignant tissue are effectively suppressed. The action of procaine is based on its interaction with the DNA itself. Molecular experiments and *in vitro* cell cultures (Villar-Garea et al. 2003) show that the methylation inhibitor interacts with GC-rich regions of DNA (higher affinity for the double-stranded macromolecule). The strongest interaction, in the gaseous phase, was with the guanine base. Procaine research continues, trying to synthesize its analogues which will be more effective. It is attractive due to the lower toxicity than that of decitabine or other nucleotide analogues (Castellano et al. 2008). This can lead to higher doses of the drug or the use of procaine as a preventive measure against cancer.

RG108 is an investigational chemical compound. RG108 is a small molecule that effectively blocks DNMT by binding to the active site of the enzyme. Classified as a non-nucleoside inhibitor of DNMT, RG108 is believed to specifically block the active site of DNMT1. When incubated at low micromoles concentrations, RG108 induces efficient demethylation of genomic DNA without any appreciable potent toxicity; RG108 thus reactivates tumor suppressor genes without impacting methylation of centromeres sequences (Brueckner et al. 2005). Thus, RG108 can be very useful in epigenetic regulation of cancer, as well as its dependent demethylation and activation of tumor suppressor genes may be one way of combating cancer. In addition, RG108 is a pioneering compound in the evolution of a new class of DNMT inhibitors in which future compounds will block the active site of the enzyme.

Sodium butyrate is an investigational chemical compound. Sodium butyrate is the sodium salt of butyric acid with likely antineoplastic activity. Butyric acid, a short chain fatty acid, competitively binds to the class I and II histone deacetylases (HDACs) zinc sites. This binding has an impact on hyperacetylation of histones that leads to a DNA conformation, which subsequently results in the uncoiling or relaxing of chromatin. Improved chromatin accessibility to transcription-regulatory complexes causes greater transcriptional activation of different epigenetically suppressed genes (Oike et al. 2014). Butyric acid, a HDAC inhibitor, provokes cell cycle arrest in G1 or G2/M phase and also increases the expression of other genes and proteins' activity associated with differentiation of cells and apoptotic signaling. Sodium phenylbutyrate approved by FDA as substances added to food (flavoring agent or adjuvant).

The antitumor activity of this drug has been investigated in many cancers, but ATRA has been shown to work only in APL, where it dissociates the PML-RARα mutant protein and allows cells to differentiate. ATRA causes complete disease remission in patients who are not resistant to this agent (Schenk et al. 2014). However, there are patients who do not respond to ATRA and the treatment should be improved with other agents such as epigenetic modifiers, i.e., inhibitors of histone deacetylases, histone methyltransferases or DNA methyltransferases.

The mechanisms of action behind therapy using ATRA are: (1) relocalization of the PML restoration of normal nuclear bodies structure and degradation of PML-RARα protein via caspase-mediated cleavage and proteosome-dependent degradation; (2) transformation of PML-RARα from a transcription repressor (CoR) into a transcription activator (CoA) under therapeutic concentration of ATRA; (3) coordinated genes expression caused by ATRA which leads to the excellent and elaborate cellular program for the commitment to differentiation. 169 genes were modulated to express, with 100 genes up-regulated and 69 down-regulated.

TSA (Trichostatin A, DB04297) is an experimental drug. TSA an inhibitor of histone deacetylases belonging to the hydroxamate class. TSA is an antifungal antibiotic obtained from Streptomyces platensis and Streptomyces hygroscopicus in 1976. The name is synonymous with Antibiotic A300. TSA is used as an anticancer medicine. TSA inhibits histone deacetylases, VEGF, cancer cell proliferation, cell migration, invasion, and also inhibits the formation of new blood vessels. TSA promotes the activation of apoptotic genes and prevents cancer cells from surviving and developing a tumor, which stops the progression of the cancer. Only one clinical trial with TSA is available so far (NCT03838926).

Z-VAD(OH)-FMK is an investigational chemical compound. Z-VAD(OH)-FMK a pan-caspase inhibitor can be used to induce necroptosis under certain stimuli. It acts as an efficient irreversible caspase inhibitor with no cytotoxic effects. Due to this it is a useful tool for studies regarding caspase activity. Drugs are applied to prevent damage to the brain or spinal cord from convulsions, ischemia, trauma, or stroke. Several of them must be administered before the event, but others may be effective for some time after. They act by different mechanisms, but frequently directly or indirectly initiate alternative cell death pathways that may provide therapeutic tools to combat cell death related with infectious diseases, neurodegenerative diseases, and ischemia-reperfusion pathologies, and may also enable the creation of alternative cytotoxic strategies in cancer treatment (Vandenabeele et al. 2006).

Zebularine (1- (β-D-ribofuranosyl) -1,2-dihydropyrimidin-2-one, DB03068) is an experimental analog of cytidine. Its synthesis was described as early as 1961. It was invented and synthesized as a cytidine deaminase inhibitor for the reason that it has no amino group at the C-4 position. Zebularine, like most cytidine analogues, was found to possess DNA methylation inhibitory properties. Zebularine became the subject of study because of its sufficient stability in acidic and neutral media. Zebularine has a lower cytotoxicity compared to 5-Aza-CR and 5-Aza-CdR—the dose of zebularine containing the same demethylation activity as 5-Aza-CR and 5-Aza-CdR is minimally toxic (Marquez et al. 2005). The need for higher doses of zebularine is influenced by its slow conversion to deoxy form by ribonucleotide reductase. This results in faster incorporation into the RNA. In addition, the liver enzyme aldehyde oxidase oxidizes zebularine at a relatively rapid rate. The cytotoxicity is reduced by the fact that the pyrimidinone ring is more stable than the 5-Aza-CR ring, i.e., zebularine forms hydrogen bonds with the guanine base (like cytidine), whereas the unstable 5-Aza-CR ring can hydrolyze and form hydrogen bonds with the cytosine base, which increases the likelihood of mutagenicity (Van Bemmel et al. 2009). Other studies have shown that zebularine is relatively selective in action, with cancer cells consuming it more rapidly than normal cells (Cheng et al. 2004),

References

Azami K, Etminani M, Tabrizian K, Salar F, Belaran M, Hosseini A, Hosseini-Sharifabad A, Sharifzadeh M (2010) The quantitative evaluation of cholinergic markers in spatial memory improvement induced by nicotine–bucladesine combination in rats. Eur J Pharmacol 636(1–3):102–107

Baldwin EL, Osheroff N (2005) Etoposide, topoisomerase II and cancer. Curr Med Chem Anticancer Agents 5(4):363–372

Biondani G, Peyron JF (2018) Metformin, an anti-diabetic drug to target leukemia. Front Endocrinol 9:446

Brockman RW, Shaddix SC, Williams M, Nelson JA, Rose LM, Schabel Jr FM (1975) The mechanism of action of 3-deazauridine in tumor cells sensitive and resistant to arabinosylcytosine. Ann N Y Acad Sci 255:501

Brueckner B, Boy RG, Siedlecki P, Musch T, Kliem HC, Zielenkiewicz P, Suhai S, Wiessler M, Lyko F (2005) Epigenetic reactivation of tumor suppressor genes by a novel small-molecule inhibitor of human DNA methyltransferases. Cancer Res 65(14):6305–6311

Campbell P, Thomas CM (2017) Belinostat for the treatment of relapsed or refractory peripheral T-cell lymphoma. J Oncol Pharm Pract 23(2):143–147

Castellano S, Kuck D, Sala M, Novellino E, Lyko F, Sbardella G (2008) Constrained analogues of procaine as novel small molecule inhibitors of DNA methyltransferase-1. J Med Chem 51(7):2321–2325

Cheng JC, Weisenberger DJ, Gonzales FA, Liang G, Xu GL, Hu YG, Marquez VE, Jones PA (2004) Continuous zebularine treatment effectively sustains demethylation in human bladder cancer cells. Mol Cell Biol 24(3):1270–1278

Cho HS, Kelly JD, Hayami S, Toyokawa G, Takawa M, Yoshimatsu M, Tsunoda T, Field HI, Neal DE, Ponder BA, et al. (2011) Enhanced expression of EHMT2 is involved in the proliferation of cancer cells through negative regulation of SIAH1. Neoplasia 13(8):676–684, IN4–IN10

Cui J, Sun W, Hao X, Wei M, Su X, Zhang Y, Su L, Liu X (2015) EHMT2 inhibitor BIX-01294 induces apoptosis through PMAIP1-USP9X-MCL1 axis in human bladder cancer cells. Cancer Cell Int 15(1):1–9

Dai Y, Chen S, Wang L, Pei XY, Kramer LB, Dent P, Grant S (2011) Bortezomib interacts synergistically with belinostat in human acute myeloid leukaemia and acute lymphoblastic leukaemia cells in association with perturbations in NF-κB and Bim. Br J Haematol 153(2):222–235

Damaskos C, Valsami S, Spartalis E, Antoniou EA, Tomos P, Karamaroudis S, Zoumpou T, Pergialiotis V, Stergios K, Michaelides C, et al. (2016) Histone deacetylase inhibitors: a novel therapeutic weapon against medullary thyroid cancer? Anticancer Res 36(10):5019–5024

Daniel M, Tollefsbol TO (2015) Epigenetic linkage of aging, cancer and nutrition. J Exp Biol 218(1):59–70

Furumai R, Matsuyama A, Kobashi N, Lee KH, Nishiyama M, Nakajima H, Tanaka A, Komatsu Y, Nishino N, Yoshida M, etal (2002) FK228 (depsipeptide) as a natural prodrug that inhibits class I histone deacetylases. Cancer Res 62(17):4916–4921

Gu B, Ding Q, Xia G, Fang Z (2009) EGCG inhibits growth and induces apoptosis in renal cell carcinoma through TFPI-2 overexpression. Oncol Rep 21(3):635–640

Havas AP, Rodrigues KB, Bhakta A, Demirjian JA, Hahn S, Tran J, Scavello M, Tula-Sanchez AA, Zeng Y, Schmelz M, et al. (2016) Belinostat and vincristine demonstrate mutually synergistic cytotoxicity associated with mitotic arrest and inhibition of polyploidy in a preclinical model of aggressive diffuse large B cell lymphoma. Cancer Biol Ther 17(12):1240–1252

Hidaka H, Watanabe M, Tokumitsu H (1990) Search for the functional substrate proteins of protein kinases and their selective inhibitors. Adv Second Messenger Phosphoprotein Res 24:485–490

Khan MA, Hussain A, Sundaram MK, Alalami U, Gunasekera D, Ramesh L, Hamza A, Quraishi U (2015) (-)-Epigallocatechin-3-gallate reverses the expression of various tumor-suppressor genes by inhibiting DNA methyltransferases and histone deacetylases in human cervical cancer cells. Oncol Rep 33(4):1976–1984

Khan N, Jeffers M, Kumar S, Hackett C, Boldog F, Khramtsov N, Qian X, Mills E, Berghs SC, Carey N, etal (2008) Determination of the class and isoform selectivity of small-molecule histone deacetylase inhibitors. Biochem J 409(2):581–589

Kramer JM (2016) Regulation of cell differentiation and function by the euchromatin histone methyltranserfases G9a and GLP. Biochem Cell Biol 94(1):26–32

Lee WJ, Shim JY, Zhu BT (2005) Mechanisms for the inhibition of DNA methyltransferases by tea catechins and bioflavonoids. Mol Pharmacol 68(4):1018–1030

Li X, Xu S, Tan Y, Chen J (2015) The effects of idarubicin versus other anthracyclines for induction therapy of patients with newly diagnosed leukaemia. Cochrane Database Syst Rev (6):179

Lochner A, Moolman JA (2006) The many faces of H89: a review. Cardiovasc Drug Rev 24(3–4):261–74. https://doi.org/10.1111/j.1527-3466.2006.00261.x

Maddocks K (2018) Update on mantle cell lymphoma. Blood J Amer Society Hematol 132(16):1647–1656

Marquez VE, Kelley JA, AGbaria R, Ben-Kasus T, Cheng JC, Yoo CB, Jones PA (2005) Zebularine: a unique molecule for an epigenetically based strategy in cancer chemotherapy. Ann N Y Acad Sci 1058(1):246–254

McPartland RP, Wang MC, Bloch A, Weinfeld H (1974) Cytidine 5'-triphosphate synthetase as a target for inhibition by the antitumor agent 3-deazauridine. Cancer Res 34(11):3107–3111

Munshi PN, Lubin M, Bertino JR (2014) 6-thioguanine: a drug with unrealized potential for cancer therapy. Oncologist 19(7):760

Nakagawa S, Sakamoto Y, Okabe H, Hayashi H, Hashimoto D, Yokoyama N, Tokunaga R, Sakamoto K, Kuroki H, Mima K, et al. (2014) Epigenetic therapy with the histone methyltransferase EZH2 inhibitor 3-deazaneplanocin a inhibits the growth of cholangiocarcinoma cells. Oncol Rep 31(2):983–988

Nandakumar V, Vaid M, Katiyar SK (2011) (-)-epigallocatechin-3-gallate reactivates silenced tumor suppressor genes, Cip1/p21 and p 16 INK4a, by reducing DNA methylation and increasing histones acetylation in human skin cancer cells. Carcinogenesis 32(4):537–544

Oike T, Ogiwara H, Amornwichet N, Nakano T, Kohno T (2014) Chromatin-regulating proteins as targets for cancer therapy. J Radiat Res 55(4):613–628

Owattanapanich W, Ungprasert P, Wais V, Kungwankiattichai S, Bunjes D, Kuchenbauer F (2019) FLAMSA-RIC for stem cell transplantation in patients with acute myeloid leukemia and myelodysplastic syndromes: A systematic review and meta-analysis. J Clin Med 8(9):1437

Puvvada SD, Li H, Rimsza LM, Bernstein SH, Fisher RI, LeBlanc M, Schmelz M, Glinsmann-Gibson B, Miller TP, Maddox AM, et al. (2016) A phase II study of belinostat (PXD101) in relapsed and refractory aggressive b-cell lymphomas: SWOG S0520. Leuk Lymphoma 57(10):2359–2369

Ren A, Qiu Y, Cui H, Fu G (2015) Inhibition of H3K9 methyltransferase G9a induces autophagy and apoptosis in oral squamous cell carcinoma. Biochem Biophys Res Commun 459(1):10–17

Schenk T, Stengel S, Zelent A (2014) Unlocking the potential of retinoic acid in anticancer therapy. Br J Cancer 111(11):2039–2045

Stuart EC, Scandlyn MJ, Rosengren RJ (2006) Role of epigallocatechin gallate (EGCG) in the treatment of breast and prostate cancer. Life Sci 79(25):2329–2336

Takashina T, Kinoshita I, Kikuchi J, Shimizu Y, Sakakibara-Konishi J, Oizumi S, Nishimura M, Dosaka-Akita H (2016) Combined inhibition of EZH 2 and histone deacetylases as a potential epigenetic therapy for non-small-cell lung cancer cells. Cancer Sci 107(7):955–962

Tan J, Yang X, Zhuang L, Jiang X, Chen W, Lee PL, Karuturi RM, Tan PBO, Liu ET, Yu Q (2007) Pharmacologic disruption of polycomb-repressive complex 2-mediated gene repression selectively induces apoptosis in cancer cells. Genes Dev 21(9):1050–1063

Thomas A, Rajan A, Szabo E, Tomita Y, Carter CA, Scepura B, Lopez-Chavez A, Lee MJ, Redon CE, Frosch A, et al. (2014) A phase I/II trial of belinostat in combination with cisplatin, doxorubicin, and cyclophosphamide in thymic epithelial tumors: a clinical and translational study. Clin Cancer Res 20(21):5392–5402

Van Bemmel DM, Brank AS, Eritja R, Marquez VE, Christman JK (2009) DNA (Cytosine-C5) methyltransferase inhibition by oligodeoxyribonucleotides containing 2-(1h)-pyrimidinone (zebularine aglycon) at the enzymatic target site. Biochem Pharmacol 78(6):633–641

Vandenabeele P, Berghe TV, Festjens N (2006) Caspase inhibitors promote alternative cell death pathways. Sci STKE 2006(358):pe44–pe44

Villar-Garea A, Fraga MF, Espada J, Esteller M (2003) Procaine is a DNA-demethylating agent with growth-inhibitory effects in human cancer cells. Cancer Res 63(16):4984–4989

Vitfell-Rasmussen J, Judson I, Safwat A, Jones RL, Rossen PB, Lind-Hansen M, Knoblauch P, Krarup-Hansen A (2016) A phase I/II clinical trial of belinostat (PXD101) in combination with doxorubicin in patients with soft tissue sarcomas. Sarcoma 2016:3126–3126

Zhou J, Bi C, Cheong LL, Mahara S, Liu SC, Tay KG, Koh TL, Yu Q, Chng WJ (2011) The histone methyltransferase inhibitor, DZNep, up-regulates TXNIP, increases ROS production, and targets leukemia cells in AML. Blood J Amer Society Hematol 118(10):2830–2839

Appendix D
Identified and Associated into Interactome Proteins

1433B	– 14-3-3 protein beta/alpha, YWHAB;
1433E	– 14-3-3 protein epsilon, YWHAE;
1433F	– 14-3-3 protein eta, YWHAH;
ABLIM1	– Actin-binding LIM protein 1;
ABLIM3	– Actin-binding LIM protein 3;
ACP1	– Low molecular weight phosphotyrosine protein phosphatase, LMW-PTPase, (PPAC);
ACTB	– Actin, cytoplasmic 1, Actin, beta;
ACTBL2	– Beta-actin-like protein 2;
ACTG1	– Actin, cytoplasmic 2, Gamma-actin;
ADAM17	– Disintegrin and metalloproteinase domain-containing protein 17;
ADAM21	– Disintegrin and metalloproteinase domain-containing protein 21, ADAM 21;
ADAMTS19	– A disintegrin and metalloproteinase with thrombospondin motifs 19;
ADAMTSL4	– ADAMTS-like protein 4;
AGO2	– Protein argonaute-2;
AIMP2	– Aminoacyl tRNA synthase complex-interacting multifunctional protein 2;
ALB	– Serum albumin;
ALDOA	– Fructose-bisphosphate aldolase A;
ALYREF	– THO complex subunit 4;
ANXA1	– Annexin A1;
ANXA13	– Annexin A13;
ANXA2	– Annexin A2;

Data from Uniprot database (https://www.uniprot.org/uniprot).

R. Navakauskienė et al., *Epigenetics and Proteomics of Leukemia*,
https://doi.org/10.1007/978-3-030-68708-3

ANXA3	– Annexin A3;
ANXA4	– Annexin A4;
ANXA5	– Annexin A5;
ANXA6	– Annexin A6;
APAF1	– Apoptotic protease-activating factor 1;
APC	– Adenomatous polyposis coli protein;
ARHGDIA	– Rho GDP-dissociation inhibitor 1;
ARHGEF2	– Rho guanine nucleotide exchange factor 2;
ARHGEF25	– Guanine nucleotide exchange factor GEFT;
ARIH2	– Protein ariadne-2 homolog;
ARPC2	– Actin-related protein 2/3 complex subunit 2;
ATF3	– Cyclic AMP-dependent transcription factor ATF-3);
B4GALT7	– Beta 1-4-galactosyltransferase 7;
BCL2L1	– Bcl-2-like protein 1;
BID	– BH3-interacting domain death agonist;
BLZF1	– Golgin-45;
BRAP	– BRCA1 associated protein;
CALD1	– Caldesmon;
CALR	– Calreticulin;
CAMP	– Cathelicidin antimicrobial peptide;
CAPG	– Macrophage-capping protein, (CAPG);
CAPN1	– Calpain-1 catalytic subunit);
CAPN10	– Calpain-10;
CAPN2	– Calpain-2 catalytic subunit;
CAPN9	– Calpain-9);
CAPZA1	– F-actin-capping protein subunit alpha-1;
CAPZA2	– F-actin capping protein alfa-subunit;
CAPZB	– F-actin-capping protein subunit beta;
CASP10	– Caspase-10;
CASP3	– Caspase-3;
CASP6	– Caspase-6;
CASP7	– Caspase-7;
CASP8	– Caspase-8;
CASP9	– Caspase-9;
CAST	– Calpastatin;
CFL1	– Coffilin 1;
CFLAR	– CASP8 and FADD-like apoptosis regulator;
CH60	– 60 kDa heat shock protein, mitochondrial, HSPD1;
CLC	– Eosinophil lysophospholipase, Galectin-10;
CLIC1	– Chloride intracellular channel protein 1;
CLIC4	– Chloride intracellular channel protein 4;
CNTN5	– Contactin-5;
COA7	– Hcp beta-lactamase-like protein C1orf163;
COF1	– Cofilin, COF1;
COMMD10	– COMM domain-containing protein 10;

CRABP1 – Cellular retinoic acid-binding protein 1;
CYCS – Cytochrome c;
DARS – Aspartate–tRNA ligase, cytoplasmic;
DES – Desmin;
DIABLO – Diablo homolog, mitochondrial;
DICER1 – Endoribonuclease Dicer;
DIS3 – Exosome complex exonuclease RRP44;
DTNB – Dystrobrevin beta;
EFHD2 – EF-hand domain-containing protein D2;
ENO1 – Alpha-enolase;
ENOA – Alpha-enolase, ENO1;
ENPL – protein Endoplasmin, gene HSP90B1;
EPS8 – Epidermal growth factor receptor kinase substrate 8;
EVC2 – Limbin;
EXOSC1 – Exosome complex component CSL4;
EXOSC2 – Exosome complex component RRP4;
EXOSC3 – Exosome complex component RRP40;
EXOSC4 – Exosome complex component RRP41;
EXOSC5 – Exosome complex component CSL4;
EXOSC6 – Exosome complex component MTR3;
EXOSC7 – Exosome complex exonuclease RRP42;
EXOSC9 – Exosome complex component RRP45;
FABP5 – Fatty acid-binding protein, epidermal;
FADD – FAS-associated death domain protein;
FAM32A – Protein FAM32A);
FAS – Tumor necrosis factor receptor superfamily member 6;
FASLG – Tumor necrosis factor ligand superfamily member 6;
FEN1 – Flap endonuclease 1;
FYCO1 – FYVE and coiled-coil domain-containing protein 1;
G6PD – Glucose-6-phosphate 1-dehydrogenase;
GAPDH – Glyceraldehyde-3-phosphate dehydrogenase;
GDI2 – Rab GDP dissociation inhibitor beta;
GDIR1 – Rho GDP-dissociation inhibitor 1, ARHGDIA;
GGTLC3 – Putative gamma-glutamyltransferase light chain 3;
GLO1 – Lactoylglutathione lyase;
GLT8D1 – Glycosyltransferase 8 domain-containing protein 1;
GNA11 – Guanine nucleotide-binding protein subunit alpha-11;
GOLGA3 – Golgin subfamily A member 3;
GPATCH3 – G patch domain-containing protein 3;
GRP75 – Stress-70 protein, mitochondrial, HSPA9;
GRP78 – Endoplasmic reticulum chaperone BiP, HSPA5;
GSN – Gelsolin;
GSTP1 – Glutathione S-transferase P;
GUCA1B – Guanylyl cyclase-activating protein 2;
H2AFV – Histone H2A.V;

H2AFZ – Histone H2A.Z;
H2BFS – Histone H2B type F-S;
H3F3B – Histone H3.3;
H3F3C – Histone H3.3C;
HIST1H1B – Histone H1.5;
HIST1H1C – Histone H1.2;
HIST1H2AJ – Histone H2A type 1-J;
HIST1H2BB – Histone H2B type 1-B;
HIST1H2BD – Histone H2B type 1-D;
HIST1H4F – Histone H4;
HIST2H2BF – Histone H2B type 2-F;
HMOX2 – Heme oxygenase;
HNRNPA1 – Heterogeneous nuclear ribonucleoprotein A1;
HNRNPA2B1 – Heterogeneous nuclear ribonucleoproteins A2/B1;
HNRNPC – Heterogenous nuclear ribonucleoproteins C1/C2;
HNRNPK – Heterogeneous nuclear ribonucleoprotein K;
HP – Haptoglobin;
HSP72 – Heat shock-related 70 kDa protein 2, HSPA2;
HSP90 – Heat shock protein HSP 90-beta, HSP90AB1;
HSP90AB1 – Heat shock 84 kDa (HSP84);
HSP90B – Heat shock protein HSP 90-beta, gene HSP90AB1;
HSP90B1 – Endoplasmin (Heat shock protein 90 kDa beta member 1), Heat shock protein Gp96 precursor;
HSPA2 – Heat shock-related 70 kDa protein 2;
HSPA5 – 78 kDa glucose-regulated protein (BiP, GRP-78);
HSPA6 – Heat shock 70 kDa protein 6;
HSPA8 – Heat shock cognate 71 kDa protein (Heat shock 70 kDa protein 8);
HSPA9 – Heat shock 70 kDa protein 9B, Stress-70 protein, (GRP75);
HSPB1 – Heat shock protein beta-1;
HSPD1 – 60 kDa heat shock protein, Chaperonin (HSP60), mitochondrial;
HSPE1 – 10 kDa heat shock protein, mitochondrial;
HSPH1 – Heat shock protein 105 kDa;
IL12RB2 – Interleukin-12 receptor subunit beta-2;
JAK3 – Tyrosine-protein kinase JAK3;
KCNA4 – Potassium voltage-gated channel subfamily A member 4 (HK1, HPCN2);
KCNA5 – Potassium voltage-gated channel subfamily A member 5;
LCN2 – Neutrophil gelatinase-associated lipocalin, (NGAL);
LCP1 – Plastin-2;
LGALS7B – Galectin-7;
LGUL – Lactoylglutathione lyase, GLO1;
LIG1 – Leucine-rich repeats and immunoglobulin-like domains protein 1;

LIMCH1	– LIM and calponin homology domains-containing protein 1;
LIMK1	– LIM domain kinase 1;
LMNA	– Lamin A/C, Splice isoform Lamin C of Lamin A/C;
LMNB2	– Lamin B2;
LPIN2	– Phosphatidate phosphatase LPIN2;
LRP1	– Prolow-density lipoprotein receptor-related protein 1;
LRRIQ4	– Leucine-rich repeat and IQ domain-containing protein 4;
LTF	– Lactotransferrin;
LYZ	– Lysozyme C;
MAB21L1	– Protein mab-21-like 1;
MARK1	– Serine/threonine-protein kinase MARK1;
MDH2	– Malate dehydrogenase;
MDP1	– Magnesium-dependent phosphatase 1;
MED10	– Mediator of RNA polymerase II transcription subunit 10;
MED14	– Mediator of RNA polymerase II transcription subunit 14;
MED17	– Mediator of RNA polymerase II transcription subunit 17;
MED18	– Mediator of RNA polymerase II transcription subunit 18;
MED20	– Mediator of RNA polymerase II transcription subunit 20;
MED22	– Mediator of RNA polymerase II transcription subunit 22;
MED29	– Mediator of RNA polymerase II transcription subunit 29;
MED6	– Mediator of RNA polymerase II transcription subunit 6;
ML12A	– Myosin regulatory light chain 12A, MYL12A;
MYL12A	– Myosin regulatory light chain 12A;
MYL4	– Myosin;
MYLPF	– Myosin regulatory light chain 2, skeletal muscle isoform;
NCL	– Nucleolin;
NDKA	– Nucleoside diphosphate kinase A, NME1;
NDUAA	– NADH dehydrogenase [ubiquinone] 1 alpha subcomplex subunit 10, mitochondrial, NDUFA10;
NDUFA10	– NADH dehydrogenase 1 alpha subcomplex subunit 10;
NDUFA2	– NADH dehydrogenase [ubiquinone] 1 alpha subcomplex subunit 2;
NDUFA6	– NADH dehydrogenase [ubiquinone] 1 alpha subcomplex subunit 6;
NDUFB10	– NADH dehydrogenase [ubiquinone] 1 alpha subcomplex subunit 10;
NDUFS1	– NADH-ubiquinone oxidoreductase 75 kDa subunit, mitochondrial;
NELFB	– Negative elongation factor B;
NELFE	– Negative elongation factor E;
NF2	– Merlin;
NME1	– Nucleoside diphosphate kinase A;
NPM	– Nucleophosmin;
NPM1	– Nucleophosmin;
NUDT3	– Diphosphoinositol polyphosphate phosphohydrolase 1;

NUP107	– Nuclear pore complex protein Nup107;
NUP133	– Nuclear Pore Complex Protein Nup133, nucleoporin 133;
NUP85	– Nuclear pore complex protein Nup85;
NUP98	– Nuclear pore complex protein Nup98-Nup96;
OASL	– 59 kDa 2'-5'-oligoadenylate synthase-like protein;
P4HB	– Protein disulfide-isomerase;
PCNA	– Proliferating cell nuclear antigen;
PDIA1	– Protein disulfide-isomerase, P4HB;
PDIA3	– Protein disulfide-isomerase A3, ER-60 (disulfide isomerase);
PDIA6	– Protein disulfide-isomerase A6;
PDK3	– Pyruvate dehydrogenase kinase, (PDH);
PEBP1	– Phosphatedyethanolamine – b.p. 1, (PEBP-1);
PFN1	– Profilin-1;
PGAM1	– Phosphoglycerate mutase 1, (PGAMI);
PHB	– Prohibitin;
PHB2	– Prohibitin;
POLA2	– DNA polymerase alpha subunit B;
POLR2M	– DNA-directed RNA polymerase II subunit GRINL1A
POTEE	– POTE ankyrin domain family member E;
POTEF	– POTE ankyrin domain family member F;
POTEJ	– POTE ankyrin domain family member J;
PPIA	– Peptidyl-prolyl cis-trans isomerase A, (PPIA);
PPT1	– Palmitoyl-protein thioesterase 1;
PRDX1	– Peroxiredoxin-1;
PSME1	– Proteasome activator complex subunit 1 ;
PSME2	– Proteasome activator complex subunit 2;
PTAFR	– Platelet-activating factor receptor;
PTK2B	– Protein-tyrosine kinase 2-beta;
RAB2A	– Ras-related protein Rab-2A, (RAB2A);
RAB2B	– Ras-related protein Rab-2B;
RABGAP1	– Rab GTPase-activating protein 1;
REL	– Proto-oncogene c-Rel;
RFC3	– Replication factor C subunit 3;
RFC4	– Replication factor C subunit 4;
RIPK1	– Recep tor-interacting serine/threonine-protein kinase 1;
RIPPLY2	– Protein ripply2;
RIT1	– GTP-binding protein Rit1;
RPL29	– 60S ribosomal protein L29;
RPL7	– 60S ribosomal protein L7;
S100A12	– Protein S100-A12, (Neutrophil S100 protein);
S100A7A	– Protein S100-A7A, (S1A7A);
S100A8	– Protein S100-A8;
S100A9	– Protein S100-A9;
SCIN	– Adseverin;
SCNM1	– Sodium channel modifier 1;

SEC13	– Protein SEC13 homolog;
SGCG	– Gamma-sarcoglycan;
SKIV2L2	– Exosome RNA helicase MTR4;
SLC12A6	– Solute carrier family 12 member 6;
SLC38A10	– Putative sodium-coupled neutral amino acid transporter 10;
SLC4A1AP	– Kanadaptin;
SLPI	– Antileukoproteinase;
SPECC1	– Cytospin-B;
SPRYD7	– Chronic lymphocytic leukemia deletion region gene 6 protein;
SREK1	– Splicing regulatory glutamine/lysine-rich protein 1;
SSX5	– Protein SSX5, (SSX5);
STAG1	– Cohesin subunit SA-1 (STAG1);
STMN1	– Stathmin;
SUN2	– SUN domain-containing protein 2;
SYNC	– Syncoilin;
SYNE1	– Nesprin-1;
SYNE2	– Nesprin-2;
TAGL2	– Transgelin-2, TAGLN2;
TAGLN2	– Transgelin-2;
TBA1B	– Tubulin alpha-1B chain, TUBA1B;
TBB5	– Tubulin beta chain, TUBB;
TBC1D10C	– Carabin;
TBC1D32	– Protein broad-minded;
THUMPD2	– THUMP domain-containing protein 2;
TIMP3	– Metalloproteinase inhibitor 3;
TMPRSS11A	– Transmembrane protease serine 11A;
TNFAIP6	– Tumor necrosis factor-inducible gene 6 protein;
TNFRSF10A	– Tumor necrosis factor receptor superfamily member 10A;
TNFRSF10B	– Tumor necrosis factor receptor superfamily member 10B;
TNFRSF25	– Tumor necrosis factor receptor;
TNRC6A	– Trinucleotide repeat-containing gene 6A protein;
TPI1	– Triosephosphate isomerase;
TPM1	– Tropomyosin 1;
TPM3	– Tropomyosin 3, Tropomyosin alpha-3 chain;
TPM4	– Tropomyosin alpha-4;
TRADD	– Tumor necrosis factor receptor type 1-associated DEATH domain protein;
TSG6	– Tumor necrosis factor-inducible gene 6 protein, TNFAIP6;
TUBA1A	– Tubulin alpha-1A chain ;
TUBA1B	– Tubulin alpha-1B chain;
TUBA1C	– Tubulin alpha-1C chain;
TUBA3C	– Tubulin alpha-3C chain;
TUBA4A	– Tubulin alpha-4A chain;
TUBA8	– Tubulin alpha-8 chain;
TUBAL3	– Tubulin alpha chain-like 3;

TUBB – Tubulin beta chain;
TUBB4A – Tubulin beta-4A chain;
TXD12 – Thioredoxin domain-containing protein 12, TXNDC12;
TXNRD2 – Thioredoxin reductase 2, mitochondrial;
UBA52 – Ubiquitin-60S ribosomal protein L40;
UBB – Polyubiquitin-B;
USP6NL – USP6 N-terminal-like protein;
VAV3 – Guanine nucleotide exchange factor;
VCL – Vinkulin;
VIM – Vimentin;
WDR81 – WD repeat-containing protein 81;
XIAP – E3 ubiquitin-protein ligase XIAP;
YARS – Tyrosine–tRNA ligase, cytoplasmic;
YWHAB – 14-3-3 protein beta/alpha;
YWHAE – 14-3-3 protein epsilon;
YWHAH – 14-3-3 protein eta, (1433F);
ZC3H13 – Zinc finger CCCH domain-containing protein 13;
ZNF12 – Zinc finger protein 12;
ZNF80 – Zinc finger protein 80.

Appendix E
Identified Proteins in NB4 Cells

Table E.1 Description of proteins identified by mass spectrometry in NB4 cells treated with HDACi BML-210 and fractionated by 2DE

Name	Acc. Nmb.	Mw (Da)/pI	Sequence cov. (%) & Peptides
1433B _HUMAN 14-3-3 protein beta/alpha	P31946	28179/4.76	30% -.MTMDKSELVQK.A, K.AVTEQGHELSNEER.N, R.VISSIEQK.T, R.YLSEVASGDNK.Q, K.EMQPTHPIR.L, K.DSTLIMQLLR.D, K.VFYLKMK.G K.MKGDYFR.Y
1433E _HUMAN 14-3-3 protein eta	P62258	29326/4.63	25% K.VAGMDVELTVEER.N, K.LICCDILDVLDK.H, K.AASDIAMTELPPTHPIR.L, K.DSTLIMQLLR.D, R.YLAEFATGNDR.K
1433F _HUMAN 14-3-3 protein eta	Q04917	28372/4.76	61% -.MGDREQLLQR.A, R.EQLLQR.A, R.LAEQAER.Y, R.YDDMASAMK.A, K.AVTELNEPLSNEDR.N, K.ELETVCNDVLSLLDKFLIK.N, K.VFYLKMK.G, K.MKGDYYR.Y, R.YLAEVASGEK.K, K.NSVVEASEAAYK.E, K.DSTLIMQLLR.D, R.NLLSVAYK.N, K.NVVGARR.S, R.VISSIEQK.T, K.TMADGNEK.K, K.EAFEISK.E, K.EAFEISKEQMQPTHPIR.L
CALR _HUMAN Calretic-ulin	P27797	48142/4.29	36% .LKEEEEDK.K, K.VHVIFNYK.G, K.DKQDEEQR.L, K.FYGDEEKDK.G, K.GQTLVVQFTVK.H R, K.EQFLDGDGWTSR.W, K.HEQNIDCGGGYVK.L, K.IKDPDASKPEDWDER.A, K.KVHVIFNYK.G, K.DKGLQTSQDAR.F

(continued)

R. Navakauskienė et al., *Epigenetics and Proteomics of Leukemia*,
https://doi.org/10.1007/978-3-030-68708-3

Table E.1 (continued)

Name	Acc. Nmb.	Mw (Da)/pI	Sequence cov. (%) & Peptides
ACTB _HUMAN Actin, cytoplasmic 1	P60709	42052/5.29	46% K.AGFAGDDAPR.A, R.AVFPSIVGRPR.H, R.HQGVMVGMGQK.D, K.IWHHTFYNELR.V, K.DLYANTVLSGGTTMYPGIADR.M, R.EIVRDIK.E, K.SYELPDGQVITIGNER.F, K.CDVDIR.K, R.VAPEEHPVLLTEAPLNPK.A, K.RGILTLK.Y, K.EITALAPSTMK.I, K.IKIIAPPER.K, R.DLTDYLMK.I, R.GYSFTTTAER.E, R.MQKEITALAPSTMK.I
CH60 _HUMAN 60 kDa heat shock protein, mitochondrial	P10809	61187/5.70	56% -.MLRLPTVFR.Q, R.LPTVFR.Q, R.QMRPVSR.V, R.AYAKDVK.F, R.TVIIEQSWGSPK.V, K.YKNIGAK.L, R.KPLVIIAEDVDGEALSTLVLNR.L, K.LVQDVANNTNEEAGDGTTTATVLAR.S, K.GANPVEIR.R, R.RGVMLAVDAVIAELK.K, K.EIGNIISDAMK.K, R.KGVITVK.D, K.TLNDELEIIEGMK.F, K.KISSIQSIVPALEIANAHR.K, K.ISSIQSIVPALEIANAHR.K, R.LKVGLQVVAVK.A, K.VGLQVVAVK.A, K.APGFGDNR.K, K.APGFGDNRK.N, K.VGEVIVTK.D, K.AQIEKR.I, R.GYISPYFINTSK.G, K.CEFQDAYVLLSEK.K, K.IPAMTIAK.N, K.NAGVEGSLIVEK.I, K.GIIDPTKVVR.T, K.RIQEIIEQLDVTTSEYEK.E, R.IQEIIEQLDVTTSEYEK.E, K.LNERLAK.L, K.LSDGVAVLK.V, K.DRVTDALNATR.A, R.VTDALNATR.A, R.AAVEEGIVLGGGCALLR.C, K.IGIEIIK.R, K.IGIEIIKR.T
CLIC1 _HUMAN Chloride intracellular channel protein 1	O00299	27248/5.09	70% KLCPGGQLPFLLYGTEVHTDTNKI, R.EEFASTCPDDEEIELAYEQVAK.A, K.LAALNPESNTAGLDIFAK.F, K.VLDNYLTSPLPEEVDETSAEDEGSQR.K, R.KFLDGNELTLADCNLLPK.L, K.LHIVQVVCK.K, R.GFTIPEFR.G, R.YLSNAYAR.E, K.IEEFLEAVLCPPR.Y, K.GVTFNVTTVDTK.R, K.NSNPALNDNLEK.G
COF1 _HUMAN Cofilin-1	P23528	18503/8.2	74% K.VFNDMKVR.K, K.SSTPEEVK.K, K.AVLFCLSEDK.K, K.MLPDKDCR.Y, R.YALYDATYETK.E, K.EDLVFIFWAPESAPLK.S, K.EILVGDVGQTVDDPYATFVK, K.MIYASSKDAIK.K, K.HELQANCYEEVK.D, K.LGGSAVISLEGKPL.- K.KNIILEEGK.E
EFHD2 _HUMAN EF-hand domain-containing protein D2	Q96C19	26795/5.15	35% R.ADLNQGIGEPQSPSR.R, R.DGFIDLMELK.L, K.AAAGELQEDSGLCVLAR.L, K.VQAINVSSR.F, R.FEEEIKAEQEER.K, K.EVDEDFDSK.L, R.LSEIDVSSEGVK.G
ENOA _HUMAN Alpha-enolase	P06733	47481/7.01	16% R.GNPTVEVDLFTSK.G, R.AAVPSGASTGIYEALELR.D, K.VVIGMDVAASEFFR.S, R.YISPDQLADLYK.S, K.LAQANGWGVMVSHR.S

(continued)

Table E.1 (continued)

Name	Acc. Nmb.	Mw (Da)/pI	Sequence cov. (%) & Peptides
ENPL _HUMAN Endoplasmin (Heat shock protein 90 kDa beta member 1)	P14625	92469/4.76	20% R.EEEAIQLDGLNASQIR.E, K.FAFQAEVNR.M, K.LIINSLYK.N, R.ELISNASDALDK.I, R.LSLNIDPDAK.V, R.GLFDEYGSK.K, K.GVVDSDDLPLNVSR.E, R.FQSSHHPTDITSLDQYVER.M, K.AEASSPFVER.L, K.DISTNYYASQK.K, K.SGTSEFLNK.M, R.GTTITLVLK.E, K.EEASDYLELDTIK.N, K.EVEEDEYK.A, R.SGYLLPDTK.A
GDIR1 _HUMAN Rho GDP-dissociatio inhibitor 1	P52565	23250/5.02	9% K.EGVEYR.I, K.YIQHTYR.K, R.FTDDDK.T
GNA11 _HUMAN Guanine nucleotide-binding protein subunit alpha-11	P29992	42124/5.5	26%R.GFTKLVYQNIFTAMQAMIR.A, R.TIITYPWFQNSSVILFLNK.K, -.MTLESMMACCLSDEVKESK.R, K.KDLLEDK.I, R.MVDVGGQR.S, K.YYLTDVDR.I, K.DLLEDK.I, K.ANALLIR.E
GRP75 _HUMAN Stress-70 protein, mitochondrial	P38646	73920/5.87	25% R.TTPSVVAFTADGER.L, R.LVGMPAKR.Q, K.NAVITVPAYFNDSQR.Q, K.DAGQISGLNVLR.V, K.STNGDTFLGGEDFDQALLR.H, R.ETGVDLTKDNMALQR.V, K.SDIGEVILVGGMTR.M, K.VQQTVQDLFGR.A, K.LLGQFTLIGIPPAPR.G, R.AQFEGIVTDLIR.R, R.VREAAEK.A, R.LVGMPAK.R\, K.YAEEDR.R, K.EEISKMR.E, R.ELLARK.D, K.KMASER.E, R.NTTIPTK.K
GRP78 _HUMAN 78 kDa glucose-regulated protein	P11021	72402/5.07	36% R.ITPSYVAFTPEGER.L, R.TWNDPSVQQDIK.F, K.TKPYIQVDIGGGQTK.T, K.KVTHAVVTVPAYFNDAQR.Q, K.VTHAVVTVPAYFNDAQR.Q, K.DAGTIAGLNVMR.I, R.IINEPTAAAIAYGLDKR.E, R.IEIESFYEGEDFSETLTR.A, K.FEELNMDLFR.S, K.VLEDSDLKK.S, K.KSDIDEIVLVGGSTR.I, K.SDIDEIVLVGGSTR.I, K.DNHLLGTFDLTGIPPAPR.G, K.IEWLESHQDADIEDFK.A, R.ALSSQHQAR.I, R.AKFEELNMDLFR.S, R.IPKIQQLVK.E, K.EFFNGKEPSR.G, K.VYEGERPLTK.D, K.ITITNDQNR.L, R.LTPEEIER.M, R.VEIIANDQGNR.I
HSP72 _HUMAN Heat shock-related 70 kDa protein 2	P54652	70022/5.6	7% R.ARFEELNADLFR.G, K.STAGDTHLGGEDFDNR.M, R.FEELNADLFR.G, R.TTPSYVAFTDTER.L

(continued)

Table E.1 (continued)

Name	Acc. Nmb.	Mw (Da)/pI	Sequence cov. (%) & Peptides
HS90B _HUMAN Heat shock 84 kDa (HSP84)	P08238	83264/4.97	32% K.HNDDEQYAWESSAGGSFTVR.A, R.ADHGEPIGR.G, K.EDQTEYLEER.R, K.FENLCK.L, K.LGLGIDEDEVAAEEPNAAVPDEIPPLEDEDASR.M, K.EQVANSAFVER.V, K.YIDQEELNK.T, R.NPDDITQEEYGEFYK.S, K.SLTNDWEDHLAVK.H, R.ALLFIPR.R, R.GVVDSEDLPLNISR.E, K.LGIHEDSTNR.R, K.EGLELPEDEEEK.K, K.EIFLR.E, R.ELISNASDALDK.I, R.YESLTDPSK.L, K.IDIIPNPQER.T, K.ADLINNLGTIAK.S, K.SLVSVTK.E, K.SIYYITGESK.E
HSPB1 _HUMAN Heat shock protein beta-1	P04792	22783/6.0	40% R.VPFSLLR.G, K.DGVVEITGK.H, R.QLSSGVSEIR.H, R.LFDQAFGLPR.L, R.VSLDVNHFAPDELTVK.T, K.LATQSNEITIPVTFESR.A, R.DWYPHSR.L, R.GPSWDPFR.D
LGUL _HUMAN Lactoyl-glu-tathione lyase	Q04760	20992/5.12	18% R.VLGMTLIQK.C, R.GFGHIGIAVPDVYSACK.R, K.RFEELGVK.F
ML12A _HUMAN Myosin regulatory light chain 12A	P19105	19839/4.67	29% K.EAFNMIDQNR.D, R.LTGDAFR.KACFDEEATGTIQEDYLR.E, K.GNFNYIEFTR.I, R.FTDEEVDELYR.E
NDKA _HUMAN Nucleo-side diphos-phate kinase A	P15531	17149/5.8	20% R.NIIHGSDSVESAEK.E, R.VMLGETNPADSKPGTIR.G, R.TFIAIKPDGVQR.G
NDUAA _HUMAN NADH dehydro-genase 1 alpha sub-complex subunit 10	O95299	41067/8.67	28% K.LAATSASARVVAAGAQR.V, R.SIFSDFVFLEAMYNQGFIR.K, R.TLYHLRLLVQDK.F, R.GIHSSVQCKLR.Y, R.SRVITVDGNICTGK.G, K.FYDDPRSNDGNSYR.L, R.VLHQFRELPGR.K, R.LLKLAATSASAR.V
PCNA _HUMAN Proliferat-ing cell nuclear antigen	P12004	29092/4.57	40% R.NLAMGVNLTSMSK.I, R.AEDNADTLALVFEAPNQEK.V, K.LMDLDVEQLGIPEQEYSCVVK.M, R.DLSHIGDAVVISCAK.D, R.LVQGSILK.K, R.SEGFDTYR.C, R.YLNFFTK.A, K.YYLAPK.I, K.MPSGEFAR.I

(continued)

Table E.1 (continued)

Name	Acc. Nmb.	Mw (Da)/pI	Sequence cov. (%) & Peptides
NPM _HUMAN Nucle-ophosmin	P06748	32726/4.64	62% K.ADKDYHFK.V, K.DYHFK.V, K.VDNDENEHQLSLR.T, R.TVSLGAGAK.D, K.DELHIVEAEAMNYEGSPIK.V, K.VTLATLK.M, K.MSVQPTVSLGGFEITPPVVLR.L, K.LLSISGK.R, K.DSKPSSTPR.S, K.DSKPSSTPRSK.G, R.MTDQEAIQDLWQWR.K, K.RSAPGGGSK.V, K.SIRDTPAK.N, K.NAQKSNQNGK.D, K.SNQNGKDSKPSSTPR.S, R.SKGQESFK.K, K.GQESFKK.Q, K.QEKTPK.T, K.GPSSVEDIK.A, K.AKMQASIEK.G, K.MQASIEK.G, K.GGSLPKVEAK.F, K.FINYVK.N
PDIA1 _HUMAN Protein disulfide-isomerase	P07237	57480/4.76	54% R.KSNFAEALAAHK.Y, K.SNFAEALAAHK.Y, K.ALAPEYAK.A, K.LKAEGSEIR.L, K.AEGSEIR.L, K.VDATEESDLAQQYGVR.G, K.FFRNGDTASPK.E, R.NGDTASPK.E, R.EADDIVNWLK.K, K.YQLDKDGVVLFK.K, K.HNQLPLVIEFTEQTAPK.I, K.SVSDYDGKLSNFK.T, K.TAAESFK.G, K.ILFIFIDSDHTDNQR.I, R.LITLEEEMTK.Y, K.YKPESEELTAER.I, R.ITEFCHR.F, K.NFEDVAFDEK.K, K.QLAPIWDK.L, K.LGETYK.D, K.LGETYKDHENIVIAK.M, K.VHSFPTLK.F, K.IFGGEIK.T, K.THILLFLPK.S, K.SVSDYDGK.L, R.ILEFFGLK.K, K.KEECPAVR.L, K.EECPAVR.L, K.KFDEGR.N, R.NNFEGEVTK.E, K.ENLLDFIK.H, K.FFPASADR.T, K.DGVVLFK.K, R.TVIDYNGER.T
PDIA3 _HUMAN Protein disulfide-isomerase A3	P30101	57146/5.98	58% R.LALFPGVA, R.DGEEAGAYDGPR.T, LLLAAAR.L, R.LAPEYEAAATR.L, K.YGVSGYPTLK.I, K.KQAGPASVPLR.T, K.QAGPASVPLR.T, K.KFISDK.D, K.DASIVGFFDDSFSEAHSEFLK.A, K.AASNLRDNYR.F, K.DLLIAYYDVDYEK.N, R.KTFSHELSDFGLESTAGEIPVVAIR.T, K.GEKFVMQEEFSR.D, K.VVVAENFDEIVNNENK.D, K.DPNIVIAK.M, K.MDATANDVPSPYEVR.G, R.GFPTIYFSPANK.K, R.EATNPPVIQEEKPK.K, K.KAQEDL.-, R.TEEEFK.K, K.FVMQEEFSR.D, R.DGKALER.F, R.FLQDYFDGNLK.R, K.SEPIPESNDGPVK.V, R.VMMVAKK.F, K.KFLDAGHK.L, K.FLDAGHK.L, K.LNFAVASR.K, K.KYEGGR.E, R.ELSDFISYLQR.E
PDIA6 _HUMAN Protein disulfide-isomerase A6	Q15084	48490/4.95	22% K.NRPEDYQGGR.T, R.TGEAIVDAALSALR.Q, K.LAAVDATVNQVLASR.Y, K.FALLKGSFSEQGINEFLR.E, K.GSFSEQGINEFLR.E, R.GSTAPVGGGAFPTIVER.E, R.EPWDGR.D, K.GESPVDYDGGR.T, K.MKFALLK.G, R.ELSFGR.G
PHB _HUMAN Prohibitin	P35232	29843/5.57	40% -.MAAKVFESIGK.F, K.VFE R.AVIFDR.F, SIGK.F, R.IFTSIGEDYDER.V, K.AAIISAEGDSK.A, R.KLEAAEDIAYQLSR.S, K.DLQNVNITLR.I, R.ILFRPVASQLPR.I, R.FDAGELITQR.E, K.EFTEAVEAK.Q, K.QVAQQEAER.A

(continued)

Table E.1 (continued)

Name	Acc. Nmb.	Mw (Da)/pI	Sequence cov. (%) & Peptides
TAGL2 _HUMAN Transgelin-2	P37802	22548/8.41	44% K.QM, EQISQFLQAAER.Y, R.YGINTTDIFQTVDLWEGK.N, R.TLMNLGGLAVAR.D, R.DDGLFSGDPNWFPK.K, R.NFSDNQLQEGK.N, R.GPAYGLSR.E, K.NVIGLQMGTNR.G
TBA1B _HUMAN Tubulin alpha-1B chain	P68363	50804/4.94	39%R.AVFVDLEPTVIDEVR.T, K.EDAANNYARGHYTIGK.E, R.NLDIERPTYTNLNR.L, R.FDGALNVDLTEFQTNLVPYPR.I, R.AFVHWYVGEGMEEGEFSEAR.E, R.IHFPLATYAPVISAEK.A, R.HGKYMACCLLYR.G, K.YMACCLLYR.G, K.DVNAAIATIKTK.R, R.AVCMLSNTTAIAEAWAR.L, K.EIIDLVLDR.I, R.LSVDYGK.K, K.CDPRHGK.Y, R.LISQIVSSITASLR.F, K.EDAANNYAR.G
TBB5 _HUMAN Tubulin beta chain	P07437	50095/4.78	52% R.EIVHIQAGQCGNQIGAK.F, R.ISVYYNEATGGK.Y, R.AILVDLEPGTMDSVR.S, R.SGPFGQIFRPDNFVFGQSGAGNNWAG, K.GHYTEGAELVDSVLDVVR.K, K.IREEYPDR.I, K.LTTPTYGDLNHLVSATMSGVTTCLR.F, R.LHFFMPGFAPLTSR.G, R.ALTVPELTQQVFDAK.N, K.NSSYFVEWIPNNVK.T, K.TAVCDIPPR.G, .ISEQFTAMFR.R, R.KLAVNMVPFPR.L, K.LAVNMVPFPR.L, R.GSQQYR.A, K.NMMAACDPR.H, R.YLTVAAVFR.G, R.GRMSMK.E, R.FPGQLNADLR.K, R.FPGQLNADLRK.L, R.EEYPDR.I
TSG6 _HUMAN Tumor necrosis factor-inducible gene 6 protein	P98066	31204/6.3	10% K.IGFHVCAAGWMAK.G, K.YVAMDPVSKSSQGK.N
TPM4 _HUMAN Tropomyosin alpha-4	P67936	28619/4.67	46%M.AGLNSLEAVKR.K, R.ELDGER.E, R.RIQLVEEELDR.A, R.KLVILEGELER.A, R.AEVSELK.C, K.NVTNNLK.S, K.SLEAASEK.YT, K.YEEVAR.K, K.YSEKEDK.Y, K.LKEAETR.A, R.AEFAER., R.LATALQK.L, K.AADESER.G
TXD12 _HUMAN Thioredoxin domain-containing protein 12		19365/5.24	21%K.VHPEIINENGNPSYKYFYVSAEQVVQGMK.E, R.LTGDAFR.K, K.KHLEDEL.-, R.TLEDGKK.E

Index

R. Navakauskienė et al., *Epigenetics and Proteomics of Leukemia*,
https://doi.org/10.1007/978-3-030-68708-3

Q

R